Lecture Notes in Mechanical Engineering

Series Editors

Francisco Cavas-Martínez, Departamento de Estructuras, Universidad Politécnica de Cartagena, Cartagena, Murcia, Spain

Fakher Chaari, National School of Engineers, University of Sfax, Sfax, Tunisia

Francesco Gherardini, Dipartimento di Ingegneria, Università di Modena e Reggio Emilia, Modena, Italy

Mohamed Haddar, National School of Engineers of Sfax (ENIS), Sfax, Tunisia

Vitalii Ivanov, Department of Manufacturing Engineering Machine and Tools, Sumy State University, Sumy, Ukraine

Young W. Kwon, Department of Manufacturing Engineering and Aerospace Engineering, Graduate School of Engineering and Applied Science, Monterey, CA, USA

Justyna Trojanowska, Poznan University of Technology, Poznan, Poland

Lecture Notes in Mechanical Engineering (LNME) publishes the latest developments in Mechanical Engineering—quickly, informally and with high quality. Original research reported in proceedings and post-proceedings represents the core of LNME. Volumes published in LNME embrace all aspects, subfields and new challenges of mechanical engineering. Topics in the series include:

- Engineering Design
- Machinery and Machine Elements
- Mechanical Structures and Stress Analysis
- Automotive Engineering
- Engine Technology
- Aerospace Technology and Astronautics
- Nanotechnology and Microengineering
- Control, Robotics, Mechatronics
- MEMS
- Theoretical and Applied Mechanics
- Dynamical Systems, Control
- Fluid Mechanics
- Engineering Thermodynamics, Heat and Mass Transfer
- Manufacturing
- Precision Engineering, Instrumentation, Measurement
- Materials Engineering
- Tribology and Surface Technology

To submit a proposal or request further information, please contact the Springer Editor of your location:

China: Dr. Mengchu Huang at mengchu.huang@springer.com
India: Priya Vyas at priya.vyas@springer.com
Rest of Asia, Australia, New Zealand: Swati Meherishi at swati.meherishi@springer.com
All other countries: Dr. Leontina Di Cecco at Leontina.dicecco@springer.com

To submit a proposal for a monograph, please check our Springer Tracts in Mechanical Engineering at http://www.springer.com/series/11693 or contact Leontina.dicecco@springer.com

Indexed by SCOPUS. All books published in the series are submitted for consideration in Web of Science.

More information about this series at http://www.springer.com/series/11236

Ranganath M. Singari · Kaliyan Mathiyazhagan ·
Harish Kumar
Editors

Advances in Manufacturing and Industrial Engineering

Select Proceedings of ICAPIE 2019

Volume 1

Editors
Ranganath M. Singari
Department of Mechanical Engineering
Delhi Technological University
New Delhi, India

Kaliyan Mathiyazhagan
Department of Mechanical Engineering
Amity School of Engineering and
Technology
Noida, India

Harish Kumar
Department of Mechanical Engineering
National Institute of Technology Delhi
New Delhi, India

ISSN 2195-4356   ISSN 2195-4364   (electronic)
Lecture Notes in Mechanical Engineering
ISBN 978-981-15-8541-8   ISBN 978-981-15-8542-5   (eBook)
https://doi.org/10.1007/978-981-15-8542-5

© The Editor(s) (if applicable) and The Author(s), under exclusive license to Springer Nature
Singapore Pte Ltd. 2021, corrected publication 2021
This work is subject to copyright. All rights are solely and exclusively licensed by the Publisher, whether
the whole or part of the material is concerned, specifically the rights of translation, reprinting, reuse of
illustrations, recitation, broadcasting, reproduction on microfilms or in any other physical way, and
transmission or information storage and retrieval, electronic adaptation, computer software, or by similar
or dissimilar methodology now known or hereafter developed.
The use of general descriptive names, registered names, trademarks, service marks, etc. in this
publication does not imply, even in the absence of a specific statement, that such names are exempt from
the relevant protective laws and regulations and therefore free for general use.
The publisher, the authors and the editors are safe to assume that the advice and information in this
book are believed to be true and accurate at the date of publication. Neither the publisher nor the
authors or the editors give a warranty, expressed or implied, with respect to the material contained
herein or for any errors or omissions that may have been made. The publisher remains neutral with regard
to jurisdictional claims in published maps and institutional affiliations.

This Springer imprint is published by the registered company Springer Nature Singapore Pte Ltd.
The registered company address is: 152 Beach Road, #21-01/04 Gateway East, Singapore 189721,
Singapore

Contents

A Review on the Fabrication of Surface Composites via Friction Stir Processing and Its Modeling Using ANN 1
Kartikeya Bector, Aranyak Tripathi, Divya Pandey, Ravi Butola, and Ranganath M. Singari

A Statistical Study of Consumer Perspective Towards the Supply Chain Management of Food Delivery Platforms 13
Gangesh Chawla, Keshav Aggarwal, N. Yuvraj, and Ranganath M. Singari

Design of an Auxiliary 3D Printed Soft Prosthetic Thumb 27
Akash Jain, Deepanshika Gaur, Chinmay Bindal, Ranganath M. Singari, and Mohd. Tayyab

Prediction of Material Removal Rate and Surface Roughness in CNC Turning of Delrin Using Various Regression Techniques and Neural Networks and Optimization of Parameters Using Genetic Algorithm 39
Susheem Kanwar, Ranganath M. Singari, and Vipin

Finding Accuracies of Various Machine Learning Algorithms by Classification of Pulsar Stars 51
Abhishek Seth, Arjun Monga, Urvashi Yadav, and A. S. Rao

Robust Vehicle Development for Student Competitions using Fiber-Reinforced Composites 61
Nikhil Sethi, Prabhash Chauhan, Shashwat Bansal, and Ranganath M. Singari

Study and Applications of Fuzzy Systems in Domestic Products 77
Vatsal Agarwal, Sunakshi, Rani Medhashree, Taruna Singh, and Ranganath M. Singari

Study of Process Parameters in Synergic MIG Welding a Review 89
Rajat Malik and Mahendra Singh Niranjan

Comparative Study of Tribological Parameters of 3D Printed ABS and PLA Materials 95
Keshav Raheja, Ashu Jain, Chayan Sharma, Ramakant Rana, and Roop Lal

Study of Key Issues, Their Measures and Challenges to Implementing Green Practice in Coal Mining Industries in Indian Context 109
Gyanendra Prasad Bagri, Dixit Garg, and Ashish Agarwal

A Brief Review on Machining with Hybrid MQL Methods 123
Rahul Katna, M. Suhaib, Narayan Agrawal, and S. Maji

Analysis of Interrelationship Among Factors for Enhanced Agricultural Waste Utilization to Reduce Pollution 135
Nikhil Gandhi, Abhishek Verma, Rohan Malik, and Shikhar Zutshi

Enhancement of Mechanical Properties for Dissimilar Welded Joint of AISI 304L and AISI 202 Austenitic Stainless Steel 145
Yashwant Koli, N. Yuvaraj, Vipin, and S. Aravindan

Effect of 3D Printing on SCM 157
Shallu Bhasin, Ranganath M. Singari, and Harish Kumar

Seasonal Behavior of Trophic Status Index of a Water Body, Bhalswa Lake, Delhi (India) 165
Sumit Dagar and S. K. Singh

Seasonal Variation of Water Quality Index of an Urban Water Body Bhalswa Lake, Delhi (India) 179
Sumit Dagar and S. K. Singh

Material Study and Fabrication of the Next-Generation Urban Unmanned Aerial Vehicle: Aarush X2 191
Rishabh Dagur, Krovvidi Srinivas, Vikas Rastogi, Prakash Sesha, and N. S. Raghava

Intelligent Transport System: Classification of Traffic Signs Using Deep Neural Networks in Real Time 207
Anukriti Kumar, Tanmay Singh, and Dinesh Kumar Vishwakarma

Fabrication of Aluminium 6082–B$_4$C–Aloe Vera Metal Matrix Composite with Ultrasonic Machine Using Mechanical Stirrer 221
Manish Kumar Chaudhary, Ashutosh Pathak, Rishabh Goyal, Ramakant Rana, and Vipin Kumar Sharma

A Fuzzy AHP Approach for Prioritizing Diesel Locomotive Sheds a Case Study in Northern Railways Network 231
Reetik Kaushik, Yasham Raj Jaiswal, Roopa Singh, Ranganath M. Singari, and Rajiv Chaudhary

Operation of Big-Data Analytics and Interactive Advertisement for Product/Service Delineation so as to Approach Its Customers 247
Harshmit Kaur Saluja, Vinod Kumar Yadav, and K. M. Mohapatra

Effect of Picosecond Laser Texture Surface on Tribological Properties on High-Chromium Steel Under Non-lubricated Conditions 257
Sushant Bansal, Ayush Saraf, Ramakant Rana, and Roop Lal

A Statistical Approach for Overcut and Burr Minimization During Drilling of Stir-Casted MgO Reinforced Aluminium Composite 269
Anmol Gupta, Surbhi Lata, Ramakant Rana, and Roop Lal

Study and Design Conceptualization of Compliant Mechanisms and Designing a Compliant Accelerator Pedal 285
Harshit Tanwar, Talvinder Singh, Balkesh Khichi, R. C. Singh, and Ranganath M. Singari

Numerical Study on Fracture Parameters for Slit Specimens for Al2124 and Micro-alloyed Steel 297
Pranjal Shiva and Sanjay Kumar

Different Coating Methods and Its Effects on the Tool Steels: A Review 307
Sourav Kumar, Kanwarjeet Singh, Gaurav Arora, and Swati Varshney

Evaluation of Work-Related Stress Amongst Industrial Workers 315
Anuradha Kumari and Ravindra Singh

Exergoeconomic and Enviroeconomic Analysis of Flat Plate Collector: A Comparative Study 329
Prateek Negi, Ravi Kanojia, Ritvik Dobriyal, and Desh Bandhu Singh

Lead–Lag Relationship Between Spot and Futures Prices of Indian Agri Commodity Market 339
Raushan Kumar, Nand Kumar, Aynalem Shita, and Sanjay Kumar Pandey

Learnify: An Augmented Reality-Based Application for Learning 349
Himanshi Sharma, Nikhil Jain, and Anamika Chauhan

Thermal Analysis of Friction Stir Welding for Different Tool Geometries 361
Umesh Kumar Singh, Avanish Kumar Dubey, and Ashutosh Pandey

Analysis of Electrolyte Flow Effects in Surface Micro-ECG 371
Dhruv Kant Rahi, Avanish Kumar Dubey, and Nisha Gupta

Investigate the Effect of Design Variables of Angular Contact Ball Bearing for the Performance Requirement 381
Priya Tiwari and Samant Raghuwanshi

Effect of Flow of Fluid Mass Per Unit Time on Life Cycle Conversion Efficiency of Double Slope Solar Desalination Unit Coupled with N Identical Evacuated Tubular Collectors 393
Desh Bandhu Singh, Navneet Kumar, Anuj Raturi, Gagan Bansal, Akhileshwar Nirala, and Neeraj Sengar

Micro-milling Processes: A Review 403
Kriti Sahai, Audhesh Narayan, and Vinod Yadava

Strategic Enhancement of Operating Efficiency and Reliability of Process Steam Boilers System in Industry 413
Debashis Pramanik and Dinesh Kumar Singh

A Step Towards Responsive Healthcare Supply Chain Management: An Overview 431
Shashank Srivastava, Dixit Garg, and Ashish Agarwal

Designing of Fractional Order Controller Using SQP Algorithm for Industrial Scale Polymerization Reactor 445
D. Naithani, M. Chaturvedi, P. K. Juneja, and V. Joshi

Additive Manufacturing in Supply Chain Management: A Systematic Review 455
Archana Devi, Kaliyan Mathiyazhagan, and Harish Kumar

A Step Towards Next-Generation Mobile Communication: 5G Cellular Mobile Communication 465
Ayush Kumar Agrawal and Manisha Bharti

Efficacy and Challenges of Carbon Trading in India: A Comparative Analysis 477
Naveen Rai and Meha Joshi

Micro-structural Investigation of Embedded Cam Tri-flute Tool Pin During Friction Stir Welding 485
Nadeem Fayaz Lone, Arbaz Ashraf, Md Masroor Alam, Azad Mustafa, Amanullah Mahmood, Muskan Siraj, Homi Hussain, and Dhruv Bajaj

Design and FEM Analysis of Connecting Rod of Different Materials 493
Sanjay Kumar, Vipin Verma, and Neelesh Gupta

Numerical Study on Heat Affected Zone and Material Removal Rate of Shape Memory Alloy in Wire Electric Discharge Machining 509
Deepak Kumar Gupta, Avanish Kumar Dubey, and Alok Kumar Mishra

A Hybrid Multi-criteria Decision-Making Approach for Selection of Sustainable Dielectric Fluid for Electric Discharge Machining Process 519
Md Nadeem Alam, Zahid A. Khan, and Arshad Noor Siddiquee

Preference Selection Index Approach as MADM Method for Ranking of FMS Flexibility 529
Vineet Jain, Mohd. Iqbal, and Ashok Kumar Madan

Impact of Additive Manufacturing in Value Creation, Methods, Applications and Challenges 543
Rishabh Teharia, Gulshan Kaur, Md Jamil Akhtar, and Ranganath M. Singari

3D Printing: A Review of Material, Properties and Application 555
Gulshan Kaur, Rishabh Teharia, Md Jamil Akhtar, and Ranganath M. Singari

Effect of Infill Percentage on Vibration Characteristic of 3D-Printed Structure 565
Pradeep Kumar Yadav, Abhishek, Kamal Singh, and Jitendra Bhaskar

Study of Slender Carbon Fiber-Reinforced Columns Filled with Concrete 575
Utkarsh Roy, Shubham Khurana, Pratikshit Arora, and Vipin

Factors Affecting Import Demand in India: A Principal Component Analysis Framework 585
Khyati Kathuria and Nand Kumar

Theoretical and Statistical Analysis of Inventory and Warehouse Management in Supply Chain Management—A Case Study on Small-Scale Industries 597
Mahesh R. Latte and Channappa M. Javalagi

Evaluation of Separation Efficiency of a Cyclone-Type Oil Separator 609
Ujjwal Suri, Shraman Das, Utkarsh Garg, and B. B. Arora

Energy Analysis of Double Evaporator Ammonia Water Vapour Absorption Refrigeration System 619
Deepak Panwar and Akhilesh Arora

Blockchain Technology as a Tool to Manage Digital Identity: A Conceptual Study 635
Ruchika Singh Malyan and Ashok Kumar Madan

Evolution in Micro-friction Stir Welding 649
Nadeem Fayaz Lone, Md Masroor Alam, Arbaz Ashraf,
Amanullah Mahmood, Nabeel Ali, Dhruv Bajaj, and Soumyashri Basu

**Traffic Noise Modelling Considering Traffic Compositions
at Roundabouts** 657
Anupam Thakur and Ramakant Rana

**Hydrogen Embrittlement Prevention in High Strength Steels
by Application of Various Surface Coatings-A Review** 673
Sandeep Kumar Dwivedi and Manish Vishwakarma

**Commencement of Green Supply Chain Management Barriers:
A Case of Rubber Industry** 685
Somesh Agarwal, Mohit Tyagi, and R. K. Garg

**Estimation of Critical Key Performance Factors of Food Cold Supply
Chain Using Fuzzy AHP Approach** 701
Neeraj Kumar, Mohit Tyagi, and Anish Sachdeva

A Short Review on Machining with Ultrasonic MQL Method 713
Rahul Katna, M. Suhaib, Narayan Agrawal, and S. Maji

**Professional Values and Ethics: Challenges, Solutions
and Different Dimension** 721
Shalini Sharma

**Design of 3D Printed Fabric for Fashion and Functional
Applications** 729
Arpit Singh, Pradeep Kumar Yadav, Kamal Singh, Jitendra Bhaskar,
and Anand Kumar

**Fabrication and Characterization of PVA-Based Films Cross-Linked
with Citric Acid** 737
Naman Jain, Gaurang Deep, Ashok Kumar Madan, Madhur Dubey,
Nomendra Tomar, and Manik Gupta

Characterization of Bael Shell (*Aegle marmelos*) Pyrolytic Biochar ... 747
Monoj Bardalai and D. K. Mahanta

**Metal Foam Manufacturing, Mechanical Properties and Its
Designing Aspects—A Review** 761
Rahul Pandey, Piyush Singh, Mahima Khanna, and Qasim Murtaza

Emerging Trends in Internet of Things 771
Yash Agarwal and K. A. Nethravathi

**Selection of Best Dispatching Rule for Job Sequencing Using
Combined Best–Worst and Proximity Index Value Methods** 783
Shafi Ahmad, Ariba Akher, Zahid A. Khan, and Mohammed Ali

Thermal Performance Investigation of a Single Pass Solar Air Heater 793
Ovais Gulzar, Adnan Qayoum, and Rajat Gupta

Modelling of Ambient Noise Levels in Urban Environment 807
S. K. Tiwari, L. A. Kumaraswamidhas, and N. Garg

Development and Characterizations of ZrB$_2$–SiC Composites Sintered Through Microwave Sintering 815
Ankur Sharma and D. B. Karunakar

Characterization of Ni-Based Alloy Coating by Thermal Spraying Process 825
Manmeet Jha, Deepak Kumar, Pushpendra Singh, R. S. Walia, and Qasim Murtaza

A Review on Solar Panel Cleaning Through Chemical Self-cleaning Method 835
Ashish Jaswal and Manoj Kumar Sinha

Investigations on Process Parameters of Wire Arc Additive Manufacturing (WAAM): A Review 845
Mayank Chaurasia and Manoj Kumar Sinha

A State-of-the-Art Review on Fused Deposition Modelling Process 855
Kamal Kishore and Manoj Kumar Sinha

3D Modelling of Human Joints Using Reverse Engineering for Biomedical Applications 865
Deepak Kumar, Abhishek, Pradeep Kumar Yadav, and Jitendra Bhaskar

Institutional Distance in Cross-Border M&As: Indian Evidence 877
Sakshi Kukreja, Girish Chandra Maheshwari, and Archana Singh

Synthesis and Characterization of PVDF/PMMA-Based Piezoelectric Blend Membrane 889
Ashima Juyal and Varij Panwar

Comparative Study of Retrofitted Columns Using Abaqus Software 897
Geeta Singh, Tarun Shokeen, and Vidrum Gaur

Optimal Pricing and Procurement Decisions for Items with Imperfect Quality and Fixed Shelf Life Under Selling Price Dependent and Power Time Pattern Demand 907
Sonal Aneja and K. K. Aggarwal

Experimental Analysis of Portable Optical Solar Water Heater 925
Hasnain Ali, Ovais Gulzar, K. Vasudeva Karanth, Mohammad Anaitullah Hassan, and Mohammad Zeeshan

CO_2 Laser Micromachining of Polymethyl Methacrylate (PMMA): A Review 939
Shrikant Vidya, Reeta Wattal, Lavepreet Singh, and P. Mathiyalagan

Design of Delay Compensator for a Selected Process Model 947
Oumayma Benjeddi, M. Chaturvedi, P. K. Juneja, G. Yadav, V. Joshi, and R. Mishra

Fly Ash, Rice Husk Ash as Reinforcement with Aluminium Metal Matrix Composite: A Review of Technique, Parameter and Outcome 953
Jagannath Verma and Harish Kumar

Optimization of CNC Lathe Turning: A Review of Technique, Parameter and Outcome 963
Vivek Joshi and Harish Kumar

Innovations and Future of Robotics 975
Ayush Kumar Agrawal, Pritam Pidge, Manisha Bharti, M. Prabhat Dev, and Prashant Kaduba Kedare

Optimization of EDM Process Parameters: A Review of Technique, Process, and Outcome 981
Akash Gupta and Harish Kumar

Impact Behavior of Deformable Pin-Reinforced PU Foam Sandwich Structure 997
Shivanku Chauhan, Mohd. Zahid Ansari, Sonika Sahu, and Afzal Husain

Sensitivity Improvement of Piezoelectric Mass Sensing Cantilevers Through Profile Optimization 1007
Shivanku Chauhan, Mohd. Zahid Ansari, Sonika Sahu, and Afzal Husain

Current Status, Applications, and Factors Affecting Implementation of Additive Manufacturing in Indian Healthcare Sector: A Literature-Based Review 1015
Bhuvnesh Chatwani, Deepanshu Nimesh, Kuldeep Chauhan, Mohd Shuaib, and Abid Haleem

System Optimization for Economic and Sustainable Production and Utilization of Compressed Air (A Case Study in Asbestos Sheet Manufacturing Plant) 1031
Debashis Pramanik and Dinesh Kumar Singh

Investigating the Prospects of E-waste and Plastic Waste as a Material for Partial Replacement of Aggregates in Concrete 1047
Abhishek Singh, Ahmad Sahibzada, Deepak Saini, and Susheel Kumar

Investigation of Combustion, Performance and Emission of Aluminium Oxide Nanoparticles as Additives in CI Engine Fuels: A Review 1055
Manish Kumar, Naushad A. Ansari, and Samsher

A Review of CI Engine Performance and Emissions with Graphene Nanoparticle Additive in Diesel and Biodiesel Blends 1065
Varun Kr Singh, Naushad A. Ansari, and Akhilesh Arora

Synthesis and Study of a Novel Carboxymethyl Guar Gum/ Polyacrylate Polymeric Structured Hydrogel for Agricultural Application 1073
Khushbu, Ashank Upadhyay, and Sudhir G. Warkar

Artificial Neural Network (ANN) for Forecasting of Flood at Kasol in Satluj River, India 1085
Abhinav Sharma and Anshu Sharma

Sewage Treatment Using Alum with Chitosan: A Comparative Study 1095
Jaya Maitra, Athar Hussain, Mayank Tripathi, and Mridul Sharma

Automatic Plastic Sorting Machine Using Audio Wave Signal 1111
S. M. Devendra Kumar, S. Prashanth, and Rani Medhashree

GSM Constructed Adaptable Locker Safety Scheme by Means of RFID, PIN Besides Finger Print Expertise 1123
S. M. Devendra Kumar, B. Manjula, and Rani Medhashree

Low-Voltage Squarer–Divider Circuit Using Level Shifted Flipped Voltage Follower 1131
Swati Yadav and Bhawna Aggarwal

Memristor-Based Electronically Tunable Unity-Gain Sallen–Key Filters 1141
Bhawna Aggarwal, Manshul Arora, Marsheneil Koul, and Maneesha Gupta

Influence of Target Fields on Impact Stresses and Its Deformations in Aerial Bombs 1153
Prahlad Srinivas Joshi and S. K. Panigrahi

A Review of Vortex Tube Device for Cooling Applications 1161
Sudhanshu Sharma, Kshitiz Yadav, Gautam Gupta, Deepak Aggrawal, and Kulvindra Singh

Implementation of Six-Sigma Tools in Hospitality Industry: A Case Study 1171
Nishant Bhasin, Harkrit Chhatwal, Aditya Bassi, and Shubham Sharma

Impact of Integrating Artificial Intelligence with IoT-Enabled Supply Chain—A Systematic Literature Review 1183
Ranjan Arora, Abid Haleem, P. K. Arora, and Harish Kumar

Experimental Study for the Health Monitoring of Milling Tool Using Statistical Features 1189
Akanksha Chaudhari, Pavan K. Kankar, and Girish C. Verma

PVT Aware Analysis of ISCAS C17 Benchmark Circuit 1199
Suruchi Sharma, Santosh Kumar, Alok Kumar Mishra, D. Vaithiyanathan, and Baljit Kaur

Correction to: Advances in Manufacturing and Industrial Engineering C1
Ranganath M. Singari, Kaliyan Mathiyazhagan, and Harish Kumar

About the Editors

Dr. Ranganath M. Singari is a Professor in the Department of Mechanical, Production & Industrial Engineering and heads the Department of Design, Delhi Technological University, India. He is a graduate in Industrial Production Engineering from Karnataka University. He completed his M.Tech in Computer Technology & Applications and Ph.D. from the Department of Production Engineering from University of Delhi, India. He has more than 60 international publications in conference and reputed journals. He is also a reviewer for reputed journals. Dr. Singari has organised several international conferences, seminars/workshops, industry-institute interactions and 6 FDP/SDP/STTP. He also serves as Chairman, Production Engineering, Skill India Programme, DTTE, Delhi. He is an expert member of several selection committees for technical, teaching and administrative positions. His research interest is materials, manufacturing, industrial management, production management, CAD/CAM, supply chain management, multi-criteria decision making and sustainable lean manufacturing. He has 25 years of research and teaching experience.

Dr. Kaliyan Mathiyazhagan is currently working as an Associate Professor in the Department of Mechanical Engineering, Amity University, India. He pursued his Ph.D. from the Department of Production Engineering, National Institute of Technology, Tiruchirappalli, Tamil Nadu. He was also a visiting research fellow at the University of Southern Denmark. He has more than 60 international publications in reputed journals and one of his papers received the best paper award in NCAME 2019, NIT Delhi. Dr. K. Mathiyazhagan is an associate editor of Environment, Development and Sustainability. He is also an editorial member of more than 5 international journals. He has served as a guest editor for several special issues in international journals and is an active reviewer of more than thirty reputed international journals. His research interest is green supply chain management, sustainable supply chain management, multi-criteria decision making, third party logistic provider, sustainable lean manufacturing, public distribution system, and Lean Six Sigma. He has more than 10 years of research and teaching experience.

Dr. Harish Kumar is currently working as an Assistant Professor at the National Institute of Technology, Delhi. He has more than 15 years of research and academic experience and has served as a scientist at different grade in CSIR - National Physical Laboratory, India (NPLI). He has been an active researcher in the area of mechanical measurement and metrology. He has worked as a guest researcher at the National Institute of Standards and Technology, USA in 2016. He has been instrumental in the ongoing redefinition of the kilogram in India. He has authored more than 70 publications in peer reviewed journals and conferences. He is an active reviewer of many reputed journals related to measurement, metrology and related areas. He has served as a guest editor of different peer reviewed journals.

A Review on the Fabrication of Surface Composites via Friction Stir Processing and Its Modeling Using ANN

Kartikeya Bector, Aranyak Tripathi, Divya Pandey, Ravi Butola, and Ranganath M. Singari

Abstract Friction Stir Processing (FSP) is a surface modification and surface composite fabrication technique that was first theorized and demonstrated in 2002. Since then, it has grown exponentially in the industry due to its efficiency, ease of usability and various other advantages over conventional surface modification and fabrication processes. Artificial Neural Networks provide a computational model that can handle complex relationships between the various determinants involved in FSP and the effect they have on the final output. ANN has been used extensively to predict the impact of various FSP determinants and hence model the most efficient values of these determinants for various base metals. This paper has tried to encapsulate the plethora of research done on the optimization of FSP determinants using ANN architecture.

Keywords FSP · MMCs · ANN

1 Introduction

Most metal matrix composites, reinforced with ceramic phases or another desired metal, exhibit a higher elastic modulus, higher strength, higher Vickers hardness and better resistance to fatigue, creep and wear than the base metal. This makes them suitable for the aerospace and automobile industries. Since the ceramic reinforcement materials introduced in the base metal matrix are non-deformable and brittle, one of the drawbacks of such composites, especially in the case of such ceramic additives, is the loss of crucial properties of the base material—ductility and toughness. Thus, the composites formed have a limited application. The solution for this is sought from the

---

K. Bector · A. Tripathi · D. Pandey · R. Butola (✉)
Departments of Mechanical Engineering, Delhi Technological University, New Delhi, Delhi 110042, India
e-mail: Ravibutola33855@gmail.com

R. M. Singari
Departments of Production and Industrial Engineering, Delhi Technological University, New Delhi, Delhi 110042, India

© The Author(s), under exclusive license to Springer Nature Singapore Pte Ltd. 2021
R. M. Singari et al. (eds.), *Advances in Manufacturing and Industrial Engineering*,
Lecture Notes in Mechanical Engineering,
https://doi.org/10.1007/978-981-15-8542-5_1

fact that for a majority of applications, the surface properties of the composite play a major role. Following this, Friction Stir Processing (FSP) was ideated and pioneered by Mishra et al. [1] in 2002. This technology stems from Friction Stir Welding (FSW) that was actualized back in 1991. It was derived from the process of using friction to weld joints even in aluminium, titanium and various other alloys. FSP is generally utilized for lightweight and flexible metals like aluminium and magnesium, but only after the base metal has experienced FSP to acquire certain desirable properties and make up for their absence of strength and hardness. Chaudhary et al. [2] studied the consequences of using FSP on different alloys like Mg_4Y_3Nd(WE43), $Mg-ZrSiO_4-N_2O_3$, Al–Si hypoeutectic A356 alloy, 5210 steel (WC-12% CO coated). The difference in the properties of friction stir processed alloys were observed as the processing was executed at different angles and speeds. A variety of ingenious materials may be used as reinforcement while preparing friction stir processed composites. The effects and properties of reinforcing materials such as Silicon Carbide, Graphite, Fly ash, Rice husk ash and boron halide were examined in detail by Butola et al. [3] for the preparation of surface composites using FSP. Properties like corrosion behaviour, tensile strength, hardness, wear-resistance and so forth were studied rigorously and summarized.

There exist several surface modification techniques like Laser Surface Engineering [4], high-energy electron beam irradiation [5], high-energy laser melt treatment [6], plasma spraying [7], stir casting [8], etc. The above-described techniques for the formation of surface composites rely on liquid-phase processing at high temperatures. Because of the nature of the processing, it is inescapable that an interfacial reaction occurs between the reinforcement and the metal matrix. Some other impeding phases may also be formed. The aforementioned problems may be mitigated by carrying out the processing at a temperature below the melting point of the substrate.

Artificial neural network is an innovative prediction model that utilizes existing data to train an intuitive network of neurons so that accurate complex predictions can be made. Okuyucu et al. [9] pioneered the use of ANN in friction stir welding. An ANN model was created for the simulation and analysis of the interrelationship of FSW determinants and the mechanical properties of the aluminium plates. Similarly, ANN models considering various FSP determinants were developed and implemented on friction stir processed alloys of aluminium and magnesium.

2 Principle and Effect of Parameters of FSP

Friction Stir Processing involves the heating and plasticization of the substrate and reinforcement material due to the friction created by the tool (with or without the protruding mandrel). FSP can be carried out on a conventional FSW machine [10]. During FSP, a non-consumable tool turns at a high RPM and gradually slides into the workpiece while applying a power pivotally until the shoulder of the instrument interacts with the outside of the workpiece, which brings about erosion. The rotating

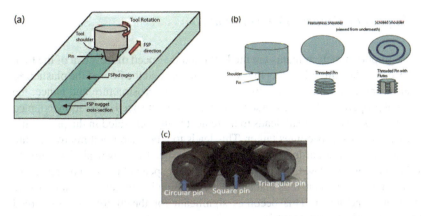

Fig. 1 FSP for frictional modification of surface layers, **a** process diagram, **b** tool design [2], **c** a variety of mandrel designs of tools for FSP [11]

tool then moves along the workpiece in the desired direction of FSP. A substantial amount of heat is engendered due to traction between the shoulder of the tool and the workpiece. The plasticized and heated material is forced along the modification line and underneath the back-up rim to the end of the tool, where it's compacted and blended because of severe deformation before it cools. FSP is a versatile method that can be used for manufacturing, modification as well as fabrication of materials with special properties (Fig. 1).

Moreover, FSP is an eco-friendly process as the heat energy required is generated through friction [12]. This leads to the formation of a dynamically recrystallized fine grain structure. Friction stir processed areas can be generated to the depths of 0.5–50 mm, with a progressive evolution from a fine-grained, thermodynamically worked microstructure to the elementary original microstructure [13]. The processing region in FSP, like FSW, is usually categorized into a thermo-mechanically affected zone (TMAZ), a stir zone (SZ), heat-affected zone (HAZ) and base metal zone(BM) [14, 15]. The SZ undergoes acute plastic deformation and primarily consists of homogeneously refined grains which are equiaxial and whose dimensions are contracted monumentally in comparison to the principal metal. During recrystallization, the formation of grains is promoted by the particle-stimulated nucleation when particles of the reinforcement material are introduced to the metal matrix. During the dynamic recrystallization process, the uniform dispersion of fine particles can inhibit grain growth. This is in accordance with the Zener–Holloman mechanism. This occurs due to the pinning action on the grain boundaries leading to a substantial amelioration of the microstructure [16]. The factors affecting the FSP modified substrate are tool traversal speed, tilt angle of tool, RPM, tool plunge depth and dimensions of the tool [12].

2.1 Impact of Process Determinants

The most important determinants are the RPM and the speed of traversal of the tool. This is because they directly affect the addition of heat and the flow of the plasticized material during FSP. This drastically influences the microstructure and hence, the mechanical properties of the processed material. A higher speed of rotation, coupled with a lower speed of traversal, leads to more heat being generated in the processing region, which in itself becomes larger. This leads to a better-refined microstructure and an increase in the hardness [17–20]. A lower speed of traversal also helps to control unusual grain growth [21]. Moreover, a higher speed of rotation of the tool or lower speed of traversal ensures that a higher augmentation of heat and more plastic deformation is achieved. This becomes significant for the mixture of reinforced particles and the base metal matrix.

Multi-pass FSP has been proven to better the material properties by aiding the plastic deformation of processed materials. Increasing the number of passes and reversing the direction of rotation of the tool with every subsequent pass ensures that the composites manufactured by multi-pass FSP have a progressively uniform phase dispersion and strengthening because of thoroughly propagated in situ reactions as compared to single-pass FSP [22]. The procedure may be designed accordingly by selecting the percentage of overlap viz. 5, 10, 25, 50, 75% and thus desirable dimensions of the modified surface may be obtained [12].

2.2 Impact of Tool (Pin) Geometry

Tools used to stir the material in FSP may be with or without a mandrel and are usually non-consumable [23]. A stir tool is usually used with the mandrel and the mandrel-less tools are used for either modifying the surface of the material or processing of reinforcement material into the native metal during the fabrication of composites. The tool size and the geometry of the tool pin significantly affect the amount of heat produced as well as the flow of the material during processing. A large shoulder diameter of the tool results in the frictional heat being more concentrated. Subsequently, the particles in the second phase are refined better and thus the microstructure is also more stable. Thus, the influence of pin profile is crucial and this results in the temperature being the lowest when a conical pin is plunged into the material. Three different pin profiles were studied by Butola et al. [11] by observing their effect on SiC, RHA and B_4C-reinforced composites with AA7075 as the native metal. The coda showed that in the stir zone, which had also decreased in size, the most homogeneous distribution of reinforcement particles was seen in the case of a square mandrel.

2.3 Fabrication of Surface Composites

Many methods have been explored by various researchers to fabricate composites containing reinforcement particles. High modulus of elasticity, high strength, etc. are some of the inherent properties possessed by metal matrix composites. Surface composites may be fabricated from these to enhance the surface properties of these composites. The main challenge affecting this has been the introduction of these particles in the base metal during FSP. Apart from casting [24], a number of other methods were also tried. The SiC powder was mixed with methanol to form a paste, which was, in turn, applied evenly onto the surface of the workpiece before FSP. But, the particles slipped easily and were splashed out of the surface due to the rotation of the tool. This leads to an irregular distribution of the reinforced phases and inefficient use of material [1]. Some other ways to incorporate the reinforced particles in the matrix include making grooves on the surface of the base metal plates and adding the particles in it. To prevent splashing, the grooves are pre-processed with a pin-less tool using FSP [25, 26]. Similarly, blind holes may be drilled into the surface of the workpiece [27]. In both of these cases, final processing via FSP ensures that the particles are uniformly dispersed in the metal matrix and the homogeneity in the dispersion of the strengthening phases vastly improves the properties of the surface composite hence formed.

Liquid-phase techniques like laser cladding and plasma spraying are often used for this purpose. Butola et al. [8] used stir casting for the introduction of natural fibres like bagasse, banana and jute to form metal matrix composites and studied their effect on the mechanical properties of the base metal. In another study, Butola et al. [28] used stir casting and ball milling to fabricate and refine MMCs and study the effect of Groundnut Shell Ash (GSA), Rice Husk Ash (RHA) and ash-forms of some other natural fibres as reinforcement. Due to the formation of a liquid-phase in the above-mentioned techniques, the deleterious reactions mentioned at the beginning of the research may happen. Using FSP for the same will ensure that a finely distributed phase of strengthening particles may be obtained while keeping the SZ in a solid-state. This effectively prevents the formation of detrimental phases and any unwanted interfacial reactions. The commonly added reinforcement materials in FSP include SiC, B_4C, GNPs and Al_2O_3.

3 Artificial Neural Networks

Artificial neural network, popularly known as ANN, is a biology-inspired architecture of nodes (neurons) that are extensively interconnected [29]. It is a complex learning model that trains on a set of data, analyzes and learns the pattern followed in it and then predicts the result of a similar dataset. It processes the information in the datasets using a connectionist approach and multiple functions are run on it simultaneously. Synapses link neurons and a weight factor is associated with each of

Table 1 The inputs and outputs used by some previous works [31]

Inputs	Outputs	References
Welding speed (WS) Rotational speed (RPM)	Yield strength (YS), Length variation	[9]
WS, RPM	Tensile shear force, Hardness	[32]
WS, RPM	Tensile strength (TS), YS, Elongation	[33]
WS, RPM, Axial force (F)	TS	[34]
WS, RPM, Tool shoulder diameter	TS	[29]

them. ANNs are data processing models that mirror the role of the biological matrix, made out of neurons and are utilized to understand convoluted capacities in different applications by determining the nonlinear relationship between the involved, influential determinants and the output(s) obtained. The model has three layers viz. input, hidden and output layers. The input layer comprises of all the input factors. Data, via the input layer, is then processed through one or more hidden layers and the corresponding output vector is calculated in the final layer. One of the most popular learning algorithms is the backpropagation algorithm [30]. One of the primary hurdles while constructing an ANN model is to choose an appropriate network framework, which includes the activation function and the number of neurons in the hidden layer. Largely, tentation is used for the same. Since its introduction, ANN has been used by various researchers to study the effect of various determinants on a process under scrutiny. Friction Stir Processing is one such process. Since the advent of this processing method, the variables that control the resultant surface composite formed, have been closely studied. Since there is no explicit correlation for estimating target determinants based on the input factors, usually target determinants are modeled by a three-layer perceptron ANN using data obtained from mechanical and microstructural experiments (Table 1).

Neural architecture with the following [35]:

- Input has **r** determinants
- Output has **s** determinants
- **p**—inputs
- **w**—weight matrix
- **b**—bias vectors
- **f**—transfer function in neurons
- **a**—transfer functions in outputs.

Training of a neuron is carried out by multiplying the input vector with a vector of weights, followed by the addition of a bias vector. The result of this processing is then fed into the hidden layer. The sum of all the inputs is then fed into a transfer function. The output thus obtained is the output of the neural network. This output is then compared with the corresponding experimental values obtained. Due to the difference in the expected and practical values, an error vector is generated. In case this error value exceeds the acceptable error limit, the output is propagated back

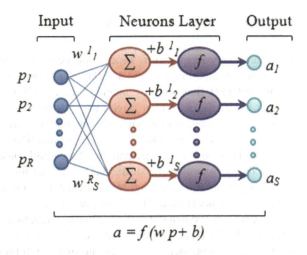

Fig. 2 Tangent sigmoid (Tansig) transfer function [31]

through the network and appropriate corrections in the weights and biases are made till the desirable values are attained.

Logarithmic sigmoid (Logsig) transfer function:

$$\psi(x) = \frac{1}{1+e^{-x}} \quad (1)$$

Linear transfer function (Fig. 2):

$$\chi(x) = \text{linear}(x) \quad (2)$$

3.1 Training of ANN

The dataset that is available is usually split into two parts, usually in a 3:1 ratio [36]. The bigger of the two datasets are used to train the ANN, while the other one acts as the testing dataset. The final outputs obtained are compared with the expected values, error calculated and in case the error exceeds the permitted limit, the output is sent back through the network and required adjustments in the weights and biases are made. The aim of the feed-forward backpropagation (BP) algorithm is to minimize the sum of the mean squared errors obtained between the calculated and practical values of output and the minimization is achieved via the gradient descent method. BP is one of the most efficient algorithms for the optimization of the weights and biases of a multi-layer supervised feed-forward network.

3.2 Implementation of ANN

There are many input factors in a neural network, changing which will, in turn, change the method of operation and the overall precision and processing speed of the network [37]. Some of these include but are not limited to the number of neurons in each hidden layer, hidden layers, the bias used and the rate of training of the network. As the count of hidden layers and nodes(neurons) in each layer are crucial in the overall functioning and performance of the ANN (act as the primary processing entity of the network), they are chosen after careful consideration. They are usually chosen by tentation because of the lack of a fixed formula to determine the same. Increasing them does not always lead to better performance in terms of the speed and accuracy of the network. In fact, it increases the complexity of the network, which in turn, tends to slow down the network after a certain limit. An unstable network may be obtained if the rate of training is increased or decreased beyond a certain limit. All the input and output determinants are normalized to prevent them from scattering. This means that the values of these determinants are divided by the maximum value and hence reduced to a value between 0 and 1. This decreases the scattering of the determinants.

3.3 Performance Evaluation of ANN

A lot of statistical models are at one's disposal to determine the performance of any neural network [26]. The most common ones include the Pearson coefficient of correlation (PCC) and mean relative error (MRE). Their equations are as follows:

Where,

$$\text{PCC} = \frac{\sum_{i=1}^{n}(f_{\text{EXP},i} - F_{\text{EXP}})(f_{\text{ANN},i} - F_{\text{ANN}})}{\sqrt{\sum_{i=1}^{n}\left((f_{\text{EXP},i} - F_{\text{EXP}})^2 (f_{\text{ANN},i} - F_{\text{ANN}})^2\right)}} \quad (3)$$

$$\text{MRE} = \frac{1}{n}\sum_{i=1}^{n}\frac{|f_{\text{ANN},i} - f_{\text{EXP},i}| \times 100}{f_{\text{EXP},i}} \quad (4)$$

$$F_{\text{EXP}} = \frac{1}{n}\sum_{i=1}^{n} f_{\text{EXP},i}, \quad F_{\text{ANN}} = \frac{1}{n}\sum_{i=1}^{n} f_{\text{ANN},i},$$

$$f_{\text{EXP}} = \text{Experimental}, \quad f_{\text{ANN}} = \text{Predicted} \quad (5)$$

The methodology of constructing an optimum ANN architecture is as follows:

1. Start
2. Normalization of data (inputs and outputs)
3. Feeding the normalized data to the hidden layer of the ANN

4. Determining the optimum values of the determinants involved
5. Executing the training algorithm of the network
6. Obtaining the Pearson coefficient
7. If PCC is atleast 0.99, continue. If it is less than that, go back to the optimization step (step 4)
8. Till the convergence of experimental and predicted data is not obtained, the processing is to be continued
9. Weights and biases vectors are obtained
10. Analysis is done based on the function used in the model
11. Final error is calculated
12. End.

4 Summary

Friction Stir Processing has a wide variety of applications which include, but are not restricted to, surface modification and fabrication. There are numerous ways to incorporate reinforcement material into the surface of the base metal and these have been optimized over time. Friction Stir Processing can be used for achieving superplasticity, producing alloys that possess special properties, improving the fatigue strength of welded joints, etc. ANN has been used extensively and successfully to figure out the most efficient parameter values for FSP by various researchers.

5 Future Scope

Friction Stir Processing has rapidly made a niche for itself in the industry. As it is a surface modification and fabrication technique that overrides the drawbacks of the existing technologies by a mile, the acceptance of this technique has exponentially grown. Moreover, the fact that FSP can be implemented on existing CNC milling machines, it is being adapted quickly far and wide through the industry. In addition to this, the varied applications of FSP, combined with its arsenal of advantages and easy adaptability, have made FSP very promising. Since artificial neural networks are also increasingly being used all through the world, the ease of determination of the relationship between the various determinants and variables associated with FSP in a mathematical way, which can accurately predict the impact of every parameter on the resultant MMC, has not only drastically reduced the cost of experimentation and research but also made it far more accessible than before.

References

1. Mishra RS, Ma ZY, Charit I (2003) Friction stir processing: a novel technique for fabrication of surface composite. Mater Sci Eng, A 341:307–310
2. Chaudhary A, Dev AK, Goel A, Butola R, Ranganath MS (2018) The mechanical properties of different alloys in friction stir processing: a review. Mater Today: Proc 5:5553–5562
3. Butola R, Singari RM, Bandhu A, Walia RS (2017) Characteristics and properties of different reinforcements in hybrid aluminium composites: a review. Int J Adv Prod Ind Eng IJAPIE-SI-MM 511:71–80
4. Pantelis D, Tissandier A, Manolatos P, Ponthiaux P (1995) Formation of wear resistant Al–SiC surface composite by laser melt–particle injection process. Mater Sci Technol 11:299
5. Lee CS, Oh JC, Lee S (2003) Improvement of hardness and wear resistance of (TiC, TiB)/Ti-6Al-4V surface-alloyed materials fabricated by high-energy electron-beam irradiation. Metall Mater Trans A 34:1461. https://doi.org/10.1007/s11661-003-0258-y
6. Ricciardi G, Cantello M, Mollino G, Varani W, Garlet E (1989) Proceedings of 2nd international seminar on surface engineering with high energy beam, science and technology, CEMUL-IST, Lisbon, Portugal, pp 415–423
7. Gui MC, Kang SB (2000) 6061Al/Al–SiCp bi-layer composites produced by plasma-spraying process. Mater Lett 46:296
8. Butola R, Pratap C, Shukla A, Walia RS (2019) Effect on the mechanical properties of aluminum-based hybrid metal matrix composite using stir casting method. Mater Sci Forum 969:253–259
9. Okuyucu H, Kurt A, Arcaklioglu E (2007) Artificial neural network application to the friction stir welding of aluminium plates. Mater Des 28:78–84
10. Li K, Liu X, Zhao Y (2019) Research status and prospect of friction stir processing technology. Coatings 9:129
11. Butola R, Ranganath MS, Murtaza Q (2019) Fabrication and optimization of AA7075 matrix surface composites using Taguchi technique via friction stir processing (FSP). Eng Res Express **1**, 025015 (2019)
12. Weglowski M (2014) Friction stir processing technology—new opportunities. Weld Int 28
13. Khalkhali A, Ebrahimi-Nejad S, Malek NG (2018) Comprehensive optimization of friction stir weld parameters of lap joint AA1100 plates using artificial neural networks and modified NSGA-II. Mater Res Express 5(6). Published 6 June 2018 • © 2018 IOP Publishing Ltd.
14. Padhy GK, Wu CS, Gao S (2018) Friction stir based welding and processing technologies—processes, parameters, microstructures and applications: a review. J Mater Sci Technol 34:1–38
15. Tamadon A, Pons DJ, Sued K, Clucas D (2017) Development of metallographic etchants for the microstructure evolution of A6082–T6 BFSW welds. Metals 7:423
16. Khodabakhshi F, Gerlich AP, Svec P (2017) Fabrication of a high strength ultra-fine grained Al-Mg-SiC nanocomposite by multi-step friction-stir processing. Mater Sci Eng A Struct Mater Prop Microstruct Process 698:313–325
17. Sathiskumar R, Murugan N, Dinaharan I, Vijay SJ (2013a) Role of friction stir processing parameters on microstructure and microhardness of boron carbide particulate reinforced copper surface composites. Sadhana Acad Proc Eng Sci 38:1433–1450
18. Tamadon A, Pons DJ, Sued K, Clucas D (2018) Formation mechanisms for entry and exit defects in bobbin friction stir welding. Metals 8:33
19. Sun P, Wang K, Wang W, Zhang X (2008) Influence of process parameter on microstructure of AZ31 magnesium alloy in friction stir processing. Hot Work Technol 37:99
20. Barmouz M, Givi MKB, Seyfi J (2011) On the role of processing parameters in producing Cu/SiC metal matrix composites via friction stir processing: investigating microstructure, microhardness, wear and tensile behavior. Mater Charact 62:108–117
21. Jana S, Mishra RS, Baumann JA, Grant G (2010) Effect of process parameters on abnormal grain growth during friction stir processing of a cast Al alloy. Mater Sci Eng A Struct Mater Prop Microstruct Process 528:189–199

22. Mehta KP, Badheka VJ (2016) Effects of tilt angle on the properties of dissimilar friction stir welding copper to aluminum. Mater Manuf Process 31:255–263
23. Butola R, Murtaza Q, Singari RM (2019) Advances in computational methods in manufacturing. In: Narayanan R, Joshi S, Dixit U (eds) Lecture Notes on Multidisciplinary industrial engineering. Springer, Singapore, pp 337–348
24. Yang R, Zhang ZY, Zhao YT, Chen G, Guo YH, Liu MP, Zhang J (2015) Effect of multi-pass friction stir processing on microstructure and mechanical properties of Al 3 Ti/A356 composites. Mater Charact 106:62–69
25. Ahmadkhaniha D, Fedel M, Sohi MH, Hanzaki AZ, Deflorian F (2016) Corrosion behavior of magnesium and magnesium-hydroxyapatite composite fabricated by friction stir processing in Dulbecco's phosphate buffered saline. Corros Sci 104:319–329
26. Sathiskumar R, Murugan N, Dinaharan I, Vijay SJ (2013b) Characterization of boron carbide particulate reinforced in situ copper surface composites synthesized using friction stir processing. Mater Charact 84:16–27
27. Reddy GM, Rao AS, Rao KS (2013) Friction stir surfacing route: effective strategy for the enhancement of wear resistance of titanium alloy. Trans Indian Inst Met 66:231–238
28. Butola R et al (2019) Experimental studies on mechanical properties of metal matrix composites reinforced with natural fibres. Ashes SAE 2019-01-1123 1–11 Technical Paper 2019-01-1123
29. Ghetiya N, Patel K (2014) Prediction of tensile strength in friction stir welded aluminium alloy using artificial neural network. Procedia Technol 14:274–281
30. Ahmad S, Singari R M, Mishra R S (2020) Modelling and optimisation of magnetic abrasive finishing process based on a non-orthogonal array with ANN-GA approach. Trans IMF 98:186–198
31. Zeidabadi SRH, Daneshmanesh H (2017) Fabrication and characterization of in-situ Al/Nb metal/intermetallic surface composite by friction stir processing. Mater Sci Eng A Struct Mater Prop Microstruct Process 702:189–195
32. Shojaeefard MH, Abdi R, Akbari M, Besharati MK, Farahani F (2013) Modeling and pareto optimization of mechanical properties of friction stir welded AA0704/AA4703 butt joints using neural network and particle swarm optimization algorithm. J Mater Design 44:190–198
33. Yousif Y, Daws K, Kazem B (2008) Prediction of friction stir Welding characteristic using neural network . Jordan J Mech Ind Eng 2:151–155
34. Lakshminarayanan A, Balasubramanian V (2009) Comparison of RSM with ANN in predicting tensile strength of friction stir welded AA7039 aluminium alloy joints. Trans Nonferrous Met Soc China 19:9–18
35. Maleki E (2015) Artificial neural networks application for modeling of friction stir welding effects on mechanical properties of 7075–T6 aluminum alloy. IOP Conf Ser Mater Sci Eng (Online) 103(1):15
36. Niyati M, Moghadam AME (2009) Estimation of products final price using bayesian analysis generalized poisson model and artificial neural networks. J Ind Eng 2:55–60
37. Maleki E, Sherafatnia K (2016) Investigation of single and dual step shot peening effects on mechanical and metallurgical properties of 18CrNiMo7-6 steel using artificial neural network. Int J Mater Mech Manuf 4:100–105

A Statistical Study of Consumer Perspective Towards the Supply Chain Management of Food Delivery Platforms

Gangesh Chawla, Keshav Aggarwal, N. Yuvraj, and Ranganath M. Singari

Abstract Supply Chain deals with fulfilling the customer request be it directly or indirectly. It involves forming a link between customers, warehouses, manufactures, retailers and suppliers. Customer experience is one factor that the company should keep in mind when deciding its supply chain network. Smart Logistics, in recent times, has a huge role to play in making the supply chain efficient and responsive. With the combination of logistics and technology, the transparency in the whole process would be increased which would lead to reduced turnaround time and ensuring safety, quality and privacy of the items delivered. This research paper aims to analyze the supply chain of food delivery business and based on those device strategies which ensure maximum participation of the customer at each stage in food delivery supported by feedback from over 120 frequent users of food delivery business from different age brackets. The entire research happens in a step-by-step manner in which the first part is based on understanding the industry, finding the key parameters and then collecting various survey opinions and responses to identify the issues faced during food delivery and how it affects the entire supply chain and lastly making a list of the key considered challenges and prioritizing them using Decision-Making Trial and Evaluation Laboratory (DEMATEL) method.

Keywords Supply chain · Turnaround time · Transparency · Smart logistics · Minimum capital investment · New strategies · DEMATEL approach

---

G. Chawla (✉) · K. Aggarwal
Department of Production and Industrial Engineering, Delhi Technological University, New Delhi, Delhi 110042, India
e-mail: gangeshchawla.dce.pie@gmail.com

N. Yuvraj
Department of Mechanical Engineering, Delhi Technological University, New Delhi, Delhi 110042, India

R. M. Singari
Department of Design, Delhi Technological University, New Delhi, Delhi 110042, India

© The Author(s), under exclusive license to Springer Nature Singapore Pte Ltd. 2021
R. M. Singari et al. (eds.), *Advances in Manufacturing and Industrial Engineering*, Lecture Notes in Mechanical Engineering,
https://doi.org/10.1007/978-981-15-8542-5_2

1 Introduction

India is unarguably one of the largest consumer marketplaces across the globe today with a population exceeding 1.3 billion. This young and new India's appetite is one of the major factors for the ever-increasing demand in the beverage and food industry on the whole [1]. Indian market is currently experiencing a boom in the food delivery segment. Since this industry is on a rise, it is important to have an efficient supply chain to further facilitate expansion. Swiggy, till October 2018 had its operations in nearly 28 cities across India. Now, it has covered a larger part of India with expanding its operations by introducing its services to 16 cities which were earlier not covered [2]. The global prospects are also very promising for this market. The total revenue in the online food delivery market totals up to US$7730 million in 2019 across the globe as shown in Fig. 1 and by 2023 the total revenue of the online food delivery market is expected to touch the US$12,536 million [3].

Today, in the era of globalization the cost of any product is enormously impacted by the efficiency of the logistical setup applied during the various phases of its life. Smart logistics deals with smoothing all these phases as well as cutting the cost at the same time. With the help of various technological setups in accordance with Industry 4.0 principles [4], e.g., Physical Internet (based on the IoT), UAVs/Drones, Blockchain, Data Analytics, Robotics and Automation and Autonomous Vehicles we can simplify most of the stages involved in the logistical setup (Fig. 2).

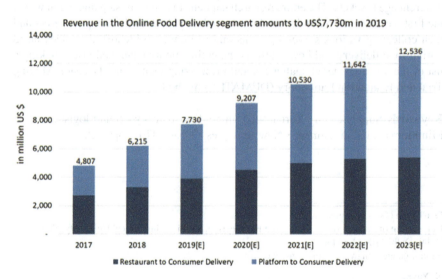

Fig. 1 Year-wise revenue comparison between restaurant to consumer delivery and platform to consumer delivery in the world

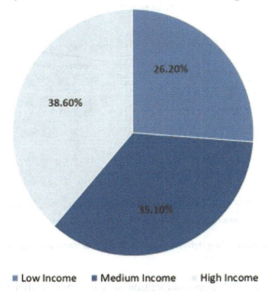

Fig. 2 Income wise classification of food delivery platform users in the world

Information Technology plays an important part in the modern supply chain ecosystem. The implementation of IT should be in accordance with the organization's strategies and goals so that there is harmony throughout the firm [5]. Logistics department could also be made more efficient with IT.

Consumers in the logistics industry comprise of both B2C and B2B segments and their satisfaction is the key driving force of the industry. Business model for B2B segments basically comprises of Carriers, Logistics and Service Providers (LSP), Courier/Express /Parcel (CEP), etc. and in the B2C segment comprises of CEP only. EBIT margins usually vary from −1 to 8% for the industry [6]. Companies are making use of big data technology from customer interaction and processes to digitalize their business and have an efficient supply chain. Internet of Things provides visibility of assets and resources [7]. Big data has become an asset for the organization enabling better supply chain performance. Big data has been characterized by 5Vs: volume, variety, velocity, veracity and value [8].

With the rise in an economy that is highly on-demand driven, food delivery has proved to be a major pillar in driving the opening of a number of restaurants. With the help of delivery service, the restaurant operators are presented with an exciting new opportunity of significantly increasing their guest engagement, building their brand loyalty and adding to the restaurant's revenue stream [9]. Online food delivery firms have increased in large numbers recently. According to a report by BCG, the Indian food market is expected to touch about Rs. 42 lakh crores by the end of 2020, reports BCG [10] (Fig. 3).

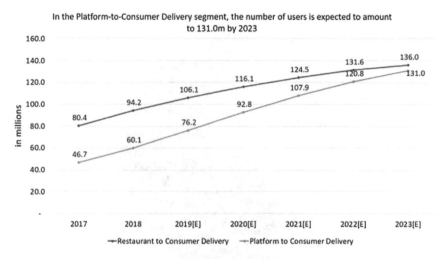

Fig. 3 No. of users that are availing delivery services from restaurants and platforms in the world

As with any industry that faces competition once it grows, similarly in the online food delivery industry also the competition is fierce causing friction for new players, and posing the risk of losing sales for existing restaurants/platforms.

In this research paper, we will first try to state and discuss the current problems and challenges in front of the industry. Then we will put forward our research which will help us to understand current demands of the customer from the industry and finally we will suggest technological solutions that will show how smart logistics can be applied in the food delivery business to improve its functioning in a cost-effective manner. The purpose of this study is to explore different types of challenges faced during the food delivery and use DEMATEL approach to analyze the problem. The DEMATEL approach helps us to understand both direct as well as indirect criteria and helps to filter out the best criterion.

2 Problem Statement

Food industry has proved out to be a profitable marketplace for all the stakeholders involved, be it suppliers, manufacturers, or even the distributors and users. In India, smartphones penetration has increased thus increasing the use of the internet. Due to this, there are major changes in the food industry about how things work and hence the online food delivery service is of utmost requirement [11].

The customer's decision after having the food or after dining-in for choosing the restaurant again or not is the final moment of truth for the customer, rather than a simple decision to go to the restaurant [12]. Similarly, in the online food ordering segment, every aspect needs to go right hence once the customer finishes his/her meal it is the time of the moment of truth whether they will use the service again or not therefore everything needs to go right. In any delivery process, the main objective is to fulfill the demand of the customer in a time-efficient and cost-effective manner while ensuring the quality of the delivered goods at the same time. Outsourcing of various operations creates many possibilities of success for the organization, but it faces various challenges that need to be dealt intelligently to ensure higher profitability [13]. Selection of a wrong supplier by the organization may result in a deep gap between them and suppliers; internal cooperation and collaboration will also be affected, and therefore productivity and efficiency of the work will decrease.

Every supply chain aims at delivering the commodity as swiftly as possible to the customer for a better customer experience. We want to analyze the food delivery business so as to find the market gaps and based on that device strategies which ensure maximum participation of the customer at each stage so as to make sure that a relationship is built with the customer base. This includes part of after-sale services till the customer receives the delivery, i.e., the transportation part, tracking of delivery to the service point preventing tampering of the delivered good and eventually enhancing the customer experience during the whole process with minimal efforts from the customer side.

Also, as of 2017, 77.7% of the users that order food online fall in the age groups of 0–24 years (33.1%) and 25–34 years (44.6%) as shown in Fig. 4 hence our survey majorly focuses on these two groups only. Given the competition in the food delivery segment and the options available to the millennials today regarding the food delivery, it is important we understand the right issues and prioritize those issues so that we spend our energy in the right direction to get the maximum gain from the situation.

When it comes to food delivery, the situation is no different. Maximum participation of the customer during the whole process ensures greater transparency and greater trust to the end consumer. With the help of modern food delivery tracking

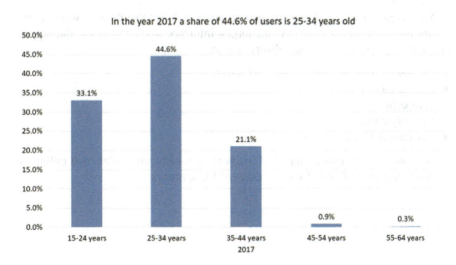

Fig. 4 Age-wise classification of food delivery platform users around the world

systems available, it is now very easy for these business owners to properly systemize and streamline their entire operations. The available software is fully customized basis the business type and its needs. Due to this, it is ensured that the whole process of ordering and delivering food is efficient and the user is delighted in return while ensuring that all the stakeholders are happy too with the efficient use of technology. It is a highly recognized fact now that the whole world today revolves around the internet and mobile solutions. All of the solutions and needs are available to the common man with the help of mobile applications and the internet. Due to this, a lot of conveniences have been created in the lives of the common man, both personally and professionally. Consumers are complaining consistently regarding the quality of the food delivered. Food-Tech start-ups like Swiggy, Zomato, Foodpanda, etc. have to face with the banter of the customers. Recently there was a very big issue created in the market regarding food delivery by Zomato. Its delivery executive opened the item to be delivered mid-way delivery, ate it, and then packed it again to finally deliver it to the customer. Here the privacy, quality of food and the trust of the customer on the delivery outlet is compromised.

When it comes to making decisions and evaluating the situation at hand currently, the certain criterion can pose some major problems for companies and organizations and make the decision-making difficult. Therefore, it is important for the decision and policymakers to choose and identify the right criteria. For instance, when dealing with selected problems like maximizing timely delivery, minimizing cost or increased quality the company finds itself in rattrap which can impose a high cost to its entire system. Therefore, with the help of mathematical and statistical techniques such as DEMATEL methods, this situation can be effectively controlled and would make the lives of the decision-makers easier. Thus, through this research, we aim to find an effective and efficient solution for the implementation of smart logistics in the food delivery business.

We talked to number of industry experts and concluded that the following issues are the most critical ones and have the largest influence when it comes to the supply chain management in the food delivery sector:

- Delivery within the initial estimated time frame
- Packaging of food
- Tampering of food
- Delivery charges
- Locating delivery address.

Now we surveyed more than 120 frequent customers who order food online to prioritize the above factors by using DEMATEL approach.

3 Literature Review

A considerable number of studies have investigated the supply chain challenges in different industries, and analyzing them by using different approaches.

Shieh [14] presented an approach of DEMATEL method wherein they recognized key performance indicator that determines service quality of a hospital by finding the key criteria and then applying DEMATEL method on them.

Kaushik [15] describes DEMATEL as a flexible and efficient approach for decision-making. This approach is based on the values obtained from a review algorithm with emphasize on compromise solutions in hybrid decision-making methods as well as criteria interrelationship studies.

Si [16] describes DEMATEL as an efficient method for the identification of cause and effect chain components of a sophisticated system. The study suggested various scenario-based upgradations to the original DEMATEL approach to obtain the optimized solution to the given problem statement.

Turker [17] uses the fuzzy DEMATEL technique in finding an effective critical input and output factors in successful efficiency measurement of university departments. In this empirical research study, the case university department's efficiency measurement factors are examined through 15 critical input and 8 output factors. According to evaluation results, several inferences about efficiency measurement factors were established by the researchers.

Sivakumar [18] discusses the barriers in the sustainable end of life practices for used plastic parts. In the paper, he uses DEMATEL approach to form a causal relationship diagram through which influential strengths of different barriers were determined.

Chawla [19] applied DEMATEL approach to find out the impact of smart logistics on the supply chain management of Telecom industry. Research was aimed to find out the key challenges in the industry, which can be eliminated by implementing smart logistics.

Jacob [20] applied Diffusion of Innovation (DOI) theory to find out the consumer perception of online food delivery apps in Kochi. The results suggested that majorly students and working population use online ordering, thus establishing its increasing popularity in youth.

Mehta [21] researched the evolving online food ordering and delivering industry to gain insights into how they interact with the customers through various social media channels. The study also highlights how these companies are trying to improve customer loyalty and engagement.

Das [22] conducted an empirical study on consumer perception of online food ordering and delivery services. The study applied statistical, mathematical and computational techniques on the collected primary data to find out the factors which encourage and discourage the decision of the users to order food online.

Gupta [23] conducted research on influence of online food delivery platforms on dynamics of restaurant business with special preference to Swiggy and Zomato. The secondary research approach concluded that the businesses who keep their image and offers disruptive & change them in due course of time will grab the major portion of market.

Manocha [24] researched that with efficient integration of information technology in transportation, warehousing, security, cargo tracking and verification have benefited the logistics companies in terms of competitive advantage, customer satisfaction, fast and timely delivery of goods minimizing the logistics costs and results in greater financial gains.

Niranjanamurthy [25] discussed E-commerce Security measures, Security Issues, Security Threats, Digital E-commerce cycle/online shopping and guidelines for safe and secure online shopping through shopping web sites and other online platforms, e.g., food delivery platforms, etc.

4 Research Methodology

DEMATEL Approach: Decision-Making Trial and Evaluation Laboratory (DEMATEL) method helps in separating the involved variables of the ecosystem into cause and effect groups hence enabling better and efficient decision-making. This technique, in accordance with the characteristics of objective affairs, can confirm the interdependent relationships amongst the variables/attributes and can cause resistance in the relationship that shows the important properties with an essential system and development trend. The end result of the DEMATEL process is a visual or graphical representation that helps the user organize their thoughts and act accordingly. The existing steps of DEMATEL method are given as:

Step 1: Generating the Direct Relation Matrix (A)
Initially, each of the respondents was asked to evaluate the direct influence between any two factors by an integer score from 0 to 4, where 0 represents no influence and 4 represents very high influence.

Data collected in the form of expert opinion which was regarding the impact of ith criteria over the jth criteria. For $i = j$, the diagonal elements are set to zero. For $i \neq j$, the value of the element is the average of all the data points collected for the impact of ith criteria over the jth criteria. Hence, the matrix A is given as:

$$A = \frac{1}{n} \sum_{k=1}^{n} A_{ij}^{k}$$

Step 2: Normalizing the direct relation matrix
Normalized form of matrix A is given by matrix X:

$$X = k.A$$

where $k = \dfrac{1}{\max\limits_{1 \le i \le n} \sum_{j=1}^{n} a_{ij}}$, $i, j = 1, 2, \ldots, n$.

Step 3: Calculate the total relation matrix
Total relation matrix T is calculated by the following formula:

$$T = X(I - X)^{-1}$$

where I is the identity matrix.

Step 4: Producing a casual diagram
r_i is the sum of each row in the total relation matrix. Similarly, c_j is the sum of each column in the total relation matrix. Next, we calculate $(r_i + c_j)$, $(r_i - c_j)$ for each criterion. $(r_i + c_j)$ reveals how much importance the criterion has, i.e., the degree of the relation of each criterion with other criterion. Higher $(r_i + c_j)$ value, higher relationship among two variables. Lower the value of $(r_i + c_j)$, lower will be the relationship between the two criteria.

$(r_i - c_j)$ depicts the kind of relationship between criteria. If the value is positive, the criterion belongs to "cause-group", i.e., this criterion influences other criteria. If the value is negative, the criterion belongs to the "effect-group", i.e., it gets influenced by other criteria.

Finally, a casual diagram is obtained with the help of the scatter plot. $(r_i - c_j)$ values are represented by the vertical axis while the $(r_i + c_j)$ values are represented by the horizontal axis.

Step 5: Set up a threshold value (optional)
The threshold value (α) helps to identify all the criteria which have a negligible effect over other criteria. The threshold value (α) is computed by the average of the elements in matrix T. This calculation aims to eliminate some minor effects elements in matrix T.

$$\alpha = \dfrac{\sum_{i=1}^{n} \sum_{j=1}^{n} t_{ij}}{N}$$

where N is the total number of elements in the matrix T.

5 Numerical Illustration

See Tables 1, 2, 3 and 4.

Table 1 Average direct relation matrix

	Delivery time	Packaging of food	Tampering of food	Delivery charges	Locating address
Delivery time	0	1.9823	2.6637	2.9557	3.0000
Packaging of food	1.8938	0	3.1150	2.6017	1.3893
Tampering of food	2.761	3.2654	0	1.9203	1.3716
Delivery charges	2.8053	2.654	1.7079	0	1.8672
Locating address	3.3451	1.2212	1.4513	1.9026	0

Table 2 Normalized direct relation matrix

	Delivery time	Packaging of food	Tampering of food	Delivery charges	Locating address
Delivery time	0	0.1869	0.2512	0.2787	0.2829
Packaging of food	0.1786	0	0.2938	0.2454	0.1310
Tampering of food	0.2604	0.3080	0	0.1811	0.1293
Delivery charges	0.2646	0.2503	0.1610	0	0.1761
Locating address	0.3155	0.1151	0.1368	0.1794	0

Table 3 Total relation matrix

	Delivery time	Packaging of food	Tampering of food	Delivery charges	Locating address
Delivery time	1.5277	1.5029	1.5247	1.5994	1.3911
Packaging of food	1.4929	1.1975	1.4049	1.4148	1.1414
Tampering of food	1.5899	1.4730	1.2225	1.4191	1.1804
Delivery charges	1.5570	1.3927	1.3218	1.2282	1.1865
Locating address	1.4666	1.1789	1.1844	1.2617	0.9449

A Statistical Study of Consumer Perspective Towards the Supply ... 23

Table 4 The sum of on criteria

	r_i	c_j	$r_i + c_j$	$r_i - c_j$
Delivery time	7.5460	7.6343	15.1804	−0.0882
Packaging of food	6.6518	6.7453	13.3971	−0.0934
Tampering of food	6.8851	6.6585	13.5436	0.2265
Delivery charges	6.6865	6.9234	13.6099	−0.2369
Locating address	6.0367	5.8446	11.8813	0.1921

6 Result

After conducting the survey and applying the DEMATEL method, the casual diagram was plotted as shown in Fig. 5. The survey was filled by more than 120 people of varying age groups ranging from 15 to 40 and above. The results of the DEMATEL method showed that factors "Tampering of food" along with "Locating address" belonged to the cause-group as the $(r_i - c_j)$ value comes out to be positive. The values for both the parameters are 0.226568582 and 0.1921499983 respectively. This means these two criteria influence the other criteria. Since the value of $(r_i - c_j)$ is greater for the factor "Tampering of food" than "Locating address", thus the former factor has a greater influence on the entire ecosystem of factors. It has a higher level of relationship with other variables. Also, factors like "Delivery time", "Packaging of food" and "Delivery Charges" belong to the effect-group as the value of $(r_i - c_j)$ comes out to be negative. The values for the three factors are −0.0882744801, −0.09345937299 and −0.2369847272 respectively. This means that these factors are influenced by the factors of the cause-group. Among the cause-group, as the value of $(r_i - c_j)$ for the factor "Tampering of food" is greater than the factor of "Locating address", the former factor has higher influence on the effect-group and thus becomes the governing factor of our analysis. Also, based on the values of $(r_i + c_j)$ we can see that "Delivery Time" has the highest level of relationship with other variables followed by "Delivery Charges" and "Tampering of Food". The $(r_i + c_j)$ values of these three factors are 15.18040915, 13.60999497 and 13.54368328 respectively. We can also see that "Locating Address" does not have deep relationships with other

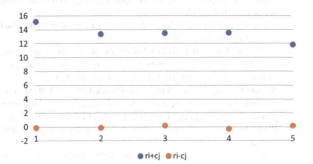

Fig. 5 Casual diagram

variables hence we can say that this factor does not really change the dynamics of the food delivery industry. Since the difference in the values of $(r_i + c_j)$ for both "Delivery Charges" and "Tampering of Food" is very small, we can say that both these factors have strong relationships with other factors and since "Tampering of Food" is the major factor that influences the entire ecosystem, based on the results of the research, it would not be wrong to say that this factor is one of the most important factors of the food delivery business.

7 Conclusion

Food industry is a very dynamic industry. The popularity of food ordering apps is on rise in countries where the majority is affluent middle class. In urban India, overwhelming reach of mobile application and a great number of unemployed youth having a scope of employment as delivery valets, have increased the popularity of online food ordering businesses.

One of the most important aspects of the food delivery industry is customer loyalty. If the customer sticks to your platform/restaurant, then you have rolling revenue up your sleeve. Timely delivery complemented with exclusive discounts is also another important factor that will ensure the success of your business. Customers like their food delivered to them quickly and in a good condition. Customers want their food to be warm and properly packed. They want total privacy regarding their food, and also proper tracking of food so that they know when it will arrive. These are methods that also increase customer loyalty towards a particular platform/restaurant. Hence, efforts need to be put to improve packaging and tracking of the food items that are out for delivery.

References

1. AIMS Institutes Homepage, https://theaims.ac.in/resources/online-food-service-in-india-an-analysis.html. Last accessed 2019/12/15
2. imarc Homepage, https://www.imarcgroup.com/india-online-food-delivery-market. Last accessed 2019/12/15
3. Statista Homepage, https://www.statista.com/outlook/374/119/online-food-delivery/india. Last accessed 2019/12/15
4. Verma R, Singh B (2017) Challenges of supply chain management in implementation of Industry 4.0. In: Proceedings of 2nd international conference on advanced production and industrial engineering. IJAPIE, New Delhi, pp 373–379
5. Kumar S, Tyagi M, Aggarwal S (2016) Identification of parameters for IT enabled supply chain performance measurement system. In: Proceedings of 1st international conference on advanced production and industrial engineering. IJAPIE, New Delhi, pp 428–433
6. PWC Homepage, https://www.pwc.com/sg/en/publications/assets/future-of-the-logistics-industry.pdf. Last accessed 2019/12/15
7. Roy D (2018) Impact of analytics and digital technologies on supply chain performance. AIMA J Manage Res 12(1)

8. Nguyen T, Zhou L, Spiegler VK, Ieromonachou P, Lin Y (2018) Big data analytics in supply chain management: a state-of-the-art literature review. Comput Oper Res 98:254–264
9. olo Homepage, https://www2.olo.com/hubfs/Olo_Delivery_Ops_Guide.pdf. Last accessed 2019/12/15
10. Parashar N, Ghadiyali MS (nd) A study on customer's attitude & perception towards digital food application services. Amity J Manage 5(1)
11. Singh A, Aditya R, Kanade V, Pathan S (2018) Online food ordering system. Int Res J Eng Technol 5(6)
12. Kivela J, Inbakaran R, Reece J (1999) Consumer research in the restaurant environment, Part 1: a conceptual model of dining satisfaction and return patronage. Int J Contemp Hosp Manage 11(5):205–222
13. Kumar P, Ashish S (2017) An analysis of global sourcing in Indian auto-component sector. In: Proceedings of 2nd international conference on advanced production and industrial engineering. IJAPIE, New Delhi, pp 393–410
14. Shieh J.-I, Wub H-H, Huang K-K (2010) A DEMATEL method in identifying key success factors of hospital service quality. Knowl-Based Syst 23(3):277–282
15. Kaushik S, Somvir (2015) DEMATEL: a methodology for research in library and information science. Int J Librariansh Adm 6(2):179–185
16. Si SL, You XY, Liu HC, Zhan P (2018) DEMATEL technique: a systematic review of the state-of-the-art literature on methodologies and applications. Math Probl Eng
17. Turker T, Etoz M, &AltunTurker Y (2016) Determination of effective critical factors in successful efficiency measurement of university departments by using fuzzy DEMATEL method. Alphanumeric J 4(1)
18. Sivakumar K, Jeyapaul R, Vimal KEK, Ravi P (2018) A DEMATEL approach for evaluating barriers for sustainable end-of-life practices. J Manuf Technol Manage 29(6):1065–1091
19. Kaushik R, Chawla G, Singh S, Singari RM (2019) An integrated analysis using DEMATEL approach for exploring the impact of smart logistics on SCM in telecom industries. Int J Adv Sci Res 4(1):12–14
20. Jacob AM, Sreedharan NV, Sreena K (2019) Consumer perception of online food delivery apps in Kochi. Int J Innov Technol Expl Eng 8(7S2):302–305
21. Mehta G, Iyer T, Yadav J (2019) A study on adoption of social media by food ordering & delivering companies for successful development of relationship marketing. Int J Res Anal Rev 6(2):53–63
22. Das J (2018) Consumer perception towards "online food ordering and delivery services": an empirical study. J Manage 5(5):155–163
23. Gupta M (2019) A study on impact of online food delivery app on restaurant business special reference to zomato and Swiggy. Int J Res Anal Rev 6(1)
24. Manocha T, Sahni MM, Nayyar P (2017) Next G-logistics: unmanned aerial vehicles or drones. Int J Market Technol 7(4)
25. Niranjanamurthy M, Chahar D (2013) The study of E-commerce security issues and solutions. Int J Adv Res Comput Commun Eng 2(7)

Design of an Auxiliary 3D Printed Soft Prosthetic Thumb

Akash Jain, Deepanshika Gaur, Chinmay Bindal, Ranganath M. Singari, and Mohd. Tayyab

Abstract Advancement in intelligent prosthetics and increased popularity for wearable technology has paved the way for the evolution of robotic appendages. The paper presents the design of an additional prosthetic thumb which seeks to extend human capabilities by improving hand manipulation and grasping ability. The device is engineered to be lightweight so that it is completely portable and comfortable to wear throughout the day. Thermoplastic polyurethane (TPU) is selected as the material for fabrication due to its high abrasion and chemical resistance, thus ensuring excellent durability. 3D printing is chosen as the manufacturing process as it is cheap, quick, and easily accessible. Moreover, it allows easy customization, thus allowing the user to modify the prosthetic thumb according to their preference. The device presents a wide range of applications from playing an extra note on the piano to improving grip on large objects. It aims to revolutionize the use of wearable robotics to upgrade human capabilities and open new avenues for enhanced human–robot integration.

Keywords Additional prosthetic thumb · Sixth finger · Soft wearable robot · 3D printed prosthetic · Human enhancement

1 Introduction

Prosthetics are often associated with artificial limbs or organs used to replace damaged parts of the body. However, its application is not limited to conventional meaning and can be used to extend or supplement human capabilities. The word *prosthesis* originates in Greek from prostithenai, meaning 'to add in place' [1]. This essence of addition is explored in our work through an extra thumb seeking to enhance gripping and manipulation ability of the hand.

A. Jain (✉) · D. Gaur · C. Bindal · R. M. Singari · Mohd. Tayyab
Delhi Technological University, New Delhi, Delhi 110042, India
e-mail: akash.jain9398@gmail.com

© The Author(s), under exclusive license to Springer Nature Singapore Pte Ltd. 2021
R. M. Singari et al. (eds.), *Advances in Manufacturing and Industrial Engineering*, Lecture Notes in Mechanical Engineering,
https://doi.org/10.1007/978-981-15-8542-5_3

Humans have evolved to have five digits on both hands along with intricate wrist mechanism to carry out various complex tasks. An additional thumb will change the way we perceive and interact with our environment. It will open new avenues for enhanced human integration with robots and promote the use of prosthetics to advance human capabilities. From playing an extra note on the guitar to picking up large objects with ease, the possibilities are endless.

Several robotic appendages have been developed as human enhancements [2–4]; however, they suffer from lack of accessibility, due to their complex nature. The prosthetic thumb is designed to be simple and easily manufacturable. With the rise in popularity for additive manufacturing and an increase in the number of people with desktop 3D printers, the prosthetic thumb would be easily accessible. It will allow personalized customization and facilitate effortless integration in people's lives.

2 Requirements and Challenges

The prosthetic thumb should feel natural and comfortable when worn. There are numerous factors which dictate a successful design. The following requirements are summarized as a result of consultation with doctors and literature review of prosthetics.

1. *Portability*: The device should be lightweight and completely portable, thus allowing the user to freely move around while wearing the prosthetic thumb.
2. *Wearability*: The system developed must be comfortable to wear and the user should be able to easily put-on and take-off the device when required.
3. *Non-restrictiveness*: There must not be any constraints on the mobility of joints, hence allowing the user to freely use their hand.
4. *Durability*: The prosthetic thumb must be resistant to the environment so that it can be used easily in day-to-day activities.
5. *Safety*: The device must be completely safe to wear and must not pose any risk to human health.

3 Design Conceptualization

A basic outline of the prosthetic thumb is first formulated which is then further refined to produce a more elaborate system. This section presents the whole conceptualization process including the general structure of the device and the actuation mechanism employed.

Design of an Auxiliary 3D Printed Soft Prosthetic Thumb

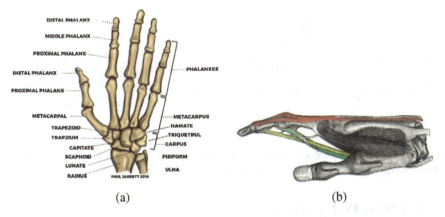

Fig. 1 Hand and wrist anatomy, **a** skeletal structure, **b** flexor and extensor tendons

3.1 Structure

To simulate natural movement, the structure of the prosthetic thumb is based on the anatomy of the human hand. The thumb comprises two narrow bones called phalanges which precede a metacarpal bone leading to the carpal bones in the wrist (see Fig. 1a). Ligaments form connective tissue between the bones and tendons serve the purpose of connecting muscles with the bones for movement [5]. Accordingly, the prosthetic thumb is designed to have three short segments corresponding to the metacarpal and the phalanges. Flexible filament serves the purpose of the ligament at the joints, while the tendon wires connecting to servomotors correspond to muscle operation. Angles provided at the joint facilitate the motion of the thumb.

3.2 Movement

Fingers comprise flexor and extensor tendons to carry out bending and stretching motion, respectively (see Fig. 1b) [6]. However, flexion and extension require two sets of tendons, thus increasing complexity along with the number of motors required. The problem is solved through the use of elastic material at the joints such that extension is the natural state of the thumb. Flexion is carried out through tension in the string, and its magnitude dictates the position of the thumb along with gripping strength. Release of tension causes the thumb to return to its original position, thus allowing a wide range of motion through the use of a single motor.

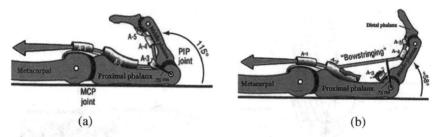

Fig. 2 Importance of the flexor pulley system. **a** The function of annular ligaments as pulleys. **b** Bowstringing effect leading to the reduction of power and loss of movement

3.3 Flexor Pulley System

Flexor tendons are tethered to the bones through annular ligaments acting as pulleys, thus providing efficient pulling force (see Fig. 2a). Damage to annular pulley leads to a bowstringing effect (see Fig. 2b), therefore causing a reduction of power along with the restriction of normal hand movement [7]. Tubular holes are cut passing through each segment of the prosthetic thumb to provide passage for the flexor tendon. This ensures that the tendon runs along with the thumb and efficient transmission of power takes place. Each of the segments thus acts as an annular pulley while simultaneously serving the purpose of providing structure to the whole thumb.

3.4 Mounting

The prosthetic thumb must be interfaced with the hand in a non-restrictive manner. To warrant this, the thumb is provided with a base that snugly fits into the side of the palm. A piece of soft fabric with elastic properties is used to secure the device. This ensures proper blood circulation and minimal restriction in normal hand movement. Servomotor along with electronics and battery is mounted on the wrist in the form of a watch. A Bowden cable running from the tip of the thumb connects to the servomotor at the wrist. The outer sheath of the cable prevents damage to the user during thumb operation. The whole setup is designed to be lightweight and portable so that it can be comfortably worn in daily life.

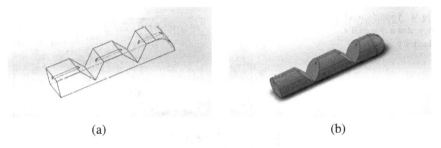

Fig. 3 Development of CAD model, **a** basic model, **b** final design

4 Design Conceptualization

4.1 CAD Model

Figure 3 illustrates the conceptualized design of the prosthetic thumb. It consists of three short segments with angles provided at the joint to facilitate the movement of the thumb. Tubular holes are cut through the device which acts as a flexor pulley system, thus ensuring the efficient transmission of tension. They also provide a path for the tendon and facilitate flexion of the prosthetic thumb. The device is provided with a flat base and a structured design to ensure quick printing and easy customization. The edges are smoothened out with a fillet to provide a more natural feel. The mounting base is not included in the diagram as it can be separately designed uniquely to each individual.

4.2 Material and Manufacturing

Each customer has a specific demand, and catering to the needs of every individual has given rise to customized solutions. 3D printing is one of the most powerful tools for providing personalized products as it allows easy modification to meet customer demands. Moreover, the process is quick, cheap and highly accessible. It is used in medical applications for custom implants and prosthetics due to its high adaptability. 3D printing is chosen as the process for the manufacturing of the prosthetic thumb as it allows the device to be tailored according to the specific needs of the user.

The material for the prosthetic thumb is required to have high elasticity in order to facilitate movement and excellent durability to allow usage in daily life. Rubber is the ideal choice for application; however, it cannot be 3D printed. Referring to the chart in Fig. 4, it can be observed that elastomers and polymers fall under the category of flexible materials. A combination of thermoplastics and elastomers produce a new 3D printable class of materials called thermoplastic elastomers (TPE) [8]. They have several variations with elasticity depending on the type of TPE and chemical

Fig. 4 Young's modulus indicates a material's elasticity

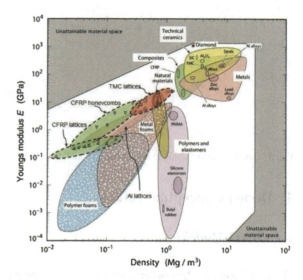

formulation used. Thermoplastic polyurethane (TPU) is a commonly used printable grade of TPE and is selected for prosthetic thumb due to its high elastic range and excellent printability [8]. Properties such as fatigue, abrasion, and chemical resistance ensure long operational life, whereas its soft and flexible nature guarantees comfort while preventing accidental harm to the user.

4.3 Actuator

The position of the prosthetic thumb along with gripping strength depends on the magnitude of the tension in the string. The actuator employed must, therefore, provide positional control to facilitate the movement of the thumb. Apart from this, the actuator must be small enough to fit comfortably in the wrist of the user and shall provide sufficient torque for the general gripping application. Both the servo and the stepper motors meet the requirements and are ideal for use as an actuator for the prosthetic thumb. 28BYJ-48 is a commonly available 5 V stepper motor that fits the required dimensions [9]. MG90S is a metal-geared 5 V servomotor of similar dimensions [10]. Comparing the datasheets of both the motors, it can be noticed that MG90S servomotor provides greater operating torque and hence is chosen as the actuator for the prosthetic thumb.

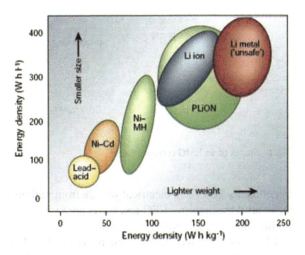

Fig. 5 Graphical representation of various types of batteries

4.4 Battery

The power system for the prosthetic thumb must be lightweight, rechargeable, durable, and medically safe for use in prosthetic devices. Figure 5 presents a visual representation of different characteristics of various batteries. As we move higher up on the graph, energy density increases along with lighter weight and smaller size. All these desirable characteristics make Li-ion an attractive choice for use in the prosthetic thumb.

They possess excellent charge density and do not display memory effort over multiple charge cycles. Moreover, they have a low self-discharge rate, require little maintenance, and are suitable for use in proximity to the human body [11]. With advancements in technology, features such as fast charging make this an excellent choice. Their small size and low weight make them suitable for mounting on the wrist with little discomfort. Considering all these factors, the Li-ion battery is chosen for use with the prosthetic thumb.

5 Control System

Buttons and switches are the most commonly employed methods for the control of prosthetic devices. However, they require to be physically operated and take away a certain degree of freedom. Flex sensors rely upon the movement of the wrist or fingers and hence do not provide independent control. The mechanism used to operate the prosthetic thumb must allow effortless control without affecting the natural movement of the user. Electro-biological signals prove to be a great solution as they can be easily detected and interfaced with the prosthetic thumb. Their low cost of implementation and relatively simple processing system make them an ideal candidate.

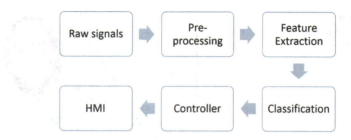

Fig. 6 Stages of an EMG control system

Intentionally generated electrical signals from various parts of the body are used as the control mechanism. EEG and EMG are two types of signals which can be used for the control of the prosthetic thumb.

Electromyographic (EMG) signals are produced due to neuromuscular activity and typically lie in the range of 100 µV to 90 mV. They display a standard behavior which makes them an ideal candidate for control systems. Moreover, their signal level can be easily distinguished due to high electric potential. Electro-encephalographic (EEG) signals, on the other hand, correspond to the brain (neural) activity and generate amplitudes below 100 µV [12]. These signals are harder to condition compared to EMG signals and are ideal for use in patients with neuromotor disabilities. Since the prosthetic thumb is intended for healthy individuals, EMG signals are selected as the control mechanism for the device.

The control system using electromyographic signals involves detection via electrodes placed on the user. The raw data is then filtered and processed before being communicated to the controller. Command signals are then sent to the motors which actuate the prosthetic thumb. The whole process is illustrated in Fig. 6. Surface EMG is used to detect the electrical signals using conductive gel electrodes which measures the complete electrical activity of a large region of muscle fibers. These electrical signals correspond to the force exerted by a muscle (or group of muscles) in real time. The common application of surface EMG includes myoelectric prosthesis, control of a limb in virtual reality and generation of biofeedback for muscular pain (Fig. 7).

Myoware muscle sensor by Advancer technologies [13] is used to detect muscle activation signals. It is appropriate for generating raw EMG signal and analog output signal for Arduino. The sensor is designed for a reliable output with low power consumption. It operates on a single power supply (+2.9 to +5.7 V) with polarity reversal protection and allows adjustments in sensitivity gain. The sensor is placed along the length of the muscle, with the electrode closest to the wire connections placed at the middle of the muscle and the second electrode on the circuit board toward to end of the muscle. The third electrode attached to the black wire is placed away from the muscle which needs to be sensed.

Our muscles generate a wide range of voltage for a simple motion of the thumb. Sample data is collected corresponding to different movements, and an appropriate signal is determined for flexion and extension of the thumb. The collected data is then processed to associate electrical signals with muscle movements. This is achieved

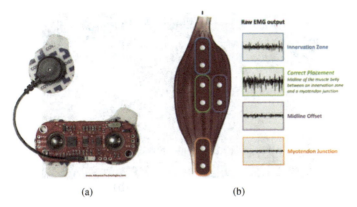

Fig. 7 **a** Myoware EMG muscle sensor, **b** correct electrode placement

using a classification technique: probabilistic neural network (PNN). The classifier gives a high level of performance accuracy for complex biological signals. Using PNN, the dataset is trained to classify the voltage range to specific thumb movements. Once training and testing are completed, a model is developed which is capable of providing the prosthetic thumb the desired movement.

6 Prototype Development

A sample prototype of the prosthetic thumb was developed by 3D printing it on a Creality CR10 3D printer with TPU as the filament material. The layer height and the wall thickness were kept at 1.2 mm and an infill of 25% was provided with cubic pattern. No supports or build plate adhesion were employed.

The printed thumb is illustrated in Fig. 8. The model is tough and durable while possessing elastic properties to facilitate the motion of the thumb. The dimensions of the prosthetic matches with other digits on the hand and hence can be comfortably integrated without any discomfort. The model is interfaced with Arduino to develop a working prototype. MG90S metal-geared servomotor is used as the actuator for controlling the tendon movement and hence the motion of the thumb. Li-ion battery is used to power the device, and Myoware muscle sensor is used to pick up signals from the forearm.

7 Conclusion

An additional prosthetic thumb for extension of human abilities was proposed, and the detailed design of the device was presented. The thumb was designed for use

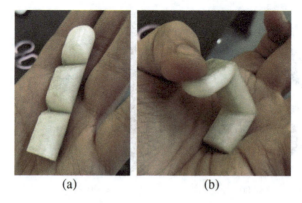

Fig. 8 Developed prototype of the prosthetic thumb. **a** Initial extension position, **b** flexion of the thumb

with healthy individuals to enhance gripping and manipulation ability. Utilizing 3D printing, the prosthetic thumb can be easily customized and manufactured in a small amount of time with little cost. TPU as the material of the thumb makes it durable and fit for use in daily life while providing elasticity in the joints to facilitate motion of the thumb. The control system utilizing EMG signals provides autonomy to the device and promotes advanced user control. The prosthetic thumb aims to promote the development of robotic devices to enhance human capabilities. It possesses vast potential in medical, military, and space applications. There is a huge scope for further research including implementation of haptics, changes in the brain due to extra digit, fatigue, and life cycle of the thumb and several others.

References

1. Prosthesis: Definition of Prosthesis by Lexico (nd) Retrieved from https://www.lexico.com/en/definition/prosthesis
2. Dormehl L (2018, August 4) Want an extra arm? A third thumb? Check out these awesome robotic appendages. Retrieved from https://www.digitaltrends.com/cool-tech/amazing-extra-limb-projects/
3. Meraz NS, Tadayoshi, Hasegawa1 Y (2018, January 15) Modification of body schema by use of extra robotic thumb. Retrieved from https://robomechjournal.springeropen.com/articles/10.1186/s40648-018-0100-3.
4. Giova (2016, October 4) The robotic sixth finger. Retrieved from https://clem.dii.unisi.it/~prattichizzo/sixthfinger/
5. Hand and Wrist Anatomy (nd) Retrieved from https://www.arthritis.org/about-arthritis/where-it-hurts/wrist-hand-and-finger-pain/hand-wrist-anatomy.php
6. Structures of the Hand (nd) Retrieved from https://teachmeanatomy.info/upper-limb/misc/structures-hand/
7. Lowe W (2017, November 15) Flexor pulleys of the fingers. Retrieved from https://www.academyofclinicalmassage.com/flexor-pulleys-fingers/
8. Flexible 3D Printing Filament—Which Should You Chose? (2019, February 21) Retrieved from https://all3dp.com/2/flexible-3d-printing-filament-which-should-you-chose/

9. 28BYJ-48 Stepper Motor Pinout Wiring, Specifications, Uses Guide & Datasheet (nd) Retrieved from https://components101.com/motors/28byj-48-stepper-motor
10. MG90S (nd) Retrieved from https://www.towerpro.com.tw/product/mg90s-3/
11. What's the Best Battery? (nd) Retrieved from https://batteryuniversity.com/learn/archive/whats_the_best_battery
12. Ferreira A, Celeste WC, Cheein FA, Bastos-Filho TF, Sarcinelli-Filho M, Carelli R (2008, March 26) Human-machine interfaces based on EMG and EEG applied to robotic systems. Retrieved from https://www.ncbi.nlm.nih.gov/pmc/articles/PMC2292737/
13. MyoWare Muscle Sensor (nd) Retrieved from https://www.advancertechnologies.com/p/myoware.html

Prediction of Material Removal Rate and Surface Roughness in CNC Turning of Delrin Using Various Regression Techniques and Neural Networks and Optimization of Parameters Using Genetic Algorithm

Susheem Kanwar, Ranganath M. Singari, and Vipin

Abstract This paper aims to compare the different regression techniques in conventional turning of a cylindrical workpiece of Acetal Homopolymer Delrin and to determine the most accurate one among them for determining the MRR and SR. Three different cutting parameters, namely feed (mm/rev), depth of cut (mm) as well as speed (RPM), are varied, and the corresponding MRR and surface roughness are represented by a Taguchi L27 orthogonal array. The orthogonal array is divided into training as well as testing datasets by making use of the train_test_split functionality in python. The testing data is one-third of the entire dataset with the remaining data forming the training data. In each case, the mean square error (MSE) is determined by contrasting obtained values with the output data present in testing dataset. Out of all the functions formulated, the neural network (NN) gives the least mean square error of 0.108. Genetic algorithm (GA) is then applied to optimize the input parameter values. It took 139 generations to achieve the optimum value of 1.735 μm for SR and 827.473 mm^3/min for MRR. This combination of resulting values was obtained at 299.887 rpm, 0.59 mm/rev feed and 1.49 mm depth of cut.

Keywords Regression · Taguchi · Mean square error · Neural network · Genetic algorithm

1 Introduction

Delrin is a thermoplastic polymer trademarked by DuPont and commonly known as polyoxymethylene (POM). It has properties similar to some metals and therefore is a potential alternative. Regression is applied to the data to determine a continuous output. It is a supervised learning operation in which the algorithm is trained on data, where it figures out the correlation among input and output variables and then

---

S. Kanwar (✉) · R. M. Singari · Vipin
Delhi Technological University, New Delhi, Delhi 110042, India
e-mail: sushkanwar@gmail.com

© The Author(s), under exclusive license to Springer Nature Singapore Pte Ltd. 2021
R. M. Singari et al. (eds.), *Advances in Manufacturing and Industrial Engineering*,
Lecture Notes in Mechanical Engineering,
https://doi.org/10.1007/978-981-15-8542-5_4

makes predictions for new output values given new input values. In this paper, the algorithms are fed the rake angle, speed and feed as input and the corresponding MRR and surface roughness values as output. Higher rake angle usually gives better surface finish [1]. When graphite and polymer composites are cut orthogonally, it is found that surface finish is extremely poor when the rake angle is between 0° and 5°. On increasing the rake angle, the concavities formed on the machine surface decrease resulting in improved surface finish [2]. The rake angle generally used varies from about 6° to 20°, and it can even reach up to 30° under certain conditions [3]. We found the surface roughness to be minimized in this region.

One of the problems with machining is that built-up edge (BUE) is created. If machining parameters are not selected carefully, then there is a build-up of layer at the tool edge called BUE and at the tool-rake interface known as built-up layer (BUL) [4–7]. Especially, when the aluminium undergoes dry machining, the effect of BUE and BUL is more pronounced. However, the environmental concerns accompanying the utilization of coolants and their disposal make dry machining an attractive alternative. To offset the disadvantage of BUE and BUL, dry machining of aluminium is accompanied by the use of high rake angles (about 30°) and ultra-hard tool materials (like diamond and CBN) [8].

Apart from surface finish, MRR is also of tremendous interest to us. Rake angle does not have as major an effect on MRR as depth of cut (DOC). DOC followed by speed is more influential in determining the MRR in a machining operation [9]. However, on increasing the rake angle and keeping the other parameters constant, there is slight decrease in the MRR [10].

Another important machining parameter is the feed rate. Feed has a greater effect on SR than other machining parameters [11, 12]. Feed is related to surface roughness by the formula $R_t = f^2/8r$ [13], where f = feed rate and R_t = peak-to-valley surface roughness. Thus, feed rate and surface finish share an inverse relationship. On the other hand, the MRR is directly proportional to feed rate [14].

Speed plays a major role in determining output parameters like MRR and surface finish. As speed increases, it leads to an increase in MRR as well while reducing the surface finish [15]. Speed and feed dominate the MRR of the workpiece [15].

There are several regression techniques which were used in predicting the SR and MRR like:

1.1 Linear Regression

Linear establishes a relationship between independent variables and dependent variables in an equation [16]. It is one of the most widely used regression analysis techniques because of its simplicity and the fact that a dataset which varies linearly is easier to fit than a nonlinear dataset [17, 18]. For n points present in a dataset, linear regression models assume a linear correlation between x and y. No relation is perfect, and often there is some error/noise which gets incorporated into the equation as shown below:

$$yi = \beta 01 + \beta 1x1 + \cdots + \beta nxn + ci = x^T \beta + \varepsilon i \text{ where } i = 1, \ldots, n$$

1.2 KNN Regression

It is among the simplest algorithmic techniques in machine learning. In KNN algorithm, the output value is the average of K nearest values. KNN regression technique is a non-parametric type of regression analysis [19].

1.3 Support Vector Regression (SVR)

SVRs come under the category of supervised learning models that predict a continuous output value for a given input. They come under the category of support vector machine (SVM) or support vector network [20]. Our goal in SVR regression is to determine a function that maximizes deviations for all the data points [21]. Errors less than the threshold value of ε are considered negligible, but those greater than that are unacceptable [22]. Linear support vector regression has a general equation:

$$y = \Sigma i = 1(ai - a)\langle xi, x \rangle + b$$

1.4 Bayesian Ridge

Target value is a linear combination of input values. Bayesian regression includes a regularization parameter. Bayesian ridge regression estimates β using L2-constrained least squares [23]. Bayesian ridge regression has a greater than quadratic fit time as the number of samples makes scaling hard. In contrast to the ordinary least squares (OLS) estimator, the weights are shifted toward zeros, which lends stability. By maximizing the marginal log-likelihood over a number of iterations, estimation is achieved. There are several implementation strategies for Bayesian ridge. The implementation mentioned in this paper is taken from [24]. Moreover, better values of the regularization parameters inspired from the recommendation in [25].

1.5 Decision Tree Regression

The core algorithm called ID3 involves breaking the dataset into increasingly smaller subsets represented by a combination of decision as well as leaf nodes [26]. Numerous connections branch forth from a decision node, each of which represents an attribute while leaf nodes represent a decision. Node corresponding to the best predictor is referred to as the root node.

1.6 Gradient Boosting Regression

It is used to make predictions by using an ensemble of ML models. The various ML models are generalized by allowing optimizing on an appropriate cost function [27]. Gradient boosting model helps in the optimization of any arbitrary differentiable loss functions. Training set of the form $\{(x1, y1)$, to $(xn, yn)\}$ used to determine approximation $F(x)$ to minimize the loss function

$$L(y, F(x)) : F = \arg\min Ex, y[L(y, F(x))].$$

1.7 Neural Networks

Simulate the neurons in the human brain. Artificial neural networks consist of a minimum of three layers, viz. the input, output and hidden layer. The connection between is assigned weights which can be positive, negative or zero [28]. The NN learns the correlation present in the data through repetition. More the number of repetitions, better the NN learns. Once it has learned the relationship, it can generalize to previously unseen data and can thus predict output values for new input data.

1.8 Genetic Algorithm (GA)

GA is an evolutionary algorithm which we have utilized for optimization. In GA, we start off with an initial population which may be randomly generated. We then select the fittest members from this initial population based on various criteria and make them pass their "genes" to the next generation. This step is known selection. "Offspring" are created by mating the parent population selected in the previous step. This process keeps on repeating until the point where the offspring produced are not much different from the parents. This point is known as convergence [29–31].

2 Experimental Investigations

In this paper, three different cutting parameters, namely feed (mm/rev), depth of cut (mm) as well as speed (RPM), vary, and the corresponding MRR (mm^3/min) and surface roughness (micrometres) are represented by a Taguchi L27 orthogonal array. The train_test_split function of sklearn splits the dataset further into two categories: training and testing. Testing data represents one-third of the entire dataset, with the rest being training data.

A homopolymer Delrin rod of diameter 34 mm was selected, and CNC turning operation was performed on it. Before performing final CNC turning operation, a roughing operation was performed on it. At the end of the turning operation, the diameter was reduced from 34 to 33 mm. In the CNC turning operation, three different depths of cuts were provided: 0.5, 1.0 and 1.5 mm. The entire rod was broken into three pieces of equal length, and all the operations for a particular depth of cut were performed on each rod in succession. Various regression techniques were then applied on the dataset, and the mean square error was calculated to determine the accuracy of the ML regression models. The scikit-learn library of python was used for implementing the various regression techniques [32]. In addition to regression techniques, a neural network has also been implemented to predict the MRR and surface roughness using the Keras library [33]. Matplotlib library has been utilized to plot the experimental MRR and SR values along with the predicted MRR and SR values [33]. MATLAB optimization toolbox is used to implement multi-objective genetic algorithm to find the optimized values of the input variables [34]. CNC turning experimental data obtained was arranged in a L27 orthogonal array (Table 1).

3 Results and Discussion

It was found that the least mean square error while predicting both MRR and SR was obtained by NN. The mean square error obtained, for each regression technique, is represented in Table 2. As can be seen from the table, neural net gives the best overall results, as it has an extremely small MSE.

The above results can be confirmed by visualizing the values obtained using regression techniques with the true values of both MRR and SR on two different graphs (the graph for neural network has not been included as the values for NN have been reshaped and will have to be represented by another scale than shown in Figs. 1 and 2).

Out of the three regressions, viz. linear, SVR and Bayesian ridge regression, linear regression has the least error for both MRR and SR. For linear regression, the equation found is:

$$y = [[0.0092381 \quad -1.31666667 \quad -1.15333333] \\ [-6.41442733 \quad 886.39633333 \quad -65.8162]] * x \\ + [1.4800000000000002 \quad 3380.573266666667].$$

Table 1 L27 orthogonal array

Exp. no.	Control factors A	B	C	Speed (s) (rpm)	Feed (f) (mm/rev)	Depth of cut (d) (mm)	Surface roughness (Ra) (μm)	MRR (mm³/min)
1	1	1	1	150	0.2	0.5	1.04	1083.77
2	1	1	2	150	0.2	1.0	2.44	1784.15
3	1	1	3	150	0.2	1.5	0.56	1120.99
4	1	2	1	150	0.4	0.5	0.90	1665.85
5	1	2	2	150	0.4	1.0	1.42	1568.58
6	1	2	3	150	0.4	1.5	0.96	1898.23
7	1	3	1	150	0.6	0.5	0.86	1702.32
8	1	3	2	150	0.6	1.0	1.14	2505.56
9	1	3	3	150	0.6	1.5	1.16	1933.68
10	2	1	1	250	0.2	0.5	2.62	2140.88
11	2	1	2	250	0.2	1.0	5.38	2216.46
12	2	1	3	250	0.2	1.5	0.52	2538.14
13	2	2	1	250	0.4	0.5	2.32	1811.81
14	2	2	2	250	0.4	1.0	3.58	2469.70
15	2	2	3	250	0.4	1.5	0.64	2040.10
16	2	3	1	250	0.6	0.5	1.92	1410.84
17	2	3	2	250	0.6	1.0	3.44	1532.72
18	2	3	3	250	0.6	1.5	0.94	1579.62
19	3	1	1	300	0.2	0.5	2.04	1047.48
20	3	1	2	300	0.2	1.0	4.80	960.00
21	3	1	3	300	0.2	1.5	1.00	1085.79
22	3	2	1	300	0.4	0.5	2.24	251.998
23	3	2	2	300	0.4	1.0	3.84	461.219
24	3	2	3	300	0.4	1.5	0.80	980.251
25	3	3	1	300	0.6	0.5	2.70	143.928
26	3	3	2	300	0.6	1.0	3.76	0.0035
27	3	3	3	300	0.6	1.5	0.74	0.1408

The equation obtained on application of linear regression was taken, and genetic algorithm (GA) was applied to it for optimizing the input variables. In order to implement GA, the optimization toolbox in MATLAB was used. In this experiment, the MATLAB genetic algorithm was selected in the optimization toolbox. The following parameters are used during the optimization: An initial population of 50 with feasible population as the function and tournament type with a crossover of 0.8. We also selected a single-point crossover and mutation which is adaptive feasible.

Table 2 Mean square error obtained for MRR and SR, which is calculated using various regressors

	Mean square error
Linear regression	0.46
Support vector regression	2.15
KNN regression	0.275
Bayesian ridge	1.981
Decision tree regression	0.647
Gradient boosting regression	0.86
Neural networks	0.108

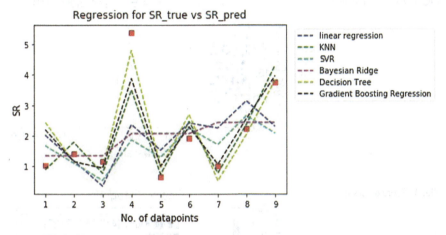

Fig. 1 Obtained and true values of surface roughness (plotted using matplotlib library in python)

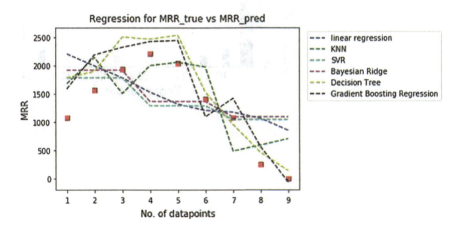

Fig. 2 Obtained and true values of material removal rate (MRR) (plotted using matplotlib library in python)

3.1 Optimization Using Genetic Algorithm (GA)

As can be seen from Fig. 3, the multi-objective GA iterates for obtaining the best solution and finds it on the 139th generation. The graph in Fig. 5 plots Objective 2 on the y-axis against Objective 1 on the x-axis giving the Pareto front. Average speed for each generation is determined in Fig. 7. The score diversity is represented in a histogram in Fig. 4 while Fig. 6 plots each individual's rank. These graphs help us determine the optimized solutions.

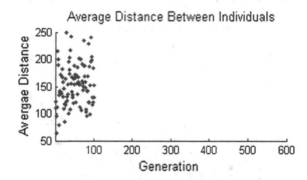

Fig. 3 Average distance

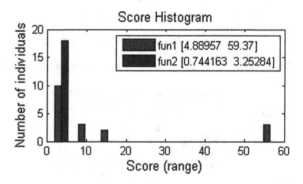

Fig. 4 Score diversity

Fig. 5 Pareto front graph

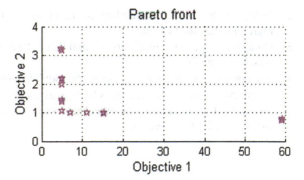

Fig. 6 Rank histogram

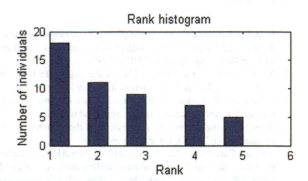

Fig. 7 Average Pareto spread

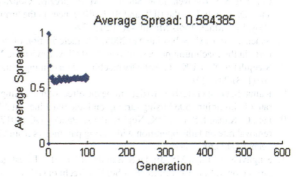

4 Conclusion

Prediction of SR and the material removal rate through regression allows us to conclude the following:

1. Neural networks give the least mean square error of 0.108 and are thus an improvement over regression.

2. In the graph showing comparison of regression techniques, it shows that KNN regression has the best fit.
3. On applying genetic algorithm, we find that optimization takes 139 generations which is quite fast.
4. The optimum cutting parameters are 150 rpm 0.6 mm/rev feed and 1.49 mm depth of cut. At this combination, SR is 0.351 μm and MRR is 1788.91 mm^3/min.
5. Increasing feed while keeping other factors constant resulted in a decrease in the surface finish of the workpiece.

References

1. Pradeesh AR, Mubeer MP, Nandakishore B, Muhammed Ansar K, Mohammed Manzoor TK, Muhammed Raees MU (2016) Effect of rake angles on cutting forces for a single point cutting tool. Int Res J Eng Technol (IRJET) 3(05)
2. Yang D, Wan Z, Xu P, Lu L (2018) Rake angle effect on a machined surface in orthogonal cutting of graphite/polymer composites. Adv Mater Sci Eng
3. Baldoukas AK, Soukatzidis F, Lontos A, Demosthenous G (2008) Experimental investigation of the effect of cutting depth, tool rake angle and workpiece material type on the main cutting force during a turning process. In: Proceedings of the 3rd international conference on manufacturing engineering (ICMEN), Ed. Ziti, Greece
4. Sanchez-Carrilero M, Marcos M (2011) SEM and EDS characterisation of layering TiO$_x$ growth onto the cutting tool surface in hard drilling processes of Ti-Al-V alloys. Adv Mater Sci Eng
5. Trent ME, Wright PK (2000) Metal cutting, 4th edn. Butterworth-Heinemann, UK
6. Carrilero MS, Bienvenido R, Sanchez JM, Alvarez M, Gonzalez A, Marcos M (2002) A SEM and EDS insight into the BUL and BUE differences in the turning processes of AA2024 Al–Cu alloy. Int J Mach Tools Manuf 42(2):215–220
7. Sánchez-Sola JM, Sebastian M (2005) Characterization of the built-up edge and the built-up layer in the machining process of AA 7050 alloy. Revista de Metalurgia 365–368
8. Sreejith PS, Ngoi BKA (2000) Dry machining: machining of the future. J Mater Process Technol 101(1–3):287–291
9. Kumar S, Gupta D (2016) To determine the effect of machining parameters on material removal rate of aluminium 6063 using turning on lathe machine. Int J Multidiscip Curr Res 4
10. Pant G, Kaushik S, Rao DK, Negi K, Pal A, Pandey DC (2017) Study and analysis of material removal rate on lathe operation with varing parameters from CNC Lathe machine. Int J Emerg Technol 8(1):683–689
11. Singhvi S, Khidiya MS, Jindal S, Saloda MA (2016) Investigation of material removal rate in turning operation. Int J Innov Res Sci Eng Technol (IJIRSET) 5(3)
12. Abdullah AB, Chia LY, Samad Z (2008) The effect of feed rate and cutting speed to surface roughness. Asian J Sci Res 1(1):12–21
13. Qu J, Shih A (2003) Analytical surface roughness parameters of a theoretical profile consisting of elliptical arcs. Mach Sci Technol 7(2):281–294
14. Gupta R, Diwedi A (2014) Optimization of surface finish and material removal rate with different insert nose radius for turning operation on CNC turning center. Int J Innov Res Sci Eng Technol (IJIRSET) 3(6)
15. Bhavani TK, Satyanarayana K, Kumar GSV, Kumar IA (2017) Optimization of material removal rate and surface roughness in turning of aluminum, copper and gunmetal materials using RSM. Int J Eng Res Technol (IJERT) 6(02)
16. Freedman D (2009) Statistical models: theory and practice, 2nd edn. Cambridge University Press, Cambridge

17. Xin Y (2009) Linear regression analysis: theory and computing, 1st edn. World Scientific, Singapore
18. Montgomery D, Peck E, Vining G (2012) Introduction to linear regression analysis, 5th edn. Wiley Publishers, New York
19. Altman NS (1992) An introduction to kernel and nearest-neighbor nonparametric regression. Am Stat 46(3):175–185
20. Corinna C, Vladimir NV (1995) Support-vector networks. Mach Learn 20(3):273–297
21. Vapnik VN (2000) The nature of statistical learning theory, 2nd edn. Springer, New York
22. Smola A, Schölkopf B (2004) A tutorial on support vector regression. Stat Comput 14:199–222
23. Pasanen L, Holmström L, Sillanpää MJ (2015) Supporting information for Bayesian LASSO, scale space and decision making in association genetics. PLoS ONE 10(4)
24. Tipping ME (2001) Sparse Bayesian learning and the relevance vector machine. J Mach Learn Res 1
25. MacKay DJC (1992) Bayesian interpolation. Comput Neural Syst 4(3)
26. Quinlan JR (1986) Induction of decision trees. Mach Learn 1:81–106
27. Brelman L (1997) Arcing the edge. Technical Report486. Statistics Department University of California, Berkeley
28. Hopfield JJ (1982) Neural networks and physical systems with emergent collective computational abilities. Proc Natl Acad Sci U S A 79(8):2554–2558
29. Goldberg DE (1989) Genetic algorithms in search optimization and machine learning, 1st edn. Addison-Wesley Publishing Company Inc., Reading, MA
30. Carroll DL (1996) Chemical laser modeling with genetic algorithms. AIAA J 34(2):338–346
31. Winter G, Cuesta P, Periaux J, Galan M, Cuesta P (1996) Genetic algorithm in engineering and computer science. Wiley, New York
32. Pedregosa F, Varoquaux G, Gramfort A, Michel V, Thirion B, Grisel O, Blondel M, Prettenhofer P, Weiss R, Dubourg V, Vanderplas J, Passos A, Cournapeau D, Brucher M, Perrot M, Duchesnay E (2011) Scikit-learn: machine learning in python. J Mach Learn Res 12:2825–2830
33. Chollet F (2015) Keras. GitHub
34. Hunter JD (2007) Matplotlib: a 2D graphics environment. Comput Sci Eng 9(3):90–95

Finding Accuracies of Various Machine Learning Algorithms by Classification of Pulsar Stars

Abhishek Seth, Arjun Monga, Urvashi Yadav, and A. S. Rao

Abstract In order to rank supervised machine learning techniques according to their accuracy, a number of them were applied on the HTRU2 dataset. Pulsars are exotic neutron stars rotating at very high RPMs which lead to a high scientific interest into recognising actual pulsars from a pool of candidates. False positives are almost indistinguishable from real positives and are generated most often due to internal and external noise and interference factors. The use of aforementioned ML techniques helps mitigate some of those problems. The raw observational data was collected by the High Time Resolution Universe Collaboration using the Parkes Observatory, funded by the Commonwealth of Australia and managed by the CSIRO. Deep investigations into the nature of exotic stars seem imminent, and a ranked list of the most accurate ML techniques presented in this paper will no doubt benefit the field of pulsar astronomy. A direct combination of future observatories and ML computation might yield unexpected results, and that is what we expect from this paper. We hope our work contributes in enabling scientific discoveries as humanity is finally becoming more and more capable of turning their heads up and understanding the mysteries of what lies beyond home.

Keywords Pulsar star · Supervised machine learning · Noise and interference · High energy radiation · HTRU2 dataset · Candidate

1 Introduction

Ever since the dawn of humanity, humans have been looking up in the sky—wondering and conjuring myths and theories to explain to ourselves and others of the happenings in the sky. This has continued till modernity, and although we have discovered far more than our ancestors ever could hope, some answers still painfully elude us. First discovered in November 1967 by Jocelyn Bell, an intermittent pulsating signal was of great interest to her and her advisor Antony Hewish. They jokingly

---

A. Seth (✉) · A. Monga · U. Yadav · A. S. Rao
Department of Applied Physics, Delhi Technological University, New Delhi, India
e-mail: abhishekseth.abhhi007@gmail.com

© The Author(s), under exclusive license to Springer Nature Singapore Pte Ltd. 2021
R. M. Singari et al. (eds.), *Advances in Manufacturing and Industrial Engineering*,
Lecture Notes in Mechanical Engineering,
https://doi.org/10.1007/978-981-15-8542-5_5

named the signal LGM-1 (for Little Green Men-1). Bell had discovered three more origins of pulsar in the sky, their characteristics were being investigated by the end of the year, and then, the duo published their findings in the February 1968 issue of Nature. Many more astronomers then started looking for similar signals, and thus, the science of pulsar astronomy was born.

A pulsar is a different kind of pulsar star which radiates periodic pulses of radio, X-ray, and Gamma-ray waves which are detectable on our planet. It forms after massive stars consume most of their fuel and thus their gravity overcomes the internal radiation pressure, collapsing the star. The outer layers are thus expelled to reveal a small core. Some pulsars are so because of their oscillation, but other discovered candidates also show two pulsar stars orbiting each other, i.e. a pulsar binary. During the first 15 years, after the first discoveries, the number of known pulsars grew to over 300. It is seen that the periods were distributed mainly from 100 ms to one second, and the rate of slowdown indicated that the pulsars with the shortest periods were the youngest.

A major forward step was made by the Gamma-ray observations that are re-examined using the periodicity found by radio, and many of the discrete sources turned out to be pulsating at GeV energies with pulse shapes similar to the radio profiles. Also, the X-ray pulsars were known to exist due to accretion in a binary system, of material from a star with a large and expanding envelope on to a condensed star, or an accretion disc, to temperatures around 10^6–10^7 K. Thus, the accretion process transferred angular momentum from the binary orbit to the star, spinning it up to a periodicity approaching one millisecond.

The rate of pulsar generation correlates with the supernova frequency which is one in every twenty-five to hundred years. Thus, calculating the typical lifetime of a pulsar yields the answer to be around ten million years. It loses its rotational kinetic energy through beam emission and does not rotate fast enough to emit powerful radiation. It then becomes unobservable. When a pulsar is ageing, it can still radiate in the radio domain with time periods measuring more than a second. The Crab pulsar has been around nine-hundred and sixty years, and thus, its period is comparatively short.

There is a total population of between 10^5 and 10^6 in the galaxy. Most of them are concentrated in the plane of the galaxy within a layer about 1 kpc thick and within a radial distance of about 10 kpc from the centre. The millisecond pulsars represent a smaller population of older pulsars. Their rate of "re-birth" in the spin-up process is much lower than the birth rate of the normal pulsars. They are found throughout the galaxy, but much less concentrated towards the plane than are the younger pulsars.

The Fermi LAT or Large Area Telescope as an instrument is especially useful to observe Gamma-ray emissions as it is very sensitive. The energetic light from such an emission reveals the location in its magnetosphere where the particles are accelerated, thus building a spatial model for the electric fields. It is the combination of these two techniques which KIPAC is using to describe the production of emissions within a pulsar.

Pulsar searching in modern science is done using several large radio telescopes in combination with signal processing algorithms and a touch of human ingenuity. The problem lies in the fact that rapid detection of pulsars is much more difficult because of the large amount of signals collected during a single observation. Apart from that, we have reached a level of scientific development which allows pulsars to be classified given enough time. Humans play a critical role in the identification process by recognising their emissions when they emerge.

The relevance of our and the preceding research lies in the fact that there are umpteen sources of man-made radio frequencies and noise which closely resemble the characteristic signals of a pulsar. This makes it very difficult to figure out which ones are the actual pulsars which in turn makes further research a hassle to say the least. While the earlier papers have done the bulk of the work in identifying features to be used in ML algorithms, we wondered if a higher accuracy could be achieved.

Our paper adds to common knowledge, not by trying to find more about pulsars, but by helping in classifying them from among a pool of "candidates". This is done by identifying features which are most relevant to pulsars and take the most different values for true and false positives. It is assumed that these features exist and those which fit the aforementioned conditions most closely were chosen. This classification is done using various machine learning techniques such as Naive Bayes, K-neighbour classifier, random forest, kernel SVM, and Keras classifier. Pulsars (or true positives) are assigned a "1" value, while the non-pulsars (or false positives) are assigned a "0" value. This, as is apparent, is a binary classification.

2 Problem Statement

Like most of the problems solved by technological developments, our problem is also one of reducing human labour and the need for efficiency and efficacy so that we can focus on other higher-order problems. In the recent decades, the field of radio astronomy has gained a lot of interest by amateurs and professionals alike. With the advent of highly sensitive telescopes and measurement instruments, the amount of data being captured from outer space on a daily basis has reached humongous proportions—numbers simply too huge for manual calculations.

The advent of the modern computer has relieved much of these pains, but that is just getting started. The combination of machine learning techniques with modern computers has the potential to do what has earlier been almost impossible. Due to the vast distances pulsar radiation travels, their radio, X-ray and Gamma-ray signals are severely dispersed and attenuated. It is hard to detect them but with the addition of noise and interference from natural and man-made sources, it becomes almost impossible to distinguish real pulsars from the total number of "candidates". This is our main problem, i.e. how to distinguish real pulsars from all possible candidates?

Machine learning, when implemented with properly chosen features and clean and scaled data, can be used to perform such a classification orders of magnitude faster than any human or team of humans. It becomes as simple as inputting eight numbers from observations and instantly receiving either a "yes" or "no" as the output.

It is ironic that the same technologies which helped us discover pulsars ended up distorting our observations, and another technology is helping us observe and identify them once again—this time with amazing speeds. Werner Becker of the Max Planck Institute for Extraterrestrial Physics said in 2006, "The theory of how pulsars emit their radiation is still in its infancy, even after nearly forty years of work." An optimistic view, such as the authors, might say that we did not have the tools then but do have them now.

All in all, in this paper we try to use various machine learning algorithms to make sense of the vast amount of available data on pulsar stars and effectively combat terrestrial and cosmic interference by intelligently choosing features which have the most differing values for pulsars and non-pulsar candidates. Although the identification and classification of pulsars is only one of the many problems, it is still the fundamental problem. Only when we observe them can we learn about them.

3 Research Methodology

After securing data from the HTRU2 dataset, we cleaned and scaled the data using the standard scalar from the Sklearn library. Then, we applied the following ML algorithms to the given data (obtained from the HTRU2 pulsar survey). It is prescient to have a look at the algorithms used for this endeavour. Following are brief summaries of the algorithms we have employed.

K-Nearest Neighbours (KNN)
The k-nearest neighbour algorithm is a nonparametric supervised machine learning method. This method makes no underlying assumptions about the data and thus has vast applications in real-life scenarios. While training, the algorithm learns the positions of the training points and their classifications. Then, during the actual testing, the algorithm simply calculates the relative distance between the point and the pre-established groups, and thus classifies the point to one of the groups. We used a KNN classifier to classify the candidates into pulsars and non-pulsars [1] (Fig. 1).

3.1 Random Forest

This method improved upon decision trees such that it avoids the problems associated with overfitting. Random forests are a way of averaging multiple deep decision trees. Random forests are examples of ensemble learning algorithms where multiple subsets of the original dataset are created, and many different classifiers may be applied to

Fig. 1 Graphical representation of KNN classification [2]

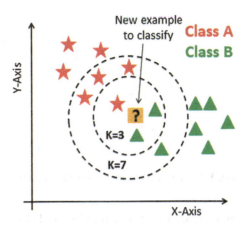

Fig. 2 Random forest algorithm [4]

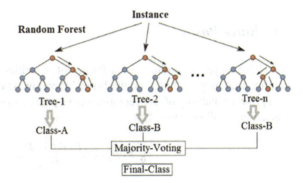

each of them. In case of random forest, multiple decision trees are used and the final output is decided after taking a vote from all trees. The aggregate prediction wins [3] (Fig. 2).

3.2 Support Vector Machine

SVM (or support vector machine) method divides the data into classes using hyperplanes. First all points are plotted onto an n-dimensional hyperspace (where n is the number of features in the dataset). Then, the hyperplane which best separates the classes such that both accuracy and robustness are maximised. To prevent misclassifications, a margin is set between the hyperplane and the closest points to it. This algorithm is also robust to outliers. Linearly separable data can be directly classified using SVM, but the nonlinear data needs extra defining features to find an appropriate

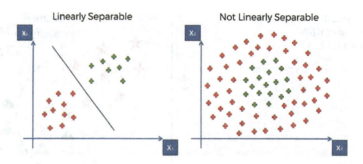

Fig. 3 Linearly and nonlinearly separable data [6]

hyperplane. The "kernel trick" is used in many types of classifiers when the data is nonlinearly separable [5] (Fig. 3).

3.3 Naive Bayes

The Naive Bayes algorithms are a family of algorithms which have one thing in common that they are based on the Bayes' theorem. The Bayes theorem states that the a priori probability of an event given the a posteriori probability of another event is given is as follows (Fig. 4):

$$P(A/B) = \frac{(P(B/A).P(A))}{P(B)}$$

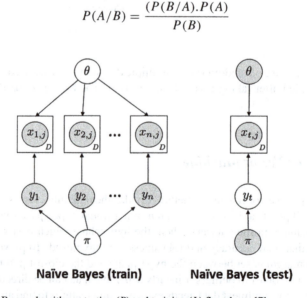

Fig. 4 Naive Bayes algorithm—testing (R) and training (L) flowcharts [7]

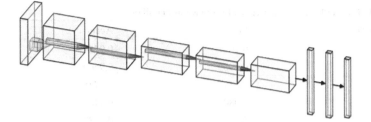

Fig. 5 Layers of a neural network [9]

The Naive Bayes assumptions are that each feature makes an *independent* an *equal* contribution to the outcome. These conditions might impose restrictions when implemented in real-life scenarios. When the independence condition holds, the Naive Bayes algorithm can be more efficient than others. It is highly scalable and gives probabilistic predictions [8].

3.4 Keras Classifier (Based on Neural Networks)

Neural networks are layers of neurons connected through synapses inside the brain. To emulate the processing methods of the brain, artificial neural networks which are analogous to biological ones were developed. Each neuron receives a signal from its predecessor, and if the signal is more than its activation threshold, the signal is propagated. Learning happens during backpropagation where the weights and biases of the individual neurons and layers are tweaked such that the next iteration will give a more accurate result. Keras is a high-level neural networks API, written in Python. Its advantage is mainly in its ability to enable fast experimentation (Fig. 5).

4 Results

After running the code for each kind of classifier, we tabulated the results. The given Table 1 shows the loss and accuracy of the model for epochs one through ten for the Keras classifier.

Plotting the accuracy and loss against number of epochs, we get the following curves (Figs. 6 and 7).

As shown, the best accuracy achieved is **97.90%** during the sixth epoch. The final accuracy achieved is **97.86%**.

Following are the final accuracies achieved using the aforementioned ML algorithms (Table 2).

Table 1 Loss and accuracy per epoch for the Keras classifier

Epoch No.	Loss	Accuracy
1	0.1001	0.9764
2	0.0905	0.9769
3	0.0847	0.9778
4	0.0808	0.9780
5	0.0789	0.9782
6	0.0770	0.9790
7	0.0762	0.9787
8	0.0752	0.9788
9	0.0744	0.9789
10	0.0740	0.9786

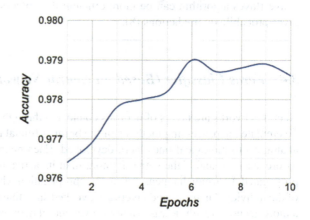

Fig. 6 Accuracy versus epochs

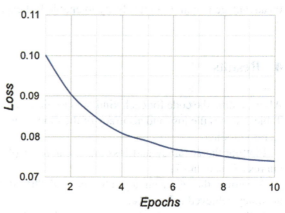

Fig. 7 Loss versus epochs

Table 2 Algorithm versus final accuracy

Algorithm implemented	Accuracy achieved (%)
K-nearest neighbours	97.944
Keras	97.900
Naive Bayes	95.061
Random forest	98.234
Support vector machine	98.122

5 Conclusion

All algorithms applied by us on the given data have given satisfactory results, which is mostly due to the high quality of the data. We found through the Keras implementation that neural networks work better for this particular set of data and parameters. As more algorithms are developed, we could hope to see higher accuracies which might ultimately do away with the need of manual classification. More study is required to understand the true nature of pulsar emissions and relates properties, but at least classification can now be done on a shorter timescale which will definitely lead to faster research.

References

1. GeeksForGeeks article. https://www.geeksforgeeks.org/k-nearest-neighbours by Anannya Uberoi
2. DataCamp article. https://www.datacamp.com/community/tutorials/k-nearest-neighbor-classification-scikit-learn
3. GeeksForGeeks article. https://www.geeksforgeeks.org/ensemble-classifier-data-mining by Avik Dutta
4. Medium article. https://medium.com/@williamkoehrsen/random-forest-simple-explanation-377895a60d2d
5. AnalyticsVidhya article. https://www.analyticsvidhya.com/blog/2017/09/understaing-support-vector-machine-example-code by Sunil Ray
6. Medium article. https://medium.com/@ankitnitjsr13/math-behind-svm-kernel-trick-5a82aa04ab04
7. TowardsDataScience article. https://towardsdatascience.com/information-planning-and-naive-bayes-380ee1feedc7
8. GeeksForGeeks article. https://www.geeksforgeeks.org/naive-bayes-classifiers
9. StackExchange article. https://datascience.stackexchange.com/questions/14899/how-to-draw-deep-learning-network-architecture-diagrams (Answered by Pablo Rivas)

Robust Vehicle Development for Student Competitions using Fiber-Reinforced Composites

Nikhil Sethi, Prabhash Chauhan, Shashwat Bansal, and Ranganath M. Singari

Abstract This paper presents a study of composite materials used in aerospace and automotive student competitions. The rationale for our research mainly arises from the increasing use of composite material technology in graduate-level student competitions. Moreover, the operational and financial constraints for research and development in student competitions are often more stringent than in the actual industry, and rigorous tradeoffs are conducted before choosing the appropriate material for the vehicles. Vehicles for three primary student competition teams corresponding to each of the authors are developed and compared to find common ground among use cases of composite material technology. The competitions included are namely Seafarer AUVSI SUAS, Shell-Eco Marathon, and Formula Student. An in-depth analysis of the material use-cases, properties, and the vehicle fabrication process is followed. The research is concluded by giving critical observations on points like optimization and sustainability.

Keywords Composites · UAV · Formula student · AUVSI SUAS · Supermileage vehicle

1 Introduction

Material science is an ever-evolving field of research with no boundaries. One of the recent entrants in its scope is fiber-reinforced composite materials. Fiber-reinforced composites, specifically carbon fiber-reinforced plastics (CFRP), are substituting conventional materials in the industry because they offer massively better strength-to-weight ratio. A study of various types of CFRP composites shows how versatile these materials are. Various arrangements of fiber planes can give the user access

---

N. Sethi (✉) · P. Chauhan · S. Bansal · R. M. Singari
Delhi Technological University, New Delhi, Delhi 110042, India
e-mail: scthi.nirvil@gmail.com

R. M. Singari
e-mail: ranganathdce@gmail.com

© The Author(s), under exclusive license to Springer Nature Singapore Pte Ltd. 2021
R. M. Singari et al. (eds.), *Advances in Manufacturing and Industrial Engineering*,
Lecture Notes in Mechanical Engineering,
https://doi.org/10.1007/978-981-15-8542-5_6

to a vast amount of applications ranging from the development of commercial super mileage vehicles to well-balanced, high-performing aerospace, and motorsport vehicles. Such prototyping and testing the material for its applicability before its production are a widely acknowledged industry practice. Research at various levels, from core researchers, industry practitioners to college graduates is important for innovation to succeed.

The present study exploits the varied experiences of multiple authors to present a review on fiber-reinforced composites, their properties, and their applications. In particular, the applications of composite materials in the automotive (Eco-Marathon), racing (FSAE), and aerospace (AUVSI SUAS)-related student competitions are covered in the current work. The cycle of development is also given a closed loop by highlighting the various methods of composite disposal and assessing the effects of the same on the environment.

2 Literature Review

In this section, an analysis of previous work conducted by other teams in similar student competitions is carried out. The competitions are taken part by teams from all over the world. As a consequence, the different student teams are often offset in their goals and developmental strategies owing to the variable amount of funding, developmental team size, technical support, research capacity, etc. These factors (in reference to the motive of this paper) affect the composite type, the production process, and the overall finished quality of the vehicle.

For instance, in [1], the authors use a combination of carbon fiber and Kevlar while employing a vacuum infusion process to develop the vehicle fairing. The mold was made through hot wiring foam. New research in fabrication methods also gives rise to advanced lighter frames, especially important for UAVs such as the case in [2], in which the authors use a hollow carbon fiber fabrication method. Some teams with well-established research facilities and significant funding also make use of autoclaves and pre-preg carbon fiber to develop their vehicles [3]. A particularly interesting use case of laminated bamboo fiber composite is used in a supermileage car in [4]. While not being the strongest of all materials, the paper presents a critical analysis of strength and composition.

Only the vehicle body is not limited to composite materials. In [5], exhaustive use of fiber composites for wheel rims and other small parts is documented for the motorsports vehicles. In [6], a chassis designed using composite materials is supported by rigorous structural analysis for multiple deflection cases. In [7], the authors have made an attempt to document the production of the control systems in an FSAE vehicle which includes its pedal box. Considering the degree of risk the regime of motorsports possess, authors in [8] have made an attempt to understand the applicability of composite fibers in the toughest form of racing, F1.

3 Materials

Prior to underlining the types of materials used and their properties, it is necessary to develop a rigorous classification of the various engineering materials and see where composites lie in the hierarchy (Fig. 1, using Refs. [9, 10]).

3.1 Glass Fiber-Reinforced Polymer

It is similar to CFRP with the exception of thin glass strands being used as the fibers. It has been used heavily as a reinforcement fiber. It consists of silica, alumina, calcium oxide, and boron oxide in varied proportions to get specific properties. It is a particularly lost-cost fiber and has vast applications in numerous fields, i.e., civil construction, automotive, marine, sporting goods, etc. [11]. Glass fiber is mainly used as reinforcement as it increases tensile strength, impact resistance, heat and chemical resistance, and dimensional stability. The use of glass fiber has the disadvantages of high viscosity of the melt, low surface quality, and damages to the tool and structure due to abrasion. It is readily available in the industry as continuous fiber rovings, chopped strands, and staple fiber [11]. It has a lower strength-to-weight ratio than carbon fiber but is much cheaper in comparison.

3.2 Fiber-Metal Laminates (FML)

FMLs are a class of metallic materials which are made by sandwiching layers of composite fiber strands among several thin metal layers with epoxy as the matrix.

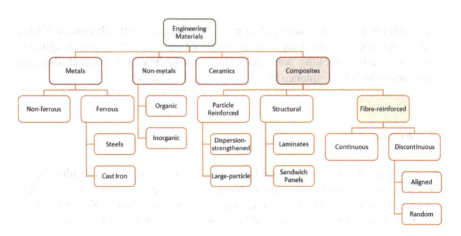

Fig. 1 Classification of materials

Some common FMLs are:

- Aramid-reinforced aluminum laminate (ARALL), based on aramid fibers
- Glass-reinforced aluminum laminate (GLARE), based on high-strength glass fibers
- CentrAl, which surrounds a GLARE core with thicker layers of aluminum
- Carbon-reinforced aluminum laminate (CARALL), based on carbon fibers.

3.3 Metal Matrix Composites

MMCs are relatively new composite materials which are made by dispersing a reinforcing material into a matrix of another metal. This material can be either metal or nonmetal. Compared to other reinforced metals, these composites have a high specific strength, greater stiffness, good wear resistance, and high operating temperatures. Their properties can also be tailored for individual applications.

3.4 Ceramic Matrix Composites (CMC)

It is a new type of material that has ceramic fibers embedded in a ceramic matrix. These are also known as CFRCs. The use of ceramics as reinforcement enhances the fracture toughness of the whole system while still exhibiting high strength and Young's modulus that is characteristic of the ceramic matrix.

3.5 Aramid Fiber

Aramid fiber is made of aromatic polyamides. They characteristically consist of large phenyl rings that are linked together by amide groups. The fibers are straight and rigid, having a high melting point and exhibit insolubility [12]. Aramid fibers are spun to make high-performance fibers like Kevlar and Nomex.

3.6 Kevlar

These are aramid fibers which have benzene as the rings in long molecular chains. The fibers have an excellent strength-to-weight ratio, high thermal stability, and high cutting resistance. The fibers have poor resistance to abrasion. Kevlar has been heavily used in knife proof armor and bulletproof vests for the past 25 years [13].

3.7 Nomex

Nomex is a high-performance aramid fiber and is also known as meta-aramid. Nomex is very low flammability and is self-extinguishing in nature. The fabric acts as a barrier to heat and fire and feels like normal textiles and is comfortable in use [14]. It is heavily used in making heat resistant garments and fabrics.

3.8 Carbon Fiber-Reinforced Polymer

It is a strong and light fiber-reinforced plastic containing strands of carbon fibers. CFRPs are expensive to produce but are commonly used in places where high strength-to-weight ratio and stiffness are essential, such as aerospace, superstructures of ships, automotive, civil engineering, sports equipment, and an increasing number of consumer and technical applications. Pure carbon fiber is characteristic of containing at least 92% carbon in it [15]. Carbon fiber is manufactured by the controlled pyrolysis process.

PAN carbon fiber—Polyacrylonitrile CF contains 68% carbon. It is polymerized by inhibitors like azo compounds or peroxides. Virgin textile PAN is not suitable for production as it takes a long time in stabilization and undergoes uncontrolled oxidation. Most of the commercially available carbon fiber used in the industry is provided by TORAY and HEXCEL groups.

Carbon Fiber Patterns
Unidirectional—This is the simplest pattern with only unidirectional fabrics. It is used when almost all the force is coming from a single axis.

1×1 Pattern—It is the easiest pattern looking like checkboxes also known as plain weave pattern which is the most standard pattern (Fig. 2).

2×2 Twill—This pattern is elastic and good for different shapes as the weave is loose. It is the most popular pattern and widely used in production (Fig. 3).

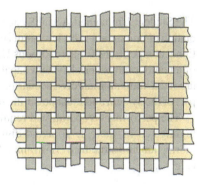

Fig. 2 1 × 1 Pattern

Fig. 3 2 × 2 Twill

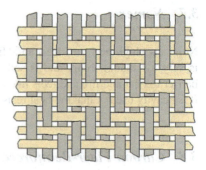

Triaxial Pattern—Triaxially woven fibers have much better shear resistance, stain resistance, and bursting resistance. The resulting fabrics do not easily crimp and are structurally better than rectangular woven fabrics.

Carbon Fiber Grades

MSI (million pounds per square inch)—This grade is based on tensile modulus which is a measure of stiffness. 33MSI is the most commonly used carbon fiber. 42MSI fiber is known as intermediate modulus fiber. Fibers with MSI 55 or higher are known as high modulus fibers and are costly and dense [15].

nK Fibers—This denotes the bundle sizes that fiber is available in 1, 3, 6, 12, 24, and 50 K are the commonly available ranges. These bundles are woven into fabrics although this has little to do with the quality of fiber.

GSM (grams per square meter)—Often, carbon fiber is sold with pre-impregnation of epoxy. This is commonly known as pre-peg. Pre-pegs are unidirectional, which means they have to be spread out-the thickness of which is indicated by GSM number.

4 Case Studies: Student Competitions

4.1 Seafarer AUVSI SUAS

The Student Unmanned Aerial Systems (SUAS) competition organized by the association of unmanned vehicle systems international which is one of the largest international competitions for autonomous UAVs. The competition entails the development of an autonomous aerial system capable of performing a complex mission centered around a humanitarian/civil scenario. The airframe used for the UAS covers a major part of the system and must be optimized and tuned to specific mission requirements.

The performance/operational requirements that drive the design and process are highlighted in Table 1. To avoid digressing from the scope of the paper, only the fabrication and structural constraints that propel the choice of material are discussed (Fig. 4).

Table 1 Design requirement

Parameter	Objective	Threshold
GTOW	<21.5 kg	<27.5 kg
Load factor	≥2	>1.5
Turning radius	<30 m	<45 m
Wind tolerance	>10.28 m/s	>7.7 m/s

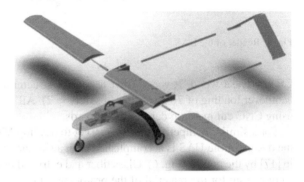

Fig. 4 Exploded CAD

Fig. 5 Removal from the mould

Owing to a short takeoff distance, high payload-to-empty weight ratio, a small turning radius, and consequently a high wing loading [16], the airframe needs to be strong enough to sustain the loads while being light enough to meet the operational requirements.

After the development of the CAD and rigorous structural analysis on ABAQUS, a fuselage featuring a monocoque shell was proposed. For initial testing, sandwiched laminates of carbon fiber, glass fiber, and balsa sheet were made, and their strengths were experimentally determined which would be further used for wing and tail skins. The results favored 200 GSM glass fiber and balsa sheets with a 45° orientation owing to the much greater load transfer capability and shear strength. The calculations for the wing spar gave the load and bending moment as the output for the selection of the length and cross section. The wing spar was built using 400 GSM unidirectional

Fig. 6 Render of the UAS

CF and balsa wood. It was further tested for structural strength with a maximum cantilever loading of 76 kg (factor of safety = 2). All the molds were manufactured using CNC cut medium density fiber boards.

For a similar competition, with a target to develop VTOL tricopter hybrid UAS, the design of the UAS is accomplished with a similar procedure as done previously in [17] by the author (Fig. 6). Glass fiber and extruded polystyrene (XPS) composite were chosen for the material of the prototype. This was partly because of a critical lack of funding and resources which is a common issue faced by many student teams.

Manufacturing Process

The fabrication of the complex blended wing body geometry was accomplished by hot wiring a block of foam along the three projected sketches on the orthographic planes (to avoid a lot of finishing) and finish it by sanding the extra corners with high grit emery paper. A similar procedure was followed for the wing. The process used for composite manufacturing is the hand epoxy layup and vacuum bagging process. The quantity of epoxy follows a general rule of 1.6 times the composite weight. This value was obtained from previous experience on similar materials.

A ratio of 1:9 for hardener: resin is used in our particular case. The different layers including the separator and breather for a uniform flow of air are placed. This was followed by adding two layers of 100 GSM glass fiber at the top and a layer each of 100 and 200 GSM glass fiber at the bottom owing to the tensile forces at the bottom (for the wing) and belly landing capability for the fuselage. The composites were

Fig. 7 XPS foam bod

Fig. 8 Vacuum bagging

Fig. 9 Final RTF prototype

cut according to shape and an epoxy layup using hands, and spatula was carried out before placing it on the mold. The vacuum was maintained for 6 h at a pressure of 150 psi with approximately 1 h of curing time.

4.2 Supermileage Eco-marathon

Eco-Marathon organized by Shell is one of the biggest supermileage competitions in the world. The competition requires the development of high mileage vehicles with specific safety and mensuration parameters. Here, only the specifics of material use and fabrication process have been documented. The vehicle was designed considering the competition requirements, i.e., improved driver safety, minimized aerodynamic drag, reduced vehicle downforce, and reduced frontal surface area of the body.

The body was also designed to comply with the following requirements:

- Allow room for multiple drivers to fit in individually.
- Provide space for the working components to operate safely.
- Maintain minimum ground clearance and provide a smooth aerodynamic flow.
- Balance out the frame with the driver and all the components in it.
- Provide a horizontal parting line for easier manufacturing of the body in two parts.

Manufacturing Process

The body was manufactured by the team members using 4 × 4 twill carbon fiber. The construction began with two plugs that were created from chopped glass fiber. The plugs were then sanded and treated with a series of primer coats, paint coats, and finally clear coats to achieve a smooth finish. The smooth male mold was then coated in Gelcoat, a material used to achieve a high-quality surface finish on a composite material. After sanding, mold release agent was applied in the mold, and two layers of carbon fiber were laid inside. The layers of carbon fiber were impregnated with a mixture of epoxy and hardener taken in the ratio of 10:3 respectively by weight (optimum being 65:35 by weight, but this increases the stiffness a lot plus would lead to inaccuracy in measurement) (Fig. 11).

Then, layers of peel ply and breather were kept on top of the layers of laminate. The whole setup (the carbon fiber, epoxy, resin, breather, and peel ply) was connected to a vacuum pump via vacuum lines and a vacuum bag. The vacuum bag was sealed tight in place using butyl tape to the flanges that were provided to the plug. This process of vacuum bagging was done, so that the excess epoxy resin gets absorbed by the breather under atmospheric pressure and also gets a smoother finish on the outside. Reinforcements and flanges were laid into the carbon fiber body during the layup process, utilizing thin, flexible aluminum sheet sandwiched inside carbon fiber as the reinforcement material. The body was allowed to cure five days while letting the pump apply pressure for 5 h.

Fig. 10 Impregnated carbon fiber

Fig. 11 Prototype vehicle

4.3 Formula Student FS FSAE

Formula Society of Automobile Engineers (FSAE) is a community of engineering undergraduate and graduate students, monitored by renowned race car design engineers. Student-run teams participate in competitions on a yearly basis all across the globe, where they design and build race cars following a certain set of rules and regulations.

Aerodynamic devices are primarily used in motorsports vehicles to generate negative lift (downforce) and to minimize drag. This gives the vehicle a competitive advantage providing superior control and performance by increasing the traction levels and reducing resistance. Components designed must be lightweight, easy to manufacture, and easy to handle and service. Several theoretical design iterations were modeled to arrive at a successful design.

A **multi-element rear wing** is designed to maximize rear-end downforce enhancing the understeering capacity of the vehicle. This increases the traction levels at the driven wheels, and hence, the capacity to accelerate increases enormously (Figs. 12 and 13).

The **front wing** is designed to achieve satisfactory levels of front-end downforce so as to balance the downforce produced at the rear end. The front wing is also the

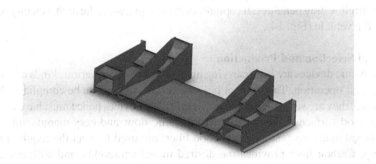

Fig. 12 Front wing

Fig. 13 Undertray

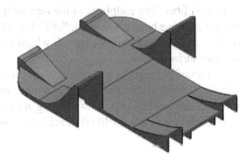

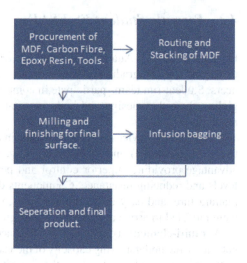

Fig. 14 Production flow Chart

onset of fluid flow in the vehicle, and hence, it also ensures that the other aerodynamic devices receive a clean flow of air.

The **undertray** is composed of three regions. The entry, middle, and the exit. Undertray achieves the goal of providing the maximum amounts of downforce levels with negligible drag penalties. It capitalizes the low pressure, the high-velocity region below the vehicle (Fig. 14).

Material Selection and Production
Aerodynamic devices are load-carrying units which undergo various kinds of loading during their operation. They cannot fail structurally and cannot be compliant. At the same time, they are supposed to be lightweight. The devices hence must have a strong core, a good surface to have better aerodynamic flow, and easy manufacturability. Fiber-based materials (essentially carbon fiber) are used to meet the required functionality. Carbon fiber provides the desired manufacturability and stiffness at the least weight. 3 k 2 × 2 twill carbon fiber weighing 200 GSM was chosen because of its good drapability, relatively low cost, high strength and stiffness, versatility, and low weight. The body panels of the car were made of a resin-infused single layer of carbon fiber. The endplates of the rear wing and front wing and mid plates of the front wing have been made using flat Rohacell 71 foam as a core between two layers of 2 × 2 200 GSM twill weave CF which improves bending stiffness and gives a pristine surface finish.

Layup of plane parts was done on sheets of glass, so that surfaces have a glass-like surface finish. For the curved contours, molds made of MDF were used. Resin infusion and wet layup vacuum bagging are used as a production process. The resin-to-fiber ratio was kept to 40:60 to achieve satisfactory results.

Robust Vehicle Development for Student Competitions using … 73

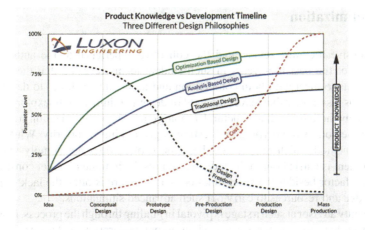

Fig. 15 Production versus design timeline [18]

Fig. 16 Mould preparation

Fig. 17 Glass finishing

Fig. 18 Final FSAE vehicle

5 Optimization

An extensive literature review and the collective experience of the authors after working for the past four years in such student teams revealed that the pressure for students to meet with both competition and academic deadlines to develop the vehicles as fast as possible leaves the field of design optimization unexplored. Thus, tools and methodologies like structural, topology, and shape optimization are avoided and the method of prototyping and testing is given a higher priority. While this is a viable method which has its own benefits, it often does not comply with how development is carried out in the industry and results in tighter budget constraints.

A key factor besides the strict timelines and pressure is often the lack of proper knowledge and resources to carry out such advanced simulations.

A faculty advisor at such a stage is pivotal in guiding through the process. However, this always leaves scope for biased designs among underprivileged teams. Figure 15 [18] shows how optimization can accelerate the design process and yield better products.

6 Sustainability

In today's world, sustainability plays a major role in determining use of materials for manufacturing and use of technology. Since the EU directive of minimum 85% of the weight fraction should be reusable or recyclable, the use of CFRP composite which is incinerated at end of life cycle is not possible [19]. Efficiency is a major driver in the automobile industry and hence, besides attention to innovation in the propulsion system, weight reduction is also an important field of experimentation. We need to access the life cycle sustainability of CFRP composite.

6.1 Manufacturing

Fabrication involves specialized techniques and tools as well as labor-intensive processes where automation has been difficult to achieve. The process usually involves the use of multiple raw materials and chemical components that are costly and time intensive in nature. Still, the development of manufacturing techniques is underway as the use of carbon fiber and other CFRP materials present an opportunity to move towards sustainable products and manufacturing.

6.2 Material Production

The first steps include material extraction, i.e., fossil fuels from the earth. Carbon fiber is made from organic polymers, which consist of long strings of molecules held together by carbon atoms. Most carbon fibers (about 90%) are made from polyacrylonitrile. CFRPs are produced using chemical-intensive energy consuming methods. Carbon fiber has high energy content compared to the rest of the fibers and is a high cost product [20]. This is a barrier against the use of CFRPs even though it has many useful properties and specifications that help drive innovation and better performance.

6.3 End of Life

Waste treatment policies aim to limit the impact of waste on the environment by promoting the use of recyclable and reusable materials. CFRPs pose a challenge in this domain as they mostly consist of multiple components that are fused together, and hence, recyclability provides a challenge. Methods like pyrolysis, hydrolysis, chemical recycling, regrinding, and incineration are often used to recycle FRCMs. Also, products are specifically fabricated for custom requirements, hence hindering widespread reusability [21].

7 Conclusion

Student competitions are a great medium for undergraduate-level students to indulge in the experimental side of the course syllabus and explore the industry. Moreover, a competitive environment makes room for innovation and sets the path for aspiring engineers and researchers. Three such competitions were studied in the current work, and the following key points were noted in comparison:

1. A different mission and size of the vehicle drives the optimum mixture of resin and hardener for the epoxy matrix.
2. The manufacturing techniques and range of materials is closely linked to the availability of resources and budget for development. Being from the same university and country, this is very similar for the three teams in our case.
3. The design process can vary from vehicle to vehicle depending on the high- and low-level requirements. However, the testing procedure and timelines are very similar for all teams worldwide. This is primarily attributed to the similar timeline of the college academics and frequency of examinations.

Acknowledgements The authors would like to thank Prof. Ranganath M. Sangari for his continuous support throughout the project. Both the UAVs in Sect 4.1 were manufactured at UAS-DTU laboratory, Delhi Technological University and the authors acknowledge the help of the team members for the same.

References

1. Faron H, Marcinkowski W, Prusak D (2015) Composite bodywork design and creation process in FSAE. Case study AGH racing. Int J Mech Eng (IJME) 4(6):13–20
2. Cornell University Unmanned air systems, Orion Technical Design Report, Seafarer AUVSI SUAS 2019
3. Flow Design Team, University of Split, Technical Design Report, Seafarer AUVSI SUAS 2019
4. Tomar P, Khandelwal H, Gupta A, Boora G (2017) Efficient design of a super mileage low cost vehicle frame using natural bamboo. Mater Today: Proc 4(9):10586–10590
5. Yay K, Murat Ereke I, Fatigue strength of a rim model with FEM using a new approximation technique. SAE Technical Paper Series 2001-01-3339
6. Olsen EV, Lemu HG (2016) Mechanical testing of composite materials for monocoque design in formula student car. Int J Mech Mechatron Eng 10(1)
7. Composite Parts for a FSAE Racecar: Monocoque Chassis, Pedal Box & Aerodynamic Undertray Project Number: P09221 Multi-Disciplinary Senior Design Conference, Kate Gleason College of Engineering
8. Savage G (2003) Composite materials technology in formula 1 motor racing Honda racing F1 team 2003
9. Callister WD, Rethwisch DG (2009) Materials science and engineering: an introduction, 8th edn. Wiley, New York
10. Chapter 4: Metal matrix composites, Advanced Materials by Design, Stati Uniti d'America, Congress. Office of technology assessment
11. Yang HM (2000) Fiber reinforcements and general theory of composites in comprehensive composite materials
12. Ash RA (2016) Vehicle armor in lightweight ballistic composites, 2nd edn.
13. De Araújo M (2016) Natural and man-made fibres: physical and mechanical properties. Fibrous Compos Mater Civ Eng Appl. https://doi.org/10.1533/9780857095583.1.3
14. Hearle JWS (2001) Textile fibers: a comparative overview. Encycl Mater Sci Technol
15. Huang X (2009) Fabrication and properties of carbon fibers. Materials (Basel) 2(4):2369–2403
16. Megson THG (2010) Introduction to aircraft structural analysis
17. Dagur R, Singh V, Grover S, Sethi N, Arora BB (2018) Design of flying wing UAV and effect of winglets on its performance. Int J Emerg Technol Adv Eng 8(3)
18. Luxon Engineering, Design Optimization, https://luxonengineering.com/services-optimization.html. Accessed 24 Nov 2019
19. Duflou JR, De Moor J, Verpoest I, Dewulf W (2009) Environmental impact analysis of composite use in car manufacturing. CIRP Ann-Manuf Technol 58:9–12
20. Song YS, Youn JR, Gutowski TG (2009) Life cycle energy analysis of fiber-reinforced composites. Compos Part A 40:1257–1265
21. Witik RA, Teuscher R, Michaud V, Ludwig C, Månson J-AE (2013) Carbon fibre reinforced composite waste: an environmental assessment of recycling. Energy Recov Landfill 49:89–99. https://doi.org/10.1016/j.compositesa.2013.02.009

Study and Applications of Fuzzy Systems in Domestic Products

Vatsal Agarwal, Sunakshi, Rani Medhashree, Taruna Singh, and Ranganath M. Singari

Abstract In the recent past, it has been observed that many domestic products have been produced with fuzzy systems. This paper explains the various programming made for the effective utilization of domestic products. This paper also explains about fuzzy and its models. To initiate with the products which are otherwise not utilizing fuzzy systems, this paper suggests a way of incorporating the fuzzy techniques in such products. With the help of this, the products will save time and human energy.

Keywords Fuzzy logic · Fuzzy logic controller (FLC) · Automated oven · Automated iron and temperature control system

1 Introduction

With increased workload and hectic living schedules, time spent and energy expended by an individual in domestic operations are becoming a strain. The need of the hour now is to adopt some superior appliances in domestic operations that can reduce human efforts and increased efficiency. The application of fuzzy logic on domestic products is now deemed to be a future solution. Fuzzy welcomes a value between 0 and 1 and uses the human linguistic variables to more accurately define the current state of input which uses predefined rules to determine the output.

V. Agarwal (✉) · Sunakshi
Mathematics and Computing Engineering, Delhi Technological University, New Delhi, Delhi 110042, India
e-mail: vatsalgrwl@gmail.com

Sunakshi
e-mail: sunaxee71@gmail.com

R. Medhashree
Electronics and Communication Engineering, Delhi Technological University, New Delhi, Delhi 110042, India
e-mail: ranims1999@gmail.com

T. Singh · R. M. Singari
Department of Design, Delhi Technological University, New Delhi, Delhi 110042, India

© The Author(s), under exclusive license to Springer Nature Singapore Pte Ltd. 2021
R. M. Singari et al. (eds.), *Advances in Manufacturing and Industrial Engineering*,
Lecture Notes in Mechanical Engineering,
https://doi.org/10.1007/978-981-15-8542-5_7

The paper is arranged as follows—Sect. 2 provides a brief study about fuzzy logic, fuzzy logic toolbox available in MATLAB and the advantages of fuzzy logic controllers over conventional PID controllers [1, 2]. In Sect. 3, we discussed the various domestic products and applications of fuzzy logic in those products that are currently in practice namely—microwave oven [3], domestic flour mill [4], gas heater [5]. Section 4 gives an extended study for automated domestic iron where fuzzy can be incorporated to deliver better results. Section 5 gives our conclusion on the basis of various studies between the PID controller and the fuzzy logic controller.

2 Fuzzy Logic

Fuzzy logic is an approach of making a decision based on computing the "degree of truthfulness" rather than the usual "true or false" (0, 1). The idea of fuzzy logic was first introduced by Dr. Lotfi Zadeh in the 1960s. While working on Natural Language Processing, a branch of machine learning, he found that natural language that human uses cannot be easily converted into a binary of 0 or 1. Many states between these two are frequently used and needed to be addressed with proper care. While dealing with fuzzy logic, 0 and 1 are taken as extreme values that show either completely false or completely true. In between these, it includes various states of truth, e.g., how much a piece of cloth is dirty can be recorded as "0.78 degrees of dirtiness." These intermediate values help any device to set output parameters according to requirements [6].

In application, fuzzy logic is defined basically as a set of if-then rules, based on linguistic variables instead of numerical variables, i.e., variables have value words rather than numbers.

Fuzzy logic since it is formulated during the 1960s has gradually developed into one of the most significant subjects of the cutting-edge world. It is valuable for individuals engaged with innovative work, mathematicians, regular researchers (science, material science, biology, and earth science), social researchers (financial matters, political theory, the board, and brain science), programming designers, examiners, architects, and therapeutic scientists. Earlier, it was considered to be a topic of mathematical concern, but now its applications can be found in varied fields such as facial pattern recognition, anti-skid braking systems, unmanned helicopters, weather forecasting, split air conditioners, stock trading, medical help, domestic flour mill, gas heater, wastewater management [7–14].

2.1 Fuzzy Logic Toolbox

Application of fuzzy logic involves the use of MATLAB fuzzy logic toolbox, "a software which interprets fuzzy logic as a theory which relates to a class of objects with unsharp boundaries which uses membership functions having values between

0 and 1." MATLAB functions and Simulink block are provided by the fuzzy logic toolbox that helps in building an environment for building and analyzing the fuzzy model. The tool stash lets you model complex framework practices utilizing straightforward rationale rules, and afterward actualize these principles in a fluffy derivation framework [2, 15].

2.2 PID Controller Versus Fuzzy Controller

Proportional derivative controller is the traditional controller that we have been using for times and takes a substantial portion of our domestic market. The goal for a PID controller is to change output when a certain setpoint is coordinated. It is commonly utilized when the contribution to the framework changes much of the time and the controller is relied upon to counterbalance it [16]. Fuzzy control is an incredible recipient in those zones of industry where because of the absence of appropriate details among input and output relations conventional PID controller strategy cannot be employed. In contrast with PID controllers, fuzzy controllers utilize a semantic framework practically identical to human observation and verbal.

A PID controller is complex and costlier than a fuzzy logic controller. Also, FLCs are more robust as they cover a wide range of operations without human intervention, and it also improves systems response time. While fuzzy logic has a well-defined set of if-then rules, a PID controller has only a fixed model, using which it outputs the result. An FLC has zero steady-state error [1, 17, 18].

3 Domestic Product Methodologies

3.1 Microwave Oven

The use of a microwave oven is prevalent in many households nowadays. A large variety of food items can be cooked and processed in it. But the amount of cooking time expected for various types and quantity of food items fluctuates considerably over a certain span of time. It makes it difficult to get the desired results every time we use it. We may have to check again and again if food is properly cooked or not, and it would lead to a decrease in efficiency and an increase in electricity consumption, in addition to extra efforts of repeating it again and again. So, we need to define a fixed set of rules for a different set of inputs and corresponding output times and implement it using a fuzzy controller to get better results (Fig. 1).

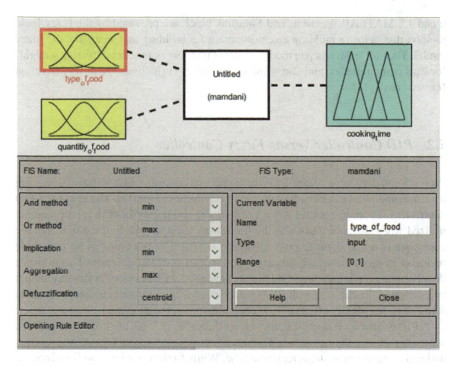

Fig. 1 Inputs and output for the model (microwave oven)

Working Principle. Fuzzy logic is already in use in the current microwave ovens. A model for automated microwave oven using intuitionistic fuzzy inference system was developed by Afsghan and Shiny Jose in their paper "Intuitionistic Fuzzy Logic Control for Microwave Ovens." Using fuzzy inputs of type of food (raw, semi-cooked, fully cooked) and quantity of food (large, medium, regular), an output regarding the cooking time of the oven (very long, long, very medium, medium, very short, short) was anticipated. Although fuzzy logic is already in practice for a microwave oven, the paper pointed out that the use of intuitionistic fuzzy will further improve the efficiency of the microwave oven [3] (Figs. 2 and 3).

3.2 Domestic Flour Mill

A flour mill is a machinery of the flour mill industry in which grain is converted into flour. Flour mills can be commercial as well as domestic. It takes raw wheat as input and gives edible wheat as output. In a flour mill, the wheat is passed between rollers, where it is ground into coarse particles of flour of varying sizes. These particles are

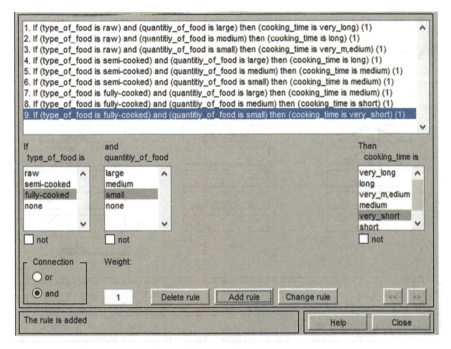

Fig. 2 If-then rules for inputs and corresponding output (microwave oven)

passed through sieves of various sizes. This process is repeated multiple times, and each time impurities are removed and flour gets finer, whiter and better in quality.

Working Principle. K. Sudha in the paper Automation of Domestic Flour Mill Using Fuzzy Logic Control proposed automation of domestic flour mill. A fuzzy logic model is proposed for the grinding process using MATLAB. Quantity (small, medium, large) and material type (soft, medium, hard) were taken as two input variables, while range (slow, intermediate, fast) is taken as an output variable. A comparison between FLC and PID controller showed that FLC gave faster response with a smaller overshoot compare to PID. This model helped in the faster and easier calculation of "range" [4].

3.3 Gas Heater

The traditional gas heater, routinely utilized for washing at home where gas and water are accessible, is a famous home machine. It fills in as follows: cool water streams into the warmer with temperature t, and stream rate u, goes through the warmth exchanger and streams out as constant boiling water with temperature t' through the heated water pipe. Right off the bat, working water pressure has a wide

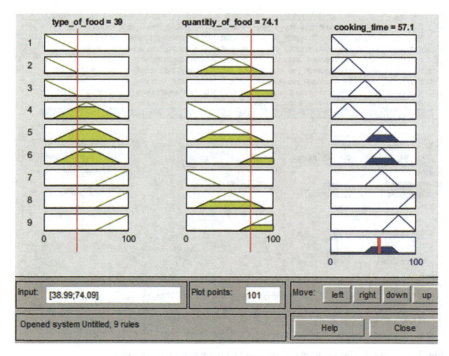

Fig. 3 Rule view of the model (microwave oven)

range. Also, the water stream rate is promptly influenced by neighboring water use, which is a major issue in urban communities of developing nations. Because of these inconsistencies, water that will turn out will have a fluctuating temperature. The absence of versatility to manage different breakdowns brought about by antagonistic working conditions has raised extraordinary issues about well-being issues.

Working Principle. In "Application of Fuzzy Logic in Home Appliance: Gas Heater Controller Design," a model dependent on the utilization of fuzzy logic alongside the PID controller was proposed. It uses temperature error and error change rate as input variables, and the open position of the gas valve was predicted as the output [5]. When analyzed against the use of only the PID controller, it gave increasingly more secure and open to washing conditions for clients with its "human brain" control approach and status checking innovation. Furthermore, the MCU enfolded in the gas warmer empowers engineers to execute a flexible and fast structure with full development highlights using contemporary innovation.

4 Extended Study: Domestic Iron

Among all domestic appliances, iron is most frequently used. Domestic iron is used to press over any fabric in order to remove its crease and wrinkles. Heat to iron plate is usually provided through coal or electricity. For domestic purpose, we consider electric iron. Electric iron has various assemblies with varying applications.

4.1 Inside Mechanism

The electric current passes through the coil present inside electric iron and raises the temp of the coil. That heat further passes to the lower plate and so pass to the clothes on pressing against it. We know that ironing is result of temperature and pressure. But overheating is an issue as it can ruin the fabric and consume energy. To avoid this, we use thermostat. A thermostat is a major component of an electric iron. As the name suggests "Thermo" means "Heat" and "stats" comes from "static", so the functioning of a thermostat is to maintain heat as constant.

4.2 Bimetallic Strip

A bimetallic strip is used in a thermostat. A bimetallic strip is a strip formed by two metal and having different coefficients of expansion, which means, on different temp, they expand by different dimensions. Both the strips are used as a component of the circuit. At normal temperature, strip stays normal and does not disturb the formation of the circuit, so the temperature can still rise. But after a particular increase when strip will encounter its expansion temperature, it starts expanding and bend toward the small pin. Small pin gets deviated and hence breaks the circuit and flow of current [6].

After some time, temperature will fall automatically, strip regain its shape. Small pin well relocates its original position and flow of current will start. Until power is switched off this cycle repeat itself.

4.3 Model Proposed

After the study of domestic iron and it's working, we proposed a model for automated domestic iron using fuzzy if-then rules. This model sets the temperature of the thermostat according to the given input parameters and hence reduces individual efforts while boosting the performance of the iron. We took the type of material, dampness of cloth and degree of wrinkles as input parameters. Type of materials

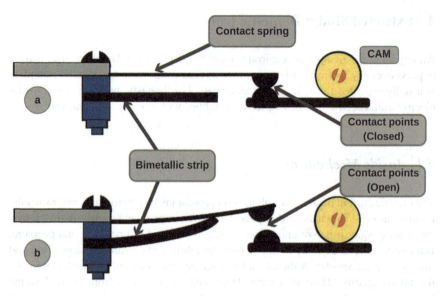

Fig. 4 **a** Under normal temperature and **b** when the iron becomes too hot

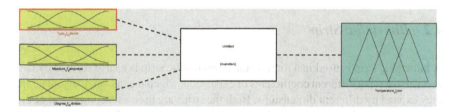

Fig. 5 Various inputs and the output of the model

used are—silk, cotton, wool. Measures of the dampness of cloth are—too moist, moist, dry and degree of wrinkles are high, low. We created a fuzzy model using MATLAB fuzzy logic toolbox and the temperature of iron is drawn as output. We draw graphs, scales and membership functions regarding how the three input parameters affect the output parameters. If a cloth has embroidery or sticky material on it, then its care has to be taken manually as done in regular iron. This model will set the temperature according to the material of the rest of the cloth piece (Figs. 4, 5, 6 and Table 1).

5 Results and Conclusion

In this paper, we made a study regarding the automation of various domestic products using fuzzy logic and its various components. These domestic products precisely include microwave oven, domestic flour mill and gas heater. Fuzzy programming by

Study and Applications of Fuzzy Systems in Domestic Products

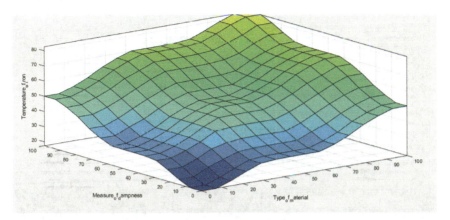

Fig. 6 A curve which portrays the relationship between various inputs and output

Table 1 Depicting rules for fuzzy model

Type of material	Dampness of cloth	Degree of wrinkle	Temperature of iron
Synthetic	Dry	Low	Very low
Synthetic	Dry	High	Low
Synthetic	Moist	Low	Low
Synthetic	Moist	High	Medium
Synthetic	Very moist	Low	Medium
Synthetic	Very moist	High	Slightly High
Wool	Dry	Low	Low
Wool	Dry	High	Medium
Wool	Moist	Low	Medium
Wool	Moist	High	Slightly high
Wool	Very moist	Low	Slightly high
Wool	Very moist	High	High
Cotton	Dry	Low	Medium
Cotton	Dry	High	Slightly high
Cotton	Moist	Low	Slightly high
Cotton	Moist	High	High
Cotton	Very moist	Low	High
Cotton	Very moist	High	Very high

the use of MATLAB fuzzy logic toolbox for different domestic products has been shown. After that, we made an extended study on domestic iron and made an attempt to propose a model for automated iron. With the help of this study, we were able to find that the use of fuzzy reduces human efforts and improvises efficiency (Fig. 7).

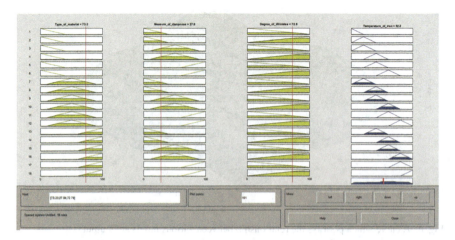

Fig. 7 Ruler representation points that at an input set of (73.2, 27.8, 72.8%) for (a type of material, a measure of dampness, degree of wrinkles), the output iron temperature is 52.2%

References

1. Langari R (1999) Past, present and future of fuzzy control: a case for application of fuzzy logic in hierarchical control. In: 18th international conference of the North American fuzzy information processing society—NAFIPS (Cat. No. 99TH8397), pp 760–765
2. Mathworks Products—Fuzzy Logic Toolbox, https://in.mathworks.com/products/fuzzy-logic.html. Last accessed 2019/10/09
3. Afshan SJ (2019) Intuitionistic fuzzy logic control for microwave ovens. Int J Math Trends Technol 65(3):134–143
4. Sudha K (2015)Automation of domestic flour mill using fuzzy logic control. Middle East J Sci Res 23(Sens Signal Process Secur):243–248
5. Rongming Z, Bian T, Qiantu W, Guaozhong D (1997) Application of fuzzy logic in home appliance: gas heater controller design. In: 1997 IEEE international conference on intelligent processing systems (Cat. No. 97TH8335), Beijing, China, vol 1, pp 373–376
6. Search Enterprise AI Fuzzy Logic definition page, https://searchenterpriseai.techtarget.com/definition/fuzzy-logic. Last accessed 2019/10/07
7. Singh H, Gupta M, Meitzler T, Hou Z-G, Garg K, Solo A, Zadeh L (2013) Real-life applications of fuzzy logic. Adv Fuzzy Syst. https://doi.org/10.1155/2013/581879
8. Mahajan R (2015) Application of fuzzy logic in automated lighting system in a university: a case study
9. Kılıç B (2017) Optimisation of refrigeration system with two-stage and intercooler using fuzzy logic and genetic algorithm. Int J Eng Appl Sci 9:42–42. https://doi.org/10.24107/ijeas.290336
10. Wakami N, Araki S, Nomura H (1993)Recent applications of fuzzy logic to home appliances. In: Proceedings of IECON '93—19th annual conference of IEEE industrial electronics, Maui, HI, USA, vol 1, pp 155–160
11. Henry N, Dahlan AA, Nasib AM, Aziz AA (2015) Performance of a variable speed of the split unit air conditioning system using fuzzy logic controller. Lecture Notes in Engineering and Computer Science, vol 1, pp 253–257
12. Kaler S, Gupta R (2017) Design of fuzzy logic controller for washing machine with more features
13. Vijayaraghavan G, Jayalakshmi M (2015) A quick review on applications of fuzzy logic in waste water treatment. Int J Res Appl Sci Eng Technol (IJRASET) 3:421–425

14. Khalid M, Omatu S, Fuzzy logic control
15. Mathswork Fuzzy Logic Documentation, https://in.mathworks.com/help/fuzzy/what-is-fuzzy-logic.html. Last accessed 2019/10/07
16. Omega Resources—How does a PID controller work? https://www.omega.com/en-us/resources/how-does-a-pid-controller-work. Last accessed 2019/10/05
17. Kaur A (2012) Comparison between conventional PID and fuzzy logic controller for liquid flow control: performance evaluation of fuzzy logic and PID controller by using MATLAB/Simulink
18. Gouda MM, Danaher S, Underwood CP (2000) Fuzzy logic control versus conventional PID control for controlling indoor temperature of a building space. IFAC Proc Vol 33:249–254. https://doi.org/10.1016/S1474-6670(17)36900-8

Study of Process Parameters in Synergic MIG Welding a Review

Rajat Malik and Mahendra Singh Niranjan

Abstract Welding is defined as one of the most efficient manufacturing and fabrication processes which exist in the category of formation of permanent metal to metal joints used widely in industries and production applications. The basic difference between conventional MIG welding and synergic MIG welding is that in synergic MIG welding, each pulse is responsible for the detachment of unit liquid metal into the puddle. Synergic MIG welding is in category of pulse-type MIG welding. A synergic MIG welding setup is designed to be a spatter less welding process that will consume lower-power heat consumption than spray or globular processes. An arc formation occurs between wire metal electrode and workpiece resulting the melting of metals at higher temperature. Combination of these melts mix together to form a single piece. Input parameters such as welding speed (v), welding voltage (V) and gas flow rate (gfr) are the important factors which directly affect the response parameters like microstructure, micro-hardness, tensile and yield strength and weld bead geometry of welded specimen. Present research aims to study various effects of input process parameters to response parameters and optimize them using Taguchi analysis of design of experiment (DOE) approach.

Keywords Synergic MIG · Welding parameters · Joint strength DOE · Taguchi

1 Introduction

1.1 Process Description

Synergic MIG welding is in the category of pulse-enabled MIG welding. Synergic welding is used to perform spatter less welding that will consume lower-power heat input than spray and globular processes. Welding each pulse is responsible for the

---

R. Malik · M. S. Niranjan (✉)
Department of Mechanical, Production and Industrial Engineering, Delhi Technological University, New Delhi, Delhi 110042, India
e-mail: mahendraiitr2002@gmail.com

© The Author(s), under exclusive license to Springer Nature Singapore Pte Ltd. 2021
R. M. Singari et al. (eds.), *Advances in Manufacturing and Industrial Engineering*,
Lecture Notes in Mechanical Engineering,
https://doi.org/10.1007/978-981-15-8542-5_8

detachment of unit liquid metal into the puddle, and metal transfer occurs through the arc with the principle of one pulse to one droplet. MIG accepts the varieties with the use of different electrode, shielding gas, and process variables optimized for a particular process.

1.2 Advantages

- Increases the net productivity.
- All direction welding is possible.
- Uniform welding is obtained with the electronically controlled parameters (Fig. 1).

Experiments were performed for finding the optimum value of rate of wire feed with pulse-enabled metal inert gas welding of Al sheets. Feed rates were varied in range of 0.5, 1 and 1.5 m/min, and the weld bead properties were examined. Measurements of weld bead were taken along the shape factor. It was observed that with the increase in wire feed rate, the weld bead becomes wider and area of cross section also increases [2] (Fig. 2).

Experiments were carried out to find out the consequences of MIG welding process parameters on the weld bead properties. Output parameters like DOP and weld bead width were measured. Dimensions of specimen were $141 \times 31 \times 10$ mm^3, and consequences of input process parameters, i.e., fraction amount of heat input and welding travel rate on weld width and depth of penetration were also observed. Welding speed varied in the range of 900–1400 mm/min. The result was obtained that with the increase in the welding travel speed, the depth of penetration shoots up

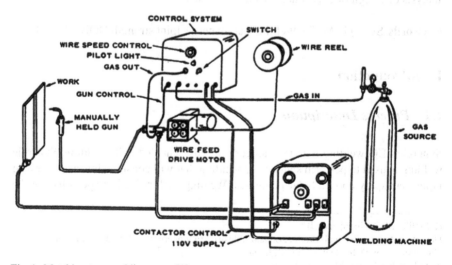

Fig. 1 Metal inert gas welding setup [1]

Study of Process Parameters in Synergic MIG Welding a Review   91

Fig. 2 Value of objective function versus wire feed speed [2]

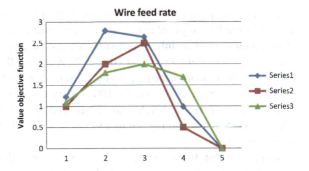

continuously, attains a maximum value at 1450 mm/min and then start decreasing beyond the optimum value [3] (Fig. 3).

Experimental study was performed on synergic MIG welding for the optimization of control parameters for achieving optimum results. Parameters were selected as gas flow rate ranging in 8–18 L/min, current 166–316 A and travel speed 27–47 cm/min. It was observed that shifting from base metal to heat affected zone, there a decrease in grain size was seen. It was also concluded that the value of micro-hardness is maximum in heat affected zone [4] (Fig. 4).

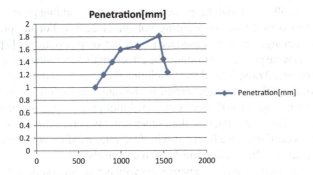

Fig. 3 Depth of penetration versus welding speed [3]

Fig. 4 Variation of micro-hardness at different regions [4]

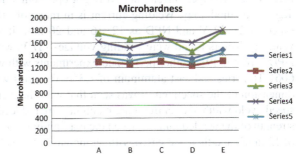

Author performed experiments using principal component analyses based on Taguchi's methodology. Material for the specimen used for the experiment is 316L steel which is used mainly in pressure vessels. Input process parameters came up with the range as welding current 100 to 124 Amp, gas flow rate 10, 15 and 20 L/min and nozzle tip to workpiece height is 9–15 mm. It was observed that the best conditions for the result were obtained at 100 amp welding current, 20 L/min gas flow rate and nozzle to workpiece height was 15 mm and the output parameters such as yield strength, UTS and percentage elongation were 322.7 MPA, 591.8 MPA and 54.53%, respectively [5].

Inspections of the characteristics of pulse on pulse frequency for the MIG welding–brazing process were carried out. Specimen was selected as two dissimilar metal plate: One is of 6061 aluminum and other is 304 stainless steel. Use of argon was done as shielding gas. This experimental study concluded as with the variation in the pulse frequency, a variation was seen in welding speed responsible for different methods or process for metal transfer. It was also seen that the use of high-energy pulse variation in the process of metal transfer is significant; whereas, using low-energy did not affect significantly [6].

Author studied the microstructural evolution of welding lap joints of two different metals, i.e., AZ31 alloy of magnesium and 2B50 wrought aluminum using metal inert gas welding. A layer of zinc material is spread between the interfaces of two metals. Specimen having dimension of 200 mm × 50 mm × 1 mm. 'ER4043' coded Al-Si wire is used as a filler material. Input parameters used in welding process—welding current is selected as 20 A, and welding voltage is 22.5 V. For the prevention of burning and cracking, zinc foil is used as a layer between two metals [7].

Experiments were performed to study the effects of factors influencing porosity of aluminum welding. Argon is used as a shielding gas. Wire feed rate was varied from 9 to 12.3 m/min. Welding voltage varied in the range from 26 to 32.0 V. Welding speed was in the range of 400–600 mm/min. Input process parameter such as gas flow rate effected more dominantly than other process parameters. It was observed that optimum conditions were found at wire feed rate, welding voltage and welding speed as 10.1 m/min, 28.5 V, 400 mm/min, respectively [8].

Author had performed several experiments with hot die steel specimen of size 200 × 100 × 5 mm^3 and observed the consequences of welding parameters like welding voltage and welding current on strength of joints of the welded metal. These input factors, i.e., voltage, current and NPD got optimum values, respectively, by using Taguchi methodology. Welding current is in range of 180–200 A, voltage taken in range of 21–27 V and NPD as 12, 16 and 20 mm. It was observed that optimal values of factors found were welding voltage—27 V, welding current—200 A and NPD—16 mm [9]. Author performed experiments with wrought aluminum alloy of 6061-T6 series and cast aluminum of A356-T6 series with 200 × 70 × 4 mm^3 as dimension of the specimen. It was seen that with the increase in the value of travel speed, weld width and reinforcement height were reduced. It was concluded that the strength of the welded joints reached the maximum value at an optimum travel speed of 12 mm/s and then decreased with an increase of travel speed [10].

Author performed experiments on specimen of 6-in. length mild steel with the objective of maximizing metal deposition rate, hardness of weld zone, penetration, reinforcement and minimization of bead height. Input process parameters, i.e., welding voltage, the traverse speed and the rate of wire feed were taken in the range of 21.1, 23.2, 24.3 V, 2.2, 3.4, 4.2 mm/min, 5.93, 7.75, 9.15 mm/min, respectively. Process used orthogonal array methodology and also SNR for the optimization of process parameters. Distance of nozzle tip and specimen has taken 17.5 mm with the use of carbon dioxide as shielded gas. Result was concluded with the increase of feed rate, the deposition rate was increased and remained constant with voltage, and with the increase in traverse speed MDR first decreases then increases. Bead height is increased as the increase in feed rate due to the fact of increase in the deposition of metal, it is decreased with the increase in welding voltage and feed rate. Hardness of metal weld region decreases first and then increases when there was an increase in welding voltage. With increase in voltage, both RFF and PSF increased. It was also resulted that the wire feed rate affected dominantly on output parameters [11].

2 Conclusions

From this review study, the following points are followed:

- Taguchi's approach is used mostly in which an orthogonal array is defined for finding the optimum parameters in synergic MIG welding.
- Different–different response parameters are affected with the change in input process parameters.
- Synergic MIG welding is better than conventional MIG welding with respect to spatter less welded joints.
- A few research work has been carried out on aluminum 5754 which offers property of great weld ability and a good machinability.
- Aluminum alloy 5754 has come up with wide application in various industries like aircraft manufacturing, shipbuilding, automobile manufacturing, etc.

References

1. https://www.google.com/search?q=mig+welding+setup+block+diagram&rlz=1C1CHBF_e nIN831IN831&sxsrf=ACYBGNQ3Ay3oj0DolUdVUH49Ns_ouE5lyQ:1575615380520& source=lnms&tbm=isch&sa=X&ved=2ahUKEwi1o7vYuKDmAhVbWysKHUq8DiQQ_ AUoAXoECA4QAw&cshid=1575615386071636&biw=1366&bih=657#imgrc=mCVDpv JQs3ibqM
2. Park HJ, Kim DC, Kang MJ, Rhee S (2008) Optimization of wire feed rate during pulse MIG welding of Al sheets. J Achiev Mater Manuf Eng 27
3. Abbasi K, Alam S, Khan MI (2012) An experimental study on the effect of MIG welding parameters on the weld-bead shape characteristics. Int J Eng Sci Technol 2

4. Meena SL, Butola R, Murtaza Q, Jayantilal H, Niranjan MS (2016) Metallurgical investigations of microstructure and microhardness across the various zones in synergic welding of stainless steel. Mater Today Proc Sci Direct
5. Ghosh N, Palb PK, Nandic G, Rudrapatid R (2018) Parametric optimization of gas metal arc welding process by PCA based Taguchi method on Austenitic Stainless Steel AISI 316L. Mater Today: Proc 5(1620–1625)
6. Kiran DV, Basu B, De A, Influence of process variables on weld bead quality in two wire tandem submerged arc welding of HSLA steel. J Mater Process Technol
7. Zhang HT, Song JQ (2011) Microstructural evolution of aluminium/magnesium lap joints welded using MIG process with zinc foil as an interlayer. Mater Lett 65:3292–3294
8. Bai Y, Gao H, Wu L, Ma Z, Cao N (2010) Influence of plasma-MIG welding parameters on aluminum weld porosity by orthogonal test. Trans Nonferrous Met Soc China 20:1392–1396
9. Kumar V, Goyal N (2018) Parametric optimization of metal inert gas welding for hot die steel by using Taguchi approach. MSRI 15:100–106. ISSN: 0973-3469
10. Nie F, Dong H, Chen S, Li P, Wang L, Zhao Z, Li X, Zhang H (2018) Microstructure and mechanical properties of pulse MIG welded 6061/A356 aluminum alloy dissimilar but joints. J Mater Sci Technol
11. Kaushik A (2017) Optimization of process parameters in synergic MIG welding of mild steel. Int J Eng Res Technol (IJERT) 6(11)

Comparative Study of Tribological Parameters of 3D Printed ABS and PLA Materials

Keshav Raheja, Ashu Jain, Chayan Sharma, Ramakant Rana, and Roop Lal

Abstract Rapid prototyping today is observed into practice and is being recognized as a significant technology for design. It demonstrates the process of design situated between conceptual design and real-world construction. It is used to automatically construct physical models from computer-aided design data or is a group of technique used to quickly fabricate a scale model of a physical part or assembly using three-dimensional computer-aided design data. The "three-dimensional printers allow designers to quickly create tangible prototypes of their designs rather than two-dimensional pictures". The 3D printer used works on fused deposition modelling (FDM) in which two materials are used to make 3D models out of the design made on CAD software. The two materials used are polylactic acid (PLA) and acrylonitrile butadiene styrene (ABS). The model of a pin will be formed using these two materials using various parameters. These parameters are—layer thickness, infill type, infill density and raster angle. In total, we will have then six parameters which include 3D printing parameters and the two materials. Using these five parameters, design of experiments will be created using the software Minitab. Various combinations will be created in the DOE with these six parameters. The pin will be then tested on the pin-on-disc machine which will help us to tell the wear rate and the coefficient of friction of the material through graphs. These graphs will help us decide the combination of parameters which have the most and the weak durability and thus can be used to create the parts we want to create for various industries. This way a comparative study of the tribological test between the two materials will be created.

Keywords 3D · PLA · ABS · DOE · Taguchi

K. Raheja · A. Jain · C. Sharma
Maharaja Agrasen Institute of Technology, New Delhi, Delhi, India

R. Rana (✉) · R. Lal
Delhi Technological University, New Delhi, Delhi 110042, India
e-mail: /ramakant@gmail.com

1 Introduction

3D printing or additive manufacturing is a process of creating a print in which each layer is laid down on the previous layers to make the final product. Hideo Kodama of Nayoga Municipal Industrial Research Institute is generally regarded to have printed the first solid object from a digital design. However, the credit for the first 3D printer generally goes to Charles Hull, who in 1984 designed it while working for the company he founded, 3D Systems Corp. Charles a Hull was a pioneer of the solid imaging process known as stereolithography and the stereolithographic (STL) file format which is still the most widely used format used today in 3D printing [1–3]. He is also regarded to have started commercial rapid prototyping that was concurrent with his development of 3D printing [4]. He initially used photopolymers heated by ultraviolet light to achieve the melting and solidification effect. Since 1984, when the first 3D printer was designed and realized by Charles W. Hull from 3D Systems Corp., the technology has evolved and these machines have become more and more useful, while their price points lowered, thus becoming more affordable [5].

Nowadays, rapid prototyping has a wide range of applications in various fields of human activity: research, engineering, medical industry, military, construction, architecture, fashion, education, the computer industry and many others. For example, for the jewellery sector, 3D printing has proved to be particularly disruptive. There is a great deal of interest and uptake based on how 3D printing can, and will, contribute to the further development of this industry [6]. From new design freedoms enabled by 3D CAD and 3D printing, through improving traditional processes for jewellery production all the way to direct 3D printed production eliminating many of the traditional steps. Also, to support new product development for the medical and dental industries, the technologies are also utilized to make patterns for the downstream metal casting of dental crowns and in the manufacture of tools over which plastic is being vacuum formed to make dental aligners. Architectural models have long been a staple application of 3D printing processes, for producing accurate demonstration models of an architect's vision. 3D printing offers a relatively fast, easy and economically viable method of producing detailed models directly from 3D CAD [7].

Minitab, the software used for creating design of experiments (DOE) is a statistical analysis software which allows the user to focus more on the analysis of data and the interpretation of results. Tribometer used to test the experiments is an instrument which measures tribological quantities, such as coefficient of friction, frictional force and wear volume, between two surfaces in contact [8–10].

There are many advantages of using 3D printing such as saving money and time in production, giving feedback of the product with less wastage, personalized items can be made, innovative ideas can be created, and the most important advantage is that once a prototype fails, it can be created again and again in less time and with less wastage [1–14].

2 Experimentation

The experimentation in this paper consists of doing tribological test on various 3D printed pins which are formed using a combination between different parameters with the help of Minitab software. These pins are then tested on pin-on-disc machine to find out their wear and coefficient of friction with respect to time.

2.1 3D Printer

The 3D printer used in making of these pins was Ender-3 by Creality, shown in Figs. 1 and 2. This 3D printer uses the fused deposition modelling (FDM) technique, shown in Fig. 3, to print the 3D product. In this technique, the model is produced by extruding small beads of material which harden to form layers. A thermoplastic filament or wire that is wound into a coil is unwinding to supply material to an extrusion nozzle head. The nozzle head heats the material up to the certain temperature and turns the flow on and off. Typically, the stepper motors are employed to move the extrusion head in the z-direction and adjust the flow according to the requirements.

The head can be moved in both horizontal and vertical directions, and control of the mechanism is done by a computer-aided manufacturing (CAM) software package running on a microcontroller [15–17]. The cylindrical pin is created in the 3D printing software according to the specifications required by the pin-on-disc machine and according to the combinations of the parameters. These pins are then run through

Fig. 1 Photographic image of the 3D printer (Ender-3 by Creality)

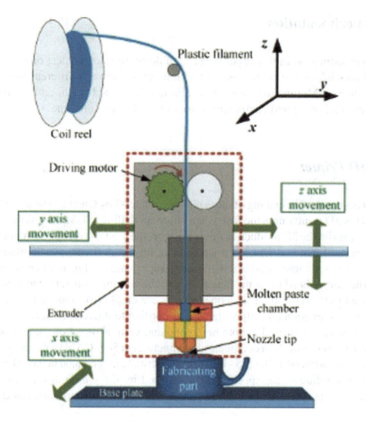

Fig. 2 Schematic representation of 3D printer

slicer which divided the model into the number of layers in which it will be formed. It is then fed to the 3D printer for the printing process.

2.2 Pin-On-Disc Machine and Tribological Test

The tribological tests were conducted by using pin-on-disc tribometer by following the ASTM G99 standard. The 3D printed pins were attached on top of the rotating disc [18]. Now the tribological tests are done on the 3D printed pins to find out the wear and coefficient of friction between the pins and the rotating disc with respect to time. The wear rate is calculated by comparing the wear volume before and after the test. The coefficient of friction is calculated by the ratio of frictional force to the loading force on the pin. The image of pin-on-disc machine is as shown in Fig. 3 [19].

Fig. 3 Pin-on-disc test rig

2.3 3D Printing Parameters

The following parameters were used in building of the 3D printed pins which determines the strength of the pins:

2.3.1 Infill Type

Infill type in a 3D print tells us the internal structure of the 3D print. This internal structure pays an important part in determining the strength of the 3D print. Figure 4a–d. shows several infill types' images that are available in the slice settings.

2.3.2 Fill Density

Infill density is the amount of filament printed inside the object, and this directly relates to the strength, weight and printing duration of the print. Figure 5a–d shows CAD images of infill density.

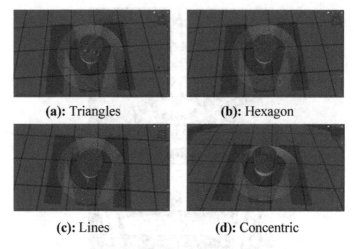

Fig. 4 a–d CAD images of various infill types

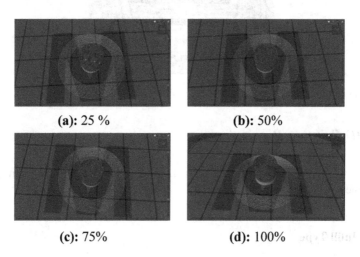

Fig. 5 a–d Cad images of fill density

2.3.3 Layer Thickness

Layer thickness in 3D printing is a measure of the layer height of each successive addition of material in the additive manufacturing or 3D printing process in which layers are stacked. The printing time required and the results of a smoother surface are also greatly determined by layer height. Layer thickness of 0.2 and 0.3 mm is used in the experiment. Figure 6a, b shows the CAD images of the used layer thickness.

Comparative Study of Tribological Parameters of 3D Printed ... 101

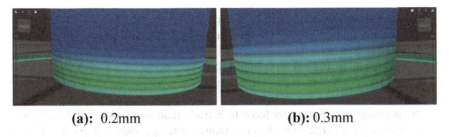

(a): 0.2mm   (b): 0.3mm

Fig. 6 a, b Cad images of layer thickness

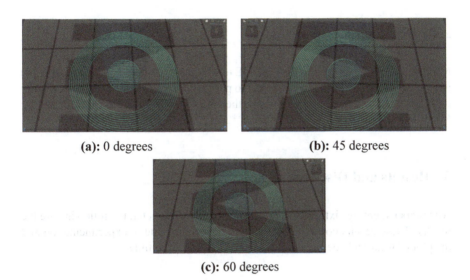

(a): 0 degrees   (b): 45 degrees

(c): 60 degrees

Fig. 7 a–c CAD images of raster angle

2.3.4 Raster Angle

Raster angle in the slicer sets the orientation of the first infill with respect to the heat bed. It determines the strength of the print at the end surface. Figure 7a–c shows the CAD images of raster angle.

2.3.5 Materials

There will be two types of materials which will be used in this experiment.

1. **Polylactic Acid (PLA)**: Polylactic acid (PLA) (is derived from corn and is biodegradable) is another well-spread material among 3D printing enthusiasts. It is a biodegradable thermoplastic that is derived from renewable resources. As a result, PLA materials are more environmentally friendly among other plastic

materials. The structure of PLA is harder than the one of ABS, and material melts at 180–220 °C which is lower than ABS. PLA glass transition temperature is between 60 and 65 °C.

2. **Acrylonitrile Butadiene Styrene (ABS): ABS is** one of the most widely used materials since the inception of 3D printing. This material is very durable, slightly flexible and lightweight and can be easily extruded, which makes it perfect for 3D printing. It requires less force to extrude than when using PLA, which is another popular 3D filament. Its glass transition temperature is about 105 °C, and temperature about 210–250 °C is usually used for printing with ABS materials.

2.4 Design of Experiments

Minitab software was used to form a DOE in which we feed the different parameters and the number of experiments we want to perform. In this Taguchi orthogonal array design is used to make combinations in the parameters. The same DOE formed by Minitab is shown in Table 1 [20]. L16(4**2 2**3), Factors:5, Runs:16.

3 Results and Discussions

After experimenting sixteen cylindrical pins on the tribometer, the following are the results. These results consist of the effects on the disc due to experimentation and the plot of wear and coefficient of friction with respect to time.

3.1 Disc Wear

Due to the friction between the two contact surfaces, i.e. the rotating disc and the stationary loaded pin, there is wear on the disc at different track diameters. Figure 8 is the image of the disc after the experiments [14] (Table 1 and Fig. 9).

3.2 Wear on Pins

The following are the images and the graphs showing the occurrence of wear on the sixteen cylindrical pins during the experiments (Fig. 10).

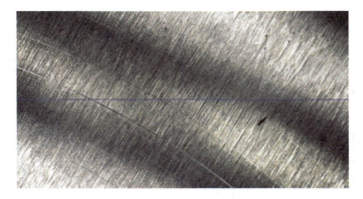

Fig. 8 Wear tracks developed on discs

Table 1 Design of Experiment (DOE) showing the average values of wear obtained in all the 16 experiments performed.

S. No.	Fill density	Infill type	Material	Layer thickness	Rester angle	Avg. wear
1	25	Triangle	PLA	0.2	45	13.41
2	25	Hexagon	PLA	0.2	45	118.13
3	25	Conc	ABS	0.3	60	71.95
4	25	Line	ABS	0.3	60	7.44
5	50	Triangle	PLA	0.3	60	8.66
6	50	Hexagon	PLA	0.3	60	5.26
7	50	Conc	ABS	0.2	45	7.52
8	50	Line	ABS	0.2	45	13.21
9	75	Triangle	ABS	0.2	60	5.26
10	75	Hexagon	ABS	0.2	60	4.96
11	75	Conc	PLA	0.3	45	2.59
12	75	Line	PLA	0.3	45	6.96
13	100	Triangle	ABS	0.3	45	5.73
14	100	Hexagon	ABS	0.3	45	8.08
15	100	Conc	PLA	0.2	60	4.37
16	100	Line	PLA	0.2	60	8.79

3.2.1 Images of Wear on Pins

See Fig. 11.

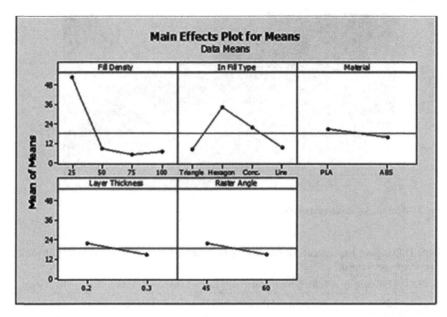

Fig. 9 Variation of wear with fill density, infill type, material, layer thickness and rester angle

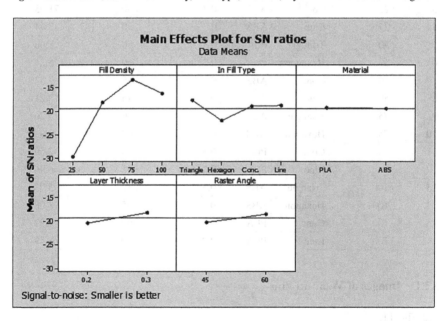

Fig. 10 Variation of SN ratio of wear with fill density, infill type, material, layer thickness and rester angle

Comparative Study of Tribological Parameters of 3D Printed ...

Fig. 11 Microscopic images of wear of all the 3D printed pins used in experimentation

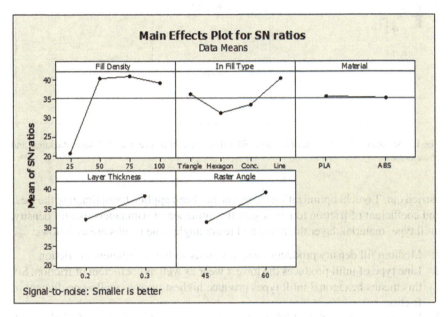

Fig. 12 Variation of SN ratio of coefficient of friction with fill density, infill type, material, layer thickness and rester angle

3.3 Coefficient of Friction

See Figs. 12 and 13.

4 Conclusions

The material undergoes wear due to sticking of the surfaces as a result of heat generated from friction. Wear test of material ABS and PLA with steel wear was

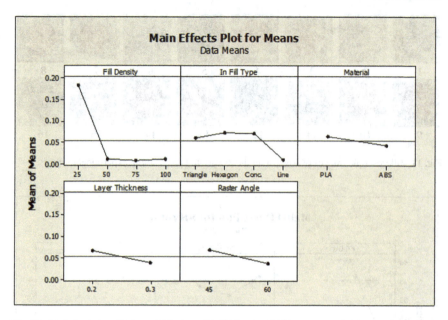

Fig. 13 Variation of coefficient of friction with fill density, infill type, material, layer thickness and rester angle

carried out. Taguchi optimization technique has been applied for optimizing the wear and coefficient of friction to investigate the influence of parameters like fill density, infill type, material, layer thickness and rester angle. The results are as follows:

1. Medium fill density produces lowest wear as well as coefficient of friction.
2. Line type of infill produces the lowest wear as well as coefficient of friction. So, this means hexagonal infill types produce highest wear as well as coefficient of friction.
3. In comparison to PLA, ABS produces less wear and coefficient of friction with respect to steel.
4. Higher layer thickness produces lower the wear and coefficient of friction.
5. Same way, higher raster angle produces lower the wear and coefficient of friction.

Acknowledgements Authors would like to acknowledge the support of Mechanical and Automation Engineering Department of Maharaja Agrasen Institute of Technology, Delhi, for allowing the use of Metrology Laboratory and it is facilitates. Authors also extend their regards to Precision Manufacturing Laboratory of Delhi Technological University, Delhi, India, for their kind help in conducting the experiments.

References

1. Sood AK, Ohdar RK, Mahapatra SS (2010) Parametric appraisal of mechanical property of fused deposition modelling processed parts. Mater Des 31:287–295
2. ASTM F2792e12a (2012) Standard terminology for additive manufacturing technologies. West Conshohocken: ASTM International
3. Yakovlev A, Trunova E, Grevey D, Pilloz M, Smurov I (2005) Laser-assisted direct manufacturing of functionally graded 3D objects. Surface Coatings Technol 190(1):15–24
4. Williams JV, Revington RJ (2010) Novel use of an aerospace selective laser sintering machine for rapid prototyping of an orbital blowout fracture. Int J Oral Maxillofac Surg 39:182–184
5. Vilaro T, Abed S, Knapp W (2008) Direct manufacturing of technical parts using selective laser melting: example of automotive application. In: Proceedings of 12th European forum on rapid prototyping
6. Webb PA (2000) A review of rapid prototyping (RP) techniques in the medical and biomedical sector. J Med Eng Technol 24(4):149–153
7. Rengier F, Mehndiratta A, Tengg-Kobligk H, Zechmann CM, Unterhinnighofen R, Kauczor U, Giesel FL (2010) 3D printing based on imaging data: review of medical applications. Int J Comput Assist Radiol Surg 5:335–341
8. Ramakant R, Kunal R, Rohit S, Roop L (2014) Optimization of tool wear: a review. Int J Mod Eng Res 4(11):35–42
9. Roop L, Singh RC (2019) Investigations of tribodynamic characteristics of chrome steel pin against plain and textured surface cast iron discs in lubricated conditions. World J Eng 16(4):560–568
10. Roop L, Singh RC (2018) Experimental comparative study of chrome steel pin with and without chrome plated cast iron disc in situ fully flooded interface lubrication. Surf Topogr Metrol Prop 6:035001
11. Singh RC, Pandey RK, Roop L, Ranganath MS, Maji S (2016) Tribological performance analysis of textured steel surfaces under lubricating conditions. Surf Topogr Metrol Prop 4:034005
12. Ramakant R, Walia RS, Surabhi L (2018) Development and investigation of hybrid electric discharge machining electrode process. Mater Today Proc 5(2):3936–3942
13. Roop L, Ramakant R (2015) A textbook of engineering drawing. Edition: 1, I.K. International Publishing House Pvt. Ltd. ISBN: 978-93-84588-68-7
14. Jain S, Aggarwal V, Tyagi M, Walia RS, Rana R (2016) Development of aluminium matrix composite using coconut husk ash reinforcement. In: International conference on latest developments in materials, manufacturing and quality control (MMQC-2016), 12–13 Feb 2016, Bathinda, Punjab India, pp 352–359
15. Ramakant R, Kumar SV, Mitul B, Aditya S (2016) Wear analysis of brass, aluminium and mild steel by using pin-on-disc method. In: 3rd international conference on manufacturing excellence (MANFEX-2016), 17–18 Mar 2016, pp 45–48
16. Ramakant R, Walia RS, Qasim M, Mohit T (2016) Parametric optimization of hybrid electrode EDM process. In: TORONTO'2016 AES-ATEMA international conference "advances and trends in engineering materials and their applications", 04–08 July 2016, Toronto, CANADA, pp 151–162
17. Ramakant R, Walia RS, Manik S (2016) Effect of friction coefficient on En-31 with different pin materials using pin-on-disc apparatus. In: International conference on recent advances in mechanical engineering (RAME-2016), 14–15 Oct 2016, Delhi, India, pp 619–624. ISBN: 978-194523970-0
18. Kaplish A, Choubey A, Rana R (2016) Design and kinematic modelling of slave manipulator for remote medical diagnosis. In: International conference on advanced production and industrial engineering, 9–10 Dec 2016

19. Khatri B, Kashyap H, Thakur A, Rana R (2016) Robotic arm aimed to replace cutting processes. In: International conference on advanced production and industrial engineering, 9–10 Dec 2016
20. Ranganath MS (2013) Vipin, optimization of process parameters in turning using Taguchi method and ANOVA: a review. Int J Adv Res Innov 1:31–45

Study of Key Issues, Their Measures and Challenges to Implementing Green Practice in Coal Mining Industries in Indian Context

Gyanendra Prasad Bagri, Dixit Garg, and Ashish Agarwal

Abstract Mining industries are the backbone of manufacturing industries because all the raw materials obtain by mining like iron, copper and brass. And the energy in the form of oil, gas and coal is achieved by mining industries and follow its significant contribution to country economies, whether their poor environment image. The objective of this study is to give an overview of the main issue of coal mining industries, their social and environmental impact, and measures to overcome these issues, in addition to identifying the main challenges to implement green supply chain management in coal mining in the Indian context.

Keywords Coal waste · Coal mining · Challenges · GSCM · SSCM · Recycling economics

1 Introduction

Mining industries were associated with various activities, from resource extraction to transportation in the marketplace in which some activity is responsible for environmental pollution. The mining industries perform such activities like extraction of raw material by drilling, blasting and cruising which create various environmental risks like air pollution, acidic drainage water, noise pollution. In a developing country, all the industries try to implement ISO 14001 certification and greener production. Mining is the backbone of all types of manufacturing industries. Mining industries

---

G. P. Bagri (✉) · D. Garg
Mechanical Engineering Department, National Institute of Technology Kurukshetra, Kurukshetra, India
e-mail: gpbagri@gmail.com

G. P. Bagri
SRM Institute of Science and Technology, NCR Campus, Delhi-NCR Campus, Delhi-Meerut Road, Modinagar, Ghaziabad, Uttar Pradesh, India

A. Agarwal
School of Engineering & Technology, Indira Gandhi National Open University, New Delhi, Delhi, India

© The Author(s), under exclusive license to Springer Nature Singapore Pte Ltd. 2021
R. M. Singari et al. (eds.), *Advances in Manufacturing and Industrial Engineering*, Lecture Notes in Mechanical Engineering,
https://doi.org/10.1007/978-981-15-8542-5_10

fulfil the requirement of raw material to manufacturing companies like all types of mineral and fossil fuel. India is the top five mineral producers in the world. Indian mining industries show rapid growth after 1947. Before 1950, India was 24 types of mineral producers; now, it has 90 mineral producers with project total value Rs. 127,662 crore, accounting for about 2.5% of the GDP [1]. Ninety-one per cent of mines are public sector units (PSUs), whereas 80% of mines have private players. By 1996–97, India had 3488 mines, of these 2271 were non-metals, 563 were coal and 654 were metals [2].

The objective of this work is to explore the key issues of coal mining industries. This study helps coal mining strategy manager to minimize the coal waste generation by waste management technique. In addition to this study, we identified the main challenges to the adoption of green-based recycling economics implementation.

2 Issues, Effects, Remedial Action and Their Possible Impact in Coal Mining Industries in India

2.1 Introduction

In the field of coal mines industry, almost all coal mining industries are public sector units. From 1975, state-owned coal mining industries come into existence. The beginning a year has a moderate coal production of around 78 million tonnes, now CIL in the field of coal production has the world's largest unit. CIL is the largest coal producer in India, which has 82 mines area with coal producing subsidiaries, and one has mine consultancy company. According to CIL report-2018 [3], during 2019 (APR-DEC), the CIL target was to produce 606.8 million tonnes of coal, which increase 39.52 MTs over the last year. Coal production in 2015–16 was 494.24 MTs, in five-year spans, and it jumped over 112.65 MTs to the current level. Coal and coal products deliver to power utilities, and it was 454.224 million tonnes in 2017–18 with a growth of 6.8 per cent over the last year (Annual Report 2018, CIL) [3]. In India, the latest technology open cast mines and underground mines are using which is fulfil the domestic as well as industrial requirement. In India, 90% of coal mines are open coal mines, which produce a huge amount of dust. G. Singh reported in their research, the present rate of extraction of coal approximately 0.8 million tonnes average daily in the country, according to this study, the coal reserves are over 100 years. Singh [4]. As per CIL Report [5], in India, total coal resources are 64,786 million tons which are using in various cooking and non-cooking applications. Table 1 shows the various application and their percentage in five years.

Table 1 Coal production in open and closed mining companies [6]

S. No.		2015		2016		2017		2018		2019 (Apr–Dec)	
		Raw coal production MTs	% of Raw coal production	Raw coal production MTs	% of Raw coal production	Raw coal production MTs	% of Raw coal production	Raw coal production MTs	% of Raw coal production	Raw coal production MTs	% of Raw coal production
1	Non Coking Coal	443.67	89.8	484.93	90.0	499.49	90.1	534.09	94.1	572.75	94.4
2	Coking Coal	50.57	10.2	53.83	10.0	54.65	9.9	33.28	5.9	34.14	5.6
3	Total	494.24	100.0	538.75	100.0	554.14	100.0	567.37	100.0	606.89	100

Table 2 SWOT of coal mining industries (expert opinion, Muduli et al. [7] and CIL report [6])

S. No.	Strength	Weakness	Opportunity	Threats
1	Most of the mining industries are PSUs (public sector)	Opencast mining	Availability of hydrotransportation method	Maximum coal reserves lie below the thick forest
2	Skilled human resources	Bureaucratic requirements	Availability of training facilities	Huge amount of gangue leaving behind Extracted superior grade of ore
3	Dedicated environmental management	Lack of clear cut specific in coal industries	The innovation of new technologies for the Extraction of coal	Frequently changing regulatory policies
4	Availability of funds for environmental expenditure	Indian coal mining industries are facing of weak productivity of employees	In the field of coal mining has opportunities to convert waste to wealth by reverse logistics	coal mining field has a lot of pressure of regularities
5	Adequate infrastructure availability	The Indian coal has a high ash content	X-ray technology can use to segregate to dry and wet coal	Emission of CH_4 trapped in coal mines
6	Most of the coal is transported by railways, which less impact on the environment	Lack of awareness of environmental impact	there is a huge demand for coal in India	In Indian coal mining industries, ending monopoly and started captive mining

(continued)

Table 2 (continued)

S. No.	Strength	Weakness	Opportunity	Threats
7	High production and huge production potential	Old mines with obsolete technology	Outsourcing of production processes	The private player is demanding to sell their product into the open market
8	Infrastructure available in almost all coal blocks	Trade Unionism	There are a lot of opportunities to value addition on their product by conversion into liquid and gas form	Mining is the emerging field in industries therefore private players can take better places with high skill employees
9	Skilled manpower available in sufficient numbers	Lack of information technology implemented in this sector		
10	The very low employee attrition rate	Poor work culture		
11	Most of Indian coal mine are mini rata category, and they have financial autonomy			

2.2 Main Issues, Effects and Best Management Practice

During the extraction of coal, coal mines are not created only air pollution but also responsible for dust; these are producing during drilling, crushing and transportation of coal from the mining area to the washing area and delivery of the end of the customers. Moreover, these processes are releases of various harmful gases like CH_4 (methane), SO_2 (sulphur dioxide), carbon monoxide and NO_x (oxides of nitrogen) where CO (carbon monoxide) and CH_4 (methane) are responsible for creating a greenhouse effect. While NO_x (oxides of nitrogen) and SO_2 (sulphur dioxide) cause for ground-level ozone and acid rain, respectively.

Acid mines drainage which produces from erosion and drained sulphide minerals present in the coal are major issues in coal mining industries (Singh 2009). Also, mining activity is responsible for global warming [8].

Air Pollution: Dust and some harmful gases such as CO, CH_4, SO_2, NO_x are produced during the extraction of coal.

Dust: During the extraction of coal, blasting and transportation are responsible for dust. In the open cast, mining is more responsible as compared to closed mines to dust production. Coal dust produced in open coal mines causes various health problems like eosinophilia, chronic bronchitis, pneumoconiosis, silicosis and several other respiratory diseases [9].

- According to the annual report 2005, it indicates by radiology studies that 12.7, 4.1% of underground mines workers and surface mines workers, respectively, suffering from pneumoconiosis.
- It is observed in the annual report 2005, the percentage of pulmonary tuberculosis disease is found in non-modernized mines which are more than the modernized mines, which one is 6.0 and 3.9%, respectively.
- Annual report 2005 shows that the percentage of tropical eosinophilia was about 8.0%.
- Coal mines' workers are also suffering from chronic bronchitis, silicosis and asbestosis [10].

Dust is generally produced by the mining extraction process and transportation to minimize dust during the extraction process using new blast less technology such as surface mining which reduces dust production. During transportation, dust produces and minimizes by the adoption of some measures like:

- To minimize coal dust spillages, avoided to overload on the vehicle.
- During transportation, coal should be wetted and covered with tarpaulin.
- Some units employ vacuum road sweepers, which one operated by mechanically to keep the haul road dust-free.
- Many companies are made service road by the dust-free material [7].

Carbon monoxide and Methane gases (CO, CH₄): These gases are responsible for the greenhouse effect, and it can overcome by using waste management techniques. Carbon monoxide (CO) can minimize by routine maintenance of the heavy earth-moving vehicle and to improve combustion efficiency. Other ways to reduce gas emission and fire fighting systems have an installed mining area as well as dump areas to prevent the fire (Singh 2009). Some of the companies adopted the sweeping and vertical mixing machine to dilute the emission of gases [7]. Methane gas is the source of energy for various industries as well as domestic purposes. Based on the principle of circular economy in mining industries, the methane gas converts waste into wealth by using various technological steps [11]. The extraction of methane gas is to construct the coal bed methane (CBM) wells in the coalfields before coal extraction [12].

SO₂ (Sulphur dioxide) and NO$_x$ (Oxide of Nitrogen): NO$_x$ (oxides of nitrogen) and SO₂ (sulphur dioxide) cause ground-level ozone and acid rain, respectively [7].

Water Pollution: During the extraction of coal, drained some water contains a high acidic chemical that pollutes the groundwater. Furthermore, the de-dusting of coal required a large amount of water; it will be 4 cubic metre water per ton of coal production (China and GAoEP 2002). Before leaving this water need for water treatment because untreated water contains heavy metals ions such as iron and mercury ions and could pollute the groundwater [12]. In the mining area, provision is to make a garland and construct a dam to store drained water for recycling and water treatment. Many companies installed effluent treatment plants (ETPs), sewage treatment plants, and silt arrestors/siltation ponds/sedimentation ponds. The treated water can be used to fulfil the water requirement of the town area instead of using freshwater [4, 7].

Noise and vibration: During the coal extraction process, due to blasting, drilling and coal transfer from the coal exploitation area to the washing area, etc., all activity is produced noise and vibration in terms of the explosion of explosive material, using heavy drilling and transportation types of machinery. Few mining industries are regularly checking noise levels and adopt preventive maintenance of heavy machinery, covering compressors and generators with acoustic materials to overcome these issues [7]. The overall effects of the mining activities likes drilling, blasting, crushing, and material transportation are producing a huge amount of noise and vibrations in the mining area, which leads to hearing loss and other health-related issues and loss of performance [1]. The noise and vibration can reduce by the use of a certain amount of explosion with control blasting techniques [4].

Fly ash and coal sludge: Coal sludge is one another solid mining waste in a coal mine. Coal sludge is produced after the washing of coal and dried. Dried coal sludge flies with the wind, which is harmful because it contained carcinogenic chemicals and toxic metals like arsenic, chromium, boron, mercury and nickel [12]. The environmental impact of the sludge overcomes by some recycling technique used like sludge which is used as low calorific value fuel in thermal power plant to produce electrical energy. The used sludge converted into fly ash in the thermal power plant which is light in weight and smaller amount as compared to sludge, and finally, it can be used in a construction company to make brick, mineral company to produce aluminium composite materials and remaining fly ash used for landfilling [12].

Solid mine waste: The mining process produces some solid waste in which coal gangue and fly ash are the main components. Coal gangue is categorized as very low coal content of topsoil and high coal content gangue. Very low coal content gangue is planned to dump in non-agricultural and non-forest land. It generally used to prepare dam or heap to retain drainage water; proper water channelling provisions are made. Besides, it is used to green belt development by geotextiles which are used to cover dump [7]. Topsoil is used to fill a low lying area with fly ash, which is a common waste management technique in coal mining industries [7]. High reach coal gangue is used in mine-mouth power generation.

Impact of Mining on Ecology: Mining practices impact the ecological system (G. Singh).

Impact of Coal Mine by Fires: Most of the coal mining industries are affected by the fires. The main reason for mine fires is causal heating by two relevant processes, for example, the interaction between oxygen and coal that is an oxidative process and the thermal process [4].

Further uncontrolled can spread over the interconnected areas. According to Singh [4], 10% of coal mines are affected by mines fires in India. Mines fire directly affect environmental and social life. Apart from that due to mines fire, impact economical of the nation shuts down the extraction of coal for some time, the release of poisoning gas, castration of coal, barriers to coal production, explosion, damage to infrastructure, etc. [4].

Impact of Coal Mine in Biodiversity: Extraction activity affects the directly or indirectly to leaving being nearby the mines. For sustainable development, industries should be balanced all three components economics, ecology and social component. The number of mining industry invests their 20% profit to sustainable development. In the field of a social component, mining industries relocate affected people nearby mines through rehabilitation action plan (RAP) and Indigenous Peoples Development Plan (IPDP) (G. Singh). Rehabilitation action plan (RAP) includes the shifting of people affected by mines, resettlement, and rehabilitation by giving land and some economically package to settle their choice places. The development plans are also having provision for training project-affected people in different trade for their economic rehabilitation. Another one is the Indigenous People Development Plan (IPDP); this plane is applicable for those villagers having within the one-kilometre area from the leasehold of the mines. Activities under IPDP are:

- Development of village roads, school, dispensary, school furniture, wells and tube wells.
- IPDP includes association activities like a youth club, Mahila Mandal, helping groups and artistic activities.
- IPDP also provides training for interpreting ownership, professional education, etc. (Fig. 1).

3 Challenges to Implementing Green Practices in Coal Mines Industries

Green practices can reduce environmental impact due to mining by using best management policy, for example, recycling waste, use the renewable product as well as possible, the increment technological innovations and the use of eco-friendly product, etc. (E. H. M. Moors, P. J. Vergragt). Therefore, the implementation of practices mentioned above could reduce environmental, social and economic impact, but industries face various challenges to implementing these practices. From the reviewing of literature and expert opinion, the following challenges are identified.

3.1 Lack of Top Management Commitment

Top management of the mining industries is *less* interested in the environmental issues and reluctance to the allocation of adequate finance, human resources, technologies to implement green practices. Top management hesitates to green practices due to a lot of inevitable documentation, etc. [13].

3.2 Size of the Firm

Small-scale mine owner has lack of technology as well as financial capability. Another problem is that much more small-scale mining is running illegally, and they are not following environmental regulation [14].

3.3 Oppose by Local Residents

Local people oppose to mechanization and modernization because of the scar of unemployment [14, 15].

3.4 Poor Legislation

Most of the Indian industries suffering from poor legislation, corruption, and poor political will relate to pollution control measures. A frequent change of environmental regulations to obstruct long-term planning of environmental issues [14, 16].

3.5 Lack of Direct Incentives

Lack of direct incentive is the major challenge in the mining industry. In the mining industries required extra fund for green practices which cannot be passed onto the consumer because the price of mining product is determined, depends upon the terminal auction market and cannot be controlled by the producer [3].

3.6 Financial Constraints

Some mining industries spend over 20% of their total revenue generated in adopting green practices like employee environmental training and upgrading equipment. The main challenges of industries to have a limited budget for green practices [16].

3.7 Technical Barriers

Many mining industries are difficult to identify new technology and expertise [16]. For green practices the need to upgrade technology like advance types of machinery, green design, green production, and waste management.

3.8 Lack of Employee Commitment

Lack of employee commitment towards green practices because mining industries do not have a proper performance evaluation system and motivating scheme for the employees to motivate them to take responsibility towards green practices. In the mining company, not properly define the responsibility of staff and not properly communicate them to their responsibility, that is why all the staff confuses what their responsibility and work are.

3.9 Lack of Awareness

Lack of awareness among product users, government personnels, and other concerned regulating bodies related to the hazardous impacts of the polluted environment [15].

3.9.1 Inappropriate Approach to Implementation

Many industries were implementing green practices without analysis, reviewing objectives and policies [17].

3.9.2 Information Gap (IG)

The information gap is one of the challenges to green practices. Lack of environmental and their impact knowledge of mining manager is a hurdle to green implementation [18]. Awareness about how to extract mining products with the use of low energy, the best utilization of waste management technique, reduction of accidental activity, less hazard to human health, etc. [19]. The addition problem creates due to the non-transparent policy of mining industries that not provided information regarding environmental issues [1, 14]. Societies, citizen and user unaware regarding the environmental impact all these are due to information Gap (IG).

3.9.3 Lack of Social Pressure (LSP)

Pressure from society, industrial association and environmental advocacy group like NGO can encourage management to take strategic action to increase their external reputation, upgradation their image in the market and reduce environmental impact. But in the case of Indian mining industries, societal pressure is very less from an advocacy group like NGOs. So that mining industries are not serious about taking corrective action towards environmental impact [15].

4 SWOT for the Coal Mining Industry

SWOT is the strategic planning technique which helps the manager to identify strengths, weaknesses, opportunities and threats into the business world. Strength and weakness: there are two dimensions used the analysis of the internal environment, whereas opportunities and a threat are used to analysis of the external environment. In the literature, many authors reported successfully implemented SWOT in the field of waste management. SWOT analysis technique can be used both as a preliminary stage and further run for the planning of strategic management [20]. This technique helps decision-maker to find out key issues and focuses on that by systematic analysis of internal and external environments [21, 22]. These tools are popular due to their clarity and simple adoption [23]. Srivastava et al. [24] successfully implement SWOT analysis in Lucknow municipality. Another study was done by Nikolaou and Evangelinos [23], in the field of mining, employed SWOT analysis in the internal and external environment of mining industries, and the author claimed that positive outcomes by the adoption SWOT. Yuan [25] employed SWOT analysis in the field of construction waste management in China (Table 2).

Now, in this study we identified strengths, weaknesses, opportunities and threats in the coal mining industry by review of the literature an expert opinion [7]. Aim of this study explores the SWOT component for coal mining industries through the literature and expert opinion. SWOT component varies with time so that industries, and SWOT analysis requires specific time intervals, for example, yearly or half-yearly.

5 Conclusions

The coal mining industries are the backbone of various industries and the main contributor to the economic growth of the country. The main objective of the above study to explore the various waste generated during coal production in mining industries from extraction to delivery. In this work we have identified the causes of waste generation and offering measures to minimise it. The various challenges for the implementation of Green Practices have also been discussed in this work. The addition of this study, identify strengths, weaknesses, opportunities and threats by expert opinion. This study helps managers making waste management strategies by recycling economy and waste management techniques and convert mining waste into wealth.

- **Key Issues in Coal Mining**
  - **Air Pollution**
    - Dust
    - Co, Ch_4
    - So_2, No_x
- **Water Pollution**
- **Noise & Vibration**
- **Coal Sludges & Fly Ash**
- **Solid Mine Waste**
- **Impact of Mining on Ecology**
- **Impact of Coal Mines fires**
- **Impact of Coal Mining in Diversity**

Fig. 1 Flow diagram: key issues in coal mining

References

1. Barve A, Muduli K (2011) Challenges to environmental management practices in Indian mining industries. 14, pp 297–301. Retrieved from https://www.researchgate.net/publication/235727022
2. Khatua S, Stanley S (2008) Integrated rural development of weaker sections in India, Ecological debt: a case study from Orissa, India. https://www.ecologicaldebt.org/publicaciones/Chapter5(125-168).pdf
3. Das A (2009) Does firm ownership differentiate environmental compliance? Evidence from Indian chromite mining industry. Munich Personal RePEc Arch 18716(18):1–26
4. Singh G (2010) Environmental issues with best management practice of coal mining in India, Feb, pp 1–6
5. CIL Report 2010
6. CIL Report 2018
7. Muduli K, Barve A, Tripathy S, Biswal JN (2016) Green practices adopted by the mining supply chains in India: a case study. Int J Environ Sustain Dev 15(2):159. https://doi.org/10.1504/ijesd.2016.076365
8. Mining India sustainably for growth Foreword (nd)

9. CIL Report 2005
10. Ghose MK (2003b) Indian small-scale mining with special emphasis on environmental management. J Clean Prod 11:159–165
11. Haibin L, Zhenling L (2010) Recycling utilization patterns of coal mining waste in China. Resour Conserv Recycl 54(12):1331–1340. https://doi.org/10.1016/j.resconrec.2010.05.005
12. Aibin L, Zhenling L (2010) Recycling utilization patterns of coal mining waste in China. Resour Conserv Recycl 54(12):1331–1340. https://doi.org/10.1016/j.resconrec.2010.05.005
13. Quazi AH (1999) Implementation of an environmental management system: the experience of companies operating in Singapore. Ind Manage Data Syst 99(7):302–311
14. Ghose MK (2003a) Promoting cleaner production in the Indian small-scale mining industry. J Clean Prod 11:167–174
15. Bowonder B (1986) Environmental management problems in India. Environ Manage 10(5):599–609
16. Hilson G (2000) Barriers to implementing cleaner technologies and cleaner production (CP) practices in the mining industry a case study of the Americas. Min Eng 13(7):699–717
17. Hale M (1995) Training for environmental technologies and environmental management. J Clean Prod 3(1–2)
18. Wu GC, Ding JH, Chen PS (2012) The effects of GSCM drivers and institutional pressures on GSCM practices in Taiwan's textile and apparel industry. Int J Prod Econ 135(2):618–636
19. Ministry of Mines (2010) Preparation of sustainable development framework for Indian Mining sector
20. Arslan O, Er ID (2008) SWOT analysis for safer carriage of bulk liquid chemicals in tankers. J Hazard Mater 154(1–3):901–913
21. Tahernejad MM, Ataei M, Khalokakaei R (2012) A strategic analysis of Iran's dimensional stone mines using SWOT method. Arab J Sci Eng 38(1):149–154
22. Agarwal A, Shankar R (2002) Analyzing alternatives for improvement in supply chain performance. Work Study 51(1):32–37. https://doi.org/10.1108/00438020210415497
23. Nikolaou IE, Evangelinos KI (2010) A SWOT analysis of environmental management practices in Greek mining and mineral industry. Resour Policy 35:226–234
24. Srivastava PK, Kulshreshtha K, Mohanty CS, Pushpangadan P, Singh A (2005) Stakeholder-based SWOT analysis for successful municipal solid waste management in Lucknow, India. Waste Manage 25(5):531–537
25. Yuan H (2013) A SWOT analysis of successful construction waste management. J Clean Prod 39:1–8. https://doi.org/10.1016/j.jclepro.2012.08.016

A Brief Review on Machining with Hybrid MQL Methods

Rahul Katna, M. Suhaib, Narayan Agrawal, and S. Maji

Abstract MQL has gained lot of attention in the past decade owing to its effectiveness over the conventional fluid delivery method. Conventional cutting fluid delivery methods have major drawback of being used in large quantity and dangerous effect on environment and workers health. MQL has emerged as a viable alternate to the conventional fluid delivery method. However, lately many more methods have been found to be more effective than the raw MQL method. Different strategies of delivering the cutting fluid directly into the cutting zone have been conceptualized and tested. These techniques are collectively coined as "Hybrid Techniques". This paper describes the various hybrid delivery methods in machining with their benefits and drawbacks.

Keywords Sustainable · Manufacturing · Minimum · Quantity · Lubrication · Green · Manufacturing

1 Introduction

Machining is very important process in the manufacturing industry. Huge amount of cutting fluids are used in machining for keeping the temperature rise under control and have a good quality product [1–4]. The various zones where heat is generated are shown in Fig. 1.

R. Katna (✉) · M. Suhaib
Department of Mechanical Engineering, Jamia Millia Islamia, New Delhi 110020 08544, India
e-mail: katnarahul@gmail.com

N. Agrawal · S. Maji
Delhi Institute of Tool Engineering, Okhla Phase 2, New Delhi 110020, India

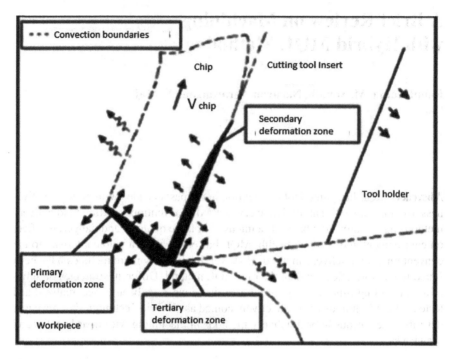

Fig. 1 Zones of heat generation in machining [4]

1.1 Conventional Cutting Fluids

However, the cutting fluids are used in large quantity and have negative effect on environment [5–7]. The high usage of cutting fluids affects the workers and also has negative impact on the environment [5, 8–10]. Dry machining is an alternative in which cutting fluids are not used and is an environment-friendly way of machining [11, 12]. However, there are many problems associated with dry cutting such as high temperature which can destroy the cutting tool and the workpiece surface [13–15]. Hence, the search is going on for a viable alternative that can be used for machining and which does not impact the environment and human [16–20]. Use of green cutting fluids has been in vogue for last many years and has proved to be better than conventional cutting fluids [21, 22]. One such alternative which has been found to be effective is MQL method. MQL stands for minimum quantity lubrication. It is a method of using straight cutting fluid, i.e. pure oil without water and atomizing it before delivering it to the machining zone during machining process. This method is effective than conventional flood delivery method in terms of providing better lubrication [23–28]. This is because the atomized oil droplets are of the order of microns and are effectively transported to the machining zone via the small cracks and also due to high delivery pressure of the carrier air. A typical MQL diagram is shown in Fig. 2. MQL method is also environment friendly as it uses very less

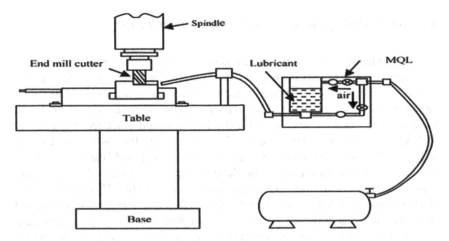

Fig. 2 MQL set-up [32]

amount if cutting fluid [29–31].

2 MQL or Minimum Quantity Lubrication

MQL has resulted in better surface finish and reduced tool wear in many experiments [33, p. 6]. However, many researchers oppose this result and have questioned the validity of the MQL in machining. Many researchers have also stated that MQL is effective in only some machining operations [34–37]. Researchers have raised serious concerns over the temperature reducing ability of MQL as it is used in very small quantity—of the order of 50–200 mL/h. It has also been reported that at very high cutting speeds, the MQL method loses its effectiveness as the oil gets vapourized before reaching the cutting zone and renders the process useless [38, 39].

Owing to the shortcomings in the conventional MQL methods, researchers have now focused on improving the effectiveness of the methods. Some research groups focus on improving the properties of the lubricant by various methods such as mixing nanoparticles and ionic liquids in the base oil. Addition of such additives greatly enhances the physiochemical properties of the oils, and machining with such lubricants has resulted in reduced friction and tool wear.

Other researchers focus on improving the delivery methods for increasing the effectiveness of the MQL method. Such methods use cooled compressed air via vortex tube, cryogenic MQL and electrostatic MQL. This paper gives an introduction to each of the methods and the current trend in their usage. In this paper, only the latest articles published in this year have been considered to give a brief overview on the applicability and the effectiveness of hybrid methods.

2.1 Minimum Quantity Lubrication in Machining

Minimum quantity lubrication is a method of cutting fluid delivery in which a very small quantity of lubricant is injected into the machining zone, mostly through the rake face of the tool. This is done by using pressurized air which atomizes the liquid lubricant and carries it with it to the machining zone. There are many benefits of MQL method over the conventional flood delivery method. It reduces the length of sticking zone of the chip on the rake face. It also reduces the friction between the chip and the tool by providing a film and thus decreases the frictional heat. High pressure also helps in chip curl and effective breaking of chips. In soft materials, use of MQL reduces the sticking tendency and thereby improves the tool life. There are two types of MQL delivery systems which are classified on the basis of mixing of compressed air with the lubricant. Figure 3 shows the different methods of mixing—external and internal. In internal mixing, the lubricant and air are mixed well before the nozzle exit, whereas in the external type of delivery, mixing is achieved just before exiting from the nozzle. The external method is far better than the internal mixing one as it ensures a homogenous flow of lubricant with the mixture. In internal mixing, the lubricant can stick to the walls of the carrying tube and can result in lumping, and the delivery is not homogenous. The pressure and the nozzle angle with respect to the cutting tool and the nozzle distance play an important role in achieving maximum efficiency from the MQL method.

3 Advancements in MQL Delivery Methods

Different ways have been researched on improving the effectiveness of MQL methods such as cryogenic MQL, cooled air MQL and electrostatic MQL. All these methods have been tested and have proved to enhance the effectiveness of the conventional MQL method. The commonly encountered problem is the air cushion or the vapour film generation which prevents the entry of the lubricant in the cutting zone.

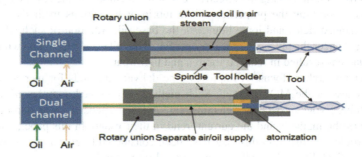

Fig. 3 Types of MQL mixing [40]

3.1 Cooled Air MQL

Cooled air MQL or CMQL is a technique in which cold air is used as lubricant carrier. Cold air carries the atomized lubricant into the machining zone. The cool air causes extra cooling effect in addition to the lubrication film provided by the lubricant.

In an attempt to improve the cooling efficiency of the conventional MQL process, Saberi et al. [41] employed a vortex tube for delivering lubricant into the machining zone. A vortex tube is a small pipe without any mechanical parts in which the input compressed air is split into two parts. One end of the vortex tube discharges cold air, and other end discharges hot air. This cold air is used for carrying the atomized lubricant. From the experiments, it was found that using this technique, the heat transfer coefficient was higher than the conventional MQL system. However, they also reported that lubricant has no effect in cooling with this method; yet, the lubricant in the MQL spray was able to reduce the grinding forces. They also reported that best surface finish is found in flood method of grinding than the MQL method of delivery because poor effectiveness of air in carrying away the heat generated and due to the circular profile of the grinding wheel which distorts the jet flow of air-lubricant mixture.

Stachurski et al. [42] used similar technique of using cold compressed air with MQL in sharpening by grinding the hobbing cutter. In a different approach, the authors used separate delivery of cold air into the machining zone in addition to the lubricant spray. They reported that among MQL, MQL + CCA and conventional flood method, MQL + CCA method gave best surface finish indicating its effectiveness over other methods and also reducing the quantity of cutting fluid required. Not only this, the cutting edges were free of defect with MQL + CCA method of delivery. The results indicate that coupling conventional MQL with cooled air approach is a better alternate than using only the conventional MQL system in machining.

Benjamin et al. [43] combined cold air through vortex tube and conventional MQL method of lubricant delivery in milling titanium alloy Ti-6Al-4V. The authors reported better lubrication in terms of reduced friction in machining with the combined method than with single MQL stream. The chips produced with the combined method of fluid delivery were also thinner and with less radius of curvature indicating less coefficient of friction at the tool chip interface. It was found from the experimental results that the combined technique was able to reduce the cutting forces by an average of 7% thereby saving on the input power requirement during machining. Too wear studies on the cutting tools showed impressive results. Tool life was found to increase by 43%. Moreover, the wear with conventional MQL was 502 microns, whereas with the combined technique, the wear was only 404 microns and no evidence of diffusion on the rake face which was found to be prominent with the conventional MQL method.

In another study done on titanium (Grade-II), Singh and Sharma [44] used vortex tube method for delivering cold and compressed air in the machining zone combined with lubricant. From the results, it was found that there is a reduction of 15% in surface roughness. Not only had this, cutting forces and tool wear also reduced with

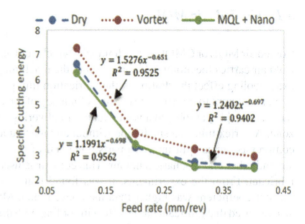

Fig. 4 Specific cutting energy with different fluid delivery method at different feed rate [46]

the combination method employed by the authors. This was due to better cooling achieved with the combination method in comparison with the MQL method alone. Salaam et al. [45] proposed the use vortex tube for generating cold air for assisting the conventional MQL proves in order to improve its effectiveness. They claim that this combined method will have higher pressure than the conventional MQL method and will also help in improving the machining efficiency. However, Ali et al. [46] reported that using hybrid method did not reduce the specific cutting energy required in the machining operation. Figure 4 clearly shows that even at different feed rate the highest specific cutting energy is with the hybrid method of fluid delivery.

Jiang et al. [47] compared the effectiveness of cold air with mist. The performance was compared with cold air without mist, and they observed a reduction in the cutting forces in the cold air with lubricant mist. Better temperature control is observed in both the cases than dry machining.

3.2 Cryogenic MQL

Cryogenic machining refers to machining done at very low temperatures. Owing to the low lubricity problem of cryogenic gases but good thermal properties, researches have use several cryogens combined with MQL method.

Shokrani et al. [13] used a double nozzle system for delivering liquid nitrogen cryogen with MQL lubricant at different cutting speeds, feed and depth of cut. The results show that cryogenic gas combined with lubricant has a significant effect on the life of the cutting tool. The life of the cutting tool was found to have increased significantly under the hybrid cooling environment. However, in this study, it was found that the lubricant also plays a role in heat carrying capacity which was not observed when machining with cold air assisted MQL. From the surface roughness data, it was found that the machining environment affected the quality of surface achieved in machining. The surface roughness achieved was comparable to that achieved with MQL. Cryogenic MQL technique thus increases the tool life in addition to providing a better surface finish and very low rise in temperature.

Hanenkamp et al. [48] used carbon dioxide in combination with atomized lubricant in milling operation of Ti6Al4V. In a strikingly different approach, the authors used different types of lubricants in addition to the carbon dioxide as cryogen. The results clearly show that the combination of cryogenic technique with MQL performs better than the conventional flood method of cooling. Comparison of surface roughness achieved with the different types of lubricants used in the machining operations revealed a striking fact that it is the presence of lubricant that controls the quality of the machined surface. This was evident because the cryogen used in each case was the same, but the lubricant employed was different, and different surface roughness values were achieved. The lowest surface roughness values were achieved with the ester + additive lubricant. Figure 5 shows the surface integrity with different methods of fluid delivery.

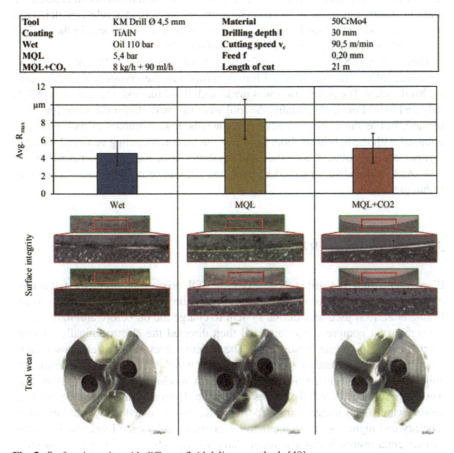

Fig. 5 Surface integrity with different fluid delivery methods [48]

Grguras et al. [49] used cryogenic MQL by premixing the lubricant in liquid carbon dioxide in machining. The experiments were done on milling TI6Al4V with carbide tool. The results indicate that cryogen oil solution was able to increase the tool life by more than 200%. The authors also reported that this method of mixing the lubricant in the carbon dioxide cryogen is also a simpler method because only a single nozzle is required in machining.

Pereira et al. [50] used carbon dioxide MQL in machining Inconel 718. They made a plug and play type adapter. From results, it is seen that the tool wear obtained with this hybrid method was close to that achieved with the flood method of cooling which indicates the effectiveness of this method as a potential alternative to the conventional flood cooling method.

Kim et al. [51] used chilly carbon dioxide gas with MQL lubricant in milling titanium alloy. The experimental results obtained clearly establish the effectiveness of hybrid MQL over the conventional MQL process. The hybrid method produced less cutting forces and resulted in lower coefficient of friction and also increased the tool life. Another outcome of this research translates the effect of lubricant delivered in the chip–work interface is that the authors found an improvement in the surface finish with an increase in the concentration of the lubricant.

Damir et al. [52] tested the effectiveness of cryogenic MQL in machining aerospace alloy. The authors used liquid nitrogen as the cryogen and a lubricant for MQL spray. The comparison was made with flood method, cryogen only, MQL only and hybrid method. Initially, the tool wear rate was almost the same for every method. But as the machining advanced towards, lowest amount of tool wear was obtained with the hybrid method. This was due to better cooling achieved with the cryogen system coupled with the MQL system. Also, the components machined with the hybrid delivery method showed less amount of residual stress than other techniques.

3.3 Electrostatic MQL

Electrostatic MQL is a strikingly new type of MQL process. In this method, fine mist of oil, either positively or negatively charged, is directed into the machining zone. Huang et al. [53] used an external system to charge the oil before atomizing them through an air compressor system and then directed the electrostatically charged oil particles into the machining zone. The benefit achieved out of electrostatically charging the oil is reduction in the wetting angle which translates into better spreadability. Also, as the oil particles are charges, it would be attracted to the neutral workpiece surface easily. Oil particles charged up electrostatically have lower surface tension which helps in easy transportation in the cutting zone thereby enabling better lubricity and higher performance than the conventionally used lubricant. From the experimental results, it was also found that positively charges oil has better lubrication properties than the negatively charges oil and the degree of polarity also affects the lubrication performance. The electrostatic MQL resulted in reduction in the wear of the cutting tool due to increase in the wetting properties, spreadability and reduction in surface tension. Similarly, in another study done on stainless steel, Xu et al. found similar results in decrease in tool wear [54]. The experimental results indicate the effectiveness of electrostatic MQL over the conventional MQL system and also indicate the viability of this method as an alternative to the conventional MQL method of cutting fluid delivery.

Jamil et al. [55] used electrostatically charges lubricant in machining aluminium alloy 6061T6. The authors used negatively charged castor oil as the lubricant in turning the aluminium 6061T6. The authors report that the negatively charged oil particles are accelerated towards the positively charged workpiece. The positive charge near the chip–workpiece region creates a suction pressure near the zone and helps in the capillary suction of the charges oil. This results in a deeper penetration into the cutting zone than the conventional MQL method and results in better performance and machining quality in terms of reduced tool wear.

Lv et al. [56] used grapheme nano platelets via the electrostatic MQL process. The focus of the researchers was to study the extent of penetration and the deposition property endowed due to the electrostatic charging. The experimental results showed an improvement in the penetration and the deposition in the metallic surface of the workpiece. The improvement in the properties was due to decrease in the wetting angle which resulted in better spreadability which is seen in other cases also. Electrostatically charged particles are able to make a better tribo film because they penetrate easily between the mating surfaces.

Xu et al. [57] used electrostatic MQL strategy in grinding operation of die steel. They reported that the dominant mechanism of action was capillary action. They also observed that 4kv electrostatic voltage was able to reduce the cutting forces significantly. This was attributed to the fact that the electrostatically charges lubricant particles were able to spread easily and resulted in better lubrication and reduced cutting temperature.

Su et al. [58] used charges nanoparticles in the lubricant and used this in the MQL system and found that electrostatically charged nanoparticles mixed lubricant with MQL is better than simply electrostatically charged lubricant with MQL. However, the drawback was that the nanoparticle charges resulted in higher mist concentration than that with the simple lubricant charges electrostatically and delivered via MQL. The authors recommend both these methods as an alternative to the conventional MQL method without electrostatic charging.

4 Conclusion and Future Scope

MQL method is no doubt a good alternative to the conventional flood method of fluid delivery. However, currently many advancements have been seen in the delivery method of cutting fluids. Following can be concluded from the literature:

1. MQL method is better than conventional flood delivery method.

2. The method of cutting fluid delivery affects the machining quality.
3. MQL still suffers from the cooling ability drawback as the volume used is very low.
4. Hybrid MQL techniques have shown better results than conventional MQL method.
5. Additional methods add up to the cost of conventional MQL as extra set-up is costly.
6. It is recommended that MQL method can be combined with flood delivery method in order to improve the heat carrying effectiveness of the method.

References

1. Adler DP, Hii W-S, Michalek DJ, Sutherland JW (2006) Examining the role of cutting fluids in machining and efforts to address associated environmental/health concerns. Mach Sci Technol 10(1):23–58
2. Astakhov VP, Joksch S (2012) Metalworking fluids (MWFs) for cutting and grinding: fundamentals and recent advances. Elsevier
3. Çakīr O, Yardimeden A, Ozben T, Kilickap E (2007) Selection of cutting fluids in machining processes. J Achiev Mater Manuf Eng 25(2):4
4. Abukhshim NA, Mativenga PT, Sheikh MA (2006) Heat generation and temperature prediction in metal cutting: a review and implications for high speed machining. Int J Mach Tools Manuf 46(7–8):782–800
5. Bartz WJ (2001) Ecological and environmental aspects of cutting fluids. Tribol Lubr Technol 57(3):13
6. Weisenberger BL (1976) Cutting oils and coolants are chief culprits in worker skin problems. Occup Health Saf Waco Tex 45(5):16
7. Grattan CEH, English JSC, Foulds IS, Rycroft RJG (1989) Cutting fluid dermatitis. Contact Dermat 20(5):372–376
8. Ellis EG (1967) LUBRICANTS: an industrial health hazard? Ind Lubr Tribol 19:141–145
9. Grijalbo L, Fernandez-Pascual M, García-Seco D, Gutierrez-Mañero FJ, Lucas JA (2013) Spent metal working fluids produced alterations on photosynthetic parameters and cell-ultrastructure of leaves and roots of maize plants. J Hazard Mater 260:220–230
10. Najiha MS, Rahman MM, Yusoff AR (2016) Environmental impacts and hazards associated with metal working fluids and recent advances in the sustainable systems: a review. Renew Sustain Energy Rev 60:1008–1031
11. King N, Keranen L, Gunter K, Sutherland J (2001) Wet versus dry turning: a comparison of machining costs, product quality, and aerosol formation. SAE Technical Paper
12. Marksberry PW, Jawahir IS (2008) A comprehensive tool-wear/tool-life performance model in the evaluation of NDM (near dry machining) for sustainable manufacturing. Int J Mach Tools Manuf 48(7–8):878–886
13. Shokrani A, Al-Samarrai I, Newman ST (2019) Hybrid cryogenic MQL for improving tool life in machining of Ti-6Al-4V titanium alloy. J Manuf Process 43:229–243
14. Klocke F, Eisenblätter G (1997) Dry cutting. CIRP Ann 46(2):519–526
15. Rubio EM, Camacho AM, Sánchez-Sola JM, Marcos M (2006) Chip arrangement in the dry cutting of aluminium alloys. J Achiev Mater Manuf Eng 16(1–2):164–170
16. Debnath S, Reddy MM, Yi QS (2014) Environmental friendly cutting fluids and cooling techniques in machining: a review. J Clean Prod 83:33–47
17. Hörner D (2002) Recent trends in environmentally friendly lubricants. J Synth Lubr 18(4):327–347

18. Katna R, Singh K, Agrawal N, Jain S (2017) Green manufacturing—performance of a biodegradable cutting fluid. Mater Manuf Process 32(13):1522–1527
19. Katna R, Suhaib M, Agrawal N (2020) Nonedible vegetable oil-based cutting fluids for machining processes – a review. Mater Manufact Process 35(1):1–32
20. Katna R, Suhaib M, Agrawal N, Singh K, Jain S, Maji S (2019) Green machining: studying the impact of viscosity of green cutting fluid on surface quality in straight turning. J Phys Conf Series 1276:012036
21. Agrawal N, Katna R (2019) Assessment of cutting forces in machining with novel neem oil-based cutting fluid. In: Applications of computing, automation and wireless systems in electrical engineering (pp 859–863). Springer, Singapore
22. Katna R, Suhaib M, Agrawal N, Jain S, Singh K, Maji S (2020) Experimental study on effect of green cutting fluid and surfactant on temperature in turning operation. In: Emerging trends in mechanical engineering (pp 437–449). Springer, Singapore
23. Dhar NR, Kamruzzaman M, Ahmed M (2006) Effect of minimum quantity lubrication (MQL) on tool wear and surface roughness in turning AISI-4340 steel. J Mater Process Technol 172(2):299–304
24. Dhar NR, Islam MW, Islam S, Mithu MAH (2006) The influence of minimum quantity of lubrication (MQL) on cutting temperature, chip and dimensional accuracy in turning AISI-1040 steel. J Mater Process Technol 171(1):93–99
25. Kamata Y, Obikawa T (2007) High speed MQL finish-turning of Inconel 718 with different coated tools. J Mater Process Technol 192–193:281–286
26. Kaynak Y (2014) Evaluation of machining performance in cryogenic machining of Inconel 718 and comparison with dry and MQL machining. Int J Adv Manuf Technol 72(5):919–933
27. Tasdelen B, Thordenberg H, Olofsson D (2008) An experimental investigation on contact length during minimum quantity lubrication (MQL) machining. J Mater Process Technol 203(1):221–231
28. Filipovic A, Stephenson DA (2006) Minimum quantity lubrication (mql) applications in automotive power-train machining. Mach Sci Technol 10(1):3–22
29. Silva LR, Corrêa ECS, Brandão JR, de Ávila RF (2013) Environmentally friendly manufacturing: Behavior analysis of minimum quantity of lubricant—MQL in grinding process. J Clean Prod
30. Chetan SG, Rao PV (2016) Environment friendly machining of Ni–Cr–Co based super alloy using different sustainable techniques. Mater Manuf Process 31(7):852–859
31. Lee P-H, Nam TS, Li C, Lee SW (2010) Environmentally-friendly nano-fluid minimum quantity lubrication (MQL) meso-scale grinding process using nano-diamond particles. In: 2010 international conference on manufacturing automation, pp 44–49
32. Murthy KS, Rajendran I (2012) Optimization of end milling parameters under minimum quantity lubrication using principal component analysis and grey relational analysis. J Braz Soc Mech Sci Eng 34(3):253–261
33. Tosun N, Huseyinoglu M (2010) Effect of MQL on surface roughness in milling of AA7075-T6. Mater Manuf Process 25(8):793–798
34. Boubekri N, Shaikh V (2015) Minimum quantity lubrication (MQL) in machining: benefits and drawbacks. J Ind Intell Inf 3(3)
35. Nguyen TK, Do I, Kwon P (2012) A tribological study of vegetable oil enhanced by nano-platelets and implication in MQL machining. Int J Precis Eng Manuf 13(7):1077–1083
36. Lawal SA, Choudhury IA, Nukman Y (2013) A critical assessment of lubrication techniques in machining processes: a case for minimum quantity lubrication using vegetable oil-based lubricant. J Clean Prod 41:210–221
37. Rotella G, Dillon OW, Umbrello D, Settineri L, Jawahir IS (2014) The effects of cooling conditions on surface integrity in machining of Ti6Al4V alloy. Int J Adv Manuf Technol 71(1):47–55
38. Khan MMA, Mithu MAH, Dhar NR (2009) Effects of minimum quantity lubrication on turning AISI 9310 alloy steel using vegetable oil-based cutting fluid. J Mater Process Technol 209(15):5573–5583

39. Sharma VS, Singh G, Sørby K (2015) A review on minimum quantity lubrication for machining processes. Mater Manuf Process 30(8):935–953
40. Tai BL, Stephenson DA, Furness RJ, Shih AJ (2014) Minimum quantity lubrication (MQL) in automotive powertrain machining. Procedia CIRP 14:523–528
41. Saberi A, Rahimi AR, Parsa H, Ashrafijou M, Rabiei F (2016) Improvement of surface grinding process performance of CK45 soft steel by minimum quantity lubrication (MQL) technique using compressed cold air jet from vortex tube. J Clean Prod 131:728–738
42. Stachurski W, Sawicki J, Wójcik R, Nadolny K (2018) Influence of application of hybrid MQL-CCA method of applying coolant during hob cutter sharpening on cutting blade surface condition. J Clean Prod 171:892–910
43. Mark Benjamin D, Sabarish VN, Hariharan MV, Samuel Raj D (2018) On the benefits of sub-zero air supplemented minimum quantity lubrication systems: an experimental and mechanistic investigation on end milling of Ti-6-Al-4-V alloy. Tribol Int 119:464–473
44. Singh G, Sharma VS (2017) Analyzing machining parameters for commercially puretitanium (Grade 2), cooled using minimum quantity lubrication assisted by a Ranque-Hilsch vortex tube. Int J Adv Manuf Technol 88(9):2921–2928
45. Salaam H, Taha Z, Tuan Ya TMYS (2012) Minimum quantity lubrication (MQL) using Ranque-Hilsch vortex tube (RHVT) for sustainable machining. Appl Mech Mater 217–219:2012–2015
46. Mahboob Ali MA, Azmi AI, Khalil ANM (2018) Specific cutting energy of Inconel 718 under dry, chilled-air and minimal quantity nanolubricants. Procedia CIRP 77:429–432
47. Jiang F, Li J, (Kevin) Rong Y, Sun J, Zhang S (2008) Study of cutting temperature in cold-air milling of Ti6Al4V alloy. In: Mitsuishi M, Ueda K, Kimura F (eds) Manufacturing systems and technologies for the new frontier. Springer, London, pp 371–376
48. Hanenkamp N, Amon S, Gross D (2018) Hybrid supply system for conventional and CO_2/MQL-based cryogenic cooling. Procedia CIRP 77:219–222
49. Grguraš D, Sterle L, Krajnik P, Pušavec F (2019) A novel cryogenic machining concept based on a lubricated liquid carbon dioxide. Int J Mach Tools Manuf 145:103456
50. Pereira O et al (2015) The Use of Hybrid CO_2 + MQL in machining operations. Procedia Eng 132:492–499
51. Kim JS, Kim JW, Lee SW (2017) Experimental characterization on micro-end milling of titanium alloy using nanofluid minimum quantity lubrication with chilly gas. Int J Adv Manuf Technol 91(5–8):2741–2749
52. Damir A, Shi B, Attia MH (2019) Flow characteristics of optimized hybrid cryogenic-minimum quantity lubrication cooling in machining of aerospace materials. CIRP Ann 68:77–80
53. Huang S, Lv T, Wang M, Xu X (2018) Enhanced machining performance and lubrication mechanism of electrostatic minimum quantity lubrication-EMQL milling process. Int J Adv Manuf Technol 94(1–4):655–666
54. Xu X, Huang S, Wang M, Yao W (2017) A study on process parameters in end milling of AISI-304 stainless steel under electrostatic minimum quantity lubrication conditions. Int J Adv Manuf Technol 90(1):979–989
55. Jamil M, Khan AM, He N, Li L, Zhao W, Sarfraz S (2019) Multi-response optimisation of machining aluminium-6061 under eco-friendly electrostatic minimum quantity lubrication environment. Int J Mach Mach Mater 21(5–6):459–479
56. Lv T, Huang S, Liu E, Ma Y, Xu X (2018) Tribological and machining characteristics of an electrostatic minimum quantity lubrication (EMQL) technology using graphene nano-lubricants as cutting fluids. J Manuf Process 34:225–237
57. Xu X et al (2019) Capillary penetration mechanism and machining characteristics of lubricant droplets in electrostatic minimum quantity lubrication (EMQL) grinding. J Manuf Process 45:571–578
58. Su Y, Lu Q, Yu T, Liu Z, Zhang C (2019) Machining and environmental effects of electrostatic atomization lubrication in milling operation. Int J Adv Manuf Technol 104(5):2773–2782

Analysis of Interrelationship Among Factors for Enhanced Agricultural Waste Utilization to Reduce Pollution

Nikhil Gandhi, Abhishek Verma, Rohan Malik, and Shikhar Zutshi

Abstract Agricultural waste is an abundant resource in India. Huge amounts of this are wasted when agriculturists burn it due to reasons ranging from lack of storage facility to financial costs incurred by said storage of the waste. The burning of agricultural waste causes major pollution problems that harm the people living in surrounding regions. The objective of this research is to identify and analyze the interrelationship among factors of using agricultural waste in a better manner to reduce the pollution levels in emerging smart cities and exploit the benefits of this resource. In this work, eight factors that affect the use of agricultural residue as a resource were identified and analyzed for the purpose of developing an ISM based model.

Keywords Agricultural waste · Smart cities · ISM · Government policies

1 Introduction

The problem of pollution due to the burning of agricultural waste has been creating a lot of shockwaves in northern India for the past few years now. Around the time when the kharif crops are harvested, tons of crop residue is produced and burned by the farmers because of the burden of storage and the financial liability that comes with it. The pollution caused adversely affects the health of the people who reside in neighboring regions. Considering the health of the people in agriculture-centric states of North India like Punjab and Haryana, this poses a major problem, whose solutions need to be urgently found.

The practice of agriculture in India began around 9000 BCE [1]. The reputation of the Indian subcontinent as a land of fertility and agricultural potential glorified the country in foreign lands. Agriculture has since been one of the most important occupational practices in India. With time though, many customs previously followed have changed and evolved to make agriculture a more profitable and environmentally

---

N. Gandhi · A. Verma · R. Malik · S. Zutshi (✉)
Delhi Technological University (Formerly Delhi College of Engineering), New Delhi, Delhi, India
e-mail: shikharzutshi@gmail.com

© The Author(s), under exclusive license to Springer Nature Singapore Pte Ltd. 2021
R. M. Singari et al. (eds.), *Advances in Manufacturing and Industrial Engineering*,
Lecture Notes in Mechanical Engineering,
https://doi.org/10.1007/978-981-15-8542-5_12

friendly trade. One of these changes is the growing use of agricultural waste. When a crop field is harvested and all the useful crop seeds and grains are procured, a lot of crop waste is left behind which, to a layman, is not as useful as the grain itself [2]. But agricultural waste is a very useful resource which has a lot of benefits. In the modern age, where the grave issues of rising pollution and environmental degradation are on the rise, a method to reuse the agricultural waste to power the nation's upcoming smart cities could prove vital to the development of India. Crop residue encompasses more than half of the world's agricultural phytomass [3]. Massive benefits can be harnessed from it, if the people using the resource have the correct means and knowledge about its benefits.

Mismanagement of crop residue has several adverse effects on the environment. The most evident being the emission of Greenhouse Gases (GHGs) [4] majorly due to the common practice of burning these residues. Some harmful effects of inefficient disposal of crop residue are that burning of residue affects the soil quality and deprives the soil of nutrients, and more broadly, its fumes contribute to pollution and ultimately respiratory problems. It would hence be appropriate to point out that the scope of research is vast in the agricultural field. More than half of the Indian population is directly or indirectly involved in agricultural production and procurement [5]. This means that possibilities of professional research in minimizing wastage of land resource and crop resource are virtually endless. The topic of agricultural waste management is important because the responsible use of the same would be a big step in upgrading the agricultural scenario in India. This work focuses on establishing such a relationship between factors that influence the constructive use of crop residue as a natural resource in order to reduce the aforementioned pollution in emerging smart cities.

2 Literature Review

Extensive research has been done until now on the uses of agricultural waste as biofuel, as manure and as fodder to the animals. Karlen and Huggins [6] have raised concerns over soil sustainability and preventing soil erosion, using crop residues. However, there has been less research until now on how to get the agricultural residue to the user at the right time, when he actually needs it. The burning of agricultural waste is very harmful to the environment because of emissions of various gases like sulfur dioxide, nitrogen dioxide, and carbon monoxide [7]. On the other hand, it is important to note that crop residue has positive uses such as packaging, wall construction, and preventing soil erosion apart from the more commonly known uses such as industrial/domestic fuel and green manure.

Crop residue can be utilized in various ways to make more meaning out of its use than GHG production by burning it. Firstly, power generation is a widely known use of crop residue. The crop residue resourced is highly dispersed and also has a low energy density [8]. This renewable resource is less polluting and will reduce global warming. Combustion-based biomass power plants can be set up to produce net conversion efficiency between 20 and 40% by consuming power from a few kilowatts up to 100 MW [9].

Secondly, crop residue can serve as an alternative source of income for farmers which generates income to farmers from rice residue and is being burnt by them. By the involvement of the rural population in the collection, transport, packing, and other actions, they generate employment. Bioethanol production is also possible from residues. The support from establishment of small energy enterprises that manufacture briquettes and pellets as a fuel of industries creates jobs and empowers rural people economically. Simultaneously, the industry would be guaranteed of decent quality and environmentally clean fuel.

Thirdly, the farmer is urged to switch to organic fertilizers made out of his own crop residue. This way, crop residue assists rural trades and enterprises to grow and succeed. The use of organic fertilizers would improve the content of soil organic carbon (SOC), increase yields and at the same time, also help employment generation by generating confined employment opportunities [10]. Crop residue management in the field helps soil become more fertile thereby resulting in savings of ₹2000 per hectare from the farmer's manure cost. To aid these uses and to promote efficient use of crop residue as a resource, the government has put in place a few policies. Android applications developed by Punjab Remote Sensing Center (PRSC) such as i-Khet Machine for assisting farmers to have access to the farming machinery/equipment for in-situ organization of crop residue, e-PEHaL for watching tree plantation, and e-Prevent to have rapid and exact information regarding instances of crop residue burning and subsidy on crop residue management machinery help put the said practice to a check, to name a few [11, 12].

Agricultural waste also finds good use as livestock feed [13]. In developing countries like India, crop residue solves the issue of feeding cattle and other domestic animals. Mostly all crop residues are fit to be fed to livestock as feed and can potentially generate profits through the increased amount of animal products.

3 Methodology Used

The ISM methodology is a structural model used to analyze the complicated socioeconomic systems [14–16]. The opinions of the selected group for the study and their practical knowledge decide whether and how the factors are interrelated and thus making it interpretive. On the basis of relationships between the enlisted factors as given in Table 1, an overall structure is portrayed in a graphical model. ISM generally has the following steps [17–19]:

- Identify factors affecting the system are noted down.
- A contextual relationship among the factors is recognized between them.
- A structural self-interaction matrix (SSIM) is established for factors, which indicates pairwise relationships among factors of the system.
- A reachability matrix is developed from the SSIM, and it is checked for transitivity.

Table 1 Agricultural waste factors

Factors	Authors
Power generation (F1)	[19–21]
Increased profits (F2)	[22, 23]
Organic fertilizers (F3)	[24]
Control soil erosion (F4)	[6, 25]
Institutional policies (F5)	[26, 27]
Livestock feed (F6)	[13, 28, 29]
Increase in SOC (F7)	[30]
Employment generation (F8)	[20, 22]

- Segregation of the reachability matrix into different levels.
- Based on the contextual relationships given in the reachability matrix, a directed graph is plotted, and the transitive links are removed.
- The outcome of the digraph is transformed into an ISM model by replacing factor nodes with detailed statements.
- The ISM model established is reconsidered to check for conceptual inconsistency.

3.1 Structural Self-interaction Matrix

Four symbols, namely V, A, X, and O are used to demonstrate the direct relation that subsists between the two sub-variables under consideration.

V: for the relation from i to j, but not in both directions
A: for the relation from j to i, but not in both directions
X: for both direction relations from i to j and j to i
O: if there is no relation from i to j and j to i

For analyzing the variables that involve in SSIM, a circumstantial relationship is taken out such that one variable leads to other. Based on this contextual relationship, SSIM has been developed and given in Table 2.

3.2 Reachability Matrix

Since there is no transitivity in this case; hence, initial reachability matrix will be used for further calculations. The driving power for each factor is the total number of factors (including itself), which may help achieve. The final reachability is developed by using an initial reachability matrix with the help of transitivity rule. The final reachability matrix is given in Table 3.

Analysis of Interrelationship Among Factors for Enhanced ...

Table 2 Agricultural waste factors

Factors	F8	F7	F6	F5	F4	F3	F2	F1
F1	V	O	O	A	O	O	V	–
F2	O	O	A	A	O	A	–	
F3	O	X	O	A	X	–		
F4	O	V	O	O	–			
F5	V	O	V	–				
F6	O	O	–					
F7	O	–						
F8	–							

Table 3 Final reachability matrix

Factors	F1	F2	F3	F4	F5	F6	F7	F8	Driving power
F1	1	1	0	0	0	0	0	1	3
F2	0	1	0	0	0	0	0	0	1
F3	0	1	1	1	0	0	1	0	4
F4	0	1	1	1	0	0	1	0	4
F5	1	1	1	1	1	1	1	1	8
F6	0	1	0	0	0	1	0	0	2
F7	0	1	1	1	0	0	1	0	4
F8	0	0	0	0	0	0	0	1	1
Dependence power	2	7	4	4	1	2	4	3	

3.3 Level Partition

The formulation of the final reachability matrix enables the reachability and antecedent set for each factor to be found. The reachability set consists of the factor itself and the other factors which it may drive, whereas the antecedent set consists of the factor itself and the other factors which may help in driving the primary factor in the first place. Thereafter, the intersection of these sets is found out for all the factors. The factors for which the reachability and the intersection sets completely intersect occupy the top level in the ISM hierarchy. A factor cannot drive any other factor above its own level in the hierarchy. This helps us find the top-level factor which once identified is separated out from the other factors. Then, the same process is repeated to find out the factors at the next level as given in Tables 4 and 5. This process is carried out until the level of each factor is found. These levels help in building the digraph and consequently the final model.

Table 4 First iteration

Factors	Reachability	Antecedent	Intersection	Level
F1	1, 2, 8	1, 5	1	
F2	2	1, 2, 3, 4, 5, 6, 7	2	One
F3	2, 3, 4, 7	3, 4, 5, 7	3, 4, 7	
F4	2, 3, 4, 7	3, 4, 5, 7	3, 4, 7	
F5	1, 2, 3, 4, 5, 6, 7, 8	5	5	
F6	2, 6	5, 6	6	
F7	2, 3, 4, 7	3, 4, 5, 7	3, 4, 7	
F8	8	1, 5, 8	8	One

Table 5 Final iteration

Factors	Reachability	Antecedent	Intersection	Level
F2	2	1, 2, 3, 4, 5, 6, 7	2	One
F8	8	1, 5, 8	8	One
F1	1	1, 5	1	Two
F3	3, 4, 7	3, 4, 5, 7	3, 4, 7	Two
F4	3, 4, 7	3, 4, 5, 7	3, 4, 7	Two
F6	6	5, 6	6	Two
F7	3, 4, 7	3, 4, 5, 7	3, 4, 7	Two
F5	5	5	5	Three

4 Results and Discussion

4.1 Formation of ISM Model

The structural model is the by-product of the initial reachability matrix. A graph is plotted indicating the relationship between the factors i and j, if a relationship exists, then it is represented by an arrow which points from i to j. This graph is known as an initially directed graph or initial digraph. After this, the transitivity is eliminated, leaving us with the final digraph. This final digraph is transformed into the ISM-based model as shown in Fig. 1. This model represents the direct relationship among different factors. It states that each LS has its own importance at its own level as shown in the model. The ISM model shows that the various policies put in place by the government to have a positive overall effect on the factors that influence the use of agricultural waste on the ground level.

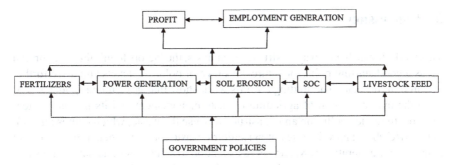

Fig. 1 ISM model

4.2 MICMAC Analysis

The objective of the MICMAC analysis is to analyze the driving and the dependence power of each variable. The variables have been classified into four quadrants or clusters as shown in Fig. 2. The first cluster consists of the autonomous factors that have weak driving power and weak dependence. The second cluster consists of the dependent factors that are weak drivers but are highly dependent. The third cluster has what are called the "linkage factors." They have strong driving power as well as strong dependence. The fourth cluster includes the independent factors having strong driving power but not so strong dependence [15]. The variables with a very strong driving power are called the key variables and falls into the category of independent or linkage factors [31].

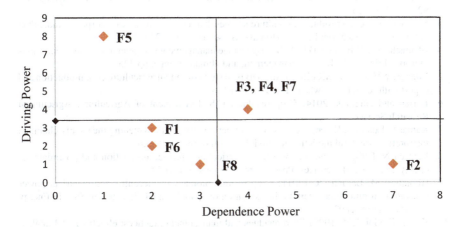

Fig. 2 MICMAC analysis

5 Conclusion

This work shows that by way of structural analysis, the factors highlighted herein can act as a good starting point for future cut-down of pollution levels and drive further research in the field. The findings of this work show that government policies (F5) drive the overall system of agricultural waste management and its use as a potent resource to reduce pollution and constructively use the waste, which comes out well in the MICMAC analysis. Uses of agricultural waste for power generation (F1) and employment generation (F8) can serve as a boon to setup futuristic smart cities and as a means to provide daily employment to the inhabitants. More farmer-centric uses of agricultural waste such as organic fertilizer production (F3), control of soil erosion (F4) in the fields in order to retain soil nutrients, increase in SOC (F7), and as livestock feed (F6) come to show that apart from cutting pollution levels and benefitting the plan of establishing smart cities, the reuse of agricultural waste augurs well at the grassroots level too. It enhances the capacity of increased profits (F2) harnessed by the agriculturalist, making it a win–win situation for all concerned. Thus, it is safe to conclude that the use of agricultural waste is a path to cutting pollution for established as well as emerging cities.

References

1. Chauhan, History of Agriculture, Department of Extension Education (2012)
2. Tripathi RP, Sharma P, Singh S (2007) Influence of tillage and crop residue on soil physical properties and yields of rice and wheat under shallow water table conditions. Soil Tillage Res 92(1–2):221–226
3. Smil V (1999) Crop residues: agriculture's largest harvest: crop residues incorporate more than half of the world's agricultural phytomass. Bioscience 49(4):299–308
4. Bhattacharyya P, Barman D (2018) Crop residue management and greenhouse gases emissions in tropical rice lands. In: Soil management and climate change, pp 323–335
5. Limbore NV, Khillare SK (2015) An analytical study of Indian agriculture crop production and export with reference to wheat. Rev Res 4(6):1–8
6. Karlen and Huggins (2014) Crop residues, U.S. Department of Agriculture: Agricultural Research Service
7. Kumar S, Luthra S, Haleem A (2013) Customer involvement in greening the supply chain: an interpretive structural modeling methodology. J Ind Eng Int 9(1):6
8. Bentsen N, Felby C, Thorsen Bo (2014) Agricultural residue production and potentials for energy and materials services. Prog Energy Combust Sci 40:59–73
9. Hiloidhari M, Baruah DC (2011) Crop residue biomass for decentralized electrical power generation in rural areas (part 1): investigation of spatial availability. Renew Sustain Energy Rev 15(4):1885–1892
10. Kumar K, Goh KM (1999) Crop residues and management practices: effects on soil quality, soil nitrogen dynamics, crop yield, and nitrogen recovery. Adv Agron 68:197–319
11. Evaluation of Crop Residue Potential for Power Generation for Indian State Punjab (2013)
12. Gadgil S, Gadgil S (2006) The Indian monsoon, GDP and agriculture. Econ Polit Wkly 4887–4895

13. Valbuena D, Erenstein O, Tui SHK, Abdoulaye T, Claessens L, Duncan AJ, Gérard B, Rufino MC, Teufel N, van Rooyen A, van Wijk MT (2012) Conservation Agriculture in mixed crop–livestock systems: scoping crop residue trade-offs in Sub-Saharan Africa and South Asia. Field Crops Res 132:175–184
14. Harary F, Norman R, Cartwright Z (1965) Structural models: an introduction to the theory of directed graphs. Wiley, New York
15. Tyagi M, Kumar P, Kumar D (2015) Analysis of interactions among the drivers of green supply chain management. Int J Bus Perform Supply Chain Model 7
16. Warfield JW (1974) Developing interconnected matrices in structural modeling. IEEE Transcr Syst Men Cybern 4(1):51–81
17. Kumar D, Jain S, Tyagi M, Kumar P, Walia RS (2016) Assessment of lean manufacturing strategies using interpretive structural modelling approach contemporary issues and challenges in management & decision sciences, pp 159–167
18. Shuaib M Khan U, Haleem A (2016) Modeling knowledge sharing factors and understanding its linkage to competitiveness. Int J Glob Bus Compet 11(1):23–36
19. Singh MD, Kant R (2008) Knowledge management barriers: an interpretive structural modeling approach. Int J Manage Sci Eng Manage 3(2):141–150
20. Gondwe KJ, Chiotha SS, Mkandawire T, Zhu X, Painuly J, Taulo JL (2017) Crop residues as a potential renewable energy source for Malawi's cement industry. J Energy S Afr 28(4):19–31
21. Nguyen TLT, Hermansen JE, Mogensen L (2013) Environmental performance of crop residues as an energy source for electricity production: the case of wheat straw in Denmark. Appl Energy 104:633–641
22. Ahmed T, Ahmad B (2014)Burning of crop residue and its potential for electricity generation. Pak Dev Rev 53:275–292
23. Dubey AK, Chandra P, Padhee D, Gangil S (2013) Energy from cotton stalks and other crop residues. CIAE, Bhopal, India
24. Baruah A, Baruah KK (2015) Organic manures and crop residues as fertilizer substitutes: impact on nitrous oxide emission, plant growth and grain yield in pre-monsoon rice cropping system. J Environ Prot 6(07):755
25. Wilhelm WW, Johnson JM, Hatfield JL, Voorhees WB, Linden DR (2004) Crop and soil productivity response to corn residue removal. Agron J 96(1):1–17
26. Cropping PIRB (2005) Crop residue management for nutrient cycling and improving soil productivity in rice-based cropping systems in the tropics. Adv Agron 85:269
27. Mittal S, Ahlgren EO, Shukla PR (2018) Barriers to biogas dissemination in India: a review. Energy Policy 112:361–370
28. Duncan AJ, Bachewe F, Mekonnen K, Valbuena D, Rachier G, Lule D, Bahta M, Erenstein O (2016) Crop residue allocation to livestock feed, soil improvement and other uses along a productivity gradient in Eastern Africa. Agr Ecosyst Environ 228:101–110
29. Owen J (1989) Department of Agricultural Economics Texas A&M University, Department of Agricultural Economics Texas A&M University
30. Wilhelm WW, Johnson JM, Karlen DL, Lightle DT (2007) Corn stover to sustain soil organic carbon further constrains biomass supply. Agron J 99(6):1665–1667
31. Ravi V, Shankar R (2004) Analysis of interactions among the barriers of reverse logistics. Technol Forecast Soc Change

Enhancement of Mechanical Properties for Dissimilar Welded Joint of AISI 304L and AISI 202 Austenitic Stainless Steel

Yashwant Koli, N. Yuvaraj, Vipin, and S. Aravindan

Abstract Welding dissimilar materials of different thicknesses always presents new problems in the process of gas metal arc welding (GMAW), when the thickness of the plate is less than 1 mm. Because of its low heat input, cold metal transfer (CMT) process is used to overcome these difficulties, which enhance the mechanical properties of the dissimilar welded joint. Thin sheets require low heat input for welding which can be possibly achieved by reducing the current and voltage or by increasing the welding speed. The aim of this study is to experimentally examine the mechanical properties of the dissimilar welded joints AISI 304L and AISI 202 using GMAW and CMT with AISI 316L filler wire. CMT prevents thin sheets from burning compared to GMAW. Therefore, the use of CMT greatly increases the tensile strength and the effects of microhardness. Residual stress of CMT samples is less as compared to GMAW and more compressive in nature at the weld bead, which helps to avoid crack nucleation.

Keywords CMT · GMAW · Dissimilar materials · Austenitic stainless steel · Mechanical properties

1 Introduction

Stainless steel is one of the most popular materials for structural applications, due to their excellent physical properties. It has many industrial applications like structural engineering such as civil construction, nuclear reactors, thermal power plants, vessels and heat exchangers due to the additional benefits and the design codes of stainless steels [1–6]. Stainless steels are of three types that is martensitic, ferritic

---

Y. Koli (✉) · N. Yuvaraj · Vipin
Delhi Technological University, New Delhi 110042, India
e-mail: yashwantkoli5@gmail.com

S. Aravindan
Department of Mechanical Engineering, Indian Institute of Technology Delhi, New Delhi, India

© The Author(s), under exclusive license to Springer Nature Singapore Pte Ltd. 2021
R. M. Singari et al. (eds.), *Advances in Manufacturing and Industrial Engineering*,
Lecture Notes in Mechanical Engineering,
https://doi.org/10.1007/978-981-15-8542-5_13

and austenitic. Out of these three, austenitic stainless steel is preferred due to easiness in welding process and its wide range of industrial applications [7]. High Cr–Ni content in AISI 304L gives better strength and corrosion resistant. For wire spool, AISI 316L is used because it has a higher level of corrosion resistance as compared to AISI 304L for its industrial applications. It has good amount of molybdenum (2.31%) that increase its resistance to pitting caused by chlorides, and AISI 316L is highly compatible with AISI 304L for welding purpose. Stainless steels can be welded by any of the welding process like manual shielded metal arc welding (SMAW), gas tungsten arc welding (GTAW), gas metal arc welding (GMAW), submerged arc welding (SAW), flux cored arc welding (FCAW) and many other resistance welding. Among the other welding techniques, GMAW is used because it can weld thin metal sheets with the help of its short-circuiting mode. This short-circuiting mode helps to reduce current and voltage at the time of short-circuiting which helps the thin sheets to be welded with minimum distortion and burn through. However, with the advancements in the arc welding of short-circuiting mode, cold metal transfer (CMT) is a new method of joining thin sheets based on conventional short-circuiting (CSC) transfer process established via "Fronius of Austria". It is an upgraded technology of gas metal arc welding (GMAW) process and is highly superior to GMAW in terms of spatter, distortion, burn through and welding cost due to its unique feature known as low heat input. Mechanism of CMT offers a liquid droplet detachment by an oscillatory motion from a filler wire. Retraction of filler wire at the time of short circuit not only causes the detachment of liquid droplet but also tends the current waveform to zero which makes low heat input possible for CMT. Low heat input causes major advantages in welding of thin sheets. CMT offers spatter-less welding, negligible distortion, avoids burn through, better mechanical properties, enhanced microstructures, low intermetallic compounds (IMCs) and lesser intermetallic layer (IML) thickness [8–11]. Sun and Karppi [2] have given the overview on joining of dissimilar metals by electron beam welding (EBW). Sudhakaran et al. [12] used particle swarm optimization (PSO) technique to curtail the angular distortion in AISI 202 austenitic stainless steel welded plates by GTAW process. He concluded that the experimental values of distortion and the values came after optimization by PSO technique are almost the same. Sathiya et al. [6] worked on finding the mechanical and metallurgical properties of AISI 304 austenitic stainless steel with the use of friction welding. An investigation is done on the tensile specimen, which concludes that as the friction time increases the joint strength decreases and fractographic observations verified that the rupture occurred frequently at the joint zone and moderately through the base metal. Sammaiah et al. [13] experimented on the mechanical properties of friction welded AA6063 and austenitic stainless steel. Results showed that as the friction pressure increases the tensile strength and toughness decrease. At the interface, the microhardness is higher for the condition of a high forge and high burn-off length. Joseph et al. [3] evaluated residual stress in dissimilar weld joint using X-ray diffraction technique. Buttering layer of Inconel-82 is applied on the dissimilar weld joint, which not only reduces the residual stress in heat-affected zone (HAZ) of ferritic steel but also avoids failure of welded joint that could happen due to residual stress. Mishra et al. [14] experimented about the tensile strength of different grades of stainless steel welded with mild steel. Results showed that AISI 202 welded with mild steel gives the best tensile strength both with TIG and with MIG. Prasad et al. [15] worked on the weld quality characteristics of welding of AISI 304L austenitic stainless steel thin plates using pulsed current micro-plasma arc welding (MPAW). Due to the presence of pulsed current, fine grains are obtained in the weld and fusion zone as compared to coarse grains in HAZ, which show that the weld zone is stronger than the parent metal.

This paper focuses on the mechanical properties of austenitic stainless steel when welded using GMAW and CMT welding methods. Both welding techniques at various process parameters are used to quantify welded joints and weld quality. In contrast to the GMAW, CMT displays superior mechanical properties.

2 Experimental Procedure

2.1 Materials and Methods

Experiments are conducted on dissimilar sheets with different thicknesses of SS304L and SS202 austenitic stainless steel. Butt joints were fabricated using GMAW and CMT processes. Fronius VR4000 TransPuls Synergic (TPS) 3200 and TPS 400i for GMAW and CMT processes, respectively, are shown in Fig. 1. SS316L is used as a

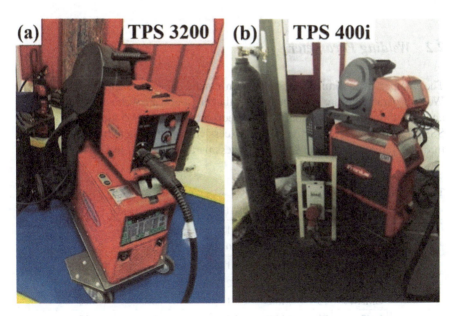

Fig. 1 Welding machine: **a** GMAW and **b** CMT

Table 1 Chemical composition in wt. (%)

Materials	C	Si	Mn	P	S	Cr	Mo	Ni
SS304L	0.0585	0.219	0.837	0.0426	0.0166	18.3	0.157	8.28
SS202	0.118	0.119	7.243	0.0967	0.0149	16.2	0.309	0.575
SS316L	0.0341	0.336	0.880	0.0451	0.0231	17.7	2.31	12.43

Table 2 Mechanical properties of substrate material

Properties	$\sigma_{0.2}$ (MPa)	σ_{UTS} (MPa)	δ (%)	$HV_{0.3N}$	Y (GPa)
SS202	225	452	75.7	325 ± 25	234.79
SS304L	200	495	95.6	238 ± 20	230.91

filler wire having 1.2-mm diameter having higher content of Cr–Ni, which prevents corrosion.

The chemical composition in wt. (%) of SS304L and SS202 substrate material and SS316L filler material used in this experiment as per ASTM standard is shown in Table 1. It shows high Cr–Ni content, which prevents corrosion. Test sample of stainless steel is cut in the dimension of size 100 mm × 50 mm × 1.5 mm and 100 mm × 50 mm × 0.8 mm for AISI 304L and AISI 202, respectively. Table 2 shows the mechanical properties of substrate material comprising tensile strength, microhardness and Young's modulus.

2.2 Welding Parameters

The number of preliminary tests is been conducted to find the best parameters. Welding parameters are set in such a way that the sheets are welded properly without any burn through or distortion. As the sheets are very thin (1.5 mm and 0.8 mm for SS304L and SS202, respectively), any excessive current or heat input leads to burn through. Table 3 shows the welding process parameters used in GMAW and CMT

Table 3 Welding process parameters

S. No	Process	I (A)	V (V)	WFR (m/min)	TS (mm/s)	CTWD (mm)
1	GMAW	65	15.1	2.1	6	10
2	CMT	65	17.6	2.0	6	10
3	GMAW	70	15.1	2.1	6	10
4	CMT	70	17.7	2.2	6	10
5	GMAW	75	15.2	2.1	6	10
6	CMT	75	17.9	2.4	6	10

I current, *V* voltage, flow rate of shielding gas = 15 L/min, *WFR* wire feed rate, *TS* travel speed

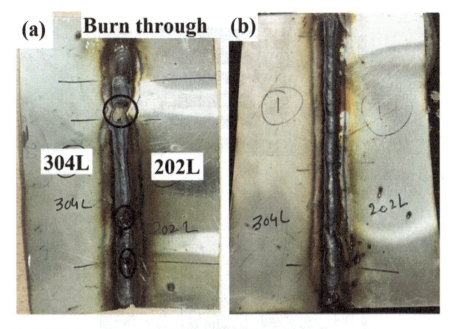

Fig. 2 Welded samples through **a** GMAW and **b** CMT

processes having current of 3 different levels while keeping a constant flow rate of shielding gas, welding speed and contact tip to workpiece distance (CTWD) for all the six samples, respectively.

Figure 2 shows burn through at three locations in sample 5 of GMAW process. Exceeding the value of current to 70A in GMAW shows burn through, while sample 6 welded with CMT shows an aesthetic bead profile while having same parameters. This is the advantage of low heat input in CMT process.

3 Result and Discussion

3.1 Tensile Test (UTM)

The tensile specimen is cut through the wire EDM as per the ASTM standard of subsize 100 mm. Large dog bone-shaped tensile specimens with a gauge length of 32 mm and a gauge width of 6 mm are shown in Fig. 3. Tinius Olsen UTM H50KS of 50 KN capacity is used to measure the tensile strength of the joint.

Ultimate tensile strength results clearly show that sample 1 which is welded at lowest current is having the highest joint strength in both the processes as shown in Fig. 4. Welded samples have comparatively higher tensile strength as compared with the substrate metal. CMT shows higher tensile strength as compared with GMAW

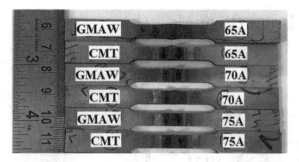

Fig. 3 Tensile specimens

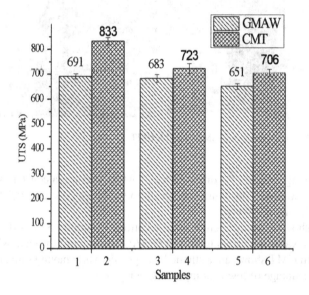

Fig. 4 Ultimate tensile strength of the joint

with 20%, 5% and 8% of strength enhancement for samples 2, 4 and 6, respectively. As the sheets were thin, the higher current results in the lower joint strength because of burn through. Joint efficiency is more than 100% for all the welded samples as fracture is taking place at the HAZ of the thinner section. Due to its low heat input characteristics in CMT, thin sheets are properly welded with higher strength and without any burn through. Figure 5 shows the fractography of the tensile sample 2. All tensile specimen experiences the same fracture point, which is HAZ of thinner side (SS202). Figure 5a shows the FESEM image of fracture surface having a crack in the middle. The EDX of crack is displayed in Fig. 5b where high amount of intermetallic compound (IMC) forms which tends to deteriorate the strength of the joint.

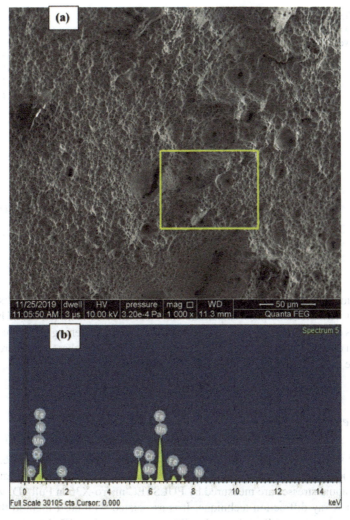

Fig. 5 Fractography of sample 2: **a** FESEM of fracture surface at HAZ of sample 2 (CMT) and **b** EDX of sample 2

3.2 Microhardness (HV)

A piece of 10 mm × 10 mm is cut through a welded sample, and its cross-sectional view is examined for microhardness. It is dry polished with emery paper to a grade of 2500. Wet polish is then proceeded with a velvet cloth using alumina powder. FISCH-ERSCOPE HM2000S is used to obtain the microhardness of the polished sample. Figure 6 shows the variation of microhardness with respect to the positions. Almost all the samples are having the same trend as the others, but it can be clearly seen that

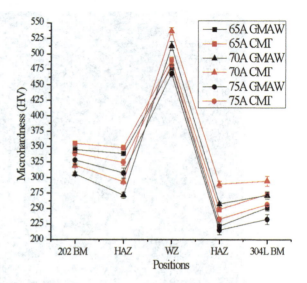

Fig. 6 Microhardness variation with the distance from the weld

CMT samples are having higher microhardness as compared with GMAW samples provided with the same process parameters. Heat-affected zone (HAZ) of both the materials experiences minimum microhardness due to the softening behaviour, which results in coarser grain structure. Weldment is showing the highest microhardness in the range of 475–550 HV due to its finer grain structure.

3.3 Residual Stress (MPa)

Residual stresses are the stresses that remain within a material or body after manufacture and material processing in the absence of external forces or thermal gradients [16]. Residual stresses are measured by PULSTEC micro-X360n Full 2D, which is based on X-ray diffraction technique. It is associated with cosα method that uses a single exposure to collect the entire diffraction cone via a 2D detector. As there are many non-destructive techniques for measuring the residual stresses, but X-ray diffraction is suitable for thin plates as its penetration is about 10 μm with spatial resolution in the range of 10 μm to 1 mm. Figure 7 shows the exposure of X-ray on sample 1 at the weldment. Standard chromium (Cr) material X-ray tube is used having collimator size of 1-mm diameter with 30 kV and 1 mA specification.

From Fig. 8, it is clearly shown that the CMT samples have the minimum residual stress, which is suitable for this thin sheet welding owing to its low heat input. Sugahara et al. [17] conclude that with lower heat input, lower residual stresses are observed and vice versa. At the weld zone, the residual stress tends to become negative (compressive) due to shrinkage of the grain size (fine grains) which can be analysed from Fig. 6, which has the highest microhardness at the weld zone. More compressive residual stress is being observed at the weldment for CMT sample to avoid any crack

Enhancement of Mechanical Properties for Dissimilar Welded … 153

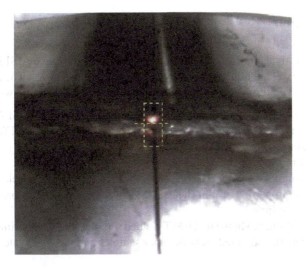

Fig. 7 X-ray exposure on sample 1 at weldment

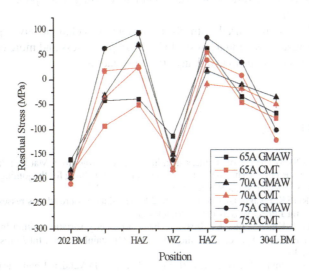

Fig. 8 Residual stress variation at different zones of welded sheets

nucleation. HAZ shows tensile residual stress in both the materials, which indicates a high tendency for a crack nucleation and thus major tensile specimens for tensile test fails at the HAZ. Due to this, mechanical properties degrade significantly.

4 Conclusions

This paper investigates the mechanical properties of AISI 304L and AISI 202 stainless steel dissimilar welded joint using the GMAW and CMT processes. The experiments were conducted on thin sheets, and the following conclusions are drawn:

1. Best suited process parameters for welding of AISI 304L and AISI 202 stainless steel of thin plates are current—65 A, welding speed—6 mm/s, CTWD—10 mm and shielding gas—pure argon. This combination of parameter gives the best results for mechanical properties that join the sheets efficiently.
2. GMAW samples experience small burn through that reduces weld quality and mechanical properties of thin sheets. CMT overcomes with this problem owing to its low heat input.
3. Tensile results have shown that CMT welded sample 2 has the maximum ultimate tensile strength compared with the other samples and the base plates of both the metals.
4. At the weldment, the Vickers microhardness value is maximum for all the samples due to the brittle nature of the weldment and welding wire spool of AISI 316L stainless steel.
5. Residual stress measured with X-ray diffraction technique gives precise and repetitive results. Residual stress of CMT samples is less and more compressive in nature which helps to avoid any crack nucleation.

References

1. Gardner L (2005) The use of stainless steel in structures. Prog Struct Mat Eng 7(2):45–55
2. Sun Z, Karppi R (1996) The application of electron beam welding for the joining of dissimilar metals: an overview. J Mater Process Technol 59(3):257–267
3. Joseph A, Rai SK, Jayakumar T, Murugan N (2005) Evaluation of residual stresses in dissimilar weld joints. Int J Press Vessels Pip 82(9):700–705
4. Jang C, Lee J, Kim JS, Jin TE (2008) Mechanical property variation within Inconel 82/182 dissimilar metal weld between low alloy steel and 316 stainless steel. Int J Press Vessels Pip 85(9):635–646
5. Muránsky O, Smith MC, Bendeich PJ, Edwards L (2011) Validated numerical analysis of residual stresses in Safety Relief Valve (SRV) nozzle mock-ups. Comput Mater Sci 50(7):2203–2215
6. Sathiya P, Aravindan S, Haq AN (2005) Mechanical and metallurgical properties of friction welded AISI 304 austenitic stainless steel. Int J Adv Manuf Technol 26(5–6):505–511
7. Sahin M (2007) Evaluation of the joint-interface properties of austenitic-stainless steels (AISI 304) joined by friction welding. Mater Des 28(7):2244–2250
8. Kumar NP, Vendan SA, Shanmugam NS (2016) Investigations on the parametric effects of cold metal transfer process on the microstructural aspects in AA6061. J. Alloy Compd. 658:255–264
9. Selvi S, Vishvaksenan A, Rajasekar E (2018) Cold metal transfer (CMT) technology-an overview. Def Technol 14:28–44
10. Zhang HT, Feng JC, He P, Zhang BB, Chen JM, Wang L (2009) The arc characteristics and metal transfer behaviour of cold metal transfer and its use in joining aluminum to zinc-coated steel. J. Mater Sci Eng A 499:111–113

11. Evangeline A, Sathiya P (2019) Cold metal arc transfer (CMT) metal deposition of Inconel 625 superalloy on 316L austenitic stainless steel: microstructural evaluation, corrosion and wear resistance properties. Mater Res Express 6:066516
12. Sudhakaran R, Murugan VV, Sivasakthivel PS (2012) Optimization of process parameters to minimize angular distortion in gas tungsten arc welded stainless steel 202 grade plates using particle swarm optimization. J Eng Sci Technol 7(2):195–208
13. Sammaiah P, Suresh A, Tagore GR (2010) Mechanical properties of friction welded 6063 aluminum alloy and austenitic stainless steel. J Mater Sci 45(20):5512–5521
14. Mishra RR, Tiwari VK, Rajesha S (2014) A study of tensile strength of MIG and TIG welded dissimilar joints of mild steel and stainless steel. Int J Adv Mater Sci Eng 3(2):23–32
15. Prasad KS, Rao CS, Rao DN (2011) A study on weld quality characteristics of pulsed current micro plasma arc welding of SS304L sheets. Int Trans J Eng Manage Appl Sci Technol 2(4):437–446
16. Wan Y, Jiang W, Li J, Sun G, Kim DK, Woo W (2017) Weld residual stresses in a thick plate considering back chipping: neutron diffraction, contour method and finite element simulation study. Mater Sci Eng 699:62–70
17. Sugahara HJ, Barros PS, Melo LG, Gonçalves IL, Rolim TL, Yadava YP, Ferreira RA (2018) Measurement of residual stresses in welded joints by DCP method. Mater Res 21(4)

Effect of 3D Printing on SCM

Shallu Bhasin, Ranganath M. Singari, and Harish Kumar

Abstract Additive manufacturing is a manufacturing science in which 3D solid object is manufactured by depositing layers by layers on the object. It is emerging as one of the innovative technologies highly effect the traditionally supply chain management. It has not only enhanced the manufacturing process but also improve the conventional supply chain management. The contribution of this article is to find out the impact of 3D printing on SCM and analyze the difference between TSCM and AMSCM.

Keywords 3D Printing · Supply chain management · Qualitative analyses

1 Introduction

Additive manufacturing is a manufacturing science in which 3D solid object is manufactured by extracting data from digital file by depositing material layer by layer in accordance with the object's 3D digital model. The various methods that are used in 3D printing are fused deposition modeling, powder bed fusion, inkjet printing and contour crafting and stereo lithography. 3D printing finds wide application in manufacturing sector-mass customization, rapid manufacturing and food, industrial application apparel, art and jewelry automation, medical bioprinting, medical devices and pills. Supply chain management (SCM) included all the processes that convert raw material into finish goods which help in managing the flow of services and goods from supplier to end customer.

S. Bhasin (✉)
Department of Mechanical Engineering, Delhi Technological University, New Delhi 110042, India
e-mail: shallubhasin2018@gmail.com

R. M. Singari
Department of Design, Delhi Technological University, New Delhi 110042, India

H. Kumar
NIT Delhi, Delhi, India

© The Author(s), under exclusive license to Springer Nature Singapore Pte Ltd. 2021
R. M. Singari et al. (eds.), *Advances in Manufacturing and Industrial Engineering*,
Lecture Notes in Mechanical Engineering,
https://doi.org/10.1007/978-981-15-8542-5_14

The main contribution of this article is to provide comprehensive review of 3D printing in context to SCM. We will find out that with the help of additive manufacturing supply chain management, there will be reduction in expenditure and time and improvement in quality as well.

Paper is divided into four sections. First section contains literature survey on 3D printing and supply chain management. Second section contains methodology we used to find the research gap. In section third, we drawn the discussion, and in forth section, conclusion has been drawn (Figs. 1, 2 and 3).

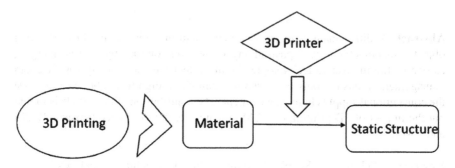

Fig. 1 3D printing

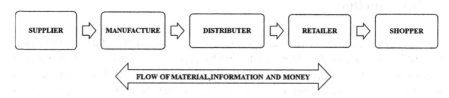

Fig. 2 Traditional supply chain management

Fig. 3 Additive manufacturing supply chain management

2 Literature Survey

To understand and analyze the 3D printing effects on supply chain, we need to explore what exactly is the 3D printing which includes different methods used in 3D printing, material used in 3D printing, application of 3D printing in manufacturing, advantages and limitations of 3D printing. Also, we need to understand what is the supply chain management—which includes process of SCM, different components of SCM and advantages of SCM. To accomplish all this, existing research work related to 3D printing and supply chain management has been reviewed.

2.1 Literature survey on 3D Printing

As indicated by Yi et al. [26], additive manufacturing is the umbrella term for assembling forms that add materials layer by layer to make parts. AM advances show various possibilities in terms of quick prototyping, tooling and direct assembling of practical parts and infer progressive advantages for assembling industries. Ngoa et al. [1] audited about 3D printing techniques, material and their development in drifting applications. The present condition of material advancement, included metal combinations, polymer composites, earthenware production and cement, was discussed alongside the advantages and downsides. Kabir et al.[2] alongside presenting brief history of 3D printing, audited the component of inserting diverse consistent filaments in to various plastics and their microstructural and mechanical property of implanting vary ent persistent strands into various plastics and their microstructural and mechanical properties including foreseeing model. Javaid et al. [3] talk about different advantages of AM in dentistry and steps utilized in make 3D printing dental model. The innovation is utilized to produce expand dental crows, spans, orthodontics props and furthermore different models, gadgets and instruments with lesser time and cost (Tables 1 and 2).

From the research review, we have drawn the following conclusion that 3D Printing is the future scope of all manufacturing process as it results in zero-wastage process, variety is free, tool less, no assembly line.Moreover, we have found the following trends in SCM-big data, collaboration, real-time information, digitalization of supply chain, agility and demand management. Also, impact of 3D printing on SCM results in mass customization, reduces complexity, changing review on resources and also changes the value adds.

3 Methodology

The additive manufacturing is a rising technique utilized for addressing the necessities of developing requests. It is a procedure of assembling parts by keeping materials

Table 1 Literature survey on supply chain management

Author	Year	Contribution
Simone Sehnema	2019	Study identifies with the development level of round economy appropriation, how well organizations oversee basic achievement factors, and the impact of chosen attributes of firms more elite classes in advancing roundabout economy
Stefan Gold	2017	Conduct the content analysis-based literature review in SCM examinations the contrast between green production network and conventional inventory network and explain the substance of green store network the board
Jiang Ying	2012	
Qinghua Zhu	2010	Look at if changed kinds of assembling ventures on ESCC exist. We likewise decide whether the Chinese producer types fluctuating in ESCC vary in their usage of the CE rehearses toward accomplishing the CE-focused on objectives on improving both natural and monetary execution

Table 2 Literature survey of 3D printing on supply chain

Author	Year	Contribution
M. Varsha Shree	2019	The quick effect of AM in SCM fuses the procedure of generation, clear creation time, shortening of materials wastage, raised adaptability and minimization of creation cost
Simone Zanoni	2018	Effect of AM advancements is not fractional or separated disturbances, yet rather envelop a scope of changes whose general results spread across various periods of SCM
Lukas Kubac	2017	Effect of AM on supply chains takes numerous structures, including disentangled generation forms, decreased costs, quicker response to request and capacity to decentralize creation
Eren Ozceylan	2017	Proposed a reproducing model for a medicinal services organization to contrast its 3DPSCM structure and TSCM. The outcomes show the solid advantages, for example, lead time and number of clients that can be accomplished by 3D PSCM contrasted with TSCM
Mohsen Attaran	2017	Development of new store network models, look at a portion of the potential advantages of AM in testing conventional as sembling imperatives, investigates its effect on the customary and worldwide production network and coordinations

which is in opposition to that of traditional. It additionally helps in accommodating profoundly complex plan, diminished material waste, reestablishing harmed portions of greater levels of popularity and so on. It has been utilized in various zones like assembling, nourishment innovation, aviation, development and so forth. The accompanying procedure has been utilized received to refresh the current utilization of 3D printing and to explore the effect of 3D imprinting on SCM.

The methodology that we used to find out the impact of 3D printing on SCM. Firstly, we find what exactly is the 3D printing materials, methods, application used in 3D printing. Secondly what is supply chain management which includes process,

Effect of 3D Printing on SCM

Table 3 Impact of AMSCM on manufacturing

Industry	Application	Benefits gain
Aerospace		
1. Design and rapid prototyping	1. Light weighting of aircraft	1. Reduce cost
2. Components manufacture	2. Engine components	2. Allow product life cycle leverages
3. Mass customization	3. Flight-certified hardware	3. Reduce lead time
Automotive		
1. Simplify production process	1. Light weight	1. Help eliminate parts
2. Component manufacture	2. Cooling system for racing cars	2. Reduce lead time
3. Design and rapid prototyping	3. BMW is using print hand tools	3. Reduce cost
4. Manufacture at requirement	4. Can be used in a backup capacity	4. Reduce labour cost
Machine tool production		
1. Design and rapid prototyping	1. Lightweight gripping system	1. Reduce inventory
2. Leaner manufacturing	2. End of arm for smarter packaging	2. Reduction in overall cost
Medical		
1. Design and rapid prototyping	1. Manufacturing human organ	1. Reduce surgery cost
2. Manufacture at requirement	2. Reconstructing bones, body parts	2. Reduce repair cost
3. Mass customization	3. Hip joints, robotic hand	3. Improve process
Architecture and construction		
1. Design and rapid prototyping	1. Generating an exact scale model of the building	1. Reduce cost
2. Manufacture at requirement	2. Printing housing components	2. Reduce Customization and time

components, application, benefits, etc., and thirdly, we study the impact of 3D printing on SCM. Further, we analyzed the difference between TSCM AND 3DSCM and drawn the conclusion form it (Table 3).

4 Discussion and research gap analyses

By studying all the parameters, we can draw the following points:

Table 4 Traditional TSC versus AMSC

Traditional supply chain	AM supply chain
Inventory required is high	Inventory required is low
Manufactured far from area of use	Manufactured nearby area of use
Supply chain is intermediation	Supply chain is disintermediation
Distribution network is complex	Distribution network is simple
Dependency on economics of scale	Dependency on economics of scope
Response to customer—lengthy	Response to customer—Quick
Challenging management of demand uncertainty	Easier management of demand uncertainty
It is push-based supply chain	It is pull-based supply chain
Lead time is more	Low lead time is there
Transportation cost is very high	Transportation cost is low
Supply chain distribution-broken machines, regional turmoil or shipping delays	Hedge against disruption

- The 3D printing finds the vast application in these industries—aerospace, automotive, machine–tool production, medical and architecture construction.
- By comparing the additive manufacturing supply chain management with the traditional supply chain management, we have drawn the following points:

So, we can say from all the research done, we can draw that with the implementation of AMSCM, there will be improved performance in the following four key areas—cost, time, quality and environment (Table 4).

5 Conclusion

3D printing, also called as additive manufacturing, is a manufacturing process in which 3D solid object is manufactured by extracting data from digital file by depositing material layer by layer in accordance with the object's 3D digital model. Supply chain management (SCM) included all the processes that convert raw material into finish goods which help in managing the flow of goods and services from supplier to end customer. By integrating 3D printing with SCM, there will be reduction in cost and lead time as well as there is improvement in quality.

References

1. Nguyen NA, Bowland CC, Naskar AK (2018) Mechanical, thermal, morphological, and rheological characteristics of high performance 3D-printing lignin-based composites for additive

manufacturing applications. Data Brief 19:936–950
2. Kabir SF, Mathur K, Seyam AFM (2019) A critical review on 3D printed continuous fiber-reinforced composites: history, mechanism, materials and proper-ties. Comp Struct 111476
3. Javaid M, Haleem A (2019) Current status and applications of additive manufacturing in dentistry: a literature-based review. J Oral Bio Craniofac Res
4. Yi L, Gläßner C, Aurich JC (2019) How to integrate additive manufacturing technologies into manufacturing systems successfully: A perspective from the commercial vehicle industry. J Manuf Syst 53:195–211
5. Ngo TD, Kashani A, Imbalzano G, Nguyen KT, Hui D (2018) Additive manufacturing (3D printing): a review of materials, methods, applications and challenges. Compos Part B: Eng 143:172–196
6. Yang W, Jian R (2019) Research on intelligent manufacturing of 3D printing/cop-ying of polymer. Adv Ind Eng Polymer Res 2(2):88–90
7. Gisario A, Kazarian M, Martina F, Mehrpouya M (2019) Metal additive manufacturing in the commercial aviation indus-try: a review. J Manuf Syst 53:124–149
8. Schneck M, Gollnau M, Lutter-Günther M, Haller B, Schlick G, Lakomiec M, Reinhart G (2019) Evaluating the use of additive manufacturing in industry applications. Procedia CIRP 81:19–23
9. Haleem A, Javaid M (2019) 3D printed medical parts with different materials using additive manufacturing. Clin Epidemiology and Global Health
10. Rajaguru K, Karthikeyan T, Vijayan V (2109) Additive manufacturing–state of art. Materials today: proceedings
11. Camacho DD, Clayton P, O'Brien WJ, Seepersad C, Juenger M, Ferron R, Salamone S (2018) Applications of additive manufacturing in the construction industry–a forward-looking review. Auto Constr 89:110–119
12. Niaki MK, Torabi SA, Nonino F (2019) Why manufacturers adopt additive manufacturing technologies: the role of sustainability. J Cleaner Prod 222:381–392
13. Fredriksson C (2019) Sustainability of metal powder additive manufacturing. Procedia Manuf 33:139–144
14. Attaran M (2017) The rise of 3-D printing: the advantages of additive manufacturing over traditional manufacturing. Bus Horizons 60(5):677–688
15. Dilberoglu UM, Gharehpapagh B, Yaman U, Dolen M (2017) The role of additive manufacturing in the era of industry 4.0. Proc Manuf 11:545–554
16. Thomas-Seale LEJ, Kirkman-Brown JC, Attallah MM, Espino DM, Shepherd DET (2018) The barriers to the progression of additive manufacture: perspectives from UK industry. Int J Prod Econom 198:104–118
17. Pereira T, Kennedy JV, Potgieter J (2019) A comparison of traditional manufacturing versus additive manufacturing, the best method for the job. Proc Manuf 30:11–18
18. Beniak J, Holdy M, Križan P, Matúš M (2019) Research on parameters optimization for the additive manufacturing process. Trans Res Proc 40:144–149
19. Pfähler K, Morar D, Kemper HG (2019) Exploring application fields of additive manufacturing along the product life cycle. Proc CIRP 81:151–156
20. Mahadik A, Masel D (2018) Implementation of additive manufacturing cost estimation tool (AMCET) using break-down approach. Proc Manuf 17:70–77
21. Kubáč L, Kodym O (2017) The impact of 3d printing technology on supply chain. MATEC web of conferences, vol 134. EDP Science
22. Sehnem S, Jabbour CJC, Pereira SCF, de Sousa Jabbour ABL (2019) Improving sustainable supply chains performance through operational excellence: circular economy approach. Res Conserv Recycl 149:236–248
23. Ying J, Li-jun Z (2012) Study on green supply chain management based on circular economy. Phys Proc 25:1682–1688
24. Zhu Q, Geng Y, Lai KH (2010) Circular economy practices among Chinese manufacturers varying in environmental-oriented supply chain cooperation and the performance implications. J Environ Manage 91(6):1324–1331

25. Wilding R, Wagner B, Seuring S, Gold S (2012) Conducting content-analysis based literature reviews in supply chain management. Supply Chain Manage Int J 17(5):544–555
26. Yi L, Gläßner C, Aurich JC (2019) How to integrate additive manufacturing technologies into manufacturing systems successfully: a perspective from the commercial vehicle industry. J Manuf Syst 53:195–211

Seasonal Behavior of Trophic Status Index of a Water Body, Bhalswa Lake, Delhi (India)

Sumit Dagar and S. K. Singh

Abstract Large and growing population and rapid pace of development have led to the degradation of natural water system. Lakes are inland bodies of water that lack any direct exchange with an ocean. Lakes may contain fresh or saltwater (in arid regions), shallow or deep, permanent or temporary lakes of all types which share many ecological and biogeochemical processes. Lake ecosystems are influenced by their watersheds, i.e., the geological, chemical, and biological processes that occur on the land and streams. Lakes play multiple roles in an urban setting. It is essential to restore and maintain the physical, chemical, and biological integrity of water bodies to achieve the required water quality, which ensure protection and propagation of fish, wildlife, plants, and also recreation in and on water. The overall goal of this study is to monitor the water quality and assess the trophic status of Bhalswa Lake in Delhi. The trophic status was assessed by using multivariate indices including Carlson Trophic Status Index (CTSI) Sakamoto, Academy and Dobson index, and USEPA-NES which primarily used total phosphorus, chlorophyll-a, and Secchi depth parameters. This study showed that the Bhalswa Lake is in moderate eutrophic condition during the monsoon and post-monsoon period.

Keywords Water quality · Lake ecosystem · Trophic status index · Phosphorus · Chlorophyll-a and secchi depth · Eutrophication

1 Introduction

The meaning of the word trophic is "related to nutrition". Trophic continuum is divided into four classes traditionally such as: (1) Oligotrophic (limited or deficient); (2) mesotrophic (medium-range); (3) eutrophic (good or great enough); and (4) hypereutophic (extensive). Multidimensional trophic concept was used before the TSI based upon the single criterion, including supply rate of organic matter to

---

S. Dagar (✉) · S. K. Singh
Environmental Engineering Department, Delhi Technological University, New Delhi 110042, India
e-mail: dagarsumit95@gmail.com

© The Author(s), under exclusive license to Springer Nature Singapore Pte Ltd. 2021
R. M. Singari et al. (eds.), *Advances in Manufacturing and Industrial Engineering*, Lecture Notes in Mechanical Engineering,
https://doi.org/10.1007/978-981-15-8542-5_15

the lake. But that criterion was doubtful because of a single parameter determination. The optimal **Trophic Status Index** (TSI) maintains the interpretation of trophic status components found in multiparameter indicators. However, TSI still has a single variable coefficient at its ease, and it can be accomplished when interrelated with the frequently used trophic criteria. The researchers reported an empirical equation to predict the concentration of phosphorus in lakes based on awareness of phosphorus loading. The link between the amount of phosphorus and microalgae production is chlorophyll-a concentration. In 1975, if a series of predictive equations could relate several of the trophic requirements frequently used, measuring all possible trophic parameters would no longer be necessary to determine trophic position. A single trophic metrics like biomass of algae, loading of nutrients, or concentration of nutrients could be the foundation for an index from which to estimate or predict other trophic requirements through the defined interactions. Alternatively, for determining the trophic condition, one of trophic requirements can also be used.

TSI is an index that would promote communication between the public and the limnologist, which is particularly vulnerable to such concerns. Algal biomass values is a difficult term, and some variables including dry and wet weight, molecule amount, chlorophyll-a, carbon amount, and Secchi disk can generally be estimated.

2 Objectives of the Study

Following are the objectives of study:-

- To characterize Bhalswa Lake on the basis of Trophic Status Index (TSI).
- To identify the probable sources of pollutants in Bhalswa Lake.
- To suggest corrective/restoration options for improvement in lake trophic status index.

3 Methodology

The methodology of present study is according to the procedure recommended in APHA (1992) and NEERI (1991) guidelines for water quality. The lake is divided in small blocks, and samples were collected from different sampling locations for the representation of the quality of water of Bhalswa Lake. The samples were collected through boats provided by Delhi Tourism.

3.1 Trophic Status Index

Trophic Status Index (TSI) is a conventional measure or instrument used to calculate a lake's trophic status or productivity. More specifically, at a particular place and moment, it is the complete living plankton weight throughout water system, Three factors estimate algal biomass separately, (a) chlorophyll, (b) Secchi depth, and (c) complete phosphorus. There is a connection between phosphorus, chlorophyll-a (concentration of algae), and Secchi depth. This implies more food is accessible for algae when phosphorus rises, so algal concentrations rise. The water becomes less transparent when algal concentrations rises and the depth of Secchi reduces. The resulting numbers cover different components and ranges from all these three calculations, so it cannot be compared or averaged directly with each other. Three trophic status index readings used an equation to standardize them in order to make them instantly comparable. A lake's overall trophic status index (TSI) is the phosphorus median TSI, the Secchi depth TSI, and thus the chlorophyll-a TSI; therefore, considering phosphorus, Secchi depth, and chlorophyll-a, it can be considered as the lake condition. It is essential to realize that in phosphorus and algal concentration, trophic status is defined divisions of a continuum. The TSI is between 0 and 100. **Oligotrophic** 0–30 is very clear water, low phosphorus, and sparse algae. 30–50 is an intermediate phase where, owing to more accessible phosphorus, the number of aquatic plant algae increases. A TSI of over 50 defines a lake that is eutrophic, with an elevated plant and algae density that may be difficult to swim in the summer at certain times. Some lakes can be eutrophic naturally, having a TSI of 50 or higher over the last 100 years. As a consequence of human operations, other lakes have gradually risen in TSI. Not necessarily, the Trophic Status Index can be interchanged with water quality. The quality of water is subjective and depends on how the water body is to be used. A lake which is great for duck hunting is not necessarily nice for water skiing. In turn, bass fishing may not be good for a lake that is good for swimming. The different attributes with respect to different parameters of TSI are mentioned in Table 1.

The relationships are not always simple between the three TSI calculations. Carlson pointed out that highly stained lakes with large amounts of organic material submerged in them can generate elevated Secchi transparency TSI scores that do not suit the model because the water color affects the depth at which the Secchi disk falls. The type and volume of the prevalent algae population might also influence the writing of Secchi the algae types may vary in the amount of pigment they carry.

3.2 TSI Values Uses

It is possible to use TSI scores to rank lake between and within geographical areas. This categorization enables water supervisors to select lakes which may require restoration or preservation. An increasing sequence in TSI standards over many decades may demonstrate degradation in the wellness of a lake.

Table 1 Trophic status index meaning and attributes (Internet sources)

TSI	Chlorophyll-a (μg/l)	Secchi depth SD (ft)	Total phosphorous (μg/l)	Attributes	Fisheries and recreation
Less than 30	Less than 0.95	More than 26.2	Less than 6	**Oligotrophy**: Clean water, oxygen all year round at the edge of the lake, very profound cold water	Fishing in the tropics dominates
In between 30 and 40	In between 0.95 and 2.6	In between 13.1 and 26.2	In between 6 and 12	Lower ponds can become anoxic (no oxygen)	Only trout fishing in profound ponds
In between 40 and 50	In between 2.6 and 7.3	In between 6.6 and 13.1	In between 12 and 24	**Mesotrophy:** Most of the summer, water is moderately evident. May be late summer "greener"	No oxygen at the lake's bottom leads to trout loss. Walleye may prevail
In between 50 and 60	In between 7.3 and 20	In between 3.3 and 6.6	In between 24 and 48	**Eutrophy**: Problems with algae and aquatic plants are feasible. Most of the year, "green" water	Fishing in warm water only. Bass may prevail
In between 60 and 70	In between 20 and 56	In between 1.6 and 3.3	In between 48 and 96	Blue–green algae dominate, scums of algae and issues with aquatic plants	Dense algae and plants of the aquatic. Low clarity of the water can prevent swimming and boating
In between 70 and 80	In between 56 and 155	In between 0.8 and 1.6	In between 96 and 192	**Hypereutrophy**: (Light productivity restricted). Dense macrophytes and algae	For recreation, water is not appropriate
More than 80	More than 155	Less than 0.8	In between 192 and 384	Scums of algae, few aquatic plants	Rough fish (carp) dominate; can kill summer fish

Table 2 Analytical parameters and methods used for TSI study (APHA Sources)

S. No.	Parameters	Method
1	Total phosphate	Stannous chloride method
2	Secchi disc depth	Diameter −25 cm
3	Chlorophyll-a	90% Acetone extraction method

3.3 Factors Influencing the TSI Values

The less possibility that a potentially harmful cyanobacteria will occur might stop more and more presence of phosphorus. It is recognized that perhaps the cyanobacteria typically accountable for producing contaminants are poor competitors of phosphorus, so the accessible levels must have been huge until they do anything in water. Phosphorus emissions can be reduced with well-designed stormwater runoff catchment processes, renovation of sewer facilities, changing householders and business practices (such as not utilizing phosphorus-rich chemicals on gardens), training and rewards, and replacing septic watershed structures with sewers.

The TSI is a useful tool for the management of lake water and also a useful science instrument of research where a trophic status standard is required. The TSI can serve as a trophic status assessment about many of biological and chemical elements of river system linked to the trophic status which can be compared. Outcome might be fuller and dynamic image of how these elements relate to each other and the ecosystems of the lake as a whole.

In this study, different analytical parameters are studied for the determination of WQI and TSI (Table 2).

3.4 Parameters Studied for the Evaluation of TSI

1. Determination of Secchi disc depth
2. Determination of total chlorophyll -a
3. Determination of total phosphate
4. Trophic Status Index
5. Trophic State Index (TSI)
6. Secchi depth transparency (SD).

4 Results

4.1 Determination of TSI (SD)

The values of TSI SD for Bhalswa Lake in the monsoon and post-monsoon seasons are represented in Table 3, and graphical rep. of comparison of TSI SD is shown in Fig. 1 mentioned below. The values of TSI SD for monsoon season are maximum at location S5, and it is minimum for location S1 and S13. And the value of TSI SD

Table 3 Comparison of TSI SD between monsoon and post-monsoon season

Sampling no	Monsoon TSI SD	Post-monsoon TSI SD
S1	94.74	91
S2	100.85	97
S3	101.84	65
S4	104.74	102
S5	105.73	103
S6	107.56	104
S7	105.4	97
S8	98.36	96
S9	97.37	95
S10	101.82	96
S11	94.74	90
S12	94.58	89
S13	94.74	86
Mean	100.19	93.15
SD	4.56	9.67

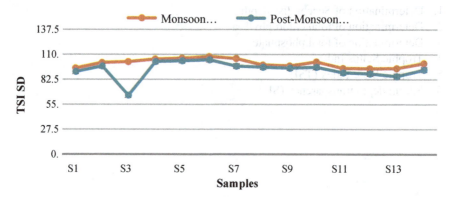

Fig. 1 Graphical representation of the comparison of TSI SD between monsoon and post-monsoon

Table 4 Comparison of TSI–TChl between monsoon and post-monsoon

Sampling no	Monsoon TSI–TChl	Post-monsoon TSI–TChl
S1	74	86
S2	30	84
S3	61	30
S4	53	30
S5	60	70
S6	67	71
S7	30	30
S8	30	30
S9	51	54
S10	75	78
S11	93	95
S12	41	30
S13	30	30
Mean	54	55
SD	19.7	25.1

for post-monsoon season is maximum at location S6, and it is minimum for location S3.

4.2 Determination of Total Chlorophyll-A (TChl)

The values of total chlorophyll-a, for Bhalswa Lake in the monsoon and post-monsoon seasons are represented in Table 4, and graphical representation of the comparison of TSI-Chl is shown in Fig. 2 as mentioned below

4.3 Determination of Total Phosphorous

The comparison of mean values of total phosphorous of all 13 locations from Bhalswa Lake was represented in Table 5 during monsoon and post-monsoon season, and graphical representation of the comparison of TSI-TP is shown below.

The graphical representation of the comparison in between the TSI-TP values in different seasons is shown in Fig. 3.

The comparison of the SD + CHL + TP/3 values for all 13 sampling locations is mentioned in Table 6 and graphical representation of the values is shown in Fig. 4.

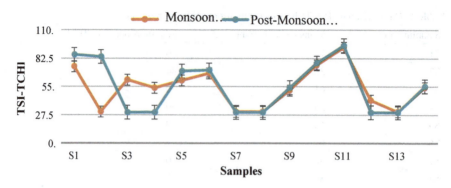

Fig. 2 Graphical representation of the comparison of TSI-TChl between monsoon and post-monsoon

Table 5 Comparison of TSI-TP between monsoon and post-monsoon

Sampling no	Monsoon TSI-TP	Post-monsoon TSI-TP
S1	133	132
S2	134	111
S3	134	111
S4	131	108
S5	136	112
S6	130	108
S7	136	112
S8	132	109
S9	135	111
S10	136	112
S11	136	112
S12	136	112
S13	133	110
Mean	134	112
SD	2	5.86

4.4 Determination of Total TSI

The TSI values for all different locations are mentioned in Table 7 and Fig. 5. The values of TSI were in between the range of 69–103, and mean values for monsoon and post-monsoon season were **96.06 ± 6.78** and **86.89 ± 10.53**, respectively. The formula used for the calculation of Carlson's Trophic Status Index is mentioned following under:

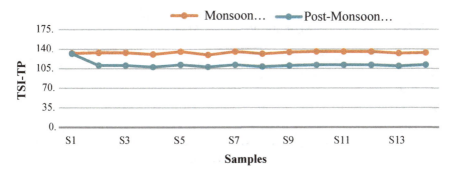

Fig. 3 Graphical representation of the comparison of TSI-TP between monsoon and post-monsoon

Table 6 Comparison of SD + CHL + TP/3 between monsoon and post-monsoon

Sampling no	Monsoon SD + CHL + TP/3	Post-monsoon SD + CHL + TP/3
S1	100.69	103.00
S2	88.47	97.33
S3	99.08	68.67
S4	96.44	80.00
S5	100.87	95.00
S6	101.80	94.33
S7	90.66	79.67
S8	86.98	78.33
S9	94.53	86.67
S10	104.54	95.33
S11	107.91	99.00
S12	90.78	77.00
S13	86.10	75.33
Mean	96.06	86.89
SD	6.781	10.539

4.5 Carlson's Trophic Status Index (C-TSI) = [TSI (TP) + TSI(CA) + TSI(SD)]/3

Mesotrophy (moderate productivity) is generally associated with index with a range in between 40–50 value; and index value more than 50 showed eutrophy status; value less than 40 are correlated with oligotrophy (low productivity). The values of the three factors, index and more than their characteristics for each parameter. Lake Bhalswa has been categorized as hypereutrophic. If it can be associated with particular occurrences within a water body, any Trophic Status Index gains value. The list of possible modifications in the body of water may happen. Some features,

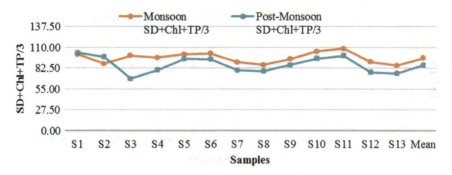

Fig. 4 Graphical representation of the comparison of SD + Chl + TP/3 between monsoon and post-monsoon

Table 7 Comparison of Carlson TSI between monsoon and post-monsoon

Sampling no	Monsoon CTSI	Post-monsoon CTSI
S1	101	103
S2	88	97
S3	99	69
S4	96	80
S5	101	95
S6	102	94
S7	91	80
S8	87	78
S9	95	87
S10	105	95
S11	108	99
S12	91	77
S13	86	75
Mean	96	87
SD	6.78	10.53

such as hypolimnitic oxygen or fish, may differ slightly and may not be linked to real TSI at times. Bhalswa Lake TSI values were discovered to be **69–103** in both monsoon season and post-monsoon season, and average TSI values were found to be **96 and 86,** respectively, for the monsoon and post-monsoon seasons. This can be ascribed to Bhalswa Lake eutrophy, where anoxic hypolimia and macrophy issues are quite feasible.

Conservation can be performed under such conditions by removing hydrophytic macrophytes and introducing hot water fish. Under these conditions, blue, green algae dominate especially that of *M. aeruginosa*, foul smell come from the water

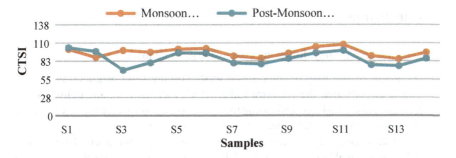

Fig. 5 Graphical representation of the comparison of Carlson TSI between monsoon and post-monsoon

and the water transparency decreases. TP increases significantly resulting in high productivity restricted, thick algae development.

5 Discussion

A close look at the lake shows that dense scums of algal, decreased macrophytes, foul smelling water lead to enormous fish murderers. It can be reduced only by draining the whole water and again refill it by harvesting rainwater. Eutrophication monitoring is a significant component of the evaluation and management of lake ecosystems. Phosphorous is initially the algae nutrient-limiting development. These results in direct and indirect biological modifications in lakes that lead to algal blooms being produced. Indirect eutrophication during the breakdown of the dead algae mass might be leads to depletion in dissolved oxygen concentration as a consequence of bacterial respiration. According to Carlson (1980), the interrelationship between the parameters can use for the assessment of quality of water for lakes. Corrective measures and lake restoration methods are crop management, live stock management, land farming, sewage water treatment, water treatment, and artificial floating islands. Suggestions from the study are that the data related to various physiochemical characteristics of water of Bhalswa Lake can be utilized as a baseline and reference value for future research work and some necessary steps can be taken on the basis of the result which is obtained from my study. Concerned authorities should take care of the lakes, necessary rules should be formed, and the violator should be charged penalty. Then only, this lake can be saved, otherwise one more lake will be lost.

6 Conclusion

Based on the above study, following conclusions can be made:-

1. The major reason may be the organic waste coming from neighboring dairies, poured directly into the Bhalswa Lake, which increases organic pollution and decreased dissolved oxygen leading to anoxic conditions in lake water, decreased DO value which is < 4, further will affect aquatic life.
2. Also lake is polluting due to inorganic waste coming from nearby landfill sites and various small factories nearby lake which are discharging their untreated waste matter directly into the lake.
3. If this process will continue, then lake will be no longer a place of recreational activity. Immediate action is needed to revive the lake. Otherwise, one more lake will be lost.

References

1. Sakshi SK, Singh Haritash AK (2017) Environmental biotechnology for control of environmental pollution. In: Proceedings of international conference on emerging areas of environmental science and engineering EAESE-2017, February 16–18, 2017
2. Shan V, Singh SK, Haritash AK (2017) Water quality Indices to determine the surface water quality of Bhindawas Wetland. In: Proceedings of international conference on emerging areas of environmental science and engineering EAESE-2017, February 16–18, 2017
3. Gour A Singh SK, Mandal A (2016) Occupational Hazard Posed to labourer operators at landfill sites. In: Proceeding of 3rd International conference on occupational and Environmental Health, 23–25 Sept 2016, New Delhi
4. Singh SK, Anunay G, Rohit G, Shivangi G and Vipul V (2016) Greenhouse gas emissions from landfills: a case of NCT of Delhi, India. J Climatol Weather Forecast
5. Gupta D, Singh SK (2015) Energy use and greenhouse gas emissions from waste water treatment plants. Int J Environ Eng-Indersci 7(1):1–10. https://doi.org/10.1504/IJEE.2015.069251
6. Jaffar MM, Shebli MK, Mussa AKA, Hadi BH, Aenab AM, Singh SK (2015) Detection of Enterobacter sakazakii from Commercial Children Dry Milk. Journal of Environmental Protection 2015 (6), 1170–1175
7. Singh SK, DhruvKatoriya DM, Sehgal D (2015) Fixed bed column study and adsorption modelling on the Adsorptim of malachite green day from wastewater using acid activated sawdust. Int J Adv Res 3(7):521–529
8. Abtahi M, Golchinpour N, Yaghmaeian K, Rafiee M, Jahangiri-rad M, Keyani A, Saeedi R (2015) A modified drinking water quality index (DWQI) for assessing drinking source water quality in rural communities of Khuzestan Province. Iran Ecol Indic 53:283–291. https://doi.org/10.1016/J.ECOLIND.2015.02.009
9. Alam M, Pathak JK (2010) Rapid assessment of water quality index of ramganga river, Western Uttar Pradesh (India) using a computer programme. Nat Sci 8:1–8
10. Avvannavar SM, Shrihari S (2008) Evaluation of water quality index for drinking purposes for river Netravathi, Mangalore, South India. Environ Monit Assess 143:279–290. https://doi.org/10.1007/s10661-007-9977-7
11. Bhargava D, Saxena B, Dewakar R (1970) A study of the geopollutants in the Godavary river basin of India. Asian Environ 78:1–4

12. BIS (1991) Water quality standards indian standard drinking water specification, Bureau of Indian Standard, Indian Standard (10500)
13. Branchu P, Bergonzini L, Ambrosi J-P, Cardinal D, Delalande M, Pons-Branchu E, Benedetti M (2010) Hydrochemistry (major and trace elements) of Lake Malawi (Nyasa), Tanzanian Northern Basin: local versus global considerations. Hydrol Earth Syst Sci Dis 7:4371–4409. https://doi.org/10.5194/hessd-7-4371-2010
14. Chandra S, Singh A, Tomar PK (2012) Assessment of water quality values in Porur Lake Chennai, Hussain Sagar Hyderabad and Vihar Lake Mumbai. India Chem Sci Trans 1:508–515. https://doi.org/10.7598/cst2012.169
15. Chauhan A, Singh S (2010) Evaluation of ganga water for drinking purpose by water quality index at Rishikesh, Uttarakhand. India Rep Opin 2:53–61
16. Devi Prasad AG (2012) Carlson's trophic status index for the assessment of trophic status of two Lakes in Mandya district. Adv Appl Sci Res 3:2992–2996
17. Dwivedi S, Tiwari I, Bhargava D (1997) Water quality of the river Ganga at Varanasi. Inst Eng Kolkota 78:1–4
18. Fadiran A, Dlamini S, Mavuso A (2008) A comparative study of the phosphate levels in some surface and ground water bodies of swaziland. Bull Chem Soc Ethiop 22:197–206
19. Goher ME, Hassan AM, Abdel-Moniem IA, Fahmy AH, El-sayed SM (2014) Evaluation of surface water quality and heavy metal indices of Ismailia Canal, Nile River Egypt. Egypt J Aquat Res 40:225–233. https://doi.org/10.1016/J.EJAR.2014.09.001
20. Joshi DM, Kumar A, Agrawal N (2009) Studies on physicochemical parameters to assess the water quality of river ganga for drinking purpose In: Haridwar District 2:195–203

Seasonal Variation of Water Quality Index of an Urban Water Body Bhalswa Lake, Delhi (India)

Sumit Dagar and S. K. Singh

Abstract Polluted water affects the human life as well as aquatic life. Some chemicals like fluoride, mercury, cadmium, arsenic, petrochemicals, nitrates, pesticides affect the human health. These chemicals are useful in the optimum concentration, but if the concentration of these chemicals increased then it will create health problems for human being. If the quantity of fluoride is < 0.5 mg/l then it is useful for dental problems, but if the quantity of fluoride is > 0.5 mg/l and exposure is more than 5 to 6 years then it causes fluorosis to the humans. Likewise, the arsenic quantity > 0.05 mg/l can cause different types of cancers; the concentration of TDS should not be above than 500–2000 mg/l; the concentration of iron (Fe) in the range of (0.3–1.0 mg/l) impacts metallic taste and brownish green stains; the concentration of manganese (Mn) should not be more than (0.1–0.3 mg/l), as it brings about bitter taste, black stains and/ brown color; the concentration of sulfate (SO_4) should not be more than (200–400 mg/l), as it causes laxative effect and deposits of scales; the concentration of nitrate (NO_3) should not be more than (45–100 mg/l); the concentration of chloride (Cl^-) should not be more than (250–1000 mg/l), as it causes high BP (blood pressure); the concentration of chromium (Cr) should not be not more than 0.05 mg/l, as it causes skin-related problems; the concentration of copper (Cu) has value more than (0.0–1.5 mg/l), as it causes blue green stains and GI irritations. The acid deposition in river, lake and pond pH lowers the concentration in surface water which further affects various plant and animal species, and the aquatic plants are so much affected. The industrial waste changes the pH of water which affects the life of aquatic plants.

Keywords Polluted water · Human health · Chemicals · Species · Plants

---

S. Dagar (✉) · S. K. Singh
Environmental Engineering Department, Delhi Technological University,
New Delhi 110042, India
e-mail: dagarsumit95@gmail.com

© The Author(s), under exclusive license to Springer Nature Singapore Pte Ltd. 2021
R. M. Singari et al. (eds.), *Advances in Manufacturing and Industrial Engineering*,
Lecture Notes in Mechanical Engineering,
https://doi.org/10.1007/978-981-15-8542-5_16

1 Introduction

Water is indeed a valuable resource for existence of life on earth. It is very difficult to substitute water with another resource, and it is genuinely a one of a blessing to human being from nature (Singh and Gupta 2017). The surface water and groundwater are generally used in various terms by man which mainly include agriculture uses, hydropower generation, for industries, fisheries, for industries and recreational exercises (Handa 1984; Singh and Gupta 2017). The freshwater in the world is just about 0.5% of the world's surface and consists of a volume of 2.84×105 km^3. Rivers comprise about 0.1% of the surface land, and nearly about 0.01% of the earth water is present in rivers. In spite of these low amounts, water scarcity is being observed last ten years (Singh and Gupta 2017).

In India, rainfall precipitation is around 3000 km^3. Precipitation in India is dependent on the northeast and southeast rainstorms (Singh and Gupta 2017). Most of the Indian rivers are perennial, and others are seasonal (Singh and Gupta 2017).

Over the most recent couple of decades, there has been a large increment in the Indian population and rapid growth of industrialization (Singh and Gupta 2017). Anthropogenic exercises identified with broad urbanization, farming practices, industrialization and population extension have drastically affected water quality over in large area of the world. Likewise, inadequate water resources have gradually restricted monitoring of water pollution and quality improvement. Contamination of water is a severe problem for the government and researchers (Singh and Gupta 2017).

2 Methodology

The methodology of the present study is according to the procedure recommended in APHA (1992) and NEERI (1991) guidelines for water quality. The lake is divided into small blocks, and samples were collected from different sampling locations for the representation of the quality of water of Bhalswa Lake. The samples were collected through boats provided by Delhi Tourism.

Objective of the research:

- To study physiochemical parameters of Bhalswa Lake.
- To find water quality index (WQI) of the lake.

Study Area:

Bhalswa Lake, Naini Lake, Hauz Khas Reservoir and Bangla Sahib Sarovar are 3 to 4 lakes in Delhi. The research selected the **Bhalswa** Lake. Complex Lake Bhalswa area is 92 ha. The lake is 350 m long (ww.dda.org.in). The location and broad view of Bhalswa Lake are shown in Fig. 1 (Image 1).

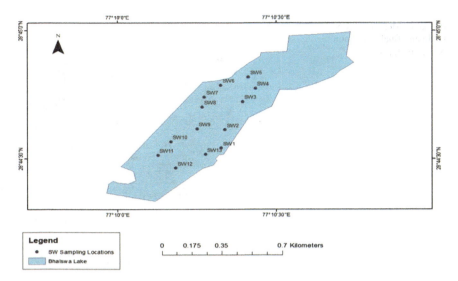

Fig. 1 Broad view and sampling location of the Bhalswa Lake

Image 1 Sampling from Bhalswa Lake

3 Experimental Analysis

The methodology of the present study is according to the procedure recommended in APHA (1992) and NEERI (1991) guidelines for water quality. The lake is divided into small blocks, and samples were collected from different sampling locations for the representation of the quality of water of Bhalswa Lake. The samples were

Image 2 Map view of New Delhi and Bhalswa region (sampling location)

collected through boats provided by Delhi Tourism. The map view of Bhalswa Lake region in New Delhi is mentioned in Fig. 1 (Image 2).

All the collected samples are immediately preserved in dark sterile boxes. In this study, thirteen (13) samples from different locations as shown in Fig. 1 of Bhalswa Lake were collected and studied for different parameters (Table 1).

Physical parameters: Temperature; hydrogen ion concentration (pH); and TDS.

Chemical parameters: Total alkalinity—TA; total hardness—TH; dissolved oxygen—DO; chloride; sulfate; nitrate; and phosphates were studied for the determination of WQI.

Methods used for the evaluation of different parameters for WQI:

Electrometric and titration methods.

Instruments/equipments used for the evaluation of different parameters for WQI:

pH meter, conductivity/TDS meter, UV–Vis spectrophotometer (Image 3).

Table 1 Analytical parameters and equipments used in the WQI study

S. No.	Parameters	Method	Instruments/equipments
1	pH	Electrometric	Multiparameter
2	Conductivity	Electrometric	Multiparameter
3	TDS	Electrometric	Multiparameter
4	Alkalinity	Titration by sulfuric acid	–
5	Hardness	Titration by EDTA	–
6	Chloride	Titration by silver nitrate	–
7	Sulfate	Barium chloride	UV–Vis spectrophotometer
8	Nitrate	Brucine-Sulfanilic acid	UV–Vis spectrophotometer
9	Phosphate	Stannous chloride Method	UV–Vis spectrophotometer

Image 3 Testing of samples collected

3.1 Determination of Temperature

A digital temperature recorder was used to measure wastewater temperature in °C.

3.2 Determination of pH

pH is defined as the negative inverse $\log_{10}$ of the H^+ concentration in the water sample and determined by the pH meter (Systronic) at room temperature. Instrument was standardized using a buffer solution with the pH range (7.4 and 9.2). It was predicted by using the formula

pH = − log[H]⁺ to calculate pH by pH meter.

3.3 *Determination of Total Dissolved Solids (TDS)*

By evaporating filtrate collected from the suspended particles, total dissolved solids were evaluated. Analyze the solids and then weigh the residue remaining in the jar that evaporates.

3.4 *Determination of Total Alkalinity (TA)*

Total alkalinity was predicted using the methyl orange indicator using the normal titration technique.

3.5 *Determination of Total Hardness (TH)*

Hardness was estimated using EDTA and EBT as an indicator using the conventional EDTA titration technique.

3.6 *Determination of Dissolved Oxygen (DO)*

Dissolved oxygen is estimated by DO meter immediately after the sample is collected. This dilution equation computes the DO in the lake after mixing of drain into lake.

3.7 *Determination of Chloride*

Based on Mohr's technique (titrant (silver nitrate); indicator (potassium chromate)), chlorides were estimated.

3.8 *Determination of Sulfate*

Sulfate concentration determination of wastewater was done with the help of turbidimetric method. Sulfate ion precipitation was done in hydrochloric acid with the barium chloride to form the uniform size crystals of barium sulfate.

3.9 Determination of Nitrate

The UV spectrophotometer was used to estimate the concentration of nitrates in different specimens at 400 nm.

3.10 Determination of Phosphate

The UV spectrophotometer was used to estimate the concentration of nitrates in different specimens at 490 nm.

3.11 Determination of Water Quality Index (WQI)

The assessment of different chemical and physical parameters was done as per CPCB Guide for the analysis of wastewater. The parameters used for the determination of quality of water are pH; TDS; chloride; TA; TH; DO; phosphate; chloride; sulfate; and nitrates used for WQI calculations at different sampling locations. WAWQI method by using most frequently measured water quality variables categorized the quality of water as per purity level. This technique has been commonly used by the different researchers (Chauhan and Singh 2010). The following equations are used to calculate the W_{QI}:

$$W_{Q_I} = \sum Q_i W_i / \sum W_i \quad (1)$$

The following equation is used to calculate rating scale (Q_i) for the determination of each parameter:

$$Q_i = 100\,[(V_i - V_o / S_i - V_o)] \quad (2)$$

For analyzing the quality of water, V_i is an estimated value for ith parameter; V_o is an optimum value of the parameter in clean water, where $V_o = 0$ (apart from the values of pH and DO (7.0 and 14.6 mg/L)), respectively; S_i is a normal value of ith parameter (Chauhan and Singh 2010).

For each parameter for the quality of water, the unit weight (W_i) is determined using the following method: $W_i = K/S_i$ at which K is the constant proportionality and could be calculated to use the equation mentioned below:

$$K = 1 / \sum (1/S_i) \quad (3)$$

Table 2 shows the rating of quality of the water by WAWQI method (Singh and Chauhan 2010).

Table 2 Rating of water quality

Value of WQI	Rating of quality of the water	Grade
0–25	Superior quality of water	a
26–50	Very good quality of water	b
51–75	Poor or not good quality of water	c
76–100	Very bad quality of water	d
Above 100	Not suitable for drinking	e

3.12 Standards of Water

World Health Organization—WHO; the Bureau of Indian Standards—BIS; and the Indian Council for Medical Research—ICMR provide the standards for different parameters for the drinking water along with its respective status classifications for the determination of WQI such as: pH (6.5–8.5)—ICMR/ BIS; TDS (500 mg/ml)—ICMR/ BIS; TA (300 mg/ml)—ICMR/ BIS; TSS (500 mg/ml)—WHO; calcium (75 mg/ml)—ICMR/ BIS; magnesium (30 mg/ml)—ICMR/ BIS; chloride (250 mg/ml)—ICMR; nitrate (45 mg/ml)—ICMR/ BIS; sulfate (150 mg/ml)—ICMR/ BIS; and dissolved oxygen (5 mg/ml)—ICMR/ BIS (Singh and Kant Kamal 2014).

Analysis of the different parameters was done for the determination for WQI as mentioned below under:

3.13 Determination of Temperature

The temperature difference (°C) of the Bhalswa Lake water in monsoon season and post-monsoon season is mentioned in Table 3.

3.14 pH Level

The average pH of all 13 sampling points was **9.0 ± 0.1** and **8.0 ± 0.3** in monsoon season and post-monsoon season, respectively. There was not an important distinction between the pH of both seasons. The pH of Bhalswa Lake water was higher than the standard parameters of pH (6.5 – 8.5)—ICMR/BIS.

Table 3 WQI of Bhalswa lake in monsoon

S. No.	TDS	TH	DO	PO$_4^{3-}$	CL$^-$	NO$_3^-$	SO$_4^{2-}$	TA	WQI
S1	1583	1050	5.2	7.6	486	42.3	123	569	380.78
S2	1856	1040	4.5	8.1	479	46.2	147	548	408.5
S3	1745	1110	4.7	8.3	486	48.9	140	604	418.11
S4	1723	1020	5.6	6.7	476	41.3	137	574	338.24
S5	1563	920	5.5	9.4	483	45.1	147	541	463.87
S6	1874	860	5.1	6.5	479	46.2	133	636	333.54
S7	1659	860	4.9	9.4	501	41.2	557	612	469.32
S8	1822	980	5.1	7.1	501	41.2	129	584	360.31
S9	1598	970	4.6	8.8	486	40.3	552	635	439.85
S10	1565	910	4.5	9.7	501	41.2	148	649	479.56
S11	1776	970	4.3	9.5	504	48.2	267	616	471.2
S12	1767	960	4.9	9.6	508	40.3	139	607	475.72
S13	1685	1340	4.7	7.573	504.1	45.7	141	598	383
Mean	1709	999	5	8	492	44	212	598	417
SD	109.6	125.7	0.4	1.1	11.3	3.1	156.2	33.4	51.6

3.15 Determination of Dissolved Oxygen (DO)

Natural water atmospheric equilibrium contains DO levels varying from 5 to 14.5 mg/l of oxygen. The DO concentration into water represents dissolution in atmosphere. DO is the variable determining whether aerobic or anaerobic organisms bring about biological modifications. Thus, the measurement of dissolved oxygen is essential for keeping procedures of aerobic therapy aimed at purifying national and industrial wastewater. A fast drop in the DO shows the river's elevated organic pollution.

3.16 Determination of Total Phosphate (TP)

Streams or other flowing water is somewhat less prone to accelerated or cultural eutrophication, so a required objective for them is a phosphate concentration below 0.3 mg/l. The required phosphate concentration should be less than 0.15 mg/l in regions where streams enter lakes or reservoirs. In conclusion, Bhalswa Lake is so much contaminated by phosphate (Table 4).

Table 4 WQI of Bhalswa lake in post-monsoon

S. No.	TDS	TH	DO	PO$_4^{3-}$	CL$^-$	NO$_3^-$	SO$_4^{2-}$	TA	WQI
S1	1375	849	4.1	5.4	370	51.5	103	462	**278.37**
S2	1676	840	3.8	6.8	346	47.4	125	484	**328.83**
S3	1404	909	3.8	5.5	398	52.5	123	512	**264.12**
S4	1595	819	4.3	4.7	347	47.1	123	486	**237.69**
S5	1370	720	3.6	5.5	385	48.2	124	412	**268.99**
S6	1639	659	3.7	5.5	365	49.6	116	578	**262.25**
S7	1493	660	4	4.5	384	45.2	345	563	**225.24**
S8	1699	779	4.3	4.4	391	48.0	118	474	**219.87**
S9	1346	769	3.6	5.4	365	45.8	369	565	**270.92**
S10	1402	709	4.1	6.1	385	45.5	129	563	**299.89**
S11	1495	769	3.9	5.5	322	53.3	246	524	**279.32**
S12	1575	760	4.2	6.4	309	49.5	124	521	**314.82**
S13	1519	1139	3.7	4.5	324	51.0	120	546	**240.87**
Mean	1507	798	**4**	**5**	**361**	**49**	**167**	**515**	**268.55**
SD	109.6	125.7	**0.3**	**0.8**	**29.1**	**2.7**	**91.7**	**49.2**	**31.62**

3.17 Determination of Chloride

The average value of chloride in all 13 sampling points was **492.0 ± 11.3 mg/l** and **361.0 ± 21.1 mg/l** in monsoon season and post-monsoon season. The chlorides (maximum) more than 250 mg/l for Class A (Singh and Kant Kamal 2014) and 600 mg/l for Classes C and E are undesirable according to water quality criteria indicated by CPCB and BIS. But, the values of chlorides were **492.0 ± 11.3 mg/l** and **361.0 ± 21.1 mg/l** in monsoon season and post-monsoon season more than the standard value **250 mg/l**.

3.18 Determination of Total Nitrate (NO$_3^-$)

The mean value of 13 samples collected from Bhalswa Lake for NO$_3^-$ concentration was **49 ± 2.7 mg/l** and **44 ± 3.1 mg/l** in monsoon season and post-monsoon season.

3.19 Determination of Total Sulfate (SO_4^{2-})

The mean value of all 13 samples of Bhalswa Lake for SO_4^{2-} concentration was **212 ± 156.2 mg/l** and **167 ± 91.7 mg/l** in the monsoon season and post-monsoon season.

3.20 Determination of Total Alkalinity (TA)

The mean value of all 13 samples of Bhalswa Lake for TA concentration was **598 ± 33.4 mg/l** and **515 ± 49.2 mg/l** in monsoon season and post-monsoon season.

3.21 Determination of Total TSI

The values of TSI were in between the ranges of 69 and 103, and mean values for monsoon and post-monsoon season were **96.06 ± 6.78** and **86.89 ± 10.53**, respectively.

4 Conclusion

Based on the above study, the following conclusions can be made:-

1. Value of WQI at all sampling sites was observed beyond 100 which represents its unsuitability.
2. The major reason may be the organic waste coming from neighbouring dairies, poured directly into the Bhalswa lake, which increases organic pollution and decreased dissolved oxygen leading to anoxic conditions in lake water, decreased DO value which is < 4, further will affect aquatic life.
3. Also, lake is polluting due to inorganic waste coming from nearby landfill sites and various small factories nearby lake which are discharging their untreated waste matter directly into the lake.
4. If this process will continue, then lake will be no longer a place of recreational activity. Immediate action is needed to revive the lake. Otherwise, one more lake will be lost.

A sewage treatment plant should be built for treating waste near the lake.

References

1. Allaa M Aenab, SK Singh et al (2016) Algae personification toxicity by GC– MASS and treatment by using material potassium permanganate in exposed basin. Egypt J Pet– Elsevier, 27 · November 2016. https://doi.org/10.1016/j.ejpe.2016.10.020
2. Katiyar A, Singh SK, Haritash AK (2017) Effect of odd even scheme to combat air pollution in Nct of Delhi. Int J Adv Res 5(2):1215–1222
3. Rustogi P, Singh SK (2017) Revival and rejuvenation strategy of water bodies in a metropolitan city: a case study of Najafgarh Lake, Delhi, India. Int J Adv Res 5(2):189–195
4. Shan V, Singh SK, Haritash AK (2017)Water quality Indices to determine the surface water quality of bhindawas wetland. In: Proceedings of international conference on emerging areas of environmental science and engineering EAESE-2017, February 16–18, 2017
5. Sorting of Glossiphondia Complanata (Linnaeus, 1758) (Rhyncobdelli: Glossiphoniidae) from three aquatic plants in river tigris within Baghdad City. Egypt J Petrol—Elsevier, 27 · November 2016. https://doi.org/10.1016/j.ejpe.2016.11.001
6. Gour A, Singh SK, Mandal A (2016) Occupational hazard posed to labourer operators at landfill sites. In: Proceeding of 3rd international conference on occupational and environmental health, 23–25 September 2016, New Delhi
7. Mehta D, Mazumdar S, Singh SK (2015) Magnetic adsorbents for the treatment of water/wastewater: a review. J Water Proc Eng 7(9):244–265 (Elsevier)
8. Jaffar MM, Shebli MK, Mussa AKA, Hadi BH, Aenab AM, Singh SK (2015) Detection of enterobacter sakazakii from commercial children dry milk. J Env Protect (6):1170–1175
9. Bansal S, Biswas S, Singh SK (2017) Fuzzy decision approach for selection of most suitable construction method of Green Buildings. Int J Sustain Built Env 6(1):122–132
10. Sawyer CN, McCarty PL, Parkin GF (2003) Chemistry for environmental engineering and science. McGraw-Hill
11. Shah KA, Joshi GS (2017) Evaluation of water quality index for River Sabarmati, Gujarat, India. Appl Water Sci 7:1349–1358. https://doi.org/10.1007/s13201-015-0318-7

Material Study and Fabrication of the Next-Generation Urban Unmanned Aerial Vehicle: Aarush X2

Rishabh Dagur, Krovvidi Srinivas, Vikas Rastogi, Prakash Sesha, and N. S. Raghava

Abstract Composite designing and airframe fabrication to develop a lightweight, robust, and high-endurance unmanned aerial vehicle (UAV) Aarush X2, next-generation urban unmanned aerial vehicle, are presented in this paper. The composite study investigated the structural feasibility of various lay-ups and fibre orientations using analytical model. The research is performed on the sandwich composite configuration of Divinycell, core, and carbon fibre-reinforced plastic (CFRP), facesheet, bonded with the epoxy matrix of AY-105 (resin) and HY-951 (hardener). The study presented five plies, [0/90/Divinycell/0/90], lay-up to sustain structural loads with the least gross take-off weight (GTOW) for manufacturing the UAV. The fabrication of the UAV was carried out using vacuum-assisted wet lay-up method (VAWLM) and demonstrated a total weight of the airframe 12.1 kg. The flight tests of around 30 h validated the structural capability of the designed sandwich composite along with the fabrication technique selected. The averaged flight time of 2.8 h on a two-stroke gasoline engine, DLE-61, with fuel tank capacity of 1.8 litres was recorded.

Keywords First-order shear deformation theory · Sandwich composite · Compressive properties · Vacuum-assisted wet lay-up

---

R. Dagur (✉) · K. Srinivas · V. Rastogi
Department of Mechanical Engineering, Delhi Technological University, New Delhi 110042, India
e-mail: rishabhdagur_bt2k14@dtu.ac.in

P. Sesha
Lockheed Martin Corporation, Hyderabad, India

N. S. Raghava
Department of Electronics & Communication Engineering, Delhi Technological University, New Delhi 110042, India

---

© The Author(s), under exclusive license to Springer Nature Singapore Pte Ltd. 2021
R. M. Singari et al. (eds.), *Advances in Manufacturing and Industrial Engineering*,
Lecture Notes in Mechanical Engineering,
https://doi.org/10.1007/978-981-15-8542-5_17

1 Introduction

In the aerospace industry, composite materials are widely used because of the high strength and lightweight characteristics offered. As follows, the mechanical prediction of the composite materials has become an active topic of research and has been studied at many diverse scales to tailor the properties. The micromechanical models such as laminate plate theory present a practical approach while designing the composite structures to investigate various fibre orientations, ply thicknesses, volume fractions, and material plies. However, designing the composite alone will not generate a strong and lightweight structure. The composite fabrication process plays a critical role in the manufacturing of a robust structure without defects.

In general, the plate deformation studies are based on the stress and displacement theories, where the broadly accepted displacement plate theory is divided into the laminate plate theory (classical laminate plate theory, CLPT) and shear deformation plate theories (first-order shear deformation theory, FOSDT, and further higher-order shear deformation theory; HOSDT) [1]. In the late nineteenth century, Kirchhoff [2] introduced the CLPT which was later improved by Love [3] in the early twentieth century. The CLPT is a quick and simple approach to predict the mechanical properties of the composite laminates [4]. However, it fits merely for the thin laminate; therefore, thick plate theories such as FOSDT and other HOSDT were formed to predict the solution [5].

The FOSDT developed by Reissner [6] and Mindlin [7] considered the effect due to the transverse shear deformation which was earlier not included in the CLPT. This addition helps FOSDT to study both thin and thick composite laminates. In these studies [6, 7], the displacement w persists to be constant along the thickness; however, the displacements u and v fluctuate linearly along the thickness of each individual layer. FOSDT is more practical than CLPT as the shear strain field is irregular at the linear interface and the displacement results allow composite collapsing. The shear correction factors ascertain the zero transverse shear conditions at the upper and lower plate surfaces. The correction factors are accountable for the accuracy between the numerical solution and the actual data [8]. Most of the structures developed using composites introduce advantages in the physical properties through fibre orientations and thicknesses. Sandhu et al. [9] demonstrated the mechanical effects on a unidirectional composite when the fibre orientations were changed for the plies. This study determined the orientation of the fibre plies for the maximum strength under a biaxial state of stress. The ability to maximise the in-plane stresses of the unidirectional composite was revealed by Brandmaier [10].

The fabrication of the composite is a critical step to construct a strong and lightweight aircraft structure. Dagur et al. [11] presented the utilisation of the vacuum-assisted wet lay-up method (VAWLM) for the sandwich composite fabrication. VAWLM eliminates the excess epoxy from the wetted lay-ups, which supports in reducing the physical weight by developing epoxy consistency in the fabricated composite. The study further claimed that the elevated temperature of the epoxy matrix impregnated in the composite thrusts the air pockets upward and simultaneously eliminated through the vacuum pump connected.

In this research, micromechanical designing and fabrication of the composite unmanned aerial vehicle were investigated. The specific objectives were: (i) to formulate FOSDT model for a sandwich composite, (ii) to optimise ply layers and orientation of the sandwich composite, (iii) to apply formulated FOSDT on fuselage and wing of Aarush X2, (iv) to fabricate Aarush X2 with the designed sandwich composite, (v) to perform flight testing of the Aarush X2.

2 Mathematical Formulation

The mathematical modelling of the Divinycell CFRP sandwich laminate was performed to calculate the optimum number of layers and fibre orientations. The entire mathematical designing of the composite is illustrated in Fig. 1. The sandwich composite is an arranged combination of three-section, two facesheets (here CFRP as top/bottom layers) and a core (here Divinycell; centre), as shown in Fig. 2. Each layer of the sandwich composite was modelled with the FOSDT and based upon the following assumptions

1. The thickness of the facesheets, $h_{t,b}$, is relatively lower than the core thickness, h_c.
2. Insignificant in-plane stresses in the core.
3. The in-plane displacements remain consistent along with the thickness of the facesheet.
4. The out-of-plane displacement is free from the z-coordinate.
5. Linear in-plane displacements of the core along z-coordinate.

2.1 Displacement Field

The displacement field expressed for the core and the facesheet in the FOSDT were as follows

Core:

$$\text{Bottom}: u = u_0 - \frac{(h_c + h_1)}{2}\psi_x, u = v_0 - \frac{(h_c + h_1)}{2}\psi_y$$

$$\text{Top}: u = u_0 + \frac{(h_c + h_2)}{2}\psi_x, u = v_0 + \frac{(h_c + h_2)}{2}\psi_y$$

$$w = w_0$$

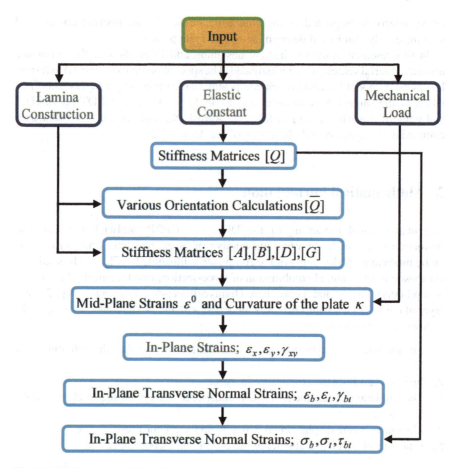

Fig. 1 FOSDT approach followed in the present study

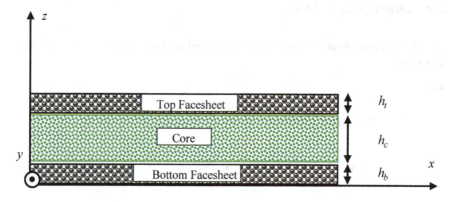

Fig. 2 Dimensions and geometry of the sandwich composite

where u, v, and w are in-plane displacements in the x, y, and z directions, respectively; u_0, v_0, and w_0 are mid-plane displacements in the x, y, and z directions, respectively; ψ_x, ψ_y are rotations of the cross sections.

The displacement field continuity at the top and bottom facesheet interface generates the displacement as shown in Fig. 2. The top and bottom here were refereed as the upper and bottom layer of the sandwich composite.

The subscript and superscript t, b, and c represent the top facesheet, bottom facesheet, and the core.

Top Facesheet

The displacement field adopted for the facesheet from Fig. 2 was defined as

$$u_t = u_c\left(\frac{h_c}{2}\right) + \left(Z - \frac{h_c}{2}\right)\psi_x^t, \quad v_t = v_c\left(\frac{h_c}{2}\right) + \left(Z - \frac{h_c}{2}\right)\psi_y^t$$

$$w_t = w_0$$

$$u_c\left(\frac{h_c}{2}\right) = u_0 + \left(\frac{h_c}{2}\right)\psi_x^c + \left(\frac{h_c^2}{4}\right)\eta_x^c + \left(\frac{h_c^3}{8}\right)\zeta_x^c$$

$$v_c\left(\frac{h_c}{2}\right) = v_0 + \left(\frac{h_c}{2}\right)\psi_y^c + \left(\frac{h_c^2}{4}\right)\eta_y^c + \left(\frac{h_c^3}{8}\right)\zeta_y^c$$

$$u_t = u_0 + \left(\frac{h_c}{2}\right)\psi_x^c + \left(\frac{h_c^2}{4}\right)\eta_x^c + \left(\frac{h_c^3}{8}\right)\zeta_x^c + \left(Z - \frac{h_c}{2}\right)\psi_x^t$$

$$v_t = v_0 + \left(\frac{h_c}{2}\right)\psi_y^c + \left(\frac{h_c^2}{4}\right)\eta_y^c + \left(\frac{h_c^3}{8}\right)\zeta_y^c + \left(Z - \frac{h_c}{2}\right)\psi_y^t$$

$$w_t = w_0$$

where ψ_y^t, ψ_x^t are top facesheet rotations in x and y directions; η_x^c, η_y^c, $\zeta_x^c \zeta_y^c$ are the higher-order relationships.

Bottom Facesheet

From Fig. 2, the displacement field for the bottom facesheet was described as

$$u_b = u_c\left(-\frac{h_c}{2}\right) + \left(Z + \frac{h_c}{2}\right)\psi_x^b, \quad v_b = v_c\left(-\frac{h_c}{2}\right) + \left(Z + \frac{h_c}{2}\right)\psi_y^b$$

$$w_b = w_0$$

$$u_c\left(-\frac{h_c}{2}\right) = u_0 - \left(\frac{h_c}{2}\right)\psi_x^c + \left(\frac{h_c^2}{4}\right)\eta_x^c - \left(\frac{h_c^3}{8}\right)\zeta_x^c$$

$$v_c\left(-\frac{h_c}{2}\right) = v_0 - \left(\frac{h_c}{2}\right)\psi_y^c + \left(\frac{h_c^2}{4}\right)\eta_y^c - \left(\frac{h_c^3}{8}\right)\zeta_y^c$$

$$u_b = u_0 - \left(\frac{h_c}{2}\right)\psi_x^c + \left(\frac{h_c^2}{4}\right)\eta_x^c - \left(\frac{h_c^3}{8}\right)\zeta_x^c + \left(Z + \frac{h_c}{2}\right)\psi_x^b$$

$$v_b = v_0 - \left(\frac{h_c}{2}\right)\psi_y^c + \left(\frac{h_c^2}{4}\right)\eta_y^c - \left(\frac{h_c^3}{8}\right)\zeta_y^c + \left(Z + \frac{h_c}{2}\right)\psi_y^b$$

$$w_b = w_0$$

2.2 Strain–Displacement

The strain and displacement associated with the sandwich composite are expressed as

Core

$$\varepsilon_x = \varepsilon_x^0 + z\kappa_x,\ \varepsilon_y = \varepsilon_y^0 + z\kappa_x,\ \gamma_{xy} = \gamma_{xy}^0 + z\kappa_{xy}$$

$$\varepsilon_x^0 = \frac{\partial u_0}{\partial x},\ \varepsilon_y^0 = \frac{\partial v_0}{\partial y},\ \gamma_{xy}^0 = \frac{\partial u_0}{\partial y} + \frac{\partial v_0}{\partial x}$$

$$\kappa_x = \frac{\partial \psi_x}{\partial x},\ \kappa_y = \frac{\partial \psi_y}{\partial y},\ \kappa_{xy} = \frac{\partial \psi_x}{\partial y} + \frac{\partial \psi_y}{\partial x},$$

$$\gamma_{xz} = \psi_x + \frac{\partial w}{\partial x},\ \gamma_{yz} = \psi_y + \frac{\partial w}{\partial y}$$

where $\varepsilon_x, \varepsilon_y, \gamma_{xy}$ are the normal strains, $\varepsilon_x^0, \varepsilon_y^0, \gamma_{xy}^0$ are the mid-core strains; γ_{xz}, γ_{yz} are the out-plane shear strains; $\kappa_x, \kappa_y, \kappa_{xy}$ are the mid-core curvatures.

Top Facesheet

$$\varepsilon_{xx}^t = \frac{\partial u_t}{\partial x} = \frac{\partial u_0}{\partial x} + \left(\frac{h_c}{2}\right)\frac{\partial \psi_x^c}{\partial x} + \left(\frac{h_c^2}{4}\right)\frac{\partial \eta_x^c}{\partial x} + \left(\frac{h_c^3}{8}\right)\frac{\partial \zeta_x^c}{\partial x} + \left(Z - \frac{h_c}{2}\right)\frac{\partial \psi_x^t}{\partial x}$$

$$\varepsilon_{yy}^t = \frac{\partial v_t}{\partial y} = \frac{\partial v_0}{\partial y} + \left(\frac{h_c}{2}\right)\frac{\partial \psi_y^c}{\partial x} + \left(\frac{h_c^2}{4}\right)\frac{\partial \eta_y^c}{\partial x} + \left(\frac{h_c^3}{8}\right)\frac{\partial \zeta_y^c}{\partial x} + \left(Z - \frac{h_c}{2}\right)\frac{\partial \psi_y^t}{\partial y}$$

$$\gamma_{xy}^t = \frac{\partial u_t}{\partial y} + \frac{\partial v_t}{\partial x} = \left(\frac{\partial u_0}{\partial y} + \frac{\partial v_0}{\partial x}\right) + \frac{h_c}{2}\left(\frac{\partial \psi_x^c}{\partial y} + \frac{\partial \psi_y^c}{\partial x}\right)$$
$$+ \frac{h_c^2}{4}\left(\frac{\partial \eta_x^c}{\partial y} + \frac{\partial \eta_y^c}{\partial x}\right) + \frac{h_c^3}{8}\left(\frac{\partial \zeta_x^c}{\partial y} + \frac{\partial \zeta_y^c}{\partial x}\right) + \left(Z - \frac{h_c}{2}\right)\left(\frac{\partial \psi_x^t}{\partial y} + \frac{\partial \psi_y^t}{\partial x}\right)$$

$$\gamma_{yz}^t = \frac{\partial w_0}{\partial y} + \psi_y^t,\ \gamma_{xz}^t = \frac{\partial w_0}{\partial y} + \psi_y^t$$

Bottom Facesheet

$$\varepsilon_{xx}^b = \frac{\partial u_b}{\partial x} = \frac{\partial u_0}{\partial x} - \left(\frac{h_c}{2}\right)\frac{\partial \psi_x^c}{\partial x} + \left(\frac{h_c^2}{4}\right)\frac{\partial \eta_x^c}{\partial x} - \left(\frac{h_c^3}{8}\right)\frac{\partial \zeta_x^c}{\partial x} + \left(Z + \frac{h_c}{2}\right)\frac{\partial \psi_x^b}{\partial x}$$

$$\varepsilon_{yy}^b = \frac{\partial v_b}{\partial y} = \frac{\partial v_0}{\partial y} - \left(\frac{h_c}{2}\right)\frac{\partial \psi_y^c}{\partial x} + \left(\frac{h_c^2}{4}\right)\frac{\partial \eta_y^c}{\partial x} - \left(\frac{h_c^3}{8}\right)\frac{\partial \zeta_y^c}{\partial x} + \left(Z + \frac{h_c}{2}\right)\frac{\partial \psi_y^b}{\partial y}$$

$$\gamma_{xy}^b = \frac{\partial u_b}{\partial y} + \frac{\partial v_b}{\partial x} = \left(\frac{\partial u_0}{\partial y} + \frac{\partial v_0}{\partial x}\right) - \frac{h_c}{2}\left(\frac{\partial \psi_x^c}{\partial y} + \frac{\partial \psi_y^c}{\partial x}\right) + \frac{h_c^2}{4}\left(\frac{\partial \eta_x^c}{\partial y} + \frac{\partial \eta_y^c}{\partial x}\right)$$

$$- \frac{h_c^3}{8}\left(\frac{\partial \zeta_x^c}{\partial y} + \frac{\partial \zeta_y^c}{\partial x}\right) + \left(Z + \frac{h_c}{2}\right)\left(\frac{\partial \psi_x^b}{\partial y} + \frac{\partial \psi_y^b}{\partial x}\right)$$

$$\gamma_{yz}^b = \frac{\partial w_0}{\partial y} + \psi_y^b, \quad \gamma_{xz}^b = \frac{\partial w_0}{\partial y} + \psi_y^b$$

2.3 Constitutive Relations for the Core and Facesheet

For the laminated sandwich composite, the stress–strain relation for the core and facesheet was defined as below

Core

By integrating the stresses over the element thickness, the resultant forces and moments for the sandwich composite core were obtained. The core was supposed as an isotropic and homogenous material with the stress–strain relation as

$$\sigma_c = \begin{Bmatrix} \sigma_{zc} \\ \sigma_{yzc} \\ \sigma_{xzc} \end{Bmatrix} = \begin{bmatrix} E_c & 0 & 0 \\ 0 & G_c & 0 \\ 0 & 0 & G_c \end{bmatrix} \begin{Bmatrix} \varepsilon_{zc} \\ \gamma_{yzc} \\ \gamma_{xzc} \end{Bmatrix}$$

The constitutive equation for the core was defined as

$$\begin{bmatrix} N \\ M \end{bmatrix} = \begin{bmatrix} [A] & [B] & [D] & [E] \\ [B] & [D] & [E] & [F] \end{bmatrix} \begin{Bmatrix} \varepsilon^{(0)} \\ \chi^{(1)} \end{Bmatrix}$$

where σ is stress; E and G are the principal modulus and shear modulus; N, Q, and M are the force, stiffness, and moment, respectively; matrices $[A]$, $[B]$, and $[D]$ are defined as extensional stiffness, coupling stiffness, and bending stiffness, respectively.

Facesheet

Both top and bottom facesheets were considered as a laminate, and the stress–strain relation of the kth layer was defined as below ($i = 1, 2$; lower face, upper face).

$$\begin{bmatrix} \sigma_x(i) \\ \sigma_y(i) \\ \sigma_z(i) \end{bmatrix}_k = \begin{bmatrix} \overline{Q}_{11} & \overline{Q}_{12} & \overline{Q}_{16} \\ \overline{Q}_{12} & \overline{Q}_{22} & \overline{Q}_{26} \\ \overline{Q}_{16} & \overline{Q}_{26} & \overline{Q}_{66} \end{bmatrix}_k \begin{bmatrix} \varepsilon_x(i) \\ \varepsilon_y(i) \\ \gamma_{xy}(i) \end{bmatrix}$$

$$\begin{Bmatrix} \sigma_{xx}^f \\ \sigma_{yy}^f \\ \tau_{yz}^f \\ \tau_{xz}^f \\ \sigma_{xy}^f \end{Bmatrix} = \begin{bmatrix} \bar{Q}_{11} & \bar{Q}_{12} & 0 & 0 & \bar{Q}_{16} \\ \bar{Q}_{21} & \bar{Q}_{22} & 0 & 0 & \bar{Q}_{26} \\ 0 & 0 & \bar{Q}_{44} & \bar{Q}_{45} & 0 \\ 0 & 0 & \bar{Q}_{54} & \bar{Q}_{55} & 0 \\ \bar{Q}_{61} & \bar{Q}_{62} & 0 & 0 & \bar{Q}_{66} \end{bmatrix}^{(k)} \begin{Bmatrix} \varepsilon_{xx}^f \\ \varepsilon_{yy}^f \\ \gamma_{yz}^f \\ \gamma_{xz}^f \\ \gamma_{xy}^f \end{Bmatrix}$$

$$f = \text{top/bottom}$$

Thus, the consecutive equations for the facesheets were expressed as

$$\begin{bmatrix} N_x \\ N_y \\ N_{xy} \end{bmatrix} = \begin{bmatrix} A_{11} & A_{12} & A_{16} \\ A_{12} & A_{22} & A_{26} \\ A_{16} & A_{26} & A_{66} \end{bmatrix} \begin{bmatrix} \varepsilon_x^0 \\ \varepsilon_y^0 \\ \gamma_{xy}^0 \end{bmatrix} + \begin{bmatrix} B_{11} & B_{12} & B_{16} \\ B_{12} & B_{22} & B_{26} \\ B_{16} & B_{26} & B_{66} \end{bmatrix} \begin{bmatrix} \kappa_x \\ \kappa_y \\ \kappa_{xy} \end{bmatrix}$$

$$\begin{bmatrix} M_x \\ M_y \\ M_{xy} \end{bmatrix} = \begin{bmatrix} C_{11} & C_{12} & C_{16} \\ C_{12} & C_{22} & C_{26} \\ C_{16} & C_{26} & C_{66} \end{bmatrix} \begin{bmatrix} \varepsilon_x^0 \\ \varepsilon_y^0 \\ \gamma_{xy}^0 \end{bmatrix} + \begin{bmatrix} D_{11} & D_{12} & D_{16} \\ D_{12} & D_{22} & D_{26} \\ D_{16} & D_{26} & D_{66} \end{bmatrix} \begin{bmatrix} \kappa_x \\ \kappa_y \\ \kappa_{xy} \end{bmatrix}$$

where

$$A_{ij} = A_{ij}(1) + A_{ij}(2)$$
$$B_{ij} = \left(\frac{h_c + h_2}{2}\right) A_{ij}(2) - \left(\frac{h_c + h_1}{2}\right) A_{ij}(1)$$
$$C_{ij} = C_{ij}(1) + C_{ij}(2)$$
$$D_{ij} = \left(\frac{h_c + h_2}{2}\right) C_{ij}(2) - \left(\frac{h_c + h_1}{2}\right) C_{ij}(1)$$

3 Load Prediction of Aarush X2

Figure 3 illustrates the subsections of the Aarush X2 which were mechanically critical and studied using the mathematical formulation developed in the present study. Both section A and section B were constructed by the facesheet of unidirectional CFRP and the core of Divinycell. Iterative calculations were performed to obtain the optimised number of layers and the fibre orientation of the facesheet required for the wing and fuselage.

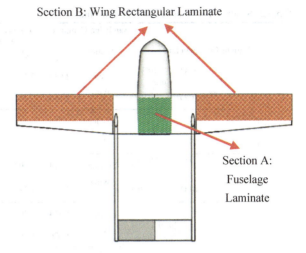

Fig. 3 Top-view of the Aarush X2, with marked zones studied using FOSDT

3.1 Fuselage Belly

The fuselage belly was assumed as a rectangular sandwich composite. The laminate was considered to be clamped at both the ends, and a unidirectional load was induced at $x = 1350$ mm from one extreme of the plate. The variation in the amount of facesheet layers and the fibre orientations were studied for the fuselage. The material property used to construct the fuselage is shown in Table 1. The boundary conditions of the laminate in the clamped arrangement were ensured by the following constraints

$$w(0, y) = \psi_x(0, y) = \psi_y(0, y) = u_0(0, y) = v_0(0, y) = 0$$

The study for the sandwich composite was performed for three, five, and seven layers, including the core ply. The different configurations, Table 2, were analysed to design the sandwich composite for the fuselage of the Aarush X2. From Table 2,

Table 1 Material properties of the CFRP and Divinycell

CFRP						
D	E_1	E_2	v_{12}	v_{23}	G_{12}	G_{23}
kg/m^3	GPa				GPa	
1580	180	9.45	0.433	0.465	6.67	3.23
Divinycell						
D	E_1	E_2	v_{12}	v_{23}	G_{12}	G_{23}
48	45	90	0.3	0.3	28	18

where D = density, E_1 = longitudinal modulus, E_2 = transverse modulus, v_{12} = in-plane Poisson ratio, v_{23} = transverse Poisson ratio, G_{12} = in-plane shear modulus, G_{23} = transverse shear modulus

Table 2 Sandwich composite configuration studied

Various Configurations				Sandwich Composite Configurations						
								FOSDT Results		
				Section	ε_x^0	ε_y^0	γ_{xy}^0	κ_x^0	κ_y^0	κ_{xy}^0
Number of Layers	3 Layers	TFO Core	[0°]	Fuselage	2.089	2.178	2.453	−1.647	−1.517	−1.803
		BFO	[0°]	Wing	2.156	2.296	2.598	−1.805	−1.665	−1.938
	5 Layers	TFO Core	[0°/0°]	Fuselage	1.514	1.590	1.784	−1.247	−1.337	−1.525
		BFO	[0°/0°]	Wing	1.672	1.711	1.893	−1.311	−1.488	−1.777
	7 Layers	TFO Core	[0°/0°/0°]	Fuselage	1.014	1.107	1.298	−0.924	−0.997	−1.129
		BFO	[0°/0°/0°]	Wing	1.155	1.244	1.338	−0.998	−1.127	−1.257
Facesheet Fibre Orientations	5 Layers	[0°/0°]		Fuselage	1.514	1.590	1.784	−1.247	−1.337	−1.525
				Wing	1.606	1.687	1.854	−1.381	−1.401	−1.679
		[0°/45°]		Fuselage	1.912	1.967	2.185	−1.480	−1.514	−1.668
				Wing	2.039	2.109	2.294	−1.699	−1.711	−1.790
		[0°/90°]		Fuselage	1.117	1.197	1.275	−1.044	−1.094	−1.158
				Wing	1.247	1.251	1.322	−1.108	−1.159	−1.213
		[45°/90°]		Fuselage	1.785	1.818	1.993	−1.380	−1.431	−1.539
				Wing	1.888	1.953	2.177	−1.483	−1.599	−1.689

TFO Top facesheet orientation, *core* Divinycell, *BFO* bottom facesheet orientation

the mid-plane strains and mid-plane curvatures of the five-layer and seven-layer sandwich composite present the possibility applicable to the fuselage; however, the seven-layer lay-up was unessentially strong and heavier than the requisite. Various fibre orientations were further studied, Table 2, and the facesheet [0°/45°] orientation presented the least mid-plane strains and curvatures, hence selected. The optimised sandwich composite, [0°/45°/Divinycell/0°/45°], stress performance is presented in Fig. 4, which denotes σ_{xx} to be around 1905 MPa.

3.2 Wing Skin

The Aarush X2 wing is shown in Fig. 3, where the wing length is 2700 mm. In order to reduce the modelling complexities, (i) the study was performed on a rectangular wing with a 500-mm width, mean of root, and tip chord, (ii) the influence due to spar and ribs was not incorporated, and (iii) single side of the wing is accounted (half of the wingspan, 1350 mm). The mechanical properties of the materials used to construct the sandwich composite are presented in Table 1. The boundary condition

Material Study and Fabrication of the Next-Generation … 201

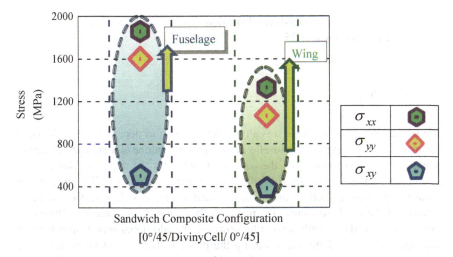

Fig. 4 Various stresses calculated for the fuselage and wing using FOSDT

of the wing was ensured for the simply supported at one end, which represented mathematically as

$$w(0, y) = M_x(0, y) = N_x(0, y) = N_{xy}(0, y) = \psi_y(0, y) = 0$$

The sandwich composite investigation was performed for various layers and further optimised for the facesheet's ply fibre orientations. Table 2 demonstrates the sandwich composite with plies 0°/45°/Divinycell/0°/45° suitable for fabricating the wing structure. The stress report of the designed sandwich composite, [0°/90°/Core/0°/90°], in the Aarush X2 wing is represented in Fig. 4, σ_{xx} approximately 1342 MPa.

4 Fabrication

VAWLM was implemented to fabricate the designed sandwich composite for the Aarush X2. The different sections of the Aarush X2, fuselage, and wing were fabricated individually initially and hereafter combined to produce the overall system. In order to maintain a clean and defect-free composite, the fabrication of the Divinycell CFRP composite was performed in a closed room chamber. The curing time of the fabrication noted was 6 h, and the resin to hardener combination, AY 105 and HY 991, was used in 10 to 1 part. To maintain a steady fabrication platform, the assemblies were fixed using jigs and fixtures. The temperature of the room was also maintained constant to avoid any sudden escalation or degradation in the curing process, which can cause inter-layer distortions and unevenness. The temperature of the room was maintained at 25 °C.

4.1 Fuselage

The fabrication of the fuselage was a two-step process, as depicted in Fig. 5, where the upper and lower part of the fuselage was fabricated alone at first. The fabrication of each part was performed on the separate moulds for the upper and lower fuselage layers, which was initiated by the surface treatment process. The cleaner and release agent was applied on the mould surface to obtain a clear surface and easy removal of the fuselage composite, respectively. The carbon fabric and Divinycell were cut in accordance with the mould; however, the additional surface area for both was trimmed to avoid shrinkage during the process. The carbon fabric layer was placed at the bottom of the mould, followed by the Divinycell and carbon fibre layer again, which were wetted with epoxy matrix before placing inside the mould. The wetted layers were followed by a separator, breather cloth, and vacuum sheet which were sealed using butyl tape. The entire assembly was attached to a 2.5 HP vacuum pump, which drops the pressure inside the sealed vacuum bag under 1 atm during the fabrication process. VAWLM maintains uniform pressure distribution across the material plies and further acts as a pseudo-forced clamp until the epoxy matrix sets the composite in the required position. The lap pressure exerted during the fabrication process further helped to remove inter-layer air pockets and manufactures a defect-free composite. Subsequently, the fabricated upper and lower layers were linked to each using VAWLM again. In the end, essential mechanical process like drilling was performed to integrate avionics, landing gear, and engine.

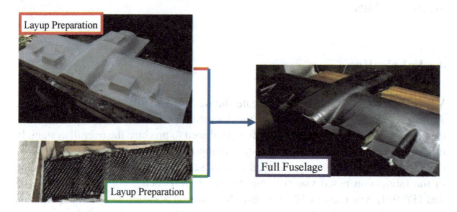

Fig. 5 Fabrication of the Aarush X2 fuselage

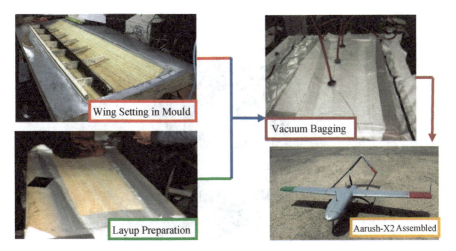

Fig. 6 Fabrication of the wing and assembly in the Aarush X2

4.2 Wing

The wings were fabricated in a similar manner as performed for the fuselage, which contains the upper and lower skin fabrication at first. The entire process is depicted in Fig. 6. The surface treatment on the upper and lower skin moulds was executed. The stacking of carbon fabric, Divinycell, separator, breather, and vacuum sheet was implemented in the same way as for the fuselage. The wetted layers on the wing mould were connected to the vacuum pump. The upper and lower skins were joined to each other by employing VAWLM in the end. The process repeated analogously for the left side of the wing, since right-side wing was fabricated before.

5 Flight Testing

The flight testing of the Aarush X2, Fig. 7, was performed in the Indian flying conditions for more than 30 h with rigorous take-offs and landings. The final inspection after each flight demonstrated UAV in healthy condition and with further prospects of flight missions. The flight time of 2.8 h was recorded on the DLE-61 engine with the fuel tank capacity of 1.8 litres. The final project flight testing was completed in Bhiwani, Haryana, and the system was inspected by Lockheed Martin officials. The UAV was found fit to be functional for the urban aerial flying conditions with the strong and shock resilient structural body.

Fig. 7 Final flight demonstration and inspection of the Aarush X2 by Lockheed Martin

6 Conclusion

This study investigated mechanical properties of the sandwich composite using FOSDT. The research demonstrated carbon fibre Divinycell sandwich composite structurally suitable and hence is used to manufacture Aarush X2. The work displayed the fabrication methodology employed to fabricate fuselage and wing of the UAV using VAWLM. The study highlights the following observations

1. The designed composite with the ply stacking of two upper carbon fibre facesheet layers, core Divinycell layer, and two bottom carbon fibre facesheet layers demonstrated stress of 1905 MPa and 1342 MPa for the fuselage and wing laminate, respectively. The orientation for each facesheet layer was $0°$ initially.
2. The different orientations, $[0°/0°]$; $[0°/90°]$; $[0°/45°]$; $[45°/90°]$, were investigated, and the results presented $[0°/45°]$ orientation to obtain maximum strength when compared to the rest.
3. The VAWLM produced sandwich composite free from the manufacturing defects with the smooth surface finish.
4. The 30-h Aarush X2 flight indicated no signs of structural failure.
5. The research displayed the selected composite designing strategy through FOSDT appropriate, complying with the structural loads estimated.

Acknowledgements The authors wish to acknowledge the technical and financial support by Lockheed Martin Corporation as a part of the Lockheed Martin Next-Generation Urban Unmanned Aerial Vehicle Project.

References

1. Aydogdu M (2009) A new shear deformation theory for laminated composite plates. Compos Struct 94–101
2. Kirchhoff G (1850) Über das Gleichgewicht und die Bewegung einer elastischen Scheibe. J für die reine und angewandte Mathematik 51–88
3. Love AEH (2013) A treatise on the mathematical theory of elasticity. Cambridge University Press, Cambridge
4. Reddy JN (1984) Energy and variational methods in applied mechanics. Wiley, NY
5. Whitney JM, Leissa AW (1969) Analysis of heterogeneous anisotropic plates. ASME J Appl Mech 261–266
6. Reissner E (1981) A note on bending of plates including the effects of transverse shearing and normal strains. Z Angew Math Phys 764–767
7. Mindlin RD (1951) Influence of rotary inertia and shear on flexural motions of isotropic elastic plates. ASME J Appl Mech 31–38
8. Wu Z, Chen R, Chen W (2005) Refined laminated plate element based on global local higher order shear deformation theory. Compos Struct 135–152
9. Sandhu RS (1969) Parametric study of tsai's strength criteria for filamentary composites. TR-68-168, Air Force Flight Dynamic Laboratory, Wright-Patterson AFB
10. Brandmaier HE (1970) Optimum filament orientation criteria. J Compos Mater 422–425
11. Dagur R, Singh D, Bhateja S, Rastogi V (2019) Mechanical and material designing of lightweight high endurance multirotor system. Elsevier Publications, materials today: proceedings

Intelligent Transport System: Classification of Traffic Signs Using Deep Neural Networks in Real Time

Anukriti Kumar, Tanmay Singh, and Dinesh Kumar Vishwakarma

Abstract Traffic control has been one of the most common and irritating problems since the time automobiles have hit the roads. Problems like traffic congestion have led to a significant time burden around the world, and one significant solution to these problems can be the proper implementation of the intelligent transport system (ITS). It involves the integration of various tools like smart sensors, artificial intelligence, position technologies and mobile data services to manage traffic flow, reduce congestion and enhance driver's ability to avoid accidents and reduce incidents or adverse weather. Road and traffic sign recognition is an emerging field of research in ITS. Classification problem of traffic signs needs to be solved as it is a major step in our journey toward building semiautonomous/autonomous driving systems. Traditionally, Mobileye had developed its first commercially deployed traffic sign recognition system with Continental AG for the BMW-7 series vehicles, but this technology has not been used much. The purpose of this work focusses on implementing an approach to solve the problem of traffic sign classification by developing a convolutional neural network (CNN) classifier using the GTSRB — German Traffic Sign Recognition Benchmark dataset. Rather than using hand-crafted features, our model addresses the concern of exploding huge parameters and data method augmentations. Our model achieved an accuracy of around 97.6% which is comparable to various state-of-the-art architectures.

Keywords Multi-class classification · Convolutional neural network · OpenCV

A. Kumar (✉) · T. Singh · D. K. Vishwakarma
Department of Information Technology, Delhi Technological University, New Delhi, India
e-mail: anu1999kriti@gmail.com

T. Singh
e-mail: tanmay_bt2k16@dtu.ac.in

D. K. Vishwakarma
e-mail: dinesh@dtu.ac.in

© The Author(s), under exclusive license to Springer Nature Singapore Pte Ltd. 2021
R. M. Singari et al. (eds.), *Advances in Manufacturing and Industrial Engineering*,
Lecture Notes in Mechanical Engineering,
https://doi.org/10.1007/978-981-15-8542-5_18

1 Introduction

With the rapid increase in population, people are aiming to look for various alternatives to lead a comfortable and easy life. Self-driving car technology is one such development toward this goal and is one of the newest inventions in the transportation system. Almost every day, new advancements in the field of driverless car technologies are taking place. However, self-driving cars are not yet legal on most of the roads. Although some companies have got permission for testing this technology, running a self-driving car is still illegal in almost all countries. According to DOT which is a US Department of Transportation and NHTSA, around 10,000 people lost their lives in 2019 due to motor vehicle traffic accidents. It also estimated that around 94% of the serious crashes are due to human error only including drunk or distracted driving cases. One of the biggest advantages of autonomous systems as these cars is that they remove such risk factors. However, there are still various challenges as they are vulnerable to mechanical issues that can cause crashes. They must know how to identify traffic signs, other vehicles, branches and other various countless objects in the vehicle's path. Based on this identification, the system must make certain decisions to avoid fatal risks and accidents by taking instantaneous actions like slowing down of the vehicle or control acceleration.

Traffic sign detection and recognition is one of the most important fields in the ITS. Based on the visual impact of traffic signs, self-driving cars can act accordingly and thus automatic recognition can avoid accidents and dangers (Fig. 1).

For this problem, our paper proposes a convolutional neural network-based architecture widely for providing high performance in image-based detection tasks. The dataset used for solving this problem is German Traffic Sign Recognition Benchmark (GTSRB) which is a multi-class traffic sign image classification dataset having around 50,000 images of various noise levels. There are several reasons for preferring this model over other state-of-the-art techniques already available. Through dataset analysis, it was observed that it consists of various challenges for which if other techniques or statistical approaches to denoising are applied, it can be computationally very expensive and hence, it is highly unsuitable for real-time applications. On the other hand, neural network-based detection and classification of the noise are computationally effective as well as achieve high performance as far as accuracy and efficiency are concerned. The paper is organized as follows: Sect. 2 explains the attempts done for a similar task. Section 3 presents the methodology used in this

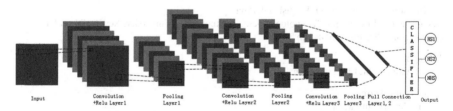

Fig. 1 Convolutional neural network (CNN)

paper for the detection as well as classification of traffic signs. Section 4 describes the evaluation metrics used for this task and the results obtained by our methodology. Section 5 refers to the conclusion and discussion of possible extensions of this research.

2 Related Work

A lot of works has already been done in detection and classification of traffic signs for future autonomous vehicle technology. Various convolutional network-based approaches have been used for this task; some of them are described here. In this paper by Garg [1], You Only Look Once (YOLOv2), single-shot detector (SSD) and faster region CNN (faster RCNN) deep learning architectures along with pretrained CNN models were compared for traffic sign detection and classification task. Various pretrained CNN models already trained on ImageNet Dataset were used; YOLOv2 combined with Coco CNN model, SSD combined with inception V2 and faster RCNN combined with ResNet pretrained CNN models were analyzed on the GTSRB dataset. The evaluation parameter used was mean average precision (mAp) and frames per second (FPS). On comparison, it was found out that YOLO is more accurate and faster than SSD and faster RCNN.

Another paper by Wang Canyong [2] proposed a novel and challenging approach of extending SSD algorithm for the traffic sign detection and identification algorithm. During the preprocessing phase, the images were normalized and fed to the VGG-16 front-end framework of the SSD algorithm. The proposed model is composed of five stacked convolution layers, three fully connected layers and a softmax layer. Using a learning rate of 0.001 and batch size of 50 and 20 for train and validation set, respectively, an accuracy rate of 96% was achieved after 20,000 iterations.

Changzhen et al. [3] proposed deep CNNs that were based on Chinese traffic sign detection algorithm using faster RCNN's region proposal network. There are seven categories of traffic signs in China and the dataset consisting of images from the Internet and roadside scenes from China. The data was augmented by motion blur and applied several levels of brightness on those images. Three different models were trained, namely VGG16, VGG_CNN_M_1024 and ZF. The ZF model had the highest detection efficiency with 60 ms as the average detection time. The model was tested on 33 video sequences captured using a mobile phone and onboard camera. The detection rate of the proposed algorithm was in real time with an efficiency of around 99%.

In another research [4], Xuehong Mao proposed a clustering algorithm based on CNN that was used to separate categories into k different subsets or families. After this, hierarchical CNN was used to train k + 1 classification CNNs, out of which one was for family classification and other k CNNs, corresponding to each family that although achieved 99.67% accuracy, this model was still computationally very expensive. Another research [5] held by Rongqiang Qian proposed an effective feature for the classification Task by using maxpooling positions. They showed how

MPP is a better feature through various experiments which indicated that MPPs demonstrate the desirable characteristics of large intraclass variance and small interclass variance in general but did not improve accuracy further.

Another research team [6] proposed a CNN-ELM model, which integrated the feature learning capacity of CNNs with extreme learning machine (ELM) because of their amazing generalization performance. In this model, firstly CNN was used for learning features and these features were then fed into the fully connected layers that were replaced by ELM for classification. The proposed model trained on GTSRB dataset achieved an accuracy of 99.4% but could not surpass the results of the state-of-the-art algorithms.

In this paper, Cireşan [7] developed a model by combining 25 different CNNs having three convolutional layers and two fully connected layers that could learn more than 88 million parameters. Although it achieved an accuracy of 99.46%, one of the biggest disadvantages of this model was that it used image augmentation due to which a reliable classification accuracy cannot be ensured for unknown data in general.

In this paper, Cireşan et al. [8] proposed a nine-layer CNN along with seven hidden layers consisting of an input layer, three convolutional layers, three maxpooling layers followed by two fully connected layers. In the preprocessing of the data, they cropped the images to equal size. Three different contrast normalization techniques were used in order to reduce high contrast variation in pictures. A grayscale representation of the original images was also produced, and the model was trained on 8 different datasets comprising original as well as sets resulting from three different contrast normalizations of color and grayscale images. Before every epoch in the training phase, images were translated, rotated and scaled based on a uniform distribution over a specified range. A recognition rate of 98.73% was achieved using CNN, and a combination of MLP and CNN achieved a 99.15% recognition rate. Both the models misclassified the 'no vehicle' traffic sign.

3 Research Methodology

In this section, we aim to discuss in detail our approach to the proposed CNN model. Section 3.1 refers to the dataset description of GTSRB and its associated statistics. Section 3.2 highlights the challenges faced with analyzing the dataset. Section 3.3 deals with the preprocessing phase overcoming the challenges mentioned previously. Section 3.4 discusses the architecture used and the hyperparameters involved in it.

Table 1 Data statistics

	No. of images	Percentage of the dataset
Training set	34,799	67.12
Validation set	12,630	24.38
Test set	4410	8.5

3.1 Dataset Description

The dataset used for training is generated at the International Joint Conference on Neural Networks (IJCNN) in 2011 for the German Traffic Sign Recognition Benchmark challenge inviting researchers to participate even without some specific domain challenge. This image dataset consists of around 43 classes representing unique traffic sign images. Training set has around 34,799 images (around 67.12%), test set has 12,630 images (around 24.38%), and validation set has 4410 images (8.5%) (Table 1).

3.2 Data Preprocessing

3.2.1 Challenges Faced

Low Image Contrast

It can occur due to several factors such as a limited range of sensor sensitivity, bad sensor transmission function and so on. This can be detected by plotting brightness histograms with the values varying from black to white on horizontal axis and number of pixels (absolute or normalized) on the vertical axis. Low image contrast will be observed if either this brightness range given is not fully used or the brightness values are concentrated around certain areas only.

Imbalanced Data

As observed from Fig. 2, the data is highly imbalanced as there exists a disproportionate ratio of images in each unique class of traffic sign. Some classes seem to have a lesser number of images than the other which causes class bias as some classes then remain underrepresented. However, there are several approaches to resolve this issue including resampling techniques (oversampling the minority class or undersampling the majority class), generating synthetic samples, changing the performance metric or the algorithm.

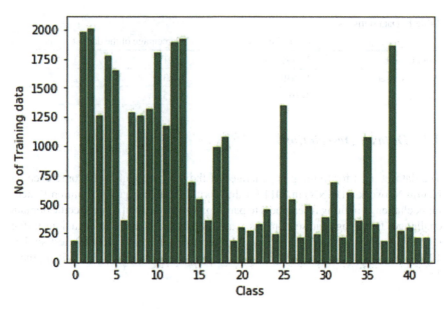

Fig. 2 Class distribution of the training set

3.2.2 Preprocessing Phase

It aims at solving the mentioned challenges obtained from the dataset by applying various techniques.

Data Augmentation

Data augmentation refers to accepting some training images in the form of batches, applying various random transformations to each image present in the batch including random rotation, changes in scale, translations, shearing, horizontal or vertical flips, replacing the original batch with the newly transformed one and finally training the CNN on this new dataset. This is done in order to recognize the target object more effectively as it increases the generalizability of our classifier. Although their appearances might change a bit, still their class labels will remain the same.

OpenCV is used for this task which is a library developed by Intel aimed at real-time computer vision. It is used for performing various image processing operations such as rotation, transformation, translation and soon (Figs. 3, 4, 5 and 6).

1. **Rotation**: Images are rotated slightly at around 10 degrees only. More rotation might cause incorrect recognition.
2. **Translation**: It will move every point in an image by some constant distance in a particular direction. It can also be considered as shifting the origin of the entire

Intelligent Transport System: Classification of Traffic ... 213

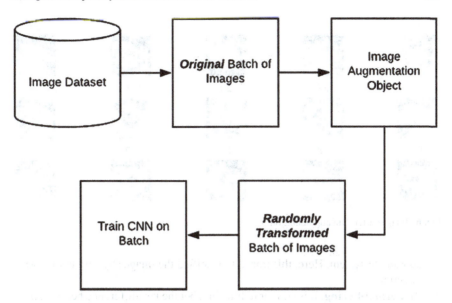

Fig. 3 Data augmentation

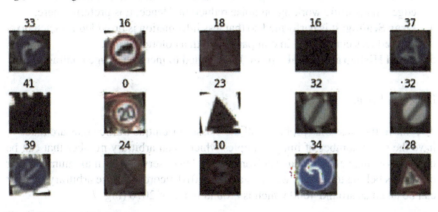

Fig. 4 Rotation of images

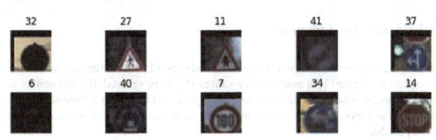

Fig. 5 Translation of image downward

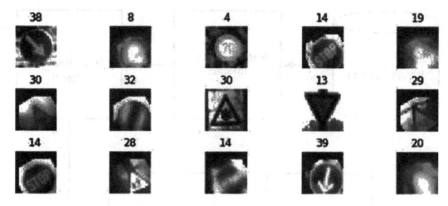

Fig. 6 Images after data augmentation

coordinate system. Here, this translation shifted the image slightly in downward direction.
3. **Bilateral Filtering**: It is similar to blurring, but the key difference between them is that blurring smoothens edges whereas a bilateral filter can keep the image's edges sharp while working on noise reduction. Hence, it is preferred here.
4. **Gray Scaling**: It is performed so that less information is provided for each pixel that reduces complexity in comparison with a colored image.
5. **Local Histogram Equalization**: It is applied to increase image contrast.

Lass Bias Fixing

To remove the class bias problem, all the classes or unique traffic signs are made to have the same number of image samples which is an arbitrary number that can be obtained on analyzing the dataset from Fig. 2. On observation, a maximum number of records belong to class 2, which is around 2010 records. So, the arbitrary number can be taken as around 4000 which is around twice of 2010 (Fig. 7).

3.3 Activation Function

Activation function is an important part of neural networks as they determine whether information received by a neuron is relevant or should be ignored. It is the nonlinear transformation which is done over the input signal, and its output is then sent to the next layer as input. They are crucial as without them, backpropagation process is not even possible.

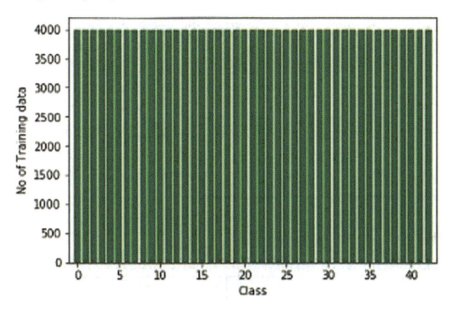

Fig. 7 Class distribution after fixing class bias issue

3.3.1 ReLU

One of the most commonly used activation functions is ReLU that is rectified linear unit. It is defined as:

$$\text{ReL} = \max(0, x)$$

One of the biggest advantages of ReLU is that it is nonlinear which makes backpropagation of errors possible and we can have multiple neuron layers activated by ReLU. Also, at a time it activates only a few neurons making the network more efficient as well as easy for computation (Fig. 8).

3.3.2 Softmax

Softmax function is another activation function that we mainly use in handling classification problems. It is applied over the final layer of the network which tells how much it is confident in its prediction. This is mainly done by performing two calculations, first exponentiating values received at each node and then normalizing this value by summing up these exponentiated values. The vector returned by the softmax function yields probability scores for each class label since they are easier for interpretation. It is represented by (Fig. 9):

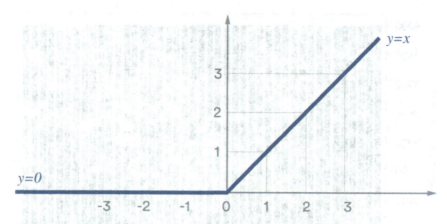

Fig. 8 ReLU activation function

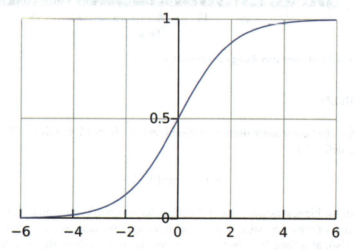

Fig. 9 Softmax activation function

$$\text{softmax}(x_i) \frac{e^{x_i}}{\sum_j e^{x_j}}$$

3.4 Model Architecture

The CNN architectures are used mostly in image processing applications as it involves processing just like our human brain. They are preferred over feed-forward neural networks as they are capable of capturing the temporal as well as special dependencies as well. In our model, we have built the deep learning model for classifying unlabeled traffic signs using CNN model architecture comprising 4 convolution layers and maxpooling layers. The kernel size is chosen as (3, 3) for these convolutional layers. First, the convolution layer takes an image as input for processing with its shape as (32, 32, 1) as the channels have been preprocessed into grayscale images.

In order to reduce the training time and overfitting, maxpooling layers are then added. Then, two fully connected layers are added which require a one-dimensional vector as input for which flattening is done. In the output layer, we have used the softmax activation function as it is a multi-class classification problem. The model architecture is shown in Fig. 10. This model is run with 700 epochs with GPU for faster processing (Table 2).

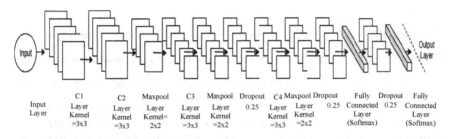

Fig. 10 Model architecture

Table 2 CNN parameters

Layer no	Layer type	Hyperparameters	Kernel size
0	Input	32 × 32 × 1	
1	Convolution (ReLU)	32 neurons	3 × 3
2	Convolution (ReLU)	64 neurons	3 × 3
3	Maxpooling		2 × 2
4	Convolution (ReLU)	64 neurons	3 × 3
5	Maxpooling		2 × 2
6	Dropout	Dropout rate = 0.25	
7	Convolution (ReLU)	128 neurons	3 × 3
8	Maxpooling		2 × 2
9	Dropout	Dropout rate = 0.5	
10	Flatten	128 neurons	
11	Fully connected (softmax)		
12	Dropout	Dropout rate = 0.5	
13	Fully connected (softmax)		

4 Evaluation and Results

Accuracy was chosen as the evaluation metric in the German Traffic Sign Recognition Benchmark challenge. Our model was tested on the validation data, and performance results were analyzed with the help of confusion matrix which in simpler terms can be described as a table depicting its confusion while making predictions, also summarizing the performance of any model. With the help of this confusion matrix obtained, accuracy can be obtained as:

$$\text{Accurancy} = \frac{TP + TN}{TP + TN + FP + FN}$$

where

TP True positive that is the observation is positive and prediction is also positive,

FN False negative that is the observation is positive but the prediction is negative,

TN True negative that is the observation is negative and prediction is also negative,

FP False positive that is the observation is negative but prediction is positive.

Using the proposed model, we have been able to reach a very high accuracy rate of around 97.6%. We also observed that our model starts saturating after 10 epochs. The number of epochs can also be reduced to 10 for decreasing the computation cost.

5 Conclusion and Future Work

In this paper, we developed a CNN architecture for the classification of unique traffic signs for self-driving car technology. We used OpenCV for image augmentation techniques for improving the model performance, and it is also suitable for real-time applications since it involves low computation at every point. For future work, we aim to identify the best architecture along with the best hyperparameters and train our proposed model on a larger dataset. We can try some other preprocessing techniques to improve the model's accuracy. We can make it a more generalistic system by first using a CNN to localize the traffic signs in realistic scenes and another one to classify them. We can also try some different architectures such as AlexNet or VGGNet and compare their performances.

References

1. Garg P, Chowdhury DR, More VN (2019) Traffic sign recognition and classification using YOLOv2, Faster RCNN and SSD. In: 10th international conference on computing, communication and networking technologies (ICCCNT) 2019

2. Canyong W (2018) Research and application of traffic sign detection and recognition based on deep learning. In: International conference on robots & intelligent system (ICRIS) 2018
3. Changzhen X, Cong W, Weixin M, Yanmei S (2017) A traffic sign detection algorithm based on deep convolutional neural network. In: IEEE international conference on signal and image processing (ICSIP) 2016
4. Mao X, Hijazi S, Casas R, Kaul P, Kumar R, Rowen C (2016) Hierarchical CNN for traffic sign recognition. In: Intelligent vehicles symposium (IV), 2016 IEEE. IEEE pp 130–135
5. Qian R, Yue Y, Coenen F, Zhang B (2016) Traffic sign recognition with convolutional neural network based on max pooling positions. In: 2016 12th international conference on natural computation, fuzzy systems and knowledge discovery (ICNC-FSKD). IEEE, pp 578–582
6. Zeng Y, Xu X, FangY, Zhao K (2015) Traffic sign recognition using extreme learning classifier with deep convolutional features with deep convolutional features. In: The 2015 international conference on intelligence science and big data engineering (IScIDE 2015) vol 9242. Suzhou, China, pp 272–280
7. Cireşan D, Meier U, Masci J, Schmidhuber J (2012) Multi-column deep neural network for traffic sign classification. In: Neural networks, Elsevier
8. Ciresan D, Meier U, Masci J, Schmidhuber J (2011) A committee of neural networks for traffic sign classification. In: Dalle Molle Institute for artificial intelligence
9. Kamal U, Das S, Abrar A, Hasan MK (2017) Traffic-sign detection and classification under challenging conditions: a deep neural network based approach. In: IEEE video and image processing cup 2017
10. Peng E, Chen F, Song X (2017) Traffic sign detection with convolutional neural networks. In: International conference on cognitive systems and signal processing
11. Zeng Y, Xu X, Fang Y, Zhao K (2015) Traffic sign recognition using deep convolutional networks and extreme learning machine. In: Intelligence science and big data engineering. Image and video data engineering (IScIDE). Springer, pp 272–280. https://doi.org/10.1007/978-3-319-23989-7_28
12. Aghdam HH, Heravi EJ, Puig D (2016) A practical and highly optimized convolutional neural network for classifying traffic signs in real-time. Int J Comput Vis
13. Zaklouta F, Stanciulescu B (2012) Real-time traffic-sign recognition using tree classifiers. IEEE Trans Intell Transp Syst 13(4):1507–1514. https://doi.org/10.1109/TITS.2012.2225618
14. Aghdam HH, Heravi EJ, Puig D (2015) A unified framework for coarse-to-fine recognition of traffic signs using bayesian network and visual attributes. In: 10th international conference on computer vision theory and applications (VISAPP), pp 87–96. https://doi.org/10.5220/0005303500870096
15. Stallkamp J, Schlipsing M, Salmen J, Igel C (2011) The German traffic sign recognition benchmark: a multi-class classification com-petition. In: International joint conference on neural networks
16. Fleyeh H, Davami E (2011) Eigen-based traffic sign recognition. IET Intel Transport Syst 5(3):190. https://doi.org/10.1049/iet-its.2010.0159

Fabrication of Aluminium 6082–B₄C–Aloe Vera Metal Matrix Composite with Ultrasonic Machine Using Mechanical Stirrer

Manish Kumar Chaudhary, Ashutosh Pathak, Rishabh Goyal, Ramakant Rana, and Vipin Kumar Sharma

Abstract The given research deals with the study of metal matrix composite in which aluminium 6082 series was selected as the matrix and boron carbide (B₄C) along with aloe vera powder were selected as reinforcement with different weight percentages. The use of aloe vera powder helps in making a green metal matrix composite as it is organic material and biodegradable product. For the fabrication of MMC with 4% or 8% of B₄C and 8% or 4% aloe vera powder, a furnace with ultrasonicator and mechanical stirrer was used. The casting was carried at 900 °C to melt the aluminium for about an hour along with the reinforcement, and then the stirrer was rotated to mix the reinforcement properly. The spectroscopy test was carried to get the different contents of the B₄C and aloe vera powder. The reinforcements were distributed homogenously, found by the optical micrographs.

Keywords Aluminium 6082 · Boron carbide · Aloe vera powder · Casting · Optical micrographs

1 Introduction

Aluminium alloy 6082 belongs to family of 6000 series in wrought aluminium–silicon–magnesium. The most popular alloy in the series is 6082, and it is formed by extrusion and rolling process. The properties like low-density and high-hardness Al6082 are greatly preferred, and there are contents of magnesium which helps to decrease the wettability property of other material like aloe vera. These materials are used in bulletproof vests, tank armour, engine sabotage powders and other industries where high hardness is required. It is often called black diamond. Aloe vera powder is made from the aloe vera plants which is organic in nature, and there uses are vastly

---

M. K. Chaudhary (✉) · A. Pathak · R. Goyal · R. Rana · V. K. Sharma
Department of MAE, Maharaja Agrasen Institute of Technology, Rohini-22, New Delhi 110086, India
e-mail: manishk0782@gmail.com

varied. In composites, it helps to decrease the wear resistance of material. It is easily available and cheap in price.

The composites are made by combining the materials which have different properties of chemical as well as physical. The composites made are far better than the base metal for almost all the cases. The composites can be formed by using different types of reinforcement materials depending upon the characteristic required. In recent days, composites are acquired quite a good place in the market due to its extraordinary properties and easily manufacturing process. The composite in which more than two or more reinforcements are used, it is termed as hybrid composite.

Raj et al. [1] have investigated on aluminium grade 6061 being reinforced with boron carbide (B_4C) with composition of 5–20% in steps of 5% volume percentage of B_4C for which stir casting technique was used. The microstructural study showed the uniform distribution of reinforcement which was carried using scanning electron microscope (SEM). Universal test machine (UTM) with 100 kN load cell was employed for tensile testing of MMC. The result from study gives the data that the hardness, yield strength and tensile strength were improved. Thangarasu et al. [2] have carried a study on the aluminium Al6082 grade which was used to make a composite using the titanium carbide (TiC). The reinforcement was used in 0, 6, 12, 18 and 24% with respect to volume fraction. The fabrication process was carried using friction stir processing (FSP) technique. For the microstructural view, optical microscope and SEM were employed. The hardness test along with tensile test was carried along with sliding wear behaviour. The study shows that the stiffness, wear resistance, etc., were enhanced by TiC. Deshpande et al. [3] have conducted the study of aluminium composite (AA7075) powder having circular shape with normal molecule size of 35 m which was chosen as lattice material. The support material utilized was pitch-based processed uncoated carbon filaments (10 μm measurement and 200-μm length). Hot squeezing was implemented for formation of powder. X-ray diffraction (XRD) technique or method was used for identifying the phase of crystals formed. Metal-coated carbon fibres were reinforced as it made the wettability property enhanced. Scanning electron microscope with the help of EDS facility was made used to check out the surface roughness. Hardness test of the composites was performed on Brinell hardness test machine. As the content of carbon fibre volume % was increased, the density of overall composites was increased. It is further reported by Authors that, at low volume, it showed quite good amount of hardness and maximum for approximate volume 20% of carbon fibre. Mohanty et al. [4] have used aluminium 7075 of cylindrical form as the matrix metal. Boron carbide of 75 μm size was used as the reinforcement in the research. They prepared a hybrid metal matrix composites by using B_4C along with coconut shell fly ash. Boron carbide was used in 0, 3, 6, 9, 12 and 15 wt.%. The process of fabrication was carried on stir casting. Stir casting being a simple and precise casting technique was implemented. Result shows that there was subsequent increment in the hardness and tensile strength for 12 wt.% of B_4C. Gireesh et al. [5], in the following research, have used aluminium 2024 as matrix and the boron carbide and graphite as the composites. For fabrication, stir casting process was involved. In this paper, the composition of boron carbide used was 5 and 10% by weight. The result from the following research shows that hardness and strength of aluminium were increased. Vardhan et al. [6], in their research, have employed aluminium A 356 as the matrix metal and the aloe vera powder as reinforcement to form a green metal matrix composite. The aloe vera has higher wettability property and contains all essential contents like magnesium (1.22%), calcium (3.58%), sodium (3.66%), %), potassium (4.06%), iron (0.1%), phosphorous (0.02%), zinc (0.02%) and copper (0.06%). The process of fabrication utilized was stir casting which is a very simple and good method of casting. The result from the paper shows that the hardness was enhanced, along with decrease in wear rate. Sankar et al. [7] investigated the microstructural and mechanical behavioural properties of copper metal matrix composite in B_4C, and crushed seashell particles were used as the reinforcement (fabrication carried using powder metallurgy). In powder form, copper is widely used in structural applications. Copper also possesses excellent electrical and thermal conductivity, and ductility, along with which it provides resistance to corrosion. B_4C comes in top three hardest known materials that also possesses excellent toughness and wear resistance. Seashells are readily available along coastal areas. By use, the following powder hardness was increased up to 5 times the native material. The use of seashell significantly improves the properties of wear resistance too. Mozammil et al. [8] have inspected the mechanical behaviour of aluminium in which copper (Cu) and titanium boride (TiB_2) have 3, 6, 9 and 12% wt.% of reinforcement. The fabrication method used was stir casting, and for testing hardness Vickers hardness test machine was used. Electron probe microscopic analysis (EPMA) was employed for checking microstructure of MMC. Thermal behaviour of MMC was also checked using differential scanning calorimetry (DSC). The result of the study shows that hardness and tensile properties were enhanced by addition of reinforcement in base metal. Sharma et al. [9] have used fly ash which is the waste left after the burning of the wood particles, and it pollutes the environment. Fly ash is one of the greatest environmental problems. In this experiment, the fly ash collected was used to enhance the property of the aluminium to certain amount. This fly ash has been reinforced with the aluminium melt. For fabricating the fly ash with aluminium melt, stir casting process has been used. For fabrication, other techniques were also used that is hot rolling, vacuum hot pressing, ball milling. The composition of fly ash used was 2, 4, 6% of weight in fixed proportion. To measure the wear and frictional force, a pin on disc set-up was used. Surface analysis of the material was performed using scanning electron microscope (SEM). As weight percentage of fly ash was increased, the wear was decreased.

From the literature review, it was observed that very few attempts have been made to fabricate a hybrid composite using B4C and aloe vera as reinforcements. So in this work, hybrid composite of B_4C and aloe vera was fabricated with the help of a stir casting technique.

2 Material

Aluminium grade of 6082 with rectangular shape has dimensions breadth 8.6 cm, width 6.5 cm and length 15 cm from Hitesh Enterprises metal shop. Due to lightweight, durability, good machinability and strength, aluminium was chosen as base material. Composition of Al6082 was determined by spectroscope testing. Table 1 gives detail of the composition of the Al6082. The reinforcements taken were boron carbide having average size of 280 microns and aloe vera powder with uniform grain sizes. Due to lightweight, durability, good machinability and high strength, aluminium was chosen as base material.

2.1 Fabrication Method

Aluminium blocks were cut into small pieces having dimensions breadth 8.6 cm, width 6.5 cm and length 7 cm, and composition for reinforcement was decided on the weight percentage. Two samples for casting were prepared. One sample contained 1.5 kg aluminium, 8% B_4C and 4% aloe vera powder, and other sample contained 1.5 kg aluminium, 4% B_4C and 8% aloe vera powder. Reinforcement was wrapped into aluminium foil, and balls of 30 g were prepared. The varying weight percentages of B_4C and aloe vera powder are, i.e. 4%, 8% and 8%, 4%, respectively, shown in Table 2.

Mechanical stirrer with ultrasonic machine as shown in Fig. 1 was inducted for the fabrication of hybrid aluminium composites. Mechanical stirrer used was made up of composite which has melting point of 3000 °C and having three blades, and length of 65 cm was utilized. Cylindrical-shaped die made up of cast iron of capacity 3 kg was selected. Cylindrical-shaped die was selected for easy machining of casting, and crucible of capacity 3 kg was introduced.

Aluminium blocks which were cut were used for fabrication, and reinforcement balls were also placed into crucible. Mechanical stirrer was mounted on casting set-up, and connection was done with control panel. Cover plate was placed on

Table 1 Chemical composition of Al6082 aluminium alloy

Element	Si	Fe	Mg	Mn	Ti	Cr	Other	Al
%age	1.0333	1.086	0.746	0.6019	0.1339	0.0894	0.1331	96.251

Table 2 Different weight percentages of B_4C and aloe vera in Al6082 alloy

Sample no	B_4C (wt.%)	Aloe vera (wt.%)	Total (wt.%)
1	8	4	12
2	4	8	12

Fabrication of Aluminium 6082–B$_4$C–Aloe Vera …

Fig. 1 Ultrasonic machine with mechanical stirrer

crucible. Connection was done, and maximum temperature limit set 900 °C. Connection switched on due to which temperature of filament started increasing gradually and preheating of the metal and reinforcement ball was done to remove the moisture. When temperature was reached at 900 °C, time was noted down and from that time metal was allowed to be kept at temperature 900 °C for melting. After one hour, metal started melting. When metal was melted completely, stirrer was brought down with the help of switch and introduced into liquid metal. Liquid melt was stirred with constant rpm for reinforcement mixing. Melt was stirred 15 min. Melt was stirred at 100 rpm for first 10 min, and at 150 rpm for last 5 min. When all reinforcement was mixed completely, stirrer was stopped and lifted up. After that, lever was pulled and liquid mixture was allowed to be poured into the die. After 10 min, die was separated and casting was taken out. Same procedure was repeated for second sample.

2.2 Machining of Casting

Casting obtained had dimension of 13 cm in length and diameter of 7.8 cm. Turning operation was performed on casting with depth of cut of 1.5 mm, and porous part of casting was removed from casting with parting operation. Facing operation was also performed on casting to make the face surface smooth. After performing above-mentioned operation, final dimension of casting was obtained. Final dimension of casting having length 9 cm and diameter 7.5 cm was cut into circular discs having diameter of 7.5 cm and thickness 5.6 mm. Six such discs were cut from each cylindrical casting, and those circular discs were proceeded for further testing.

2.3 Microstructure

The study of microstructure constituent of the composites is done through optical micrograph equipment. The mixing of reinforcement in the composites can be verified through the optical microscope. For the optical micrograph, various processes are taken under consideration. First step involves cutting of the samples into small pieces so that it can be mounted on the Bakelite with some resin. Second step involves the grinding of the sample being mounted with different grit papers, varying in nature to make the scratches remove and wash them with soapy water. In the third step, polishing of sample has been done with polishing machine to remove the scratches left in second step. Fourth step involves etching of the polished surface. This was done to remove the thin layers which were produced during the grinding and polishing work. Final step involves the taking of image with optical micrograph with different zooming lens [10–16] (Fig. 2).

3 Result and Discussion

Figure 3 shows the micrographs obtained from optical microscope which shows that the mixing of the reinforcement was proper and in uniform manner. Figure 3a, b shows the reinforcement B4C and aloe vera used 8% and 4% by wt. %, respectively, with different zooming lens of 100X and 200X. Similarly, Fig. 3c, d shows the reinforcement B_4C and aloe vera used 4% and 8% by wt.%, respectively.

Fig. 2 Polished surface of samples

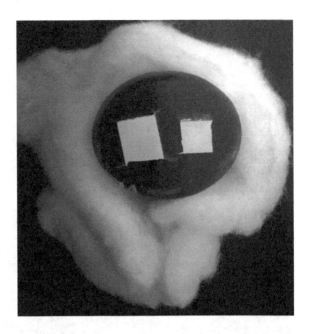

The fabrication of green hybrid metal matrix composites based on aluminium 6082 with ultrasonic machine using mechanical stirrer has been done. The test from micrograph shows the fair enough distribution of the reinforcement in the composite prepared.

4 Conclusion

In this study, a hybrid aluminium composite was fabricated using the ultrasonic-assisted stir casting technique. Two aluminium composites with 4 wt% B_4C, 8 wt% aloe vera and 8 wt% B_4C, and 4 wt% aloe vera were prepared.

To inspect the distribution of the reinforcement particles, optical microscopic images at 100X and 200X magnifications were obtained. The optical microscopic images present the equal distribution of the reinforcement in the aluminium matrix.

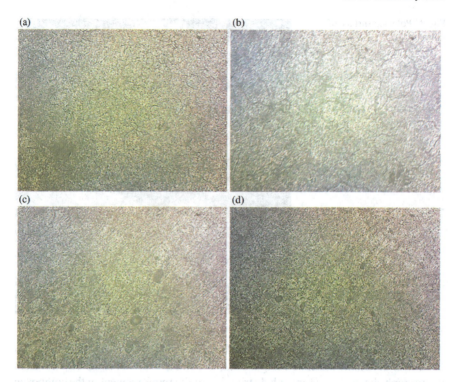

Fig. 3 Optical micrograph of Al6082 reinforced with: **a** 8% B$_4$C and 4% aloe vera with 100X zooming lens; **b** 200X zooming lens; **c** 4% B$_4$C and 8% aloe vera with 100X zooming lens; and **d** 200X zooming lens

References

1. Raj R, Thakur DG (2019) Effect of particle size and volume fraction on the strengthening mechanisms of boron carbide reinforced aluminium metal matrix composites. Proc Inst Mech Eng Part C: J Mech Eng Sci 233(4):1345–1356
2. Thangarasu A, Murugan N, Dinaharan I, Vijay SJ (2015) Synthesis and characterization of titanium carbide particulate reinforced AA6082 aluminium alloy composites via friction stir processing. Arch Civil Mech Eng 15(2):324–334
3. Deshpande M, Gondil MR, Waikar R, Mahata TS (2016) Processing of Carbon fibre reinforced Aluminium (7075) metal matrix composite. In: International conference on renewable energy and materials for sustainability
4. Mohanty RM, Balasubramanian K, Seshadri SK (2008) Boron carbide-reinforced aluminium 1100 matrix composites: fabrication and properties. Mater Sci Eng a 498(1–2):42–52
5. Hima Gireesh C, Durga Prasad K, Ramji K (2018) Experimental investigation on mechanical properties of an Al6061 hybrid metal matrix composite. J Compos Sci 2(3):49
6. Vardhan TV, Nagaraju U, Gowd GH, Ajay V (2017) Evaluation of properties of LM 25-alumina–boron carbide MMC with different ratios of compositions. Int J Appl Eng Res 12(14):4460–4467
7. Sankar M, Devaneyan P, Pushpanathan D, Myszka D (2018) Microstructural characterization and mechanical behavior of copper matrix composites reinforced by B4C and sea shell powder. J Cast Mater Eng 2:24. https://doi.org/10.7494/jcme.2018.2.1.24

8. Mozammil S, Karloopia J, Verma R, Jha PK (2019) Effect of varying TiB2 reinforcement and its ageing behaviour on tensile and hardness properties of in-situ Al-4.5% Cu-xTiB2 composite. J Alloy Compd 793:454–466
9. Sharma VK, Singh RC, Chaudhary R (2017) Effect of flyash particles with aluminium melt on the wear of aluminium metal matrix composites. Eng Sci Technol Int J 20(4):1318–1323
10. Singh R, Chaudhary R, Sharma V (2019) Fabrication and sliding wear behavior of some lead-free bearing materials. Mater Res Express 6(6):066533
11. Sharma V, Singh R, Chaudhary R (2018) Wear and friction behaviour of aluminium metal composite reinforced with graphite particles. Int J Surf Sci Eng 12(5/6):419
12. Sharma VK, Singh RC, Chaudhary R, Saxena M, Anand M (2019) Effects of flyash addition on the dry sliding tribological behavior of aluminium composites. Mater Res Express 6(8):0865f4
13. Roop L, Singh RC (2018) Experimental comparative study of chrome steel pin with and without chrome plated cast iron disc in situ fully flooded interface lubrication. Surf Topogr Metrol Prop 6:035001
14. Singh RC, Pandey RK, Roop L, Ranganath MS, Maji S (2016) Tribological performance analysis of textured steel surfaces under lubricating conditions. Surf Topogr Metrol Prop 4:034005
15. Sharma VK, Singh RC, Chaudhary R (2017) Effect of flyash particles with aluminium melt on the wear of aluminium metal matrix composites. Eng Sci Technol Internat J 20(4):1318–1323
16. Sharma VK, Singh RC, Chaudhary R (2018) Experimental study of tribological behaviour of casted aluminium-bronze. Mater Today: Proc 5(14):28008–28017

A Fuzzy AHP Approach for Prioritizing Diesel Locomotive Sheds a Case Study in Northern Railways Network

Reetik Kaushik, Yasham Raj Jaiswal, Roopa Singh, Ranganath M. Singari, and Rajiv Chaudhary

Abstract The Indian railway network is one of the largest rail networks that undergo continuous expansion. However, more reliable and safe service in-network is the subject of growing attention in India but to work on these challenges is a cumbersome task as each factor is influencing each other. Studying these vulnerable parameters is a tedious task and will contribute significantly to improving the operations and overall management of the railway network. The purpose of this research study is to identify and optimize different vulnerable factors that are affecting the smooth functioning of the railway network through the identification of hierarchical correlations among parameters and prioritizing performing a specific order for identification and improvement. The major focus is on handling imprecise and vague information with the help of a fuzzy synthesis of information. Based on an extensive review of the literature and expert opinion from the industry, 10 vulnerable parameters were identified and expert elicitation was applied to determine the correlation among factors. Based on these parameters, Diesel locomotive sheds of the northern railways network were prioritized and ranking of these systems was obtained. The research provides reliable information for decision-makers to form active strategies and management policies for a more adequate and reliable system of the network. Also, a more safe and profitable network that will promote sustainable development throughout India. This study can be a guide of the methodology to be implemented to other multiple criteria decision-making problems.

Keywords Fuzzy · AHP · Railways diesel sheds

R. Kaushik (✉) · Y. R. Jaiswal · R. Singh · R. Chaudhary
Departments of Mechanical Engineering, Delhi Technological University, New Delhi 110042, India
e-mail: reetik_bt2k16@dtu.ac.in

R. M. Singari
Departments of Production & Industrial Engineering, Delhi Technological University, New Delhi 110042, India

© The Author(s), under exclusive license to Springer Nature Singapore Pte Ltd. 2021
R. M. Singari et al. (eds.), *Advances in Manufacturing and Industrial Engineering*, Lecture Notes in Mechanical Engineering,
https://doi.org/10.1007/978-981-15-8542-5_20

1 Introduction

Every organization endeavors to enhance its productivity and always in search of tools that on implementation present limited resources and derives the best output. The study aims to deliver a simple model which when employed in India Railways, eight largest employer of the world, helps in smoothening its workflow, enhancing its reliability and availability. Indian Railways is the fourth-largest rail network and runs 20,000 passenger trains and 9200 trains in the freight segment daily. Due to its imperative nature, the economic budget of the fiscal year 2019–20 of the Indian government gave a budgetary allocation of ₹65,837 crore. The data displays the importance of this organization on daily economic, social affairs of the country, and the demand for high reliability [1].

The complex large-scale organization on which relies the functioning of a vast section of country faces disruptions, irregularities, failures. Locomotive failure is one of the most challenging and hindering tasks to the Indian Railways aiming for the smooth running of its trains around the country. Daily at an average, more than five locomotives are failed in between their journey, leading to chaos and congestion in the route and ultimately leading to heavy traffic block. Hence, there is an urgent need to analyze the functioning of the organization to reduce the cases of failures. The analyze will consider not only the technical faults of locomotives but also the functioning failures which are driven by factors such as individual technical capability, organizational structure, network topology, safety investment, education and training, emergency management plan, operating environment, rules and regulations, individual workload and stress [1, 3].

The research study attempt is to organize the complex network of the organization into a structured form, which is explicit and understandable also, which helps us to know the inter-relationship and inter-dependency of various factors. We have used an integrated Fuzzy AHP approach for creating a conceptual model with correlated parameters that are prioritized using this model.

2 Research Methodology

Fuzzy Logic is widely used in a variety of fields. The advantage of using fuzzy logic is that it helps in optimization of problems when limited or uncertain information is available to the user. Most computer systems are based on binary logic, where something is either true or false or an element is a member of a set or not, but the advantage of fuzzy logic approach is that elements can be partially part of a set. Moreover, we can use linguistic variables in addition to numeric values to make the problem more realistic and reflective of real-life situations. Thus a fuzzy control system is a control system based on fuzzy logic that is a mathematical system that analyzes input values in terms of logical variables that take any value between 0 and 1. Fuzzification operations are then carried out which convert a crisp or numeric value to linguistic variables or fuzzy subsets. This step is followed by the application of fuzzy rules between inputs and outputs which are a bunch of IF-THEN statements that specify the output obtained when a particular input condition is satisfied. The final step in the process involves Defuzzification, which gives a crisp output from fuzzy sets. These outputs are then used to help make various judgements and eliminate human error and bias [2].

Analytical hierarchy approach (AHP) is used to resolve many problems involving multi-criteria decision models. Saaty first coined this method that integrated expert opinion and forms basic hierarchy system using scores, so that the problem can be dis-integrated in lower hierarchies and priorities of different problems are evaluated [4, 5]. AHP method includes vagueness as the scores are based on personal judgement of the expert panel so to improve the quality of the method fuzzy-based AHP model is used. Triangular fuzzy numbers are rated by the experts and tabulated which are then used to form pairwise comparison of both criteria and the respective decisions alternatives. These triangular variables are termed as linguistic variables. Procedure for Fuzzy AHP model is as follows:-

Table 1 represent the relationship priority of fuzzy members that illustrates the triangular scale of the relationship among parameters, for example, is the expert chooses "Parameter 1 (F1) to be of strong influence than Parameter 2 (F2)", then on fuzzy scale triangular value [6, 7, 8] is allotted. On the other hand, if the decision-maker states F2 to be strong influence on F1 then the fuzzy value of [1/8, 1/7, 1/6] is chosen.

After scaling through all the expert opinions correlation among the factors can be determined and can be thus represented in decision-maker triangular fuzzy matrix as shown in Eq. 1, where $\widetilde{\lambda_{ij}^k}$ represents the kth expert preference of the criteria ith over the jth that is represented in a triangular fuzzy number. The "~" symbol means the decision of first expert made over parameter one and second and equals to, $\widetilde{\lambda_{12}^1} = [6, 7, 8]$.

$$\tilde{A}^k = \begin{pmatrix} \widetilde{\lambda_{11}^k} & \widetilde{\lambda_{12}^k} & \cdots & \widetilde{\lambda_{1n}^k} \\ \widetilde{\lambda_{21}^k} & \cdots & \cdots & \widetilde{\lambda_{2n}^k} \\ \cdots & \cdots & \cdots & \vdots \\ \widetilde{\lambda_{n1}^k} & \widetilde{\lambda_{n2}^k} & \cdots & \widetilde{\lambda_{nn}^k} \end{pmatrix} \qquad (1)$$

Table 1 Fuzzy influence scale

Scale	Priority	Fuzzy scale
1	Equal influence	(1, 1, 1)
3	Weak influence	(2, 3, 4)
5	Fair influence	(4, 5, 6)
7	Strong influence	(6, 7, 8)
9	Very strong influence	(9, 9, 9)

Next step in the methodology is to calculate fuzzy triangular scale value for all the decision-makers so average of all the expert preference is made and is represented as $\widetilde{\lambda}_{ij}$ as shown in Eq. 2.

$$\widetilde{\lambda}_{ij} = \frac{\sum_{k=1}^{K} \widetilde{\lambda}_{ij}^{k}}{K} \qquad (2)$$

After averaged opinions of experts are obtained, the decision matrix can be updated as Eq. 3.

$$\tilde{A} = \begin{pmatrix} \widetilde{\lambda}_{11} & \cdots & \widetilde{\lambda}_{1n} \\ \vdots & \ddots & \vdots \\ \widetilde{\lambda}_{n1} & \cdots & \widetilde{\lambda}_{nn} \end{pmatrix} \qquad (3)$$

Then geometric mean of fuzzy triangular values are calculated using Eq. 4 and is represented in triangular values as $\tilde{x}_i$.

$$\tilde{x}_i = \left(\prod_{j=1}^{n} \widetilde{\lambda}_{ij} \right)^{1/n}, \quad i = 1, 2, \ldots, n \qquad (4)$$

In the next step, the fuzzy weight of each parameter is calculated by first doing summation of each value of $\tilde{x}_i$ then the inverse of summation vector is found and placed in increasing order and finally the fuzzy weight of each parameter is incorporated by multiplying increasing order of the inverse value with respective $\tilde{x}_i$ values. The steps are shown in Eq. 5.

$$\begin{aligned}\tilde{w}_i &= \tilde{x}_i \oplus (\tilde{x}_1 \oplus \tilde{x}_2 \oplus \cdots \oplus \tilde{x}_n)^{-1} \\ &= (lw_i, mw_i, uw_i) \end{aligned} \qquad (5)$$

The de-fuzzified values are calculated using centre of area method using Eq. 6.

$$M_i = \frac{lw_i + mw_i + uw_i}{3} \qquad (6)$$

Then in the final step, the non-fuzzy numbers are normalized using Eq. 7.

$$N_i = \frac{M_i}{\sum_{i=1}^{n} M_i} \qquad (7)$$

To calculate the scores of each alternative the normalized weight of both the alternatives and criteria are calculated and tabulated. Then each weight of criteria is multiplied with related alternation and summed to produce the total score of experts. Thus, the final decision can be made based on the highest scores received and priority order among the alternatives can be found. The applicability of the above-illustrated method is shown using northern railways network diesel sheds as a real case study example in the next section.

3 Northern Railways Diesel Sheds a Case Study

Northern railways diesel are situated at five prime locations throughout northern India. The main purpose of these diesel sheds is to carry out maintenance of diesel locomotives running in the Indian Railways network. The aim of this research is to apply the methodology of Fuzzy AHP to categorize the diesel sheds of northern railways in priority order based on parameters extracted through a review of the literature and expert opinions. Previously various studies are conducted on diesel shed maintenance systems and using AHP methods but previous research was limited in terms of considering the human error in opinions of different experts [6]. Thus, a fuzzy-based approach is applied in the network to improve the rating order of diesel shed based on a more robust and effective approach. The study incorporates five diesel shed alternatives and 10 criteria based on the decision of various experts of the railways industry. Thus, there is a need to calculate normalized weight of both alternatives and criteria which are analyzed in separate parts.

3.1 Determining the Weights of Alternative

The parameters for the problem were evaluated based on a review of literature and opinions of experts from the industry. According to the survey form filled by the experts' 10 parameters were found and are shown in Tables 2, 3.

According to above-mentioned methodology next step is to calculate geometric mean of the fuzzy triangular scale values. This is done using Eq. 4. One example of F1 criteria is illustrated below and shown as Eq. 8

$$\tilde{x}_i = \left(\prod_{j=1}^{n} \tilde{\lambda}_{ij}\right)^{1/n} = \left[\left(1*6*2*\frac{1}{6}*9*1*\frac{1}{8}*2*6*6\right)^{\frac{1}{10}};\right.$$

$$\left(1*7*3*\frac{1}{5}*9*1*\frac{1}{7}*3*7*7\right)^{\frac{1}{10}};$$

$$\left.\left(1*8*4*\frac{1}{4}*9*1*\frac{1}{6}*4*8*8\right)^{\frac{1}{10}}\right]$$

$$= [1.66, 1.95, 2.23] \tag{8}$$

Table 2 Parameters influencing diesel sheds [1, 7, 8]

Symbol	Criteria	Description
F1	Individual technical capability	Technical capacity of individuals known as skills, knowledge, and experience of an individual. Capable employees can handle situations with relatively fewer mistakes
F2	Network topology	Topology of network is a dire factor as connectivity planning is necessary. For example, if one line needs repairs It should not stop the functioning of other lines or create train delays
F3	Safety investment	Indian Railways invests a substantial amount for safety as any disaster or damage causes huge amount for repairs
F4	Equipment/facility condition	The Facility and Equipment condition is very prominent for optimum output
F5	Station layout	Layout of the Railway station is very crucial as it helps in better service to passengers. It is very important in areas where space is scarce
F6	Emergency management plan	Emergency management plan is necessary for disaster situation to prevent losses both economic and social
F7	Rules and regulations	Rules and regulations are important for smooth functioning of any company
F8	Natural environment	It is evident to avoid natural disaster-prone areas for railway facilities or networks as it can disrupt the working of Railways
F9	Individual workload and stress	Individuals work and stress is directly related to performance of employee
F10	Education and training	Social conditions refer to employee's connectivity with each other and for passenger social conditions of the facility

Repeating the above step final geometric means of fuzzy triangular correlation values for all parameters are calculated and shown in Table 4. Following further steps of method, inversed values are also calculated and tabulated in increasing order (Table 5).

In the next step, the fuzzy weights of criteria's are calculated using Eq. 5, sample calculation on F1 criteria is shown in Eq. 9 (Table 6).

$$\tilde{w}_i = [(1.66 * 0.08); (1.95 * 0.09); (2.23 * 0.11)] \tag{9}$$

In the final step of the methodology, the relative non-fuzzy values of each parameter (M_i) are calculated using Eq. 6, and then the normalized fuzzy weights (N_i) are calculated using Eq. 7 for each parameter (Table 7).

Table 3 Averaged pairwise comparison of parameters

Q#	A. Imp	S. Imp	F. Imp	W. Imp	Criteria	Eq. Imp	Criteria	W. Imp	F. Imp	S. Imp	A. Imp
1		O			F1		F2				
2				O	F1		F3				
3	O				F1		F4		O		
4					F1		F5				
5					F1	O	F6				
6					F1		F7			O	
7				O	F1		F8				
8		O			F1		F9				
9		O			F1		F10				
10					F2	O	F3				
11					F2		F4		O		
12				O	F2		F5				
13					F2		F6				O
14			O		F2		F7				
15	O				F2		F8				
16					F2		F9	O			
17					F2		F10				O
18				O	F3		F4				
19					F3	O	F5				
20	O				F3		F6				

(continued)

Table 3 (continued)

Q#	A. Imp	S. Imp	F. Imp	W. Imp	Criteria	Eq. Imp	Criteria	W. Imp	F. Imp	S. Imp	A. Imp
21		O			F3		F7				
22					F3	O	F8				
23				O	F3		F9				
24			O		F3		F10				
25					F4		F5			O	
26					F4		F6				O
27					F4		F7	O			
28				O	F4		F8				
29			O		F4		F9				
30		O			F4		F10	O			
31	O				F5		F6				
32					F5	O	F7				
33					F5		F8	O			
34					F5		F9		O		
35					F5		F10			O	
36					F6	O	F7				
37	O				F6		F8				
38				O	F6		F9				

(continued)

Table 3 (continued)

Q #	A. Imp	S. Imp	F. Imp	W. Imp	Criteria	Eq. Imp	Criteria	W. Imp	F. Imp	S. Imp	A. Imp
39					F6		F10		O		
40		O			F7		F8				
41					F7		F9			O	
42				O	F7		F10				
43					F8	O	F9				
44			O		F8		F10				
45			O		F9		F10				

Table 4 The averaged fuzzy triangular scale matrix

Criteria	F1	F2	F3	F4	F5	F6	F7	F8	F9	F10
F1	[1, 1, 1]	[6, 7, 8]	[2, 3, 4]	[1/6, 1/5, 1/4]	[9, 9, 9]	[1, 1, 1]	[1/8, 1/7, 1/6]	[2, 3, 4]	[6, 7, 8]	[6, 7, 8]
F2	[1/8, 1/7, 1/6]	[1, 1, 1]	[1, 1, 1]	[1/6, 1/5, 1/4]	[2, 3, 4]	[1/9, 1/9, 1/9]	[4, 5, 6]	[9, 9, 9]	[1/4, 1/3, 1/2]	[1/9, 1/9]
F3	[1/4, 1/3, 1/2]	[1, 1, 1]	[1, 1, 1]	[2, 3, 4]	[1, 1, 1]	[9, 9, 9]	[6, 7, 8]	[1, 1, 1]	[2, 3, 4]	[4, 5, 6]
F4	[4, 5, 6]	[4, 5, 6]	[1/4, 1/3, 1/2]	[1, 1, 1]	[1/8, 1/7, 1/6]	[1/9, 1/9, 1/9]	[1/4, 1/3, 1/2]	[2-4]	[4-6]	[6-8]
F5	[1/9, 1/9, 1/9]	[1/4, 1/3, 1/2]	[1, 1, 1]	[6, 7, 8]	[1, 1, 1]	[9, 9, 9]	[1, 1, 1]	[1/4, 1/3, 1/2]	[1/6, 1/5, 1/4]	[1/8, 1/7, 1/6]
F6	[1, 1, 1]	[9, 9, 9]	[1/9, 1/9, 1/9]	[9, 9, 9]	[1/9, 1/9, 1/9]	[1, 1, 1]	[1, 1, 1]	[9, 9, 9]	[2, 3, 4]	[1/6, 1/5, 1/4]
F7	[6, 7, 8]	[1/6, 1/5, 1/4]	[1/8, 1/1/7, 1/6]	[2, 3, 4]	[1, 1, 1]	[1, 1, 1]	[1, 1, 1]	[6, 7, 8]	[1/8, 1/7, 1/6]	[2, 3, 4]
F8	[1/4, 1/3, 1/2]	[1/9, 1/9, 1/9]	[1, 1, 1]	[1/4, 1/3, 1/2]	[2, 3, 4]	[1/9, 1/9, 1/9]	[1/8, 1/7, 1/4]	[1, 1, 1]	[1, 1, 1]	[4-6]
F9	[1/8, 1/7, 1/6]	[2, 3, 4]	[1/4, 1/3, 1/2]	[1/6, 1/5, 1/4]	[4, 5, 6]	[1/4, 1/3, 1/2]	[6, 7, 8]	[1, 1, 1]	[1, 1, 1]	[4, 5, 6]
F10	[1/8, 1/7, 1/6]	[9, 9, 9]	[1/6, 1/5, 1/4]	[1/8, 1/7, 1/6]	[6, 7, 8]	[4, 5, 6]	[1/4, 1/3, 1/2]	[1/6, 1/5, 1/4]	[1/6, 1/5, 1/4]	[1, 1, 1]

A Fuzzy AHP Approach for Prioritizing Diesel ...

Table 5 Geometric means of fuzzy comparison values

Criteria	$\tilde{x}_i$		
F1	1.66	1.95	2.23
F2	0.58	0.66	0.75
F3	1.71	1.98	2.26
F4	0.96	1.17	1.41
F5	0.62	0.68	0.78
F6	1.12	1.18	1.25
F7	0.91	1.06	1.22
F8	0.49	0.56	0.64
F9	0.87	1.05	1.28
F10	0.57	0.67	0.79
Total	9.49	10.96	12.61
Reverse power	0.11	0.09	0.08
Increasing order	0.08	0.09	0.11

Table 6 The relative fuzzy weights of each parameter

Criteria	$\tilde{w}_i$		
F1	0.13	0.18	0.25
F2	0.05	0.06	0.08
F3	0.14	0.18	0.25
F4	0.08	0.10	0.16
F5	0.05	0.06	0.09
F6	0.09	0.11	0.14
F7	0.07	0.10	0.13
F8	0.04	0.05	0.07
F9	0.07	0.09	0.14
F10	0.05	0.06	0.09

3.2 Determining Weights of Alternatives with Respect to Criteria

The same methodology is applied to find normalized fuzzy values for alternatives. Thus, the alternative which is different diesel sheds are first compared with each other, and correlation is formed pairwise for each alternative. Then pairwise comparison is drawn with respect to each criteria that means ten more iterations are performed which will be quite handful to explain each one of them, so only the first criteria are compared with each alternative (Table 8).

Table 7 Averaged and normalized weights

Criteria	M_i	N_i
F1	0.185	0.177
F2	0.063	0.060
F3	0.188	0.180
F4	0.112	0.108
F5	0.066	0.063
F6	0.111	0.106
F7	0.101	0.096
F8	0.053	0.051
F9	0.102	0.097
F10	0.064	0.061

Table 8 Northern railways diesel sheds

Alternatives	Northern railway
T1	Kalka (KLK)
T2	Alambagh Lucknow (LKO)
T3	Ludhiana (LDH)
T4	Tughlakabad (TKD)
T5	Shakurbasti (SSB)

The pairwise comparison of first criteria with the respective diesel sheds alternatives based on which fuzzy triangular matrix is formed based on average values (Table 9).

Using the similar methods mentioned above, the relative fuzzy weights of alternatives based on each parameter are calculated then using those values, the geometric fuzzy means of each alternative are tabulated in Table 10. Finally, the normalized and non-fuzzy averaged values are obtained by the use of the Area of Center method.

Similarly, normalized non-fuzzy relative weights of each diesel sheds are calculated for each parameter. The tabulated values are shown in Table 11.

Table 9 The averaged fuzzy triangular scale matrix

	T1	T2	T3	T4	T5
T1	[1, 1 ,1]	[2, 3, 4]	[1, 1 ,1]	[1/4, 1/3, 1/2]	[1/6, 1/5, 1/4]
T2	[1/4, 1/3, 1/2]	[1, 1 ,1]	[1/8, 1/7, 1/6]	[1/9, 1/9, 1/9]	[2, 3, 4]
T3	[1, 1 ,1]	[6, 7, 8]	[1, 1 ,1]	[4, 7, 6]	[6, 7, 8]
T4	[2, 3, 4]	[9, 9, 9]	[1/6, 1/5, 1/4]	[1, 1 ,1]	[9, 9, 9]
T5	[4, 5, 6]	[1/4, 1/3, 1/2]	[1/8, 1/7, 1/6]	[1/9, 1/9, 1/9]	[1, 1 ,1]

A Fuzzy AHP Approach for Prioritizing Diesel ...

Table 10 Geometric means and fuzzy weights of alternatives with respective to "T1" Criterion

	$\tilde{x}_i$			$\tilde{w}_i$			M_i	N_i
T1	0.61	0.72	0.87	0.08	0.11	0.15	0.112	0.108
T2	0.37	0.44	0.52	0.05	0.07	0.09	0.067	0.065
T3	2.70	3.00	3.29	0.35	0.45	0.56	0.454	0.438
T4	1.93	2.17	2.41	0.25	0.33	0.41	0.329	0.317
T5	0.43	0.48	0.56	0.06	0.07	0.10	0.074	0.072
Total	6.04	6.82	7.64					
Inverse	0.17	0.15	0.13					
increasing order	0.13	0.15	0.17					

Table 11 Normalized non-fuzzy relative weights of each alternative for each criterion

	F1	F2	F3	F4	F5	F6	F7	F8	F9	F10
T1	0.216	0.175	0.150	0.224	0.108	0.176	0.187	0.218	0.237	0.236
T2	0.177	0.132	0.334	0.377	0.065	0.292	0.151	0.086	0.105	0.120
T3	0.106	0.481	0.161	0.146	0.438	0.153	0.129	0.305	0.288	0.338
T4	0.075	0.126	0.241	0.087	0.317	0.109	0.174	0.102	0.276	0.224
T5	0.426	0.085	0.114	0.166	0.072	0.270	0.358	0.289	0.094	0.082

By using Tables 7 and 11, individual scores for each diesel sheds alternative based on each parameter are presented in Table 12.

Table 12 Aggregated results for each alternative according to each criterion

Criteria	Scores of alternatives with respect to related criterion					
	Weights	T1	T2	T3	T4	T5
F1	0.177	0.216	0.177	0.106	0.075	0.426
F2	0.060	0.175	0.132	0.481	0.126	0.085
F3	0.180	0.150	0.334	0.161	0.241	0.114
F4	0.108	0.224	0.377	0.146	0.087	0.166
F5	0.063	0.108	0.065	0.438	0.317	0.072
F6	0.106	0.176	0.292	0.153	0.109	0.270
F7	0.096	0.187	0.151	0.129	0.174	0.358
F8	0.051	0.218	0.086	0.305	0.102	0.289
F9	0.097	0.237	0.105	0.288	0.276	0.094
F10	0.061	0.236	0.120	0.338	0.224	0.082
Total		0.192	0.355	0.381	0.484	0.273

4 Conclusion

The study focused mainly on categorizing the Diesel Locomotive Sheds of northern railways network based on various parameters. These parameters were determined by a review of literature and opinions of experts. Present research gives a clear picture of current situation of various diesel sheds spread across northern India. The initial phase of research was focused on collection of data and finding out critical parameters to carry out research. Collection of expert opinions through surveys and interviews played a crucial role. Based on the use of Fuzzy AHP methodology it was found that the diesel shed T4 "Tughlakabad" is highest priority alternative that means the most preferred diesel shed for maintenance and carry out repairs for the failed locomotive. But the decision of maintaining a locomotive and carrying out repairs are done based on various other factors like distance from the location, location of failed locomotive if it has been towed. The research study provides an insight on how to improve the diesel sheds of northern railways network as different ratings of parameters are done so it can be clearly improved based on these parameters for example when considering safety investment as the parameter diesel shed T5 needs improvement. Thus, the research study can be used to improve the system of working on diesel sheds based on different parameters.

Depending on various parameters, prioritizing Diesel Locomotive Sheds is a very crucial task as most of the investment and improvement for sheds are depending on these parameters. Since most of these parameters are conflicting in nature and an effective, methodology is required to observe and formulate results. Keeping in mind the need for industry Fuzzy AHP theory model is applied in this research study. As the expert decision is dependent on both intangible and tangible criteria which create vagueness in these linguistic variables. Thus fuzzy set models are applied to deal with this issue. Hence Fuzzy AHP methodology gave the priority rating of northern railways diesel shed based on 10 critical parameters (T4 > T3 > T2 > T5 > T1). As a result of the case study, it is observed that diesel shed TKD Tughlakabad outperforms others. Nevertheless, this research study can also be used to target particular diesel sheds based on different improvement parameters.

References

1. Hassan A, Purnomo M, Anugerah A (2019) Fuzzy-analytical-hierarchy process in failure mode and effect analysis (FMEA) to identify process failure in the warehouse of a cement industry. J Eng Des Technol. https://doi.org/10.1108/JEDT-05-2019-0131
2. Seker S, Zavadskas EK (2017) Application of fuzzy dematel method for analyzing occupational risks on construction sites. Sustainability 9:2083
3. Ayhan M (2013) A fuzzy AHP approach for supplier selection problem: a case study in a gear motor company. Int J Manag Value and Supply Chains 4 https://doi.org/10.5121/ijmvsc.2013.4302
4. Zhang J, Chen X, Sun Q (2019) A safety performance assessment framework for the petroleum industry's sustainable development based on FAHP-FCE and human factors. Sustainability

11:3564
5. Aslani R, Feili H, Javanshir H (2014) A hybrid of fuzzy FMEA-AHP to determine factors affecting alternator failure causes. Manage Sci Lett 4(9):1981–1984
6. Attri R, Dev N, Sharma V (2013) Interpretive structural modelling (ISM) approach: an overview. Res J Manage Sci 2:3–8
7. Rade KA, Pharande VA, Saini DR (2017) Interpretive structural modeling (ISM) for recovery of heat energy. Int J Theoret Appl Mecha 12(1):83–92
8. Jena J, Fulzele V, Gupta R, Sherwani F, Shankar R, Sidharth S (2016) A TISM modeling of critical success factors of smartphone manufacturing ecosystem in India. J Adv Manage Res 13(2):203–224

Operation of Big-Data Analytics and Interactive Advertisement for Product/Service Delineation so as to Approach Its Customers

Harshmit Kaur Saluja, Vinod Kumar Yadav, and K. M. Mohapatra

Abstract On the one hand, big-data analytics has brought an uprising in the predictive modeler by enabling the complex data sets getting ordered. On the other hand, the interactive advertisement has changed the complete framework of the advertising sector by making advertisements content structured in such a way that it is customer-central. The paper helps to widen the view to explore the emerging earnest of customization techniques in the advertising sector with interactive enablers. The paper further examines how interactive advertisement and big-data has helped to represent product/service from the view of a customer and also improved the product/service performance. In order of study, exhaustive literature reviews resulting in three hypotheses are developed to take on the above-mentioned concerns.

Keywords Big-data analytics · Interactive advertisement · Customization · Customer-central

1 Introduction

The data is generated exponentially after the two decades of the internet revolution. The data are complex, and therefore it becomes hard to analyze and categorize in the form of data. Datasets whose size is too large and therefore it makes difficult for database software tools to capture, store, manage and analyze. McKinsey Global Institute Manyika 2011.

Data is so voluminous and complex in nature that traditional data processing application software is not able to generate the actual required information. It extracts value from data and puts them to a particular size of data set. The number of methods is evolved through continuous engineering leads to better analytical software which predicts the requirement of data and maintains the data transparency. Managers

---

H. K. Saluja (✉) · V. K. Yadav · K. M. Mohapatra
School of Humanities and Social Sciences, Harcourt Butler Technical University, Kanpur, Uttar Pradesh, India
e-mail: harshmit0405@gmail.com

© The Author(s), under exclusive license to Springer Nature Singapore Pte Ltd. 2021
R. M. Singari et al. (eds.), *Advances in Manufacturing and Industrial Engineering*, Lecture Notes in Mechanical Engineering,
https://doi.org/10.1007/978-981-15-8542-5_21

looking at the outputs of various algorithms available in the market and their decisions are highly dependable on these algorithms.

With the fast changing users demand companies are continuously searching for new modes to communicate with customers. As a new edge Strategy Company focuses on two-way communication which helps in improving their product/services as well as to better understand the customer needs. Interaction with the intention penetrates the idea in a swift manner and gives birth to Interactive communication which means influencing each other. Though advertising is traditionally conceptualized as one-way communication, but the arrival of the Internet has opened newer avenues for advertising to change its way. It started becoming two-way systems from one-way i.e. interactive leading to change its media vehicle [1]. have catalyst interactive advertising as the "paid and unpaid presentation and promotion of products, services, and ideas by an identified sponsor through mediated means involving mutual action between consumers and producers of the product or service".

The basic reason for the need of interactive advertisement is the growing changes in consumer's lifestyle, attitude and preferences especially when they are online as they become demanding in their nature and requires more customized production while doing window-shopping. It is observed that Consumer's purchasing power has increased over a span of time and even the cash liquidity goes high in the market which has evolved with their desires changing with every second and with every new thing that they observe which goes online and/or offline. The need of advertisement has become to understand the consumer's all activity going on online like what ad they see? what ad they skip? what ad they view again? what ad they would not like to see again?, etc. also their mailbox, the pages they prefer viewing all this shows consumer interest and preferences which is the basic parameter of understanding consumer response towards the advertisement.

Literature highlighted that big data plays a great role in advertising research. The first one is by developing fundamental frameworks of advertising with new data sources and refining with existing theories. Second is by improving the delivery of advertisement message by targeting a customized message to the target customer. The third is by sending messages to focus groups for developing new theories [2].

Normal advertisements are the ones that are advertiser-oriented and they lack in connecting customers with them. Their main aim is to sell the product/service and generate profits. Whereas Interactive advertisements are the ones that are consumer-oriented and they try to link the customers from their ad towards their product/service. The main aim is not only for sale but also for the sustainability of their customer for the long term.

2 Literature Review

2.1 Big Data

Literature highlights that data alone is found everywhere but extracting useful information from large data requires analytical techniques. There comes the role of big data as today when we are talking about business, engineering, advertisement, etc. its huge data cannot be analyzed without big data predictable and analytical techniques [3]. Data is found, everywhere and it becomes essential to extract useful information from large data which requires predictable and analytical techniques. For it, big data analytics are required which are nowadays used in bus, engine, etc. for converting the huge data into useful forms. For industrial upgrade, it has proved itself as a measure of high growth and tool for their competencies. For scientific research activities, it helps in re-examining, scientific thinking, and methods. It is a tool also for emerging disciplines and for perceiving and predicting it is proving itself to one of the biggest tools. In ancient times, predictions were done more on institutions rather than using analytical therefore results cannot be considered as authentic. But big data analytics are reliable and also perceives human insights as to the important parameter for the result analysis [4].

It was seen that in older times to predictions were done but at that time it was more based on intuitions, and therefore these were somewhere or other considered to be biased and not so clear indicator of future possibilities. But, Big data decisions are reliable as on one-hand as it is based on evidence rather than intuition, and on the other hand it also perceives human insights as an important parameter for analysis of data which helps to give a reliable and authentic result [5]. The growing importance of big data is not hidden anymore. Big data has brought a revolution in every field it may be business, public administration, and so on [6]. Information, technology, methods, and impact are considered essential characteristics of big data [7]. Information is the main characteristic of big data as if there is no information there is no need for big data. Information urges the need for big data. Without the use of technology, the authenticity of big data results is questionable. It is technology only which has triggered the need for big data among all fields. The method if not applied to raw data does not lead to fair results. The impact is the most important feature of big data. It is the impact only which makes big data what it is today. After the characteristic now let us know its components. Big data has three V's, and they are variety, velocity, and volume where volume implies large data, velocity means with high speed, and variety as the huge sources that are available [8]. Further, the V's were expanded by adding value, verocity, variability, and visualization [9]. Furthervolatile was added to the list of V's. Big data consists of many pros but privacy and security are considered as the cons in big data as the data many times that is personal and maybe the consumer does not want to share gets leaked [10].

2.2 Interactive Advertisement

Advertising if looked back has gone through lots of eras starting from stone-age to print media to radio to television to the internet to the in-app advertisement and today being interactive in advertising has become the need for successful advertisement. Advertising has shifted from anonymity to identity and to promote products, brands, services, etc. by online and/or offline media there came the need for interactive advertising. It has changed from being advertiser-oriented to consumer-oriented. Today, in the advertisement sector consumer is considered as king and their desires are of utmost importance for the sale of any product or service. Consumers are exposed to different kinds of everyday advertising which have made them develop a sensitive understanding of the advertisement, and that is the reason that advertisements are needed to be an interactive way [11, 12]. The main emphasis of the advertisement is that speaks the desire, need, the desire of the customer. Online advertising has become the base for consumer awareness and needs for their purchase or non-purchase of the product or service. A study was conducted to compare the psychological effectiveness of interactive and non-interactive advertising, and it was concluded that the trust of a consumer does not vary whether the advertisement is interactive or non-interactive. Though it was seen that obviously, interactive advertisement makes consumers highly involved personally and after that psychologically effective [13]. The lifestyle of a consumer is the main constituent for presenting an effective advertisement in front of them which makes consumer influence towards the advertisement. Interactive advertisement is dependent according to e-lifestyle of consumer and especially the average hours spent online by consumers has become the main determinant of analyzing its online habits, needs, desires, attitudes, etc. [14]. The medium through which message gets delivered to the consumer is an important tool for analyzing the effectiveness of online advertising [15]. That is why it becomes essential to understand what media vehicle to be used for effective and efficient delivery of the message. Being clear with the message content and getting up to date knowledge by consumer is important to make interactive advertising and also to retain consumers for a lifetime [16]. The need to build up the trust of consumers towards in-app advertisement is the most favorable factor to remain consumers for the long term [17].

2.3 Big-Data Analytics and Interactive Advertisement

Big-data analytics with Interactive Advertisement help to target, personalized, and customized messages for their consumers so as to increase efficiency on one-way and save money on the other by triggering the right people with the right product or service. These analytics help the advertisers, marketers, etc. by understanding customer-relevant details like shopping patterns, interest, habits and past trends. It can be said to serve as a window for cognizing consumer's psychology.

3 Hypothesis Development

With the help of Big Data Analytics, we are able to get the concrete data from a large set of data. Therefore, we are able to know and hence target our customers. Big data analytics through their ability to collect huge data knowing consumers are able to get the But as there is no Interactive Advertisement, we do not know how to send the message to the customer. The knowledge of even what message to be sent is also ambiguous. Therefore, when there is Big Data Analytics but there is no Interactive Advertisement, we are able to know the target customer but we do not know the media vehicle through which we should target them. Therefore, only Big Data Analysis without Interactive Advertisement will only lead to knowing the product/service representation target customer (Fig. 1).

H1 Big Data Analytics helps us to know target customer for our product or service representation.

As there is no Big Data Analysis used therefore, we do not know our target customer from the large data set. The technique of Interactive Advertisement is there so we know how and when we are also aware of the media content and how we can trigger our customers towards it. But without Big Data Analysis we do not know our target customer. Therefore, Interactive Advertisement without the Big Data Analysis will only lead us towards knowing the way of message delivering, message content, etc. but we are unsuccessful as we don't know our target customer (Fig. 2).

H2 Interactive Advertisement helps us to create proper message content, understand media vehicle of delivering our product or service message to customer.

Big Data Analysis when used with Interactive Advertisement helps us to perfectly represent our product or service towards our customers. As we know are target customer on the one hand and on the other hand, we know our advertisement content, we know when, where, and how we shall appeal to our customers (Fig. 3).

Fig. 1 Role of big-data analytics. *Compilation* Own source

Fig. 2 Role of interactive advertisement. *Compilation* Own source

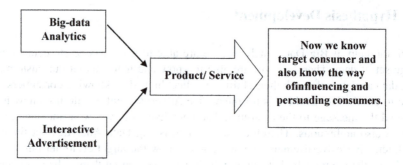

Fig. 3 Role of interactive advertisement with big-data analytics. *Compilation* Own source

H3 Big Data Analytics with Interactive Advertisement helps to represent product or service perfectly towards customers.

4 Research Methodology

To understand the role and linkage of Big data Analytics with Interactive Advertisement, we have asked few questions in terms of yes and no so as to understand that whether respondents are aware of this terminology "Big Data with Interactive advertisement" or not. For this, we have randomly selected a sample of 430 respondents from Kanpur city. Five questions in the terms of **Yes** and **No** are asked and these responses are represented below in terms of bar picturization (Fig. 4).

1. Do you know when you click your response is recorded?

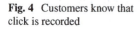
Fig. 4 Customers know that click is recorded

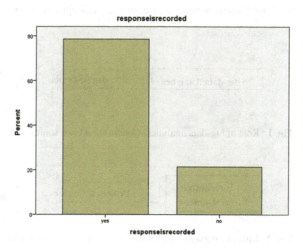

Operation of Big-Data Analytics and Interactive Advertisement … 253

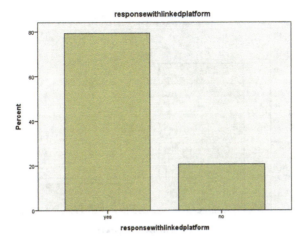

Fig. 5 Customer recorded responses are linked with various platform

From the above table, it is focused that around 78.6% (338) respondents are aware that whatever they click be it is their mobile screen or laptop or desktop, etc. that is recorded in spite of 21.4% (92) respondents are unaware (Fig. 5).

2. Do you feel that the recorded response has any relation in the next time you visit same/linked platform?

From the above table, it is found that 79.3% (341) respondents know that if they have visited any online platform once, the next time whenever they view either same or any linked platform, they will show the products or service they have tried to watch in the prior platform, rather 20.7% (99) respondents feel that there is no such relation.

3. Do you think that advertiser uses to a great extent your previous response in generating new advertisement to you?

The above table represents that around 83.5% (359) respondents feel that advertising agencies, marketers, etc. view their activity online and then similar to their interests, the advertisement is posted to them with 16.5% (71) respondents feel that they do not think the marketing agencies use their online data for showing such advertisements (Fig. 6).

4. Do you think that recorded response is an interface in your life privacy?

The data represents that 65.8%(283) respondents feel that whatever the customer view, search, buy, keep in cart, etc. they think the data recorded of them is a threat in their life privacy whereas 34.2% (147) think that it is not a threat or interface in their life privacy (Fig. 7).

5. When you feel that advertiser is using your previous data searches for convincing you to purchase. Do you find this as Fair-Trade Practice?

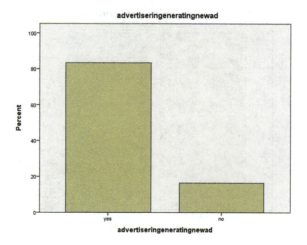

Fig. 6 Advertiser uses big-data

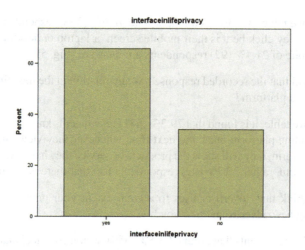

Fig. 7 Recorded response is an interface in life privacy

The above data shows that 64.9% (279) respondents feel that the advertiser uses their data searches for convincing them to purchase is a fair-trade practice whereas 35.1% (151) feels it is not a fair-trade practice (Fig. 8).

5 Conclusion

Big data and marketing analytic will help us to know everything about its customers like personality buying habits, income level, etc. and we can apply predictive analytics

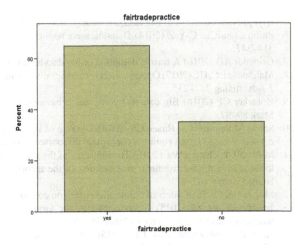

Fig. 8 Interactive advertisement with big data is a fair practice

that will allow us to customize his taste and preferences. The paper tries to show that how big data is used with interactive advertisement for these three hypotheses were taken and then few questions are asked and it was represented through bar which on one hand shows big data analytics alone can target customer and on the other hand interactive advertisement know the way of an appealing customer. At last, big data analytics when used with interactive advertisement helps us to customize through proper techniques, and therefore product or service is properly represented to the customer.

6 Future Scope

The big data with interactive advertisement can be used as predictive analysis for its customers. The hypotheses in this paper are checked through a few basic questions that give a basic explanation and show the importance of big-data analytics with interactive advertisement. So, further study can be done in a more detailed way by analyzing various aspects in detail that will enhance our knowledge.

References

1. Muhammad Awais TS (2012) valuable internet advertising and customer satisfaction cycle. Int J Comput Sci 375–380
2. Sinanc SS (2017) Big data: a review. J Advertising 227–235
3. Mauro AD (2018) A formal definition of big data based on its essential features. J Consum Mark 122–135
4. Jina X, Benjamin W, Waha XY (2015) Significance and challenges of big data research. Elsevier 59–64

5. Brynjolfsson AM (2012) Big data: the management revolution. Harvard Bus Rev 1–9
6. Philip Chen CL, C-Y Z (2014) Data-intensive applications, challenges, techniques. Elsevier 314–347
7. Grimaldi AD (2016) A formal definition of big data based on its essential features, 122–135
8. Malthouse EC, HL (2017) Opportunities for and pitfalls of using big data in advertising research. J Advertising 227–235
9. Hofacker CF (2016) Big data and consumer behaviour: Imminent Opportunities. J Consum Mark 89–97
10. Salem Mohammed S, Busen CS (2014) The role of interactive advertisements in developing consumer-based brand equity: a conceptual discourse. Procedia Soc Behav Sci 98–103
11. Millissa FY Cheung WT (2016) The influence of the propensity trust on mobile users attitude towards in-app advertisement: an extension of the theory of unplanned behaviour. Comput Human Behav 102–111
12. Mishra GK (2016) Effects of online advertising on consumers. J Human Soc Sci 35–41
13. Jinsong Huang SS (2015) Attitude towards the viral Ad: expanding traditional models to interactive advertising. J Interact Mark 36–46
14. Amir Abedini Koshksaray DF (2015) The relationship between e-lifestyle and internet advertising avoidance. Australas Mark J 38–48
15. Anusha G (2016) Effectiveness of online advertising. Int J Research- Granthaalayah, 14–21
16. Bilal Aslam HK (2017) Digital advertising around paid spaces, e-advertising industry's engine: a review and research agenda. Telematics Inform 1650–1662
17. To MF (2017) The influence of the propensity to trust on mobile users' attitudes towards in-app advertisements: an extension of the theory of planned behavior. Computers in human behavior, pp 102–111

Effect of Picosecond Laser Texture Surface on Tribological Properties on High-Chromium Steel Under Non-lubricated Conditions

Sushant Bansal, Ayush Saraf, Ramakant Rana, and Roop Lal

Abstract The objective of this paper is to analyse the significant effects of negative dot patterned laser texture on piston ring material under non-lubricated conditions. The aim is to carry out experiments to study the tribological behavioural dynamics of the working on the interface of the textured silver steel piston ring material and EN-31. The percentage of the area of the pin textured, of the total area of the pin surface available, was 41.50%. With the predefined working parameters for the experimental working, such as the load applied, sliding velocity and the track distance travelled over the course of wear for different diameters of specimens used, the values of wear, coefficient of friction and the variation in temperature throughout the experimental duration was measured. The pin and disc interface was not provided with lubrication, and the experiment was run under dry conditions. This experimental study is helpful in the better understanding of the potential of the negative texturing on the surface, leading to the reduction in friction and enhanced wear resistance for IC engines' piston and cylinder interface.

Keywords Tribology · Texture · Wear · Friction · Steel

1 Introduction

The objective of this review is to provide the analysis of the significant effects on high-chromium steel as piston ring material subjected to various tests. Several researches have been carried out to observe the variations in the effect of the working of the steel material when observed with respect to another material such as EN-31 or EN-34 or other materials for improvement in certain properties such as friction resistance, reduction in material wear, lesser heat generation that directly affect the working

---

S. Bansal · A. Saraf · R. Rana (✉)
Maharaja Agrasen Institute of Technology, New Delhi 110086, India
e-mail: 7ramakant@gmail.com

R. Rana · R. Lal
Delhi Technological University, New Delhi 110042, India

life of the material and the output of the working of the material [1–4]. Mainly, the researches that have been carried out earlier were conducted under lubricated conditions, over the material contact surface under testing. These are the prime parameters since life of an engine is dependent on its losses and wear experienced by it, during functioning [5–9].

Previous studies show the response of tribological parameters on the steel when tested under lubricated conditions without any modification in the physical surface property at the area of contact testing [10–14]. The largest source of frictional losses in the internal combustion engine is cylinder and piston system which comprises early 50% of the total frictional losses. The piston rings contribute 70–80% of friction in the cylinder and piston system. Thus, the cylinder and piston system being the paramount needs optimization to achieve high efficiency [15–19].

In order to study the possibilities of optimization of the piston cylinder, an experimental examination is important to study the aforementioned tribological characteristics. Henceforth, the pin and disc setup is considered to observe and study the effects of the materials in contact. The pin material represents the piston ring while the disc represents the cylinder lining. Wear measurements had been carried out and analysed in cylinder and piston ring during their mating with each other when the automotive engine was operated under artificially created dusty environmental conditions. Tests were conducted on pin-on-disc test apparatus with boundary-lubricated cast iron materials, and the values of specific wear rates were observed to be between 10−13 and 10−10 mm^3/mm/N. Based on the Newton–Raphson–Murty algorithm, the nonlinear finite element method was used to analyse the piston rings problems based on the theories of elasto-hydrodynamic and hydrodynamic lubrication [20–23].

The pin and disc interface subjected to and worked in worst case scenario conditions which is lack of any lubrication. This was done to ensure the properties of material chosen and the parameters which are varied for increasing the performance of the interface when used as a piston and cylinder interface stay put even if the interface is subjected to the maximum amount of mishandling and is capable of proper functioning as expected till the end of its life span. Due to lack of any kind of lubrication, the interface was subjected to a higher amount of friction than normal working conditions, and the wear observed was higher too when compared to the results observed by other researchers working in similar spectrum but with use of lubrication as well [24–26].

Parameters such as area of contact between the disc and pin had a significant impact on the amount of wear induced in on disc and pin, and a change in the frictional force was observed. Due to usage of negative dotted pattern on pin surface which was formed mainly to reduce area in contact between the pin and disc interface resulted in reduction in friction and hence a important factor in longevity of the life span of the interface when used in applications such as engine cylinders when compared to previous results borrowed from past researches done by fellow researchers [27–29].

2 Morphological Structure and Material Testing

2.1 Specimen Specifications

2.1.1 Preparation of the Specimen Disc

Plain surface finished discs of 100 mm diameter and 8 mm of thickness were produced for experimentation using the casting process, EN-31 was used as the base material for the making of discs, as shown in Fig. 1. The smoothness of the disc surface was ensured by surface grinding of disc surface using a surface grinder, which was followed by rubbing of disc surface against soft belt emery paper which further smoothened out the possibility of any debris and irregularities on the disc surface [30–34].

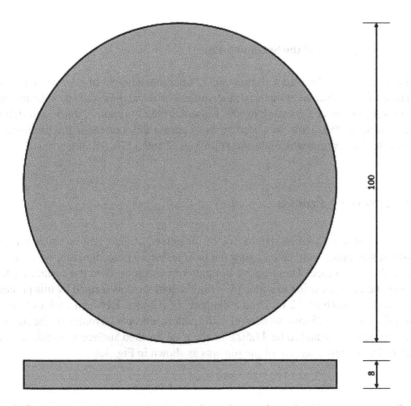

Fig. 1 Schematic diagram for specimen disc (dimensions are in mm)

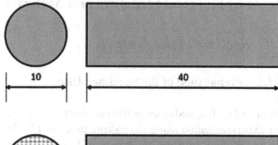

Fig. 2 Schematic diagram for specimen pin without texturing

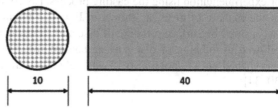

Fig. 3 Schematic diagram for specimen pin with texturing

2.1.2 Preparation of the Specimen Pins

Cylindrical pins of diameters 10 mm and 12 mm with a height of 40 mm each were produced from silver steel material; the surface of these pins was smoothened out by performing surface grinding on pin's circular face to ensure flatness and debris free surface; surface table was used so as to ensure the flatness of the pin surface schematic diagram for which are shown in Figs. 2 and 3 [26, 35, 36].

2.2 Texturing Process

The pins that were used as specimens, of diameters, 10 and 12 mm, after surface finishing was done, were taken under the laser texturing procedure that was carried out on the test surface. The negative dot pattern was grooved on the surface by the picosecond laser texturing machine [37, 38]. Each of the dots formed by this process had a surface depth of 0.2 mm and a diameter of 0.2 mm. Between each dot, there was a distance of 0.25 mm, so as to give the pattern a proper coverage of the surface area that was calculated to be 41.50% of the total finished surface available for the testing. The textured surface of the pin was as shown in Fig. 4.

3 Tribological Test

To emulate the piston and cylinder interface and evaluate the properties like wear resistance and coefficient of friction, tribological test was chosen as it provided a realistic test to simulate the same interface in the form of the pin and disc contact;

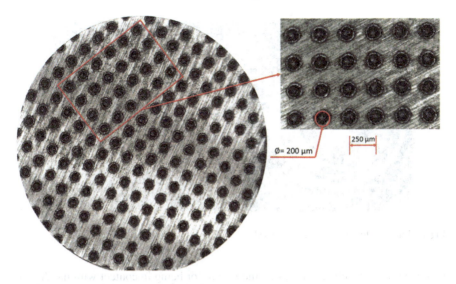

Fig. 4 Textured surface of the specimen pin

pin-on-disc apparatus was used for performing the test; schematic diagram of pin-on-disc apparatus is shown in Fig. 5 along with its top view in Fig. 6; this allowed a continuous sliding contact type testing of interface which is more accurate and effective way of ensuring the results obtained are as realistic as possible [39].

The disc was provided with such provisions so that it could be mounted horizontally onto the pin-on-disc apparatus with the help of four Allen bolts; this ensured secure mounting of the disc while it is rotated at a constant rpm during the test and to maintain it in the horizontal orientation.

Pins were mounted with the help of fixture clamps of different sizes to accommodate different diameters; pins were mounted such that a protrusion of about 4 mm of

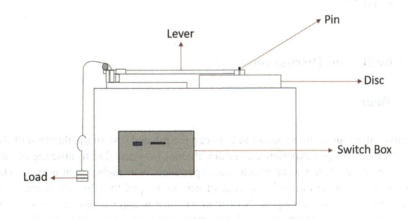

Fig. 5 Schematic diagram of pin-on-disc setup

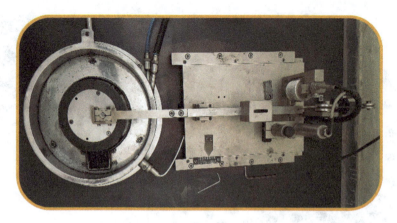

Fig. 6 Top view for pin-on-disc setup used

total length can be obtained outside the fixture for being in contact with the disc so as to avoid the possibility of buckling in pin in the duration of the experiment.

The tribometre which was used for the performance of the experiment was a high-temperature rotary type (TR-20L-PHM800-DHM850) which can rotate the disc from 300–3000 rpm range, it is also capable of providing 20–300 N range of load onto the pin and disc interface.

The test was performed under fixed-parameter conditions; parameters such as sliding velocity of the disc with respect to the pin were 6 m/s, which was kept as constant along with maintaining a constant pressure of 0.26 MPa for tribological performance studies. Observations were made by different track diameter in between the disc and pin interface 50 and 70 mm for pins with diameter 10 and 12 mm, respectively, during which the constant rpm of rotation was calculated and kept equal to 2291 and 1637 rpm for 10 and 12 mm diameters of pins, respectively. The duration of each test was set to be 500 s with total distance traversed by pin onto the disc equal to 3000 m.

4 Results and Discussion

4.1 Wear

Due to continuous sliding contact in between the pin and disc in the duration of the test, the disc and pin experienced certain amount of wear; due to absence of any kind of lubrication, the wear which was experienced was higher than normal. The pressure between the pin and disc contact was kept equal to 0.26 MPa throughout the experiment with the help 30 N of load which was applied initially as a result; the amount of wear which occurred was measured to increase for first 120–140 s

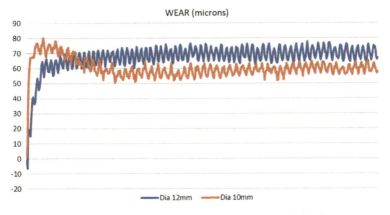

Fig. 7 Wear

from ≈1 micron to a maximum of 80 microns; after the initial time period of the experiment, the amount of wear experienced remained to be nearly constant which is approximately equal to an average of 60.027 microns. The variation of amount of wear which occurred for both specimens has been plotted and shown in Fig. 7 [39].

4.2 Coefficient of Friction (COF)

Whenever there is contact between two identical or different surfaces which are in relative motion with respect to each other, frictional force is developed between the two contact surfaces. This frictional force is governed by the value of coefficient of friction which varies for different material and is also governed by various other factors such as variation in temperature, the presence of abrasives in between contact surfaces. During the tribological test, such variation in values of coefficient of friction was observed, and this variation in the values of coefficient of friction for both the specimens has been plotted and shown in Fig. 8. The average values of coefficient of friction (COF) for silver steel pin specimen of diameters 10 and 12 mm with respect to the EN-31 specimen disc are 1.1594 and 1.1504, respectively.

4.3 Temperature Variation

The heat variation was recorded by the thermography camera, while the tribometre setup was being run for the testing of the specimen. Infrared thermography is the process of detecting, processing and visualizing the invisible infrared radiation that an object emits. The results of the thermography imaging by the imaging camera are as shown in Figs. 9 and 10 [27]. The variation in temperature is a direct result of

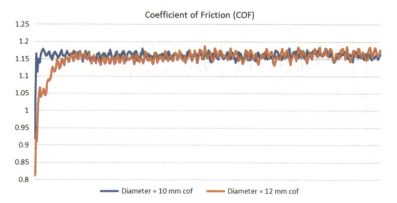

Fig. 8 Coefficient of friction

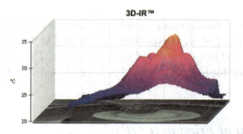

Fig. 9 Thermographic of working setup

Fig. 10 Image of working setup

the sliding phenomenon which results in generation of frictional forces between the two surfaces of the materials continuously in contact. The factor of no lubrication affects the variation of the temperature in the specimens while continuous running of the tribological test on the pin and disc setup. The image shows the temperature range of the disc during the non-lubricated run of the experiment, i.e. from 34.404 to 38.514 °C.

5 Conclusions

An attempt was made to analyse the tribological behaviour of steel by the means of an experimental investigation on the pin-on-disc test rig. The material of the stationary pin was analogous to the material of piston ring material, while the disc material was EN 31. The tribometre was used to simulate the results in terms of wear and coefficient of friction.

From the experiments, the following conclusions were drawn:

1. The increase in diameter, the pin increases the wear.
2. But, the same diameter has very negligible effect on the coefficient of friction.
3. The textures have significantly improved the thermal dispersions.

References

1. Wakuri Y, Hamatake T, Soejima M, Kitahara T (1992) Piston ring friction in internal combustion engines. Tribol Int 25:299–308
2. Nautiyal PC, Singhal S, Sharma JP (1983) Friction and wear processes in piston rings. Tribol Int 16:43–49
3. Pawlus P (1993) Effects of honed cylinder surface topography on the wear of piston-piston ring-cylinder assemblies under artificially increased dustiness conditions. Tribol Int 26:49–55
4. Childs THC, Sabbagh F (1989) Boundary-lubricated wear of cast irons to simulate automotive piston ring wear rates. Wear 134:81–97
5. Hwu CJ, Weng C (1991) Elastohydrodynamic lubrication of piston rings. Wear 150:203–215
6. Picken DJ, Hassaan HA (1983) A method for estimating overhaul life of internal combustion engines including engines operating on biogas and methane. J Agric Eng Res 28:139–147
7. Scott D, Smith AI, Tait J, Tremain GR (1975) Materials and metallurgical aspects of piston ring scuffing—a literature survey. Wear 33:293–315
8. Moore SL (1981) Piston ring lubrication in a two-stroke diesel engine. Wear 72:353–369
9. Grabon W, Pawlus P, Sep J (2010) Tribological characteristics of one-process and two-process cylinder liner honed surfaces under reciprocating sliding conditions. Tribol Int 43:1882–1892
10. Grabon W, Koszela W, Pawlus P, Ochwat S (2013) Improving tribological behaviour of piston ring–cylinder liner frictional pair by liner surface texturing. Tribol Int 61:102–108
11. Truhan JJ, Qu J, Blau PJ (2005) The effect of lubricating oil condition on the friction and wear of piston ring and cylinder liner materials in a reciprocating bench test. Wear 259:1048–1055
12. Michalski J, Woś P (2011) The effect of cylinder liner surface topography on abrasive wear of piston–cylinder assembly in combustion engine. Wear 271:582–589
13. Johansson S, Nilsson PH, Ohlsson R, Rosén BG (2011) Experimental friction evaluation of cylinder liner/piston ring contact. Wear 271:625–633
14. Olander P, Hollman P, Jacobson S (2013) Piston ring and cylinder liner wear aggravation caused by transition to greener ship transports–comparison of samples from test rig and field. Wear 302:1345–50

15. Mezghani S, Demircia I, Zahouani H, El Mansori M (2012) The effect of groove texture patterns on piston-ring pack friction. Precis Eng 36:210–217
16. Srivastava DK, Agarwal AK, Kumar J (2007) Effect of liner surface properties on wear and friction in a non-firing engine simulator. Mater Des 28:1632–1640
17. Kapsiz M, Durat M, Ferit F (2011) Friction and wear studies between cylinder liner and piston ring pair using Taguchi design method. Adv Eng Soft 42:595–603
18. Gara L, Zou Q, Sangeorzan BP, Barber GC, McCormick HE, Mekari MH (2010) Wear measurement of the cylinder liner of a single cylinder diesel engine using a replication method. Wear 268:558–564
19. Velkavrh I, Ausserer F, Klien S, Voyer J, Ristow A, Brenner J, Forêt P, Diem A (2016) The influence of temperature on friction and wear of un-lubricated steel/steel contacts in different gaseous atmospheres. Tribol Int 98:155–171
20. Tiwari S, Bijwe J and Panier S (2011) Gamma radiation treatment of carbon fabric to improve the fiber–matrix adhesion and tribo-performance of composites. Wear 271:2184–2192
21. Bijwe J, Gupta MK, Parida T, Trivedi P (2015) Design and development of advanced polymer composites as high performance tribo-materials based on blends of PEK and ABPBI. Wear 342–3:65–76
22. Djoufack MH, May U, Repphun G, Brögelmann T, Bobzin K (2015) Wear behaviour of hydrogenated DLC in a pinon- disc model test under lubrication with different diesel fuel types. Tribol Int 92:12 20
23. Truhan JJ, Qu J, Blau PJ (2005) A rig test to measure friction and wear of heavy duty diesel engine piston rings and cylinder liners using realistic lubricants. Tribol Int 38:211–218
24. Sudeep U, Pandey RK, Tandon N (2013) Effects of surface texturing on friction and vibration behaviours of sliding lubricated concentrated point contacts under linear reciprocating motion. Tribol Int 62:198–207
25. Pandey RK, Tandon N, Singh AK (2013) Fuel saving in IC engine by surface texturing of piston rings National conference on recent advancements in mechanical engineering (NCRAME-2013) (8–9 November 2013) (Itanagar, Arunachal Pradesh, India: NERIST) pp 1–4
26. Roop L, Ramakant R (2015) A textbook of engineering drawing, 1st edn I.K. International Publishing House Pvt. Ltd. ISBN: 978-93-84588-68-7
27. Jain S, Aggarwal V, Tyagi M, Walia RS, Rana R (2016) Development of aluminium matrix composite using coconut husk ash reinforcement. In: International conference on latest developments in materials, manufacturing and quality control (MMQC-2016). 12-13 February 2016, Bathinda, Punjab India, pp 352–359
28. Ranganath MS, Vipin (2013) Optimization of process parameters in turning using taguchi method and anova: a review. Int J Adv Res Innovation 1:31–45
29. Singh RC, Roop L, Ranganath MS, Rajiv C (2014) Failure of piston in IC engines: a review. Int J Mod Eng Res (IJMER) 4(9):1–10
30. Ramakant R, Kunal R, Rohit S, Roop L (2014) Optimization of tool wear: a review. Int J Mod Eng Res 4(11):35–42
31. Roop L, Singh RC (2019) Investigations of tribodynamic characteristics of chrome steel pin against plain and textured surface cast iron discs in lubricated conditions. World J Eng 16(4):560–568
32. Roop L, Singh RC (2018) Experimental comparative study of chrome steel pin with and without chrome plated cast iron disc in situ fully flooded interface lubrication Surf Topogr Metrol Prop 6:035001
33. Singh RC, Pandey RK, Roop L, Ranganath MS, Maji S (2016) Tribological performance analysis of textured steel surfaces under lubricating conditions. Surf Topogr Metrol Prop 4:034005
34. Ramakant R, Walia RS, Surabhi L (2018) Development and investigation of hybrid electric discharge machining electrode process. Mater Today: Proc 5(2):3936–3942
35. Ramakant R, Vipin Kumar S, Mitul B, Aditya S (2016) Wear analysis of brass, aluminium and mild steel by using pin-on-disc method. In: 3rd international conference on manufacturing excellence (MANFEX-2016), March 17-18, pp 45–48

36. Ramakant R, Walia RS, Qasim M, Mohit T (2016) Parametric optimization of hybrid electrode EDM process. In: TORONTO'2016 AES-ATEMA international conference "advances and trends in engineering materials and their applications", July 04 – 08, 2016, Toronto, Canada, pp 151–162
37. Ramakant R, Walia RS, and Manik S (2016) Effect of friction coefficient on En-31 with different pin materials using pin-on-disc apparatus. In: International conference on recent advances in mechanical engineering (RAME-2016), October 14–15, 2016, Delhi, India, pp 619–624. ISBN:-978-194523970-0
38. Kaplish A, Choubey A, Rana R (2016) Design and kinematic modelling of slave manipulator for remote medical diagnosis. In: International conference on advanced production and industrial engineering, 9-10 December, 2016
39. Khatri B, Kashyap H, Thakur A, Rana R (2016) Robotic arm aimed to replace cutting processes. In: International conference on advanced production and industrial engineering, 9-10 December, 2016

A Statistical Approach for Overcut and Burr Minimization During Drilling of Stir-Casted MgO Reinforced Aluminium Composite

Anmol Gupta, Surbhi Lata, Ramakant Rana, and Roop Lal

Abstract Aluminium hybrid composites are a new generation of metal matrix composites that have the potentials of satisfying the recent demands of advanced engineering applications. These demands are met due to improved mechanical properties, amenability to conventional processing technique and possibility of reducing production cost of aluminium hybrid composites. This paper attempts to develop and characterize the aluminium matrix composite using magnesium as reinforcement and aluminium (5052) as matrix. The development of aluminium metal matrix composites (MMC) reinforced with 10 wt% MgO was accomplished by stir casting process. The surface morphology of the developed composite was studied using scanning electron microscope (SEM) equipped with energy dispersive X-ray spectroscopy (EDX). X-ray diffraction (XRD) was also carried out to study the phase, texture and grain boundary conditions. Further, response surface methodology was employed to study the effect of drill diameter, speed and depth of cut on output parameters viz. burr height and overcut during the drilling of developed aluminium composite. The result inferred the optimal combination of input parameters for achieving minimum burr height and minimum overcut.

Keywords Aluminium matrix composites · SEM · EDX · Burr height · Overcut

1 Introduction

In composites, materials are combined in such a way as to enable us to make better use of their parent material while minimizing to some extent the effects of their deficiencies. The simple term 'composites' gives indication of the combinations of two or more material in order to improve the properties. In the past few years, materials

---

A. Gupta · S. Lata · R. Rana (✉)
Maharaja Agrasen Institute of Technology, New Delhi 110086, India
e-mail: 7ramakant@gmail.com

R. Rana · R. Lal
Delhi Technological University, Delhi 110042, India

development has shifted from monolithic to composite materials for adjusting to the global need for reduced weight, low cost, quality and high performance in structural materials. Aluminium matrix composites (AMCs) are extensively used in the areas of aerospace and automotive industries primarily due to their enhanced performance, economic and environmental benefits [1–5].

Aluminium matrix composites (AMC) is developed by reinforcing aluminium matrix with ceramic particles. AMC exhibits better mechanical properties than unreinforced aluminium alloys. Aluminium is only second to steel. Molten temperature of 850C was used and stirring carried out for 45 min at the rate of 200 rpm. The blade with angle 45 and 60 will give the uniform distribution. Silicon carbide particles were preheated to 200 C and introduced into the vortex created in the molten alloy [6]. The influence of stirring parameters on microstructure and mechanical properties of the composites was investigated using scanning electron microscopy (SEM), X-ray diffractometry (XRD) [7]. Kamaljit et al. [4] found that stirring speed and stirring time influenced the microstructure and the hardness of composite. Microstructure analysis revealed that at lower stirring speed with lower stirring time, the particle clustering was more. Increase in stirring speed and stirring time resulted in better distribution of particles. The hardness test results also revealed that stirring speed and stirring time have their effect on the hardness of the composite. The uniform hardness values were achieved at 600 rpm with 10 min stirring. But beyond certain stir speed the properties degraded again. Heat treatment of all composites and Al alloys improved abrasive wear resistance. Mechanical properties such as hardness and compressive strength improved. Composite containing 1.5 vol.% MgO fabricated at 850 °C showed improved properties such as hardness, strength and toughness in comparison with other specimens. Furthermore, toughness of composites, generally decreased by increasing the content of MgO [8]. With increase in temperature hardness and wear resistance also increases and hardness will increase initially with ageing time, and after a peak value, it tends to decrease. In case of the hybrid composites, the wear rate do not depend on type of reinforcement, but wear rate of the hybrid composites decreases with increasing the weight percentage of reinforcing materials. The abrasive wear resistance of Al-based hybrid composites will be controllably altered by thermal ageing. The microphotographs of the composites studied revealed the uniform distribution of the particles in the matrix system. The experimental density values were agreed with that of the theoretical density values of the composites obtained using the rule of mixture for composites [1*]. Zohoor et al. [7] found that higher degree of defects and micro-porosity is observed at composites with higher SiC content which is the result of increase in the amount of interface area. The base alloy also shows higher hardness towards the outer periphery than the inner periphery due to the refinement of primary aluminium grains and eutectic silicon phases. Similar to the hardness observations, higher tensile strength is observed towards the outer periphery [9]. Zhai yan-bo et al. [10] studied the characteristics of two Al-based functionally gradient composites reinforced by primary Si particles and Si/in situ Mg2Si particles in centrifugal casting and concluded hardness and wear resistance of Al-19Si-5 Mg tube in the inner layer are greatly higher than those in the other layers of Al-19Si-5 Mg tube and Al-19Si tube; XRD analysis that the reinforcement particles in Al-19Si tube are single kind of primary Si particles and are primary Si particles mixing with Mg2Si particles in Al-19Si-5 Mg tube [11–17].

2 Experimental Details

2.1 Preparation of Aluminium Matrix Composite

Aluminium is the most abundant element on earth surface; hence, the demand of aluminium and its alloys in automotive and aerospace industry is growing. Aluminium alloys intended for use in production of castings are generally available as ingots of varying size or in other forms suitable for re-melting. Applications of such cast materials have included the production of cast components using DRA, with stirring to suspend particles in the liquid metal prior to casting and solidification of the article. Aluminium of grade 5052 is selected as the matrix element, and its composition is given in Table 1 [18–20].

Magnesium was selected as the reinforcement due is exceptional workability with aluminium, and its various properties strips of magnesium were burned on Bunsen burner to obtain white magnesium oxide powder of or around 30–50 μm. The important role played by the magnesium during the composite synthesis is the scavenging of the oxygen from the dispersoid surface.

Ex situ process is followed in the manufacturing of AMC. 10% by weight of magnesium oxide powder is added and preheated to 650 °C in a furnace separately; the aluminium pieces were added in the crucible. The chamber of the furnace was first heated to a temperature of 850 °C this helps to melt aluminium quickly and reduces the oxidation level. The temperature of the chamber is then raised to 975 °C and stirred constantly at 540 rpm using an impeller fabricated from graphite and driven by a variable AC motor [21–24].

After stirring for about 15 min at a temperature of 850 °C, the slurry was reheated and held at a temperature of 950 °C for 15 min to ensure the slurry is in completely molten state. It is essentially required for vortex formation for the uniformly dispersion of particles. The stirrer which was used in the set-up was at 90° from the shaft and made of graphite. The reisano uniform dispersion of particles in case of no vortex formation. In the process, stirring speed was 240 rpm which was effectively producing vortex without any spattering. Stirring speed is decided by fluidity of metal speed; dispersion of particulates is not proper because of ineffective vortex. The mould was prepared using a wooden pattern and baked at 200 °C for 5 min. The slurry is then poured into the mould at uniform rate to avoid trapping of gasses. The mould is allowed to cool [25, 26].

Table 1 Composition of Al 5052

Element	Aluminium (%)	Magnesium (%)	Chromium (%)	Copper (%)	Manganese (%)	Iron (%)	Silicon (%)	Zinc (%)	Others(%)
Composition	95.8	2.2–2.8	0.15–0.35	≤0.1	≤0.1	≤0.4	≤0.25	≤0.1	0.15

Fig. 1 Microprocessor controlled furnace

Fig. 2 Drilling process on vertical milling machine

2.2 Drilling Process

Drilling operation was carried out on the fabricated MMC to drill holes with different parameters to study the surface morphology of the aluminium composite. Vertical milling machine was used with different sized drill bits to drill the holes. Holes were drilled with different values of input parameters, i.e. drill bit diameter, spindle speed and depth of cut. The drill diameters chosen for the experiment are 6, 9, 12 mm. The spindle speeds chosen for the experiment are 660, 1110 and 1320 rpm. The depth of cuts chosen for the experiment is 5, 7 and 9 mm. The Fig. 1 shows the stir casting furnace used, whereas Fig. 2 is showing the drilling process being conducted on the prepared composite.

2.3 Burr Height Measurement

Burrs and die rolls are typical defects of trimmed surfaces. Burr height is typically used as an index to measure tool wear, because it is easy to measure during production. It was measured using a puppet dial gauge. The workpiece with the drilled holes is shown in Fig. 3 and the same holes were measured for the burr height using puupy

Fig. 3 Workpiece with drilled holes

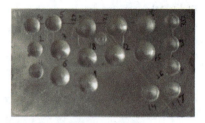

Fig. 4 Measurement of burr height using puppy dial gauge

dial gauge. The Measurement process of burr height is shown in Fig. 4. The probe of the gauge was moved around the edge of the holes over the irregular burrs, and the burr height was measured by the deflection of the dial.

2.4 Overcut Measurement

Overcuts are the defects of the machined surface. These are used as an index to measure the dimensional accuracy, surface wear and the tool wear as it is very easy to measure. A digital vernier caliper was used to measure the amount of overcut in each hole. The probe tips of the vernier caliper were inserted in the holes, and the reading was noted. The nominal diameter of the holes was subtracted from the readings, and overcut was measured.

2.5 Regression Analysis

Regression analysis is a statistical process for estimating the relationships among variables. Regression equation was generated for overcut and burr height as the dependent variables (Y). The drill diameter, spindle speed and depth of cut were chosen as the input, i.e. independent variables (X). The values of unknown parameters (β) were calculated using general regression. The final regression equation was found to be:

$$\text{Burr height} = 0.35 - 0.146 X1 + 0.00040 X2 + 0.032 X3 + 0.000043 X1 X2 \\ + 0.0109 X1 X3 - 0.000055 X2 X3 - 0.000002 X1 X2 X3$$

$$\text{Overcut} = -3.63 + 0.477 X1 + 0.00303 X2 + 0.505 X3 - 0.000446 X1 X2 \\ - 0.0627 X1 X3 - 0.000402 X2 X3 + 0.000058 X1 X2 X3$$

where

X1 Depth of cut (mm)
X2 Spindle Speed (mm)
X3 Drill Diameter (mm).

2.6 Scanning Electron Microscopy Analysis

The SEM is routinely used to generate high-resolution images of shapes of objects (SEI) and to show spatial variations in chemical compositions: (1) acquiring elemental maps or spot chemical analyses using EDS, (2) discrimination of phases based on mean atomic number (commonly related to relative density) using BSE, and (3) compositional maps based on differences in trace element 'activators' (typically transition metal and rare earth elements) using CL. The electron microscopy was performed on the prepared samples, and the following images were generated on different magnifications and depth. A total of eight images were taken at different depths and magnifications as shown below. The high-resolution SEM images shown in Figs 5 and 6 depicts the surface morphology of the prepared sample. The complete distribution of the elements was seen at some points, and other points showed incomplete distribution. The whiteness impression indicated traces of magnesium and various other elements.

3 Results and Discussion

The main effects plot for SN ratio of burr height and main effects for burr height are shown in Figs. 7 and 8 respectively. Depth of cut: The depth of cuts chosen for the experiment is 5, 7, 9 mm.

Spindle speed: Spindle speeds chosen for the experiment are 660, 1110 and 1320 rpm.

Fig. 5 SEM images of the fabricated material at 6.24KX

Fig. 6 SEM images of the fabricated material at 10.72KX

Spindle speed: Spindle speeds chosen for the experiment are 660, 1110 and 1320 rpm.

The above Fig. 8 shows the effect of depth of cut, spindle speed and drill diameter on burr height. The plot is made with help of experimental values shown in Table 2.

Depth of cut: As from the plot, it is clear that with the increase in depth of cut the burr height decreases at a uniform rate with maximum at 5 mm depth of cut.

Spindle speed: At lower speed, i.e. 660 rpm, burr height is minimum. As the speed is increased, the burr height also increases to a certain point but starts to shorten with further increase in speed, i.e. at moderate speeds more burr is formed.

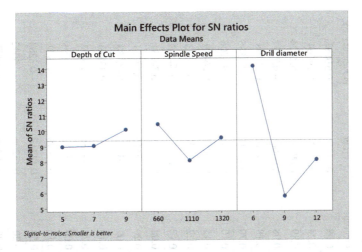

Fig. 7 Main effect plot for SN ratio of burr height

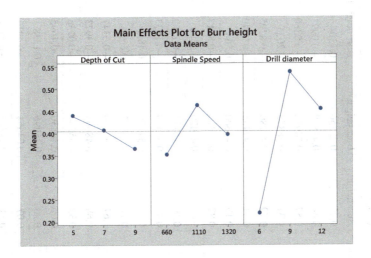

Fig. 8 Main effect plot for burr height

Drill diameter: At smaller diameter, i.e. at 6 mm, negligible burr is formed but with increase in diameter burr height goes to maximum and then shortens with increase in diameter. All the values of responses are shown in a tabular form in Table 3.

Figure 9 shows the variation of burr height with respect to depth of cut and spindle speed. The depth of cut is represented on y-axis and spindle speed on spindle speed. Burr height is shown by the shaded region, i.e. contour plot. At lower speed, i.e. 660 rpm, burr height is minimum. As the speed is increased, the burr height also increases to a certain point but starts to shorten with further increase in speed, i.e. at moderate speeds more burr is formed. With increase in spindle speed and depth

Table 2 Experimental value and theoretical values of burr height and overcut

	Independent variables			Experimental values		Theoretical value		Error		Experimental values	Theoretical value	Error
	DOC	Spindle speed	Drill diameter	Burr height		Burr height		Burr height (%)		Overcut	Overcut	Overcut (%)
1	5	660	6	0.25		0.28750		0.1304		0.03	0.0115	0.61667
2	5	660	9	0.46		0.41830		−0.099		0.24	0.34122	0.29664
3	5	660	12	0.57		0.54910		−0.0380		0.89	0.69396	−0.2824
4	5	1110	6	0.31		0.38875		0.2025		0.08	0.04608	−0.7361
5	5	1110	9	0.79		0.43180		−0.8295		0.18	0.24762	0.27308
6	5	1110	12	0.51		0.47485		−0.0740		0.70	0.44916	−0.5585
7	5	1320	6	0.19		0.43600		0.5642		0.10	0.07296	−0.3706
8	5	1320	9	0.82		0.43810		−0.8717		0.15	0.20394	0.26449
9	5	1320	12	0.07		0.44020		0.8409		0.22	0.33492	0.34313
10	7	660	6	0.19		0.16722		−0.1362		0.17	0.06072	0.64282
11	7	660	9	0.41		0.35550		−0.1533		0.08	0.26694	0.70031
12	7	660	12	0.28		0.54378		0.4850		0.20	0.47316	0.57731
13	7	1110	6	0.10		0.29637		0.6625		0.09	0.03012	0.66533
14	7	1110	9	0.64		0.39150		−0.6347		0.16	0.24174	0.33813
15	7	1110	12	0.68		0.48663		−0.3973		0.24	0.45336	0.47062
16	7	1320	6	0.39		0.35664		−0.0935		0.04	0.01584	−0.6040
17	7	1320	9	0.60		0.40830		−0.4695		0.12	0.22998	0.47822
18	7	1320	12	0.38		0.45996		0.1738		0.59	0.44412	−0.3284
19	9	660	6	0.07		0.04694		−0.4912		0.20	0.13296	−0.5042
20	9	660	9	0.31		0.29270		−0.0591		0.06	0.19266	0.68857

(continued)

Table 2 (continued)

	Independent variables			Experimental values	Theoretical value	Error	Experimental values	Theoretical value	Error
21	9	660	12	0.63	0.53846	−0.170	0.58	0.25236	0.56489
22	9	1110	6	0.18	0.20399	0.1176	0.05	0.01416	−0.7168
23	9	1110	9	0.39	0.35120	−0.1104	0.08	0.23586	0.66082
24	9	1110	12	0.58	0.49841	−0.1637	0.36	0.45756	0.21322
25	9	1320	6	0.30	0.27728	−0.0819	0.02	−0.0412	0.51456
26	9	1320	9	0.43	0.37850	−0.1360	0.17	0.25602	0.33599
27	9	1320	12	0.40	0.47972	0.1661	0.70	0.55332	−0.2650

Table 3 Rank table of burr height response

	Response table for burr height			Response table for signal to noise ratios		
Level	Depth of cut	Spindle speed	Drill diameter	Depth of cut	Spindle speed	Drill diameter
1	0.4411	0.3522	0.2200	8.998	10.464	14.188
2	0.4078	0.4644	0.5389	9.052	8.122	5.806
3	0.3656	0.3978	0.4556	10.093	9.557	8.149
Delta	0.0756	0.1122	0.3189	1.095	2.342	8.382
Rank	3	2	1	3	2	1

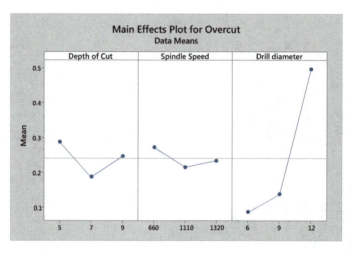

Fig. 9 Main effects plot for overcut

of cut, burr height increases. The darker region identifies the higher burr heights as shown by the dark green pattern with the quality score of the peak region between 0.4 and 0.6.

From the plot made with the help of experimental values shown in Fig. 8, we can see that at spindle speed more than 800 rpm, and at drill diameter 9 mm, maximum burr height is seen, i.e. at moderate and higher spindle speeds with greater drill diameters, maximum burr is formed. At smaller drill diameters, i.e. between 6 and 8 mm almost no burr is formed even at higher speeds. Whereas moderate burr is formed at medium drill diameters, i.e. between 8 and 9 mm and all spindle speeds. The peak region shows a quality score of more than 0.8. All the repsonses of overcut have been shown in the tabular format in the Table 4.

Table 4 Rank table of overcut response

	Response table for burr height			Response table for signal to noise ratios		
Level	Depth of cut	Spindle speed	Drill diameter	Depth of cut	Spindle speed	Drill diameter
1	0.28778	0.27222	0.08667	14.825	15.369	23.404
2	0.18778	0.21556	0.13778	16.833	16.221	17.979
3	0.24667	0.23444	0.49778	16.939	17.006	7.214
Delta	0.10000	0.05667	0.41111	2.114	1.637	16.191
Rank	2	3	1	2	3	1

3.1 Overcut Measurement

Figure 10 shows the main effect plot for SN ratio for overcut with depth of cut, spindle speed and drill diameter on overcut. The plot is made with help of experimental values presented in Table 4 and is shown in Fig. 10.

Depth of cut: As from the plot, it is clear that with the increase in depth of cut the overcut first decreases to a minimum in this case at 7 mm (depth of cut) and then increases to a mean.

Spindle speed: At lower speeds, i.e. 660 rpm, overcut is maximum. With increase in speed, overcut decreases to a constant.

Drill diameter: At smaller diameter, i.e. at 6 mm, overcut is negligible. But with increase in diameter overcut increases to very high values with maximum at 12 mm (in this experiment).

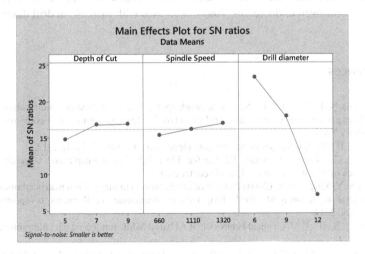

Fig. 10 Main effect plot for S-N ratios for overcut

4 Conclusion

For obtaining improved characteristics in the composite fabricated using the stir casting process, the knowledge of following process parameters is essential. The important parameters in stir casting are stirring speed, stirring temperature, preheating of reinforcement, addition of magnesium, blade pattern, preheating of mould and feed rate of reinforcement. Stirring speed promotes bonding between matrix and reinforcement and directly controls the flow pattern of molten metal. Addition of magnesium up to 10% of the metal improves wettability of the melt. The pattern of the blades of the stirrer drastically affects the flow pattern. It has been observed that the stirrer should be kept 20 mm above the base of the crucible. Stirrer with blade angle 45° and 60° and four blades will give uniform distribution of reinforcement in the metal matrix. In order to obtain uniform distribution, the feed rate of the reinforcement should be kept constant. Non-uniform feed rate promotes clustering of particles at different places leading to inclusion defect. The metal matrix composite with aluminium as matrix phase reinforced with magnesium been successfully manufactured using the stir casting method. It has been found that the operational parameters as discussed play a key role in achieving good wettability, uniform distribution and good casting of the composite material. Also the increased porosity levels in the composite material have led to a decrease in its density. The experiments were conducted on a vertical milling machine for the machining of Al–Mg MMC. The response of various parameters such as burr height and overcut was studied. It was found that spindle speed, depth of cut and drill diameter significantly affect all these parameters. In case of burr height, with increase in DOC burr height decreases. With increase in spindle speed, burr height increases. And with increase in drill diameter burr height first increases then decreases. While in case of overcut, the overcut first decreases then increases with increase in DOC. With increase in spindle speed, overcut decreases. And it increases with increase in drill diameter.

References

1. Altinkok N, Koker R (2004) Neutral network approach to prediction of bending strength and hardening behaviour of particulate reinforced (Al-Si-Mg) aluminium matrix composites, 10th February 2004
2. Cheney B (2012) Introduction to scanning electron microscopy, 3rd July 2012
3. Swayze GA, Clark RN, Sutley SJ, Hoefen TM (2006) Spectroscopic and x-ray diffraction analyses of asbestos in the world trade center dust
4. Hai Zhi Y, Xing Yang L (2004) Review of recent studies in magnesium matrix composites
5. Jaykumar A, Rangraj M (2014) Properties of aluminium metal matrix composites, 2nd February, 2014
6. Kamaljit S (2011) Mechanical behaviour of Al based MMC reinforced with aluminium oxide, July 2011
7. Kumar R, Jegan J, Initha L (2014) Processing of aluminium metal matrix composites (AMMC) through stir casting route

8. Shabani Ostad M, Mazahery A (2012) Application of GA to optimize the process condition of Al Matrix nano-composites, 17th August, 2012
9. Zhai Y, Miu CL, Wang K (2019) Characteristic of two Al based functionally gradient composites reinforced by primary Si particles and Si/in situ Mg2Si particles in Centrifugal casting, 4th May 2009
10. Rajeshkumar GB, Sonawane PM (2013) Preparation of aluminium matrix composite by using stir casting method
11. Zohoor M, Salami P, Give Besharati MK (2012) Effect of processing parameters on fabrication of Al-Mg/Cu composites via F.S.R, 7th March 2012
12. Ramakant R, Kunal R, Rohit S, Roop L (2014) Optimization of tool wear: a review. Int J Modern Eng Res 4(11):35–42
13. Roop L, Singh RC (2019) Investigations of tribodynamic characteristics of chrome steel pin against plain and textured surface cast iron discs in lubricated conditions. World J Eng 16(4):560–568
14. Roop L, Singh RC (2018) Experimental comparative study of chrome steel pin with and without chrome plated cast iron disc in situ fully flooded interface lubrication. Surf Topogr Metrol Prop 6:035001
15. Singh RC, Pandey RK, Roop L, Ranganath MS, Maji S (2016) Tribological performance analysis of textured steel surfaces under lubricating conditions. Surf Topogr Metrol Prop 4:034005
16. Ramakant R, Walia RS, Surabhi L (2018) Development and investigation of hybrid electric discharge machining electrode process. Mater Today: Proc 5(2):3936–3942
17. Ranganath MS, Vipin (2013) Optimization of process parameters in turning using taguchi method and anova: a review. Int J Adv Res Innovation 1:31–45
18. Roop L, Ramakant R (2015) A textbook of engineering drawing edn 1. International Publishing House Pvt. Ltd., I.K. ISBN: 978-93-84588-68-7
19. Ramakant R, Vipin Kumar S, Mitul B, Aditya S (2016) Wear analysis of brass, aluminium and mild steel by using pin-on-disc method. In: 3rd international conference on manufacturing excellence (MANFEX-2016), March 17–18, pp 45–48
20. Ramakant R, Walia RS, Qasim M, Mohit T (2016) Parametric optimization of hybrid electrode EDM process. In: TORONTO'2016 AES-ATEMA international conference. Adv Trends Eng Mater Their Appl July 04-08, 2016, Toronto, Canada, pp 151–162
21. Ramakant R, Walia RS, Manik S (2016) Effect of friction coefficient on En-31 with different pin materials using pin-on-disc apparatus. In: International conference on recent advances in mechanical engineering (RAME-2016), October 14–15, 2016. Delhi, India, pp 619–624. ISBN:-978-194523970-0
22. Kaplish A, Choubey A, Rana R (2016) Design and kinematic modelling of slave manipulator for remote medical diagnosis. In: International conference on advanced production and industrial engineering, 9–10 December, 2016
23. Khatri B, Kashyap H, Thakur A, Rana R (2016) Robotic arm aimed to replace cutting processes. In: International conference on advanced production and industrial engineering, 9–10 December, 2016
24. Singh RC, Roop L, Ranganath MS, Rajiv C (2014) Failure of piston in IC engines: a review. Int J Mod Eng Res (IJMER) 4(9):1–10
25. Jain S, Aggarwal V, Tyagi M, Walia RS, Rana R (2016) Development of aluminium matrix composite using coconut husk ash reinforcement. In: International conference on latest developments in materials, manufacturing and quality control (MMQC-2016), 12–13 February 2016, Bathinda, Punjab India, pp 352–359
26. Kumari N, Bhaskar HB, Kiran TS (2015) Characterization of Za-27 alloy reinforced with MgO particles by stir casting technique, April-June 2015

Study and Design Conceptualization of Compliant Mechanisms and Designing a Compliant Accelerator Pedal

Harshit Tanwar, Talvinder Singh, Balkesh Khichi, R. C. Singh, and Ranganath M. Singari

Abstract Compliant mechanisms work on the elastic body deformations of a material to transfer or/and amplify an input displacement to an output desired displacement. They are highly preferred in applications demanding friction-less and backlash-free motion with high precision. In this paper, the accelerator lever arm and the torsional spring are replaced by an equivalent distributed compliant mechanism. In this mechanism, flexibility of lever arm eliminates the use of torsional spring and additional bolted parts in the entire assembly. The traditional accelerator pedal of passenger car consists of a stationery mount and a lever arm, loaded by a torsional spring, which is directly linked to the accelerator cable operated by the driver. The compliant accelerator pedal can be fabricated as an entire monolithic piece of polypropylene using 3D printing technique. The CAD model of the assembly is modelled on SolidWorks and simulated on ANSYS. The mathematical model of accelerator pedal is developed and simulated using MATLAB. Nonlinear model results that the necessary displacement can be achieved with precision control over the pull of the accelerator cable without compromising the automobile ergonomics. This mechanism is validated and used as an accelerator pedal in an all-terrain Baja vehicle which is designed and fabricated at the university.

Keywords Compliant mechanism · Accelerator pedal · Polypropylene · Elastic segment

1 Introduction

A mechanism can be broadly classified into a rigid-body mechanism and a compliant mechanism. Former is the traditional method comprising of rigid links and joints that can be a lower pair or a higher pair. Whereas latter is the use of elasticity of materials to devise a mechanism that fulfils the intended need without the using number of linkages and joints. Compliant mechanism works by the elastic body

H. Tanwar (✉) · T. Singh · B. Khichi · R. C. Singh · R. M. Singari
Delhi Technological University, New Delhi 110042, India
e-mail: harshittanwar007@yahoo.in

deformations of material to transfer and/or amplify an input displacement and force to an output displacement and force. Compliant mechanism completely eliminates the usage of hinges and joints. There are two types of these mechanisms: Discrete compliance are those which achieve compliance through the elastic pairs connected by the same material but reduced cross-sectional area to provide flexibility, and distributed compliance are those which achieve compliance through elastic segments of uniform cross-sectional area throughout. They have advantage over the former type by providing uniform stress distribution across the length of the elastic segment. Compliant mechanism has finite but very low stiffness along the intended axis and very large stiffness along the other axis. They are extremely used [1] for small-consumer products and machinery parts like Tweel tire from Michelin, compliant bicycle sprocket and centrifugal compliant clutch [2, 3].

A cable throttle system has a throttle cable attached to the accelerator pedal on one end, and the force by the driver is applied on the other end. An electronic throttle control system has a pedal position sensor that senses the pedal position and sends it to the electronic control module to regulate the engine throttle. In both the cases, accelerator pedal of a passenger car works as a lever of first kind having pivot as a revolute joint at a stationary mount, torsional spring and force to pull the throttle cable as load and force from the drivers foot as an effort. Researchers have proposed several methods to optimize accelerator pedal [4–7]. However, improving the rigid-linkage mechanism to be frictionless and backlashless is far from possible. Also, it is difficult to reduce the cost due to production of number of parts and its entire assembly cost. The monolithic elastic mechanism replaces the need of joints and the torsional spring.

2 Preliminaries

2.1 Pseudo-Rigid-Body Modelling

Pseudo-rigid-body model is extensively used for understanding the behaviour of flexible parts because they allow flexible bodies to be modelled as rigid bodies, thus allowing application of analysis and synthesis methods from rigid-body mechanisms [8, 9]. Traditional modelling approaches for elastic bodies focus on point-by-point variations in force and displacement. PRBMs, on the other hand, describe the behaviour of whole compliant segments and hence are useful when tackling design issues at the component level. PRBM helps in solving the model by considering the given elastic segment into two parts. One becomes the rigid segment, and the other becomes the pseudo rigid segment which represents the motion of the elastic segment [10] (Fig. 1).

Study and Design Conceptualization of Compliant ... 287

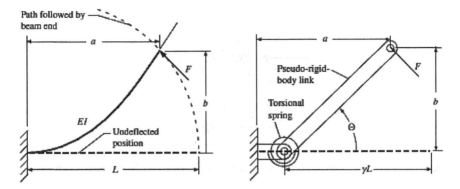

Fig. 1 PRBM model generated from a fixed-free compliant segment [8]

The pseudo rigid segment is considered to be 0.85 L or 0.83 L, where L is the length of the elastic segment, and the coefficients are known as characteristic radius factor. Former coefficient is used when θ is rotated more than 45°, and latter coefficient is used when θ is rotated less than 15°.

Extending the pseudo-rigid-body model to the four-bar linkage, we obtain the angles of three links with respect to horizontal in any configuration. Now, applying the loop closure equation given by:

$$L_2 \cos \theta + L_3 \cos \beta - L_4 \cos \varphi = L_1 \tag{1}$$

$$L_2 \sin \theta + L_3 \sin \beta - L_4 \sin \varphi = 0 \tag{2}$$

and differentiating with respect to θ gives equations in the matrix form:

$$\begin{bmatrix} -L_3 \sin \beta & -L_4 \sin \varphi \\ -L_3 \cos \beta & -L_4 \cos \varphi \end{bmatrix} \begin{bmatrix} \frac{d\beta}{d\theta} \\ \frac{d\varphi}{d\theta} \end{bmatrix} = \begin{bmatrix} L_2 \sin \theta \\ -L_2 \cos \theta \end{bmatrix} \tag{3}$$

As the deflection is known, $\Delta\theta$ angle is calculated, and the other two angles are obtained by kinematic sensitiveness equation (which is derived from the loop closure equation after differentiating it with respect to θ) given by (Fig. 2):

$$\frac{d\beta}{d\theta} = \frac{L_2 \times \sin(\varphi - \theta)}{L_3 \times \sin(\beta - \varphi)} \tag{4}$$

$$\frac{d\varphi}{d\theta} = \frac{L_2 \times \sin(\beta - \theta)}{L_4 \times \sin(\beta - \varphi)} \tag{5}$$

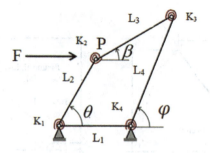

Fig. 2 Angles of links with respect to horizontal and representation of characteristic stiffness

Final angles after the deflection are given by:

$$\beta_{update} = \beta + \frac{d\beta}{d\theta}\Delta\theta \qquad (6)$$

$\beta_{update} \rightarrow \beta$ angle after the deflection

$$\varphi_{update} = \varphi + \frac{d\varphi}{d\theta}\Delta\theta \qquad (7)$$

$\varphi_{update} \rightarrow \varphi$ angle after the deflection

$$\theta_{update} = \theta + \Delta\theta \qquad (8)$$

$\theta_{update} \rightarrow \theta$ angle after the deflection

μ_i is the change in rotation experienced by the torsional spring.

$$\mu_1 = \theta_{update} - \theta \qquad (9)$$

$$\mu_2 = (\beta_{update} - \beta) - \mu_1 \qquad (10)$$

$$\mu_4 = \varphi_{update} - \varphi \qquad (11)$$

$$\mu_3 = (-\beta_{update} + \beta) + \mu_4 \qquad (12)$$

F is the force applied at the point P_i. P_x and P_x' are the x-coordinate of point P_i before and after the deflection.

$$P_x = L_2 \cos\theta \qquad (13)$$

$$P'_x = L_2 \cos \theta_{update} \tag{14}$$

$$\Delta P_x = P'_x - P_x \tag{15}$$

Now, area of cross section is calculated by putting the equation of characteristic stiffness into the elastic equilibrium equation.

$$K_i = 2.25 \frac{EI}{L_i} \tag{16}$$

As the force applied on the four-bar linkage is known and the angles after the deflection are calculated, moment of inertia of each elastic segment is determined using the elastic equilibrium equation.

$$\text{Potential Energy} = \text{Strain Energy} + \text{Work Potential} \tag{17}$$

$$PE = \frac{1}{2} \sum_{I=1}^{4} K_i \mu_i^2 - F \times \Delta P_x \tag{18}$$

Minimizing the potential energy after differentiating throughout by θ gives:

$$K_1 \mu_1 + K_2 \mu_2 \left(\frac{d\beta}{d\theta} - 1 \right) + K_3 \mu_3 \left(\frac{d\varphi}{d\theta} - \frac{d\beta}{d\theta} \right) + K_4 \mu_4 \left(\frac{d\varphi}{d\theta} \right)$$
$$= F(-L_2 \sin \theta_{update}) \tag{19}$$

3 Modelling and Simulation

3.1 Material Selection

Various factors are considered while selecting material for designing the accelerator pedal [11, 12]. These include cost, density, resilience, Young's modulus, flexural modulus and yield strength of the material. Polypropylene is, hence, selected because of its good machinability and low value of Young's modulus and resilience which means that it will absorb less energy upon loading and gives higher mechanical advantage. Several researchers have also used polypropylene as their material for compliant designs since it provides excellent surface finish (Table 1).

Table 1 Different material choices for a compliant design [8]

Material	E (GPa)	S_y (MPa)	$(S_y/E) \times 1000$	$(S_y^2/2E) \times 0.001$
Polycrystalline silcon	169	930	5.5	2600
Polyethylene (HDPE)	1.4	28	20	280
Nylon 6,6	2.8	55	20	540
Polypropylene	1.4	35	25	437.5
Kevlar (82 vol.%) in epoxy	86	1517	18	13,000
E-glass (73.3 vol.%) in epoxy	56	1640	29	24,000

3.2 Mathematical Model

Mathematical model is formulated on MATLAB using pseudo-rigid-body model equations, and the required width and height of cross-sectional elastic segment are obtained. For the operation of the accelerator pedal with the test engine, the required deflection of the accelerator pedal is 30 mm which provides the full throttle condition for the test engine. The force applied on the commercial accelerator pedal under test by the driver, given by the load cell sensor, is approximately 150 N. The force can also be calculated using the stiffness of the tension spring/torsional spring used on commercial accelerator pedal and its product with the required deflection. Lengths and initial angles of the elastic segments are provided for an ergonomic pedal design (Fig. 3).

The obtained width is 7.1 mm and height is 30 mm of cross-sectional elastic segment from the mathematical model. Height of 30 mm provides enough strength to prevent it from lateral deflection (Figs. 4, 5, 6, 7 and 8).

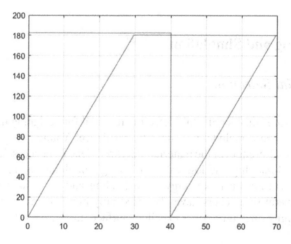

Fig. 3 Nonlinear plot of the undeformed (blue) and deformed (red) four-bar elastic model. Deformation not only causes it to translate in x-axis, but also in y-axis downwards

Study and Design Conceptualization of Compliant … 291

Fig. 4 Solid model of a compliant accelerator pedal on SolidWorks

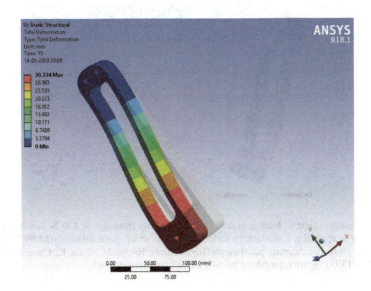

Fig. 5 Total deformation of 30.33 mm, which is the required condition for full throttle

3.3 Solid Modelling and FEA Simulation

Solid model is generated using SolidWorks, and FEA simulation is run on ANSYS static structural. A force of 150 N is applied on the footrest of the accelerator pedal

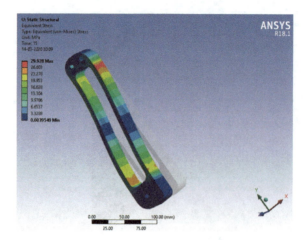

Fig. 6 Equivalent (von mises) stress is high near the juncture of two elastic segments reaching a maximum of 29.92 MPa

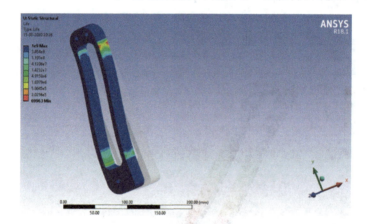

Fig. 7 A repetitive type of loading is considered that varies from 0 N to 150 N. Goodman mean stress theory is applied to calculate the fatigue life cycle which gives minimum of 69,963 cycles before the compliant accelerator pedal shows fatigue failure. Haque, M., Goda, K., Ogoe, S., Sunaga, Y. (2019, p. 139) [13] are referred for the S-N curve of polypropylene

in the simulation which gives the total deformation as 30.33 mm. In addition, the equivalent stress (von mises) generated is 29.92 MPa which is under the safe limits.

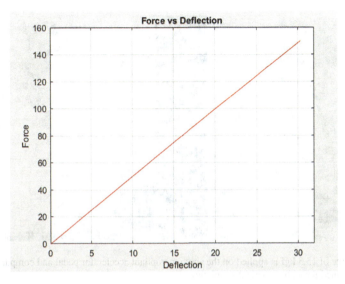

Fig. 8 Approximately linear Force vs Deflection plot shows optimal sensitivity [4] of the accelerator pedal

4 Validation

The accelerator pedal is validated on a plane surface by bolting it to a rigid frame. A digital spring balance is used to set the applied force on the pedal to 15.3 kgf. The deflection of the actual complaint accelerator pedal is compared to the one obtained through the total deformation in numerical simulation (Fig. 9).

5 Conclusion

This paper provides an alternate design for the traditional accelerator pedal. The study of compliant mechanism helps to design an accelerator pedal with distributed compliance. Pseudo-rigid-body model equations give excellent results for elastic four-bar mechanism, and hence, it is used to obtain the width and height of the cross-sectional segment through mathematical model. FEA simulation helped to verify the required deflection to operate the engine up to full throttle condition and keep the stresses induced below the material yield point. In addition, it will reduce the weight and material and assembling cost. It can be easily manufactured using CO_2 laser cut, waterjet machining, milling and 3D printing technique. Actual compliant accelerator pedal is then compared to the numerically simulated pedal. It is obtained that the deformation in the actual pedal is minutely more than that of the simulated one. This can be explained on the basis of manufacturing and mounting errors. The fatigue life of the polypropylene can be increased by making it a composite structure [13]. It

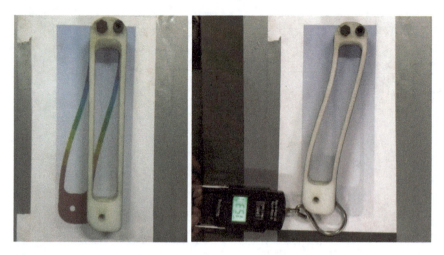

Fig. 9 Force of 15.3 kgf is applied on the actual compliant accelerator pedal and compared to the results obtained through numerical simulation

also opens the wide scope of using compliant mechanisms as an alternative to other components in the automotive industry.

References

1. Shuib S, Ridzwan M, Kadarman H (2007) Methodology of compliant mechanisms and its current developments in applications: a review. Am J Appl Sci 4:160–167
2. Roach G, Howell L (2002) Evaluation and comparison of alternative compliant overrunning clutch designs. J Mech Des 124(3):485–491
3. Crane N, Weight B, Howell L (2001) Investigation of compliant centrifugal clutch designs. In: ASME design engineering technical conferences, vol 2. ASME, pp 587–597
4. Liu P, Zhang T (2013) Modeling and acceleration response analysis for vehicle drivability. In: advanced materials research conference, vol 712–715. Trans Tech, Switzerland, pp 1473–1476
5. Deng T-M, Fu J, Shao Y, Peng J, Xu J (2018) Pedal operation characteristics and driving workload on slopes of mountainous road based on naturalistic driving tests. Safety Sci J 119:40–49
6. Mangukia N, Mangukia N (2018) Design and fabrication of brake pedal for all terrain vehicle-designed in consideration with the leverage, sitting position and brake force. Int J Eng Dev Res 6(2):562–568
7. Jain A, Pawar A, Kondhlakar G (2017) Design and analysis of accelerator pedal for four wheeler commercial vehicle. Int J Sci Res Dev 5(10):731–734
8. Howell L, Magleby S, Olsen B (2013) Handbook of compliant mechanisms. 1st end. John Wiley &Sons Ltd., United Kingdom
9. Pandiyan A, Kumar A (2016) Design methods for compliant mechanisms used in new age industries-A review. J Appl Eng Sci 14(2):223–232
10. NPTEL (2019) https://nptel.ac.in/courses/112/108/112108211/. Last accessed 23 Jan 2019
11. Sessions J, Pehrson N, Tolman K, Erickson J, Fullwood D, Howell L (2016) A material selection and design method for multi-constraint compliant mechanisms. In: 40th mechanism and robotics conference 2016, IDETC/CIE, vol 5A. ASME, North Carolina (2016)

12. Kaneko K (2008) Inelastic deformation behavior of polypropylene in large strain region and after cyclic preloadings. Def Sci J 58(2):200–208
13. Haque M, Goda K, Ogoe S, Sunaga Y (2019) Fatigue analysis and fatigue reliability of polypropylene/wood flour composites. Adv Ind Eng Polymer Res 2(3):136–142

Numerical Study on Fracture Parameters for Slit Specimens for Al2124 and Micro-alloyed Steel

Pranjal Shiva and Sanjay Kumar

Abstract This paper analyses a notched metallic specimen using the finite element method in ABAQUS CAE. The conditions of a tension test are simulated, and the specimen is subjected to uniaxial tensile loading by imposing strain rate. Contact criteria is applied between the reference node of loading pins and the specimen, and then, the load is applied on the top-loading pin keeping the bottom pin fixed. CPS8R plane stress element type is used to define the mesh. The classical metal plasticity theory is used in the present simulation, and the load–crack mouth opening displacement (CMOD) of two different materials, aluminium alloy Al2124 and micro-alloyed steel, are compared. Further, the stress developed on the specimens at the notched tip in x- and y-directions are also compared. The change observed in the behaviour of specimen upon varying its thickness is analysed by computing load–displacement diagrams. The results obtained from simulation are later compared to empirical data, and results are matched, with error within limits. Thus, the process, described in the paper, eliminates the necessity to conduct experiments as the behaviour, and properties of any material can be estimated using finite element method.

Keywords Finite element method · ABAQUS CAE · Al 2124 · Micro-alloyed steel

1 Introduction

The research article describes a novel method of analysing aluminium alloy Al2124 and micro-alloyed steel using finite element method (FEM), which is a numerical technique to solve complex problems governed by differential equations. Conventional methods of experimentation are time-consuming and costly. Moreover, in destructive testing, a new specimen needs to be attached for every experiment. The specimens are thus analysed through simulating a tensile test on ABAQUS CAE [1,

P. Shiva · S. Kumar (✉)
Department of Mechanical Engineering, Delhi Technological University (Formerly DCE), New Delhi, Delhi 110042, India
e-mail: sanjaydce2008@gmail.com

2], one of the superior software packages due to its user-friendliness and high accuracy. Aluminium Al2124 is widely used to manufacture aircraft structures [3] while micro-alloyed steel is used in automotive crankshafts [4]. The aim of the research is to obtain their properties and characterize various thin plates according to their thickness.

2 Theory

During tensile testing, the true stress developed in the specimen is greater than the nominal (engineering) stress. This is because of the lateral contraction and decrease in the area, by the law of conservation of volume. The true stress and nominal stress acting on a material given by [5] are:

$$\sigma_N = \frac{F}{A_O}; \sigma_T = \frac{F}{A}$$

where F = force applied, A_O = original cross section area, A = instantaneous cross section area.

The true and nominal stress are related by:

$$\sigma_T = \sigma_N(1 + \epsilon_N)$$

where σ_T = true stress, σ_N = nominal stress, ϵ_N = nominal strain.

The true strain and nominal strain are related by the following equation:

$$\epsilon_T = \int_{L_O}^{L} \frac{dL}{L} = \ln\left(\frac{L}{L_O}\right) = \ln(1 + \epsilon_N)$$

where

$$\epsilon_N = \frac{\Delta L}{L_O}$$

And, ΔL = change in length, L_O = initial length.

Thus, the plastic strain developed in a material is given by:

$$\epsilon_P = \epsilon_T - \frac{\sigma_T}{E}$$

where E = modulus of elasticity of the material ($\frac{\sigma_T}{\epsilon_T}$ at yield).

The typical compositions of aluminium Al2124 and micro-alloyed steel are given in Tables 1 and 2 [6]:

Table 1 Composition of aluminium Al2124

Element	Percentage (%)
Copper	5.12
Zinc	1.13
Manganese	0.592
Iron	0.565
Magnesium	0.454
Silicon	4.09
Titanium	0.112
Tin	0.112
Lead	0.044

Table 2 Composition of micro-alloyed steel

Element	Percentage (%)
Carbon	0.12
Manganese	1.6
Phosphorous	0.025
Sulphur	0.015
Copper	0.5
Nickel + Chromium + Molybdenum	0.5
Niobium + Titanium + Vanadium	0.15

3 Modelling

3.1 Design

The specimen is modelled as a two-dimensional deformable planar part with a width of 8 mm, length of 40 mm and a slit of 1.5 mm at the centre. Two holes of 1.5 mm radii, required for the loading pins, are cut at upper and lower ends as shown in Fig. 1.

3.2 Material Properties

The data for elasticity and plasticity using the stress–strain curve from uniaxial tension are used as input to define the material properties for aluminium alloy and micro-alloyed steel [6]. A solid (homogeneous) section is defined and subsequently assigned to the part.

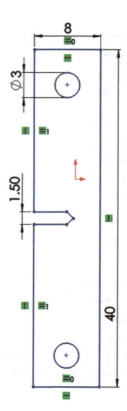

Fig. 1 Design and dimensions of the specimen

3.3 Assembly

The instance is assembled in the assembly module as an independent instance. The loading pins are modelled as discrete rigid bodies by establishing contact constraints with the reference points assembled with the specimen. Further, tie contact interaction is defined for constraining the movement between loading pin and specimen.

3.4 Step

Static, general step is defined with a time period of 1 s. Nonlinear geometry option is enabled due to high deformations, and the initial increment is lowered to 0.01.

Fig. 2 Application of load

3.5 Load and Boundary Condition

Displacement of 0.2 mm is applied on the top pin in the Y-direction, and the bottom pin is restrained by imposing pinned boundary conditions to simulate experimental conditions (Fig. 2).

3.6 Meshing

The model is meshed with a total of 2172 CPS8R elements (8-node biquadratic, reduced integration) and free element type. The mesh is improved around the notch to increase the accuracy of the results (Fig. 3).

4 Results and Discussion

The following curves are graphed for both the materials:

1. Stress–strain curve at the notched node—to discern the Young's modulus, ultimate tensile strength and yield stress.

Fig. 3 Mesh generated around the notch

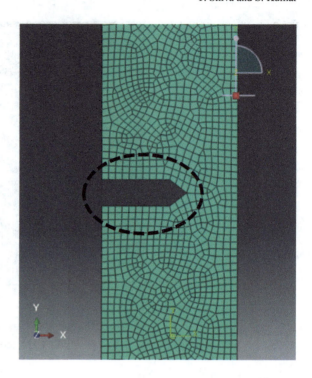

2. Stress in X- and Y-directions at the notched node—to compare the relative stresses.
3. Load and displacement curves for 2, 4 and 8 mm specimens—to visualize the effect of thickness on the reaction force (Figs. 4, 5, 6).

The probe values obtained from these curves are exported on to MATLAB programme to generate Figs. 7, 8, 9 and 10.

5 Inference

1. The reaction force developed at the fixed pin is directly proportional to the thickness of the specimen (for the same displacement) (Table 3).
2. The stress experienced in Y-direction is much higher than that in X-direction.
3. The following tables compare the values obtained from the experiment and numerical simulation (Tables 4 and 5).

Fig. 4 Von-mises stress contours around notch

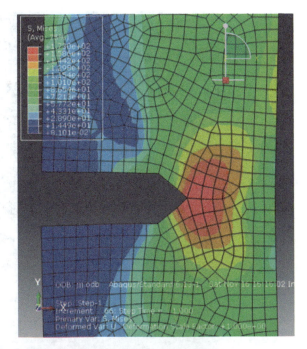

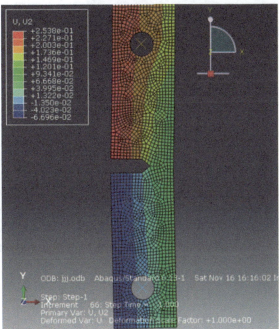

Fig. 5 Displacement in Y-direction

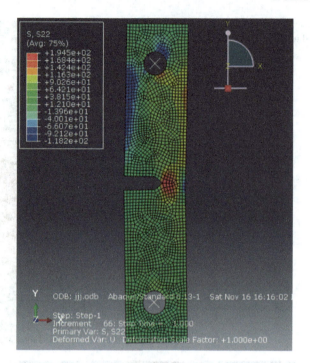

Fig. 6 Stress in Y-direction

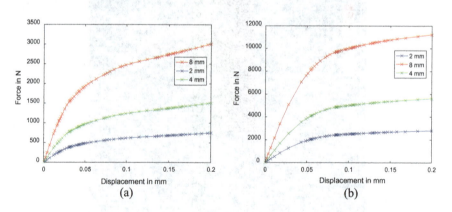

Fig. 7 Force versus displacement of **a** Aluminium Al2124, **b** micro-alloyed steel

Numerical Study on Fracture Parameters for Slit Specimens … 305

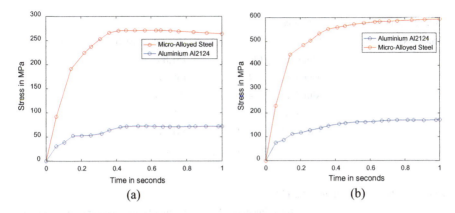

Fig. 8 Stress in **a** X-direction, **b** Y-direction for both materials

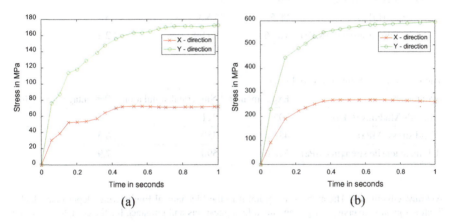

Fig. 9 Stress in X- and Y-direction **a** Aluminium Al2124, **b** micro-alloyed steel

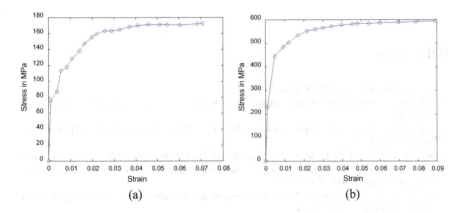

Fig. 10 Stress versus strain for **a** Aluminium Al2124, **b** micro-alloyed steel

Table 3 Values for the reaction force

Thickness (mm)	Material	Maximum reaction force (N)
2	Aluminium 2124	751.06
4	Aluminium 2124	1502.12
8	Aluminium 2124	3004.24
2	Micro-alloyed steel	2807.57
4	Micro-alloyed steel	5615.14
8	Micro-alloyed steel	11,230.3

Table 4 Values for aluminium Al2124

Parameter	Experimental	Numerical simulation	Percentage error (%)
Young's Modulus (GPa)	66	71	7.5
Yield stress (MPa)	80.5	76.3	5.2
Ultimate tensile strength (MPa)	167.6	171.4	2.2

Table 5 Values for micro-alloyed steel

Parameter	Experimental	Numerical simulation	Percentage error (%)
Young's Modulus (GPa)	197	211	7.1
Yield stress (MPa)	481.67	450	6.5
Ultimate tensile strength (MPa)	546.67	590	7.9

Acknowledgements The authors are grateful to the Mechanical Engineering Department, Delhi Technological University, for giving valuable suggestions and guidance for the completion of the project.

References

1. Abaqus/CAE User's Guide, Version 6.13.
2. Tension test tutorial, https://www.youtube.com/watch?v=pyJSoCqOvcQ&t=600s, last accessed 2019/12/01
3. Material property data, https://www.metalsuppliersonline.com/propertypages/2124.asp#top, last accessed 2019/12/03
4. Ko YS, Park JW, Park H, Lim JD Application of high strength microalloyed steel in a new automotive crankshaft. Hyundai and Kia Motor Company Research and Development Division, Whasung, Kyunggi 445-850, Korea
5. University of Cambridge. https://www.doitpoms.ac.uk/tlplib/mechanical_testing_metals/true.php, last accessed 2019/11/29
6. Pandya N, Sehgal DK, Pandey RK (2010) Determination of the fracture properties of metallic alloys using miniature specimen technique and finite element simulation, M.Tech Thesis. I.I.T, Delhi

Different Coating Methods and Its Effects on the Tool Steels: A Review

Sourav Kumar, Kanwarjeet Singh, Gaurav Arora, and Swati Varshney

Abstract Surface coatings are done on the various tool steel materials to increase their surface properties. In this paper, work done by various researchers on different coating methods and its effects on the properties of material has been reviewed. The effect of coating on microstructure, grain structure, Vickers microhardness, wearability, thermal properties, and adhesive strength of the substrate with and without the coating are also been studied in this paper. The outcome of the literature has shown a substantial improvement in the surface properties of the tool steels due to the application of the coating. This paper also represents the future scope of work that can be done by various researchers.

Keywords Coating · Tool steels · Microstructure · Vickers microhardness · Grain structure · Wearability

1 Introduction

1.1 Tool Steel

Tool Steels are widely used in the mold, die, and industries where there is a need for making products used for high load work where the wear resistance is the prime concern [1]. Tool steel contributes a huge role in engineered materials that are used by the industries [2]. Due to a significant proportion of chromium and carbon, it is used in both coal and hot working operation. Due to its exceptional properties, it is an expensive material mainly for larger jobs due to material cost and the making cost. But when there is a need for high toughness material for making the component high alloy steels can't be used due to economical factors. So, the surface of carbon steels is then coated with high alloy tool steels which increases its hardness and wear resistance making it suitable for operation and is economical in nature.

S. Kumar (✉) · K. Singh · G. Arora · S. Varshney
Delhi Institute of Tool Engineering, Okhla Phase-II, New Delhi, Delhi 110020, India
e-mail: souravkumar3694@gmail.com

© The Author(s), under exclusive license to Springer Nature Singapore Pte Ltd. 2021
R. M. Singari et al. (eds.), *Advances in Manufacturing and Industrial Engineering*,
Lecture Notes in Mechanical Engineering,
https://doi.org/10.1007/978-981-15-8542-5_26

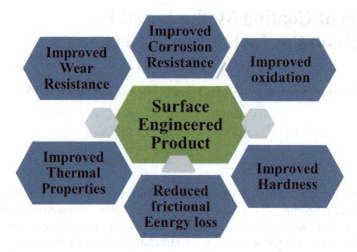

Fig. 1 Desired properties for a surface engineered product

Coatings are mainly applied to a surface to change its properties such as Wear resistance, hardness, corrosion resistance, adhesion, and wettability. These properties when improved make the substrate more stable and reliable to use. Desired properties required in a surface engineered product are shown in Fig. 1.

An important aspect to consider in the coating is the uniform thickness of the coating. To obtain uniform thickness various processes are used. Expensive machinery and operations are used to carry out the uniformity of the coatings.

The expand in the tribological field helped to achieve a better result in recent years and helped to achieve results which were unachievable earlier. The most highlighted deposition techniques seen in recent years are the thermal spray based and microwave coating methods.

Thermal spray coatings. Thermal spray refers to the group of coatings technologies used for the coatings of materials. There are mainly three types of thermal coatings, electric arc spray, plasma arc spray, and flame spray. In these procedures, the coating materials are heated to change them into the molten or semi-molten state. The molten metal is then sprayed to the surface of the substrate using the atomization jets. When the motel metal struck the surface of the substrate it forms a bond and a layer on the surface starts to build up [3].

Thermal coatings are widely used in industries because a broad number of materials can be used for the coatings as any material which doesn't putrefy on heating can be used. Another advantage is that it peels off the layer of the substrate and recoats the damaged area without changing its shape, size, and properties. The disadvantage of this coating is that it is impossible to coat a product having some cavity or having a very small size as the torch cannot reach into it [4].

Microwave hybrid coatings. Microwaves are the electromagnetic waves that have a frequency ranging 1–300 GHz. These microwaves have varying wavelengths. From a house kitchen to an industry, microwaves are used extensively. A household microwave oven has a 2.45 GHz frequency whereas the frequency of an industrial microwave is 915 MHz to 18 GHz. The basic microwave oven used in kitchens is usually used for heating food items having water content in it having a low temperature. Nowadays, the domestic microwave is being used by the researchers for higher temperature applications in industries.

2 Recent Works on Thermal Spray Coatings and Microwave Coatings

This section focuses on the researches done on thermal spray coatings and microwave coating and their effect on the substrate in recent years.

Verdon et al. [5] showed the change in the microstructure of the substrate when tungsten carbide-cobalt (WC–Co) coating was done on the substrate using HVOF thermal spray technique. The microstructure was checked by XRD and SEM methods and it was observed that new crystalline phases were formed. The grain structure became refined with the coating.

Singh et al. [4, 6] showed the development of Nickle and CeO_2 based coatings using a microwave. It was observed that optimum amount of 1wt% CeO_2, refined the microstructure of the substrate. It also enhanced the hardness of the substrate and increased its wear resistance. The hardness became 30% higher on addition of CeO_2. Similarly, in one of his studies, the effect of neodymium oxide by microwave cladding, showed that the microstructure became refined. The Vickers microhardness and the abrasive wear properties become enhanced.

Veronesi et al. [7] focused on microwave-based SHS coating of NiAl on the titanium substrate. It was observed that properties like hardness, toughness, and wear resistance of the substrate showed improvement.

Chandrasekaran et al. [8] noticed that microwave heating formation of a hotspot and thermal runaway was occurring. So, to avoid this, frequency of microwaves was varied, which resulted in a reduction of hotspots and arcing problems. They also concluded that a susceptor material is to be selected to provide constant heating at all the temperatures.

Yahaya et al. [9] discussed the benefit of microwave hybrid heating over the direct heating in the microwave. It was seen that the microwave hybrid heating is a faster process than direct microwave heating. It also eliminates the thermal runaway problem. It was seen that using susceptor offers a single energy source.

Singh et al. [10] in their study highlighted the latest achievements in the material joining and surface coating using microwave. EN-31 Tool steel was used as a substrate and it was coated using the microwave coating. XRD, SEM, Vickers hardness and pin on desk test were carried out. After the substrate was coated, it was seen that the microstructure refined, Vickers microhardness improved and the wear resistance increased.

Prasad et al. [11] carried out an experiment in which AISI 1040 steel was taken as a substrate for coating by microwave irradiation. The coating material used for the process was Ni/La$_2$O$_3$ composite. After the coating, microhardness of the substrate was tested and it was seen that microhardness increased after the microwave irradiation.

Agrawal [12] in his study discussed the growing utilization of microwave energy in different fields, including metal processing, sintering joining, cladding, recycling of used tyres, and WC–Co-based ceramic composites powder processing.

Prasad et al. [13] carried out an experiment in which temperature gradient studies of part with Co-based uni layer coating and multilayer coatings were observed by finite element simulation. Higher thermal conductivity was observed in Co-based with Cr$_3$Cr$_2$ reinforcements compared to other coatings so it is more resistant against corrosion caused due to heating. Therefore, it has been proposed that there is a development of thermal barrier coating layer on Co-based cladding.

Sharma [14] in his article discussed the experiment he carried out in which Ni-WC composite added with CeO$_2$ to form new composite Ni-WC-CeO$_2$, which were then used for coating by HVOF method on the substrate. After the coating, both the substrate having different coatings were compared and it was seen that the coating having CeO$_2$ gave better microstructure and was having better wear resistance.

Gupta et al. [15] carried out an experiment in which WC$_{10}$CO$_2$Ni cladding was done on SS-316 substrate using microwave irradiation. The clad microstructure showed structure like a composite with Iron, cobalt, nickel, and chromium forming the tough matrix, and tungsten carbide formed skeleton-like structure. This clad formed a dense and uniform structure that resulted in notably higher microhardness. As compared to SS-316 steel material with a sliding speed of 0.5 m/s these microwave clads exhibited 84 times higher wear resistance.

Shih et al. [16] conducted an experiment in which Nickle based layer on the SKD 61 tool steel was deposited using electroless plating. There was an increased coefficient of friction due to the presence of a non-diamond carbon phase which eventually obstructs the lubricity and therefore decreases the tool life. Qualitative Information on the adhesive strength between the diamond and the substrate can be provided by Tribotest.

Wu et al. [17] carried out an experiment in which a HVOF thermal coating was used. An increase of about 25% porosity was observed in the nanostructured YSZ TBC as compared to conventional TBC which is basically due to a larger amount of interspats in the coating. It was also seen that the thermal conductivity for nano-coating was 40% (0.8–1.1 W/m/EK) lower than the conventional coating showing better thermal insulation.

Mondal et al. [18] conducted an experiment where pure copper powder of different sizes and porosity were heated using microwaves. It was seen that the pure copper powder with varying size (6–383 μm) and having an initial porosity (24–44%) coupled with microwave and heated to a much higher temperature. It was observed smaller the powder size, the higher the heating rate will be. Also, compacts that have more porosity heats faster.

Oghbaei et al. [19] performed a comparative analysis of microwave and conventional sintering related to its applications and advantages. It was found that sintering materials consume lower energy with a microwave than conventional sintering. Due to enhanced mechanisms of microwave, the diffusion process intensifies. It was observed that by using microwave sintering higher heating rates can be achieved. Higher density and better grain distribution can be achieved by using microwave sintering.

Lohit et al. [20] carried out an investigation in which Ni-WC composite cladding was carried out on SS-316L steel using the microwave heating. It was observed that by microwave radiation having 2.45Ghz frequency with 900 W power, developed clad of 500 μm in 360 s. The microhardness of the clad developed on the substrate was found to be 477.5 Hv which was higher than that of uncoated substrate.

In this review article by Singh et al. [21] applications of microwave processing on manufacturing industries have been discussed. It was observed that improved microstructures can be achieved by optimization of various parameters such as temperature, heating rates temperature gradients, and heating time. It was also seen that finer microstructural development, higher densification parameters, finer average grain size, and lower porosity defect result in better properties compared to conventional techniques.

Work carried out by Gupta et al. [22] highlighted the experimental study of microstructure characteristics of microwave cladding on Austenitic stainless steel. SS-316 was taken as substrate in a powdered form of 40 μm. Nickle powder was used as the coating material. After the coating, it was observed that the phases changed in the substrate, microstructure became refined and the Vickers microhardness increased.

Gupta et al. [23] carried out an experiment to study the characterization of microwave composite cladding. (EWAC (Ni-based) + 20% $Cr_{23}C_6$ powder) was taken as the composite material and SS-316 was taken as the substrate material. Cladding was developed by using microwave heating. It was observed that the Vickers microhardness was the highest with the composite cladding and the microstructure of the substrate after the cladding was improved.

In a similar study done by Kaushal et al. [24] the composite taken was Ni-based and 20% added WC8Co. and the cladding was developed on the SS-304 substrate using the microwave heating process. It was observed that microwave heating induces material phase modification in the powder layer, carbide like chromium carbide, and W_2 clad formed. The clad surface was free from any type of micro-cracks and pores. Also, the mean microhardness was four times higher (840 ± 20 Hv) than the substrate (220 Hv).

Kaushal et al. [25] carried out an experiment in which the development of composite cladding on martensitic steel SS-420 using the microwave heating process. It was observed that the microstructure of the clad developed shows the cellular structure of the clad area. The Vickers microhardness was three times the substrate material (652 ± 90 Hv).

Vinay et al. [26] carried out an experimental study in which the development of coating was done by microwave hybrid coating method. The clads were developed using Nickle based powder with 10wt% silicon carbide (Ni–Al + 10%SiC). The developed clads shows metallurgical bonding with the martensitic steel substrate by partial mutual diffusion of elements. It was seen that the porosity in clads was very less and no visible crack formation was observed. The Vickers microhardness was found to be higher than the substrate due to the formation of carbides and intermetallic hard phases. Also, it was observed that the increase in load, results in the rate of wear increase.

Agarwal [21] in his experimental study concluded that the coatings developed by microwave hybrid heating were having better microstructure than that developed by a conventional heating method. It was also observed that the strength in the microwave heated parts was 30% higher than that of conventional heated parts.

Kaushal et al. [27] conducted an experiment in which cladding of Ni-based + 10% SiC was developed on the SS-420 martensitic steel using the microwave hybrid heating process. It was observed that the microstructure of the substrate after cladding was improved, the Vickers microhardness of the substrate after cladding became three times better. It was also seen that the wear rate of the substrate decreased.

Clark et al. [28] in his study discussed the processing of the material using microwave energy. He observed that the microwave characteristic includes high speed and uniform heating. Heating of microwave absorbing phases and components in materials enhanced physical and/or mechanical properties.

Work carried out by Toma [29] discussed the development of cermet coating using the HVOF thermal spray coating method. The coating material taken was Cr_3C_2 and 1.4571 steel was taken as the substrate. It was observed that weight loss in a highly corroded matrix was much greater than the less corroded matrix. It was also found that the addition of 4 wt% Cr improved corrosion and the erosion resistance of the WC-based coating.

3 Conclusion

- The microwave heating processing is proved to be the most efficient and time-saving process having the best results.
- The cladding on tool steel makes the tool steel more favorable to use as the surface properties after the claddings are more suitable for the heavy load applications.
- The development of composite coating by microwave hybrid heating was discussed.
- Comparisons between the microwave hybrid heating and conventional heating proved that microwave hybrid heating is more effective and stable.

4 Future Work

A very little work has been done for the development of coatings on tool steel by the microwave process. Therefore, a lot of scope is still there to investigate the effect of coatings developed by microwave heating on mechanical, microstructure, and wear properties of tool steel.

References

1. Cui C, Guo Z, Liu Y, Xie Q, Wang Z, Hu J, Yao Y (2007) Characteristics of cobalt-based alloy coating on tool steel prepared by powder feeding laser cladding. Opt Laser Technol 39(8):1544–1550. Author F, Author S (2016) Title of a proceedings paper. In: Editor F, Editor S (eds) Conference 2016, LNCS, vol 9999. Springer, Heidelberg, pp 1–13 (2016)
2. Jonda E, Labisz K, Dobrzański LA (2015) Microstructure and properties of the hot work tool steel gradient surface layer obtained using laser alloying with tungsten carbide ceramic powder. J Achievements Mater Manuf Eng 73(2):214–221. Author, F.: Contribution title. In: 9th International proceedings on proceedings. Publisher, Location, pp 1–2 (2010)
3. Davis JR (ed) (2004) Handbook of thermal spray technology. ASM international
4. Singh K, Sharma S (2018) Development of Ni-based and CeO_2-modified coatings by microwave heating. Mater Manuf Processes 33(1):50–57
5. Verdon C, Karimi A, Martin JL (1998) A study of high velocity oxy-fuel thermally sprayed tungsten carbide-based coatings. Part 1: microstructures. Mater Sci Eng: A 246(1–2):11–24
6. Singh K, Sharma S (2019) Effect of neodymium oxide on microstructure, hardness and abrasive wear behaviour of microwave clads. Mater Res Express 6(8):086599
7. Veronesi P, Leonelli C, Poli G, Casagrande A (2008) Enhanced reactive NiAl coatings by microwave-assisted SHS. COMPEL-The Int J Comput Math Electr Electron Eng 27(2):491–499
8. Chandrasekaran S, Ramanathan S, Basak T (2012) Microwave material processing—a review. AIChE J 58(2):330–363
9. Yahaya B, Izman S, Konneh M, Redzuan N (2014) Microwave hybrid heating of materials using susceptors-a brief review. Adv Mater Res 845:426–430
10. Singh S, Gupta D, Jain V (2016) Recent applications of microwaves in materials joining and surface coatings. Proc Inst Mech Eng, Part B: J Eng Manuf 230(4):603–617
11. Prasad A, Gupta D, Sankar MR, Reddy AN (2014) Experimental investigations of Ni/La2O3 composite micro-cladding on AISI 1040 steel through microwave irradiation. In All India manufacturing technology, design and research conference, vol 55, pp 1–6
12. Agrawal D (2010) Latest global developments in microwave materials processing. Mater Res Innovations 14(1):3–8
13. Prasad CD, Joladarashi S, Ramesh MR, Sarkar A (2018) High temperature gradient cobalt based clad developed using microwave hybrid heating. AIP Conf Proc 1943(1):020111
14. Sharma S (2012) Wear study of Ni–WC composite coating modified with CeO_2. The Int J Adv Manuf Technol 61(9–12):889–900
15. Gupta D, Sharma AK (2011a) Investigation on sliding wear performance of WC10Co2Ni cladding developed through microwave irradiation. Wear 271(9–10):1642–1650
16. Shih HC, Sung CP, Fan WL, Lee CK (1993) Formation and tribological application of CVD diamond films on steels. Surf Coat Technol 57(2–3):197–202
17. Wu J, Guo HB, Zhou L, Wang L, Gong SK (2010) Microstructure and thermal properties of plasma sprayed thermal barrier coatings from nanostructured YSZ. J Thermal Spray Technol 19(6):1186–1194

18. Mondal A, Agrawal D, Upadhyaya A (2008) Microwave heating of pure copper powder with varying particle size and porosity. J Microw Power Electromagn Energy 43(1):5–10
19. Oghbaei M, Mirzaee O (2010) Microwave versus conventional sintering: a review of fundamentals, advantages and applications. J Alloy Compd 494(1–2):175–189
20. Lohit RB, Bhovi PM (2017) Development of Ni-WC composite clad using microwave energy. Mater Today: Proc 4(2):2975–2980
21. Singh S, Gupta D, Jain V, Sharma AK (2015) Microwave processing of materials and applications in manufacturing industries: a review. Mater Manuf Processes 30(1):1–29
22. Gupta D, Sharma AK (2011b) Development and microstructural characterization of microwave cladding on austenitic stainless steel. Surf Coat Technol 205(21–22):5147–5155
23. Gupta D, Bhovi PM, Sharma AK, Dutta S (2012) Development and characterization of microwave composite cladding. J Manuf Processes 14(3):243–249
24. Kaushal S, Gupta D, Bhowmick H (2017a) On microstructure and wear behavior of microwave processed composite clad. J Tribol 139(6):061602
25. Kaushal S, Sirohi V, Gupta D, Bhowmick H, Singh S (2018) Processing and characterization of composite cladding through microwave heating on martensitic steel. Proc Inst Mech Eng, Part L: J Mater: Des Appl 232(1):80–86
26. Sirohi V, Gupta DG (2015) Surface modification of martenistic stainless steel through microwave hybrid heating (Doctoral dissertation)
27. Kaushal S, Gupta D, Bhowmick H (2017b) Investigation of dry sliding wear behavior of Ni–SiC microwave cladding. J Tribol 139(4):041603
28. Clark DE, Folz DC, West JK (2000) Processing materials with microwave energy. Mater Sci Eng, a 287(2):153–158
29. Toma D, Brandl W, Marginean G (2001) Wear and corrosion behaviour of thermally sprayed cermet coatings. Surf Coat Technol 138(2–3):149–158

Evaluation of Work-Related Stress Amongst Industrial Workers

Anuradha Kumari and Ravindra Singh

Abstract Work-related stress has become a significant reason for illness and a hazard to the mental and social prosperity of workers. This paper depicts the assessment of work-related stress using two physiological signals such as saturation of peripheral oxygen (SpO_2) and heart rate (HR) along with psychological scale (DASS 21). These physiological signals received from the industrial workers using mental arithmetic task (MAT)-based stimuli. Discriminant analysis and independent t test are used to distinguish the stress or normal state of the workers. The results obtained from the HR signal are accurate 84.2 and 78.2% in the case of SpO_2. Finally, the investigation concludes that physiological signals based or psychological scale-based examination gives the improved detection rate of stress.

Keywords Heart rate · Saturation of peripheral oxygen · Independent T-test

1 Introduction

American Psychological Association describes stress as "the common response of the human body when any demand put on it" [1]. Stress has become very common in the workplace; it is reported that 33% of workers in the industries are suffering from stress [2]. Stress at work is defined as the change in worker's cognitive and physical state at the workplace [3]. It usually occurs when there is a mismatch between workplace demands and worker's capability [4]. Various issues cause occupational stress, i.e. workplace harassment, extended working hours, overload and isolation, etc. [5]. Occupational stress causes various problems in workers that led to a reduction in the work performance due to injury, illness, absenteeism, violation, anger, hopelessness, indigestion, insomnia, anxiety, high blood pressure, fatigue, helplessness, low execution and emotional stability turnover cost and quality control cost [6].

A. Kumari (✉) · R. Singh
Department of Design, Delhi Technological University, New Delhi, India
e-mail: raizada53@gmail.com

Assessment of human stress is always a significant challenge for researchers; it can be assessed and measured in terms of physical, perceptual and behavioural response [7]. It causes behavioural and physiological changes that can be estimated by the psychological scale and physiological signal-based techniques [8].

In this present paper, psychological scale DASS 21 (depression, anxiety and stress) scale along with physiological signals specifically the saturation of peripheral oxygen (SpO_2) and heart rate (HR) is used to evaluate the stress amongst industrial workers.

2 Background

This section presents the background of occupational stress and covered various investigations, methods and case studies conducted for stress detection. The literature of work-related stress consists of substantial details relevant to this investigation. Table 1 gives a review of the literature related to this research.

3 Methods and Materials

The primary purpose of the investigation is to evaluate the occupational stress amongst industrial participants. The investigation was conducted in Godrej & Boyce, Punjab. The investigation was conclusive in nature, and data was gathered from all the participants through two different methodologies, i.e. psychological questionnaire-based (DASS21) and physiological signal-based (SpO_2 and heart rate signal). Discriminant analysis and independent t test were used to distinguish the stress or normal state of the workers as shown in Fig. 1.

3.1 Physiological Measures

Two physiological signals heart rate (HR), saturation of peripheral oxygen (SpO_2) have identified.

Heart rate: Physical and cognitive stress increases the movement of autonomic nervous system (ANS). When any human body comes under stress, the central nervous system orders the brain and heart to prepare for such circumstances, which leads to an increase in the heart rate. Heart rate signal is a significant stress indicator as compared to other physiological signals [15]. The proposed device to carry out this investigation is the NEULOG heart rate sensor, to measure the heart rate.

Saturation of peripheral oxygen: It is defined as the proportion of oxygen saturated haemoglobin comparative to the total hemoglobin in the blood. Ordinary oxygen saturation level in the human body blood is considered as 95–100%. Oxygen level below the 90% can influence the mental ability of the person [16]. The proposed

Table 1 Literature review related to the work-related stress assessment [9–14]

Sr. No.	Author(s), year	Methods	Findings	Contribution towards stress evaluation
1	Sujatha and Mishra (1996)	Effort–reward imbalanced scale	Sample: public and private sector of bank employees 1. Study was attempted to compare the work-related stress between the public & private sector of bank employees 2. Results indicate that public sector employees are more stressed than others	Less
2	Upadhayay and Singh (1999)	Perceived stress scale	Sample: college teachers and executives 1. The study attempt to compare the occupational stress between college teachers and executives. 2. Results indicate that executives are more than college teachers due to an increase in their responsibilities	Moderate
3	Healey and Picard (2004)	Psychological-signal based (heart rate, electrocardiogram, electromyograms, and skin conductance)	Sample: bus drivers 1. The whole experiment was designed to gathered driver(s) physiological response during the driving task 2. It showed that skin conductance and HR signals were the most significant stress indicators as compared to other collected signals	Moderate

(continued)

Table 1 (continued)

Sr. No.	Author(s), year	Methods	Findings	Contribution towards stress evaluation
4	Zhai and Barreto (2006)	Psychological-signal based (pupil diameter pressure, skin temperature, galvanic skin response, and blood volume)	Sample: computer user 1. All the four physiological signals were collected from the computer user during MAT task stimuli 2. A SVM (support vector machine) is used for the classification of stressed and non-stressed users 3. Results showed that pupil diameter was the most significant stress indicator	Moderate
5	Fernandes et al. (2008)	Perceived stress scale	Sample: young physician 1. The study was attempted to observe that How Social support impact on role stress. 2. The study found that role stress can be reduced by enhancing social support at the workplace	Less
6	Chaly et al. (2014)	DASS-21	Sample: software professionals and school teachers 1. The study was an attempt to evaluate the work-related stress amongst school teachers and software professionals at Trivandrum 2. Results indicate that 15% of school teachers and 71% of software professionals suffering from work-related stress	Moderate

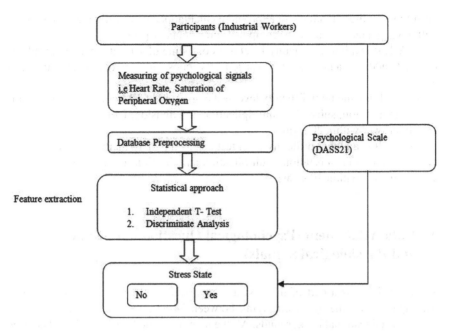

Fig. 1 Flow chart of research methodology

device to carry out this investigation is pulse oximeter (AFE4400) to measure the oxygen level in human body blood.

3.2 Psychological Questionnaire Measures

Depression, anxiety and stress scale (DASS 21) is utilized for the assessment of work-related stress. The psychological questionnaire analysis was used to verify the perception of the high, medium and low levels of stress. This scale consists of 21 questions in which workers need to rate the various circumstances that occurred with them from the last few weeks in the workplace on a scale of "3" to "0" where a "3" was used to represent "always" (applied to me) and "0" was used to represent "never" (did not apply) [17].

3.3 Task Validation/Legitimization

In the investigation, Measurement Incorporated Secure Test (MIST) procedure is used; it runs on a computer; it creates mental arithmetic problems, no special tool was used in this study; real-life situations were used to simulate stress. Participants need to solve these mathematical problem calculations using only the human mind with no assistance from a pen, computer and calculator [18].

Task 1: It is a lower difficulty level test consisting of 10 mathematical problems related to addition and substation. Participants choose the correct answer within 20 s gave.

Task 2: It is a medium difficulty level test consisting of 10 mathematical problems related to addition, substation, multiplication and division. Participants choose the correct answer within 20 s gave.

Task 3: It is a higher difficulty level test consisting of 10 mathematical problems related to addition, substation, multiplication and division. Participants choose the correct answer within 10 s gave.

4 Data Attainment (Psychological Questionnaire Data and Physiological Signals)

The carried out as a part of this research involved 50 (15 females and 35 males) participants with the age group lying between 24 and 53 years old with various social and educational backgrounds. All the members who have participated are free from the history of medication and drugs. The following steps were followed to collect the psychological questionnaire date and physiological signals data from the participant:

Step 1: DASS 21 scales were applied to the participants to reveal the stress level amongst them. It consists of 21 questions in which workers need to rate their bad situations happened at workplace from the scale of "0" to "3", where a rating of "3" was used to represent "always and "0" was used to represent "never".

Step 2: Heart rate and SpO_2 signals were chosen to collect. On the index finger of the left hand, NEULOG heart rate sensor was placed, and on the index finger of the right hand, the SpO_2 sensor was placed. HR and SpO_2 signals were collected at rest position before task; Task 1 is given to the participant during task again; these signals were collected; after completion of Task 1 again, these signals were collected at rest position.

Step 3: Task 2 is given to the participants during their task HR, and SpO_2 signals were collected; after completion of Task 2 again, these signals were collected at rest position.

Step 4: Task 3 is given to the participants during their task HR, and SpO_2 signals were collected; after completion of Task 3 again, these signals were collected at rest position. Significantly, difference can be seen in physiological signals during all the rest and tasks condition. Whole experimentation process was free from the uncertainty of human obstruction, the same light conditions and the same temperature. Here, Figs. 2 and 3 show the acquired heart rate and SpO_2 signals.

Fig. 2 It shows the acquired heart rate signal

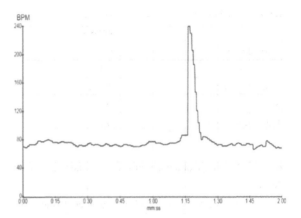

Fig. 3 It shows the acquired SpO$_2$ signal

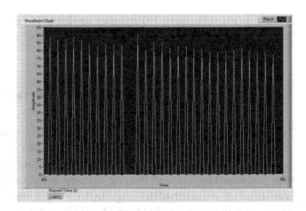

5 Results and Conclusion

This section introduces the results acquired by applying the techniques and methods described previously.

5.1 Questionnaire Analysis

DASS 21 applied to all the participants to assess the stress in terms of behaviour change. Table 2 shows the DASS21 severity scoring. If the stress score is laying between 0 and 7, the person is not under stress; if stress score lying from 8–9, 10–12, 13–16 and 17+ then, people are under mild, moderate, severe and extreme severe stress conditions [17].

Table 3 As per DASS21 questionnaire analysis, it is found that 46% of workers

Table 2 Psychological scale severity scoring

Severity	Stress score	Stress response
Normal state	0–7	No
Mild state	8–9	Yes
Moderate state	10–12	Yes
Severe state	13–16	Yes
Extreme severe state	17+	Yes

Table 3 Percentage of number of stressed or non-stressed workers

Stress	Non-stress
46% [women worker (9 out of 15) men worker (14 out of 35)]	54%

women workers (9 out of 15) and men workers (14 out of 35) are in stress condition and 54% of workers are in non stressed conditions.

5.2 Statistical Analysis of Acquired Physiological Signals

Heart rate and SpO$_2$ signals were received from all the participants. As per DASS 21 score, all participated members are divided into two groups (stress and non-stress), and statistical features were extracted from the collected data, and independent t test was performed between all the non-stressed and stressed workers to investigate the statistically distinct between two groups. Discriminant analysis was performed in all the prevailing features of SpO$_2$ and heart rate signals to recognize which variable is discriminate.

Statistical Feature Extraction: Statistical features are extracted from the collected database. These statistical features are mean, median, mode, standard deviation, variance, minima, maxima and peak-peak value.

Independent t test: On all the extracted features of the selected signals, Independent T-TEST was performed to recognize the dominating features that are statistically distinguishable between non-stressed stressed industrial workers. After that, discriminant analysis was performed on dominant features of the signals. Independent t test was performed on SPSS software.

$$t = \frac{(\overline{x_1} - \overline{x_2}) - (\mu_1 - \mu_2)}{s_{\overline{x_1} - \overline{x_2}}} \quad (1)$$

Null hypothesis H$_0$: $u_1 = u_2$, (mean of two different groups (stressed and non-stresses) are equal).

Alternative hypothesis H$_A$: $u_1 \neq u_2$ (mean of two different groups (stressed and non-stresses) are not equal).

Evaluation of Work-Related Stress Amongst Industrial Workers

However, t test is applied to all the extracted features of SpO_2 and HR signals. From all the extracted features of the signals, maxima of the SpO_2 signal and standard deviation of the HR signal is statistically significant as shown in Table 4.

Discriminate Analysis: Discriminate analysis is performed on maxima of the SpO_2 and standard deviation of the heart rate signal. It is used to determine which variables discriminate between two or more groups in which dependent variables are predictor and independent variables are groups. It also specifies which independent variable is the best predictor of the group.

Table 5 shows that the means of each group are significantly different from one another if the value of **F-Prob** is smaller than 0.05. In the present table, values (0.021, 0.18, 0.028, 0.012, 0.023, 0.001 and 0.027) were statically significantly varying with each task's difficulty level.

Table 6 shows that the means of each group are significantly different from one another if the value of **F-Prob** is smaller than 0.05. In present table values (0.0019, 0.031, 0.023, 0.029, 0.021, 0.029 and 0.138) were statically significantly varying with each task's difficulty level.

Table 4 Independent t test on dominant features of SpO_2 and heart rate signals

SpO_2 (maxima)				HR (standard deviation)			
Independent t test on stressed versus non-stressed	t-value	Df	Sig	Independent t test on stressed versus non-stressed	t-value	Df	Sig
Rest	0.52	5	0.468	Rest	1.60	12	0.135
Task 1	0.48	2	0.081	Task 1	2.62	4	0.059
Rest after Task 1	0.68	4	0.398	Rest after Task 1	1.72	11	0.115
Task 2	2.08	2	0.038	Task 2	2.96	4	0.032
Rest after Task 2	1.28	4	0.102	Rest after Task 2	1.95	9	0.091
Task 3	2.95	2	0.018	Task 3	2.25	4	0.011
After task	1.402	1	0.151	After task	1.14	5	0.204

Significance level ($a = 0.05$)

Table 5 Test of equality of different groups' mean of the heart rate signal

Situation	Wilks' lambda	F	Df1	Df2	Sig.
Rest	0.712	4.692	3	41	0.021
Task 1	0.798	3.288	3	41	0.018
Rest after Task 1	0.622	5.261	3	41	0.028
Task 2	0.862	7.061	3	41	0.012
Rest after Task 2	0.512	4.145	3	41	0.023
Task 3	0.978	0.834	3	41	0.001
Rest after Task	0.913	0.727	3	41	0.027

Table 6 Test of equality of group means (SpO$_2$ signal)

Situation	Wilks' lambda	F	Df1	Df2	Sig.
Rest	0.715	5.621	3	41	0.019
Task 1	0.676	4.342	3	41	0.031
Rest after Task 1	0.729	5.013	3	41	0.023
Task 2	0.780	5.517	3	41	0.029
Rest after Task 2	0.751	5.881	3	41	0.021
Task 3	0.813	7.132	3	41	0.029
Rest after Task	0.791	1.328	3	41	0.138

Work-related stress classification: Classification is a process of preparing a model that maps each attribute set X to one of the predefined class names Y. Here, in our case, class names Y is the stress class of industrial workers characterized by DASS 21 score (no stress, mild, moderate, severe) and attribute set X is the functions are a linear combination of two statistical features, namely standard deviation of HR, maxima of SpO$_2$ HR extracted from the physiological parameters. Function attributes used for stress classification because these functions are sufficient for the detection of stress levels of the industrial workers.

Table 7 shows that at rest phase of the task segment is observed as the no stress category. Task 1 is observed as mild-level stress category. Task 2 is considered a Moderate stress category. Task 3 is considered a severe stress category. After the task, it may or may not be in the stress category. It is presumed that the change in the stress level of the workers is solely due to the various tasks given to them.

Table 8 shows that the discriminant analysis can classify the stress and non-stress level amongst workers with an accuracy of 84.2% for HR (SD) and 78.8% for SpO$_2$ (Max) into their respective stress class according to DASS21 scale.

Table 7 Categorization of worker's task segment as per stress category and stress class

Task segment used	Stress category	Stress class
Rest	Mild	1
Task 1	Moderate	2
Task 2	Severe	3
Task 3	Extreme severe	4
Rest after task	May/may not stress	

Table 8 Discriminant analysis-based classification of stress class using function for workers on HR (standard deviation) and SpO$_2$ (maxima)

HR (standard deviation)				SpO$_2$ (maxima)			
Sr. No.	Actual stress values	Predicted stress values	Correctly classified in its class (Yes/No)	Sr. No.	Actual stress values	Predicted stress values	Correctly classified in its class (Yes/No)
1	2	2	Yes	1	2	2	Yes
2	1	1	Yes	2	2	1	No
3	2	2	Yes	3	2	2	Yes
4	3	3	Yes	4	3	3	Yes
5	1	1	No	5	1	1	No
6	1	1	Yes	6	1	1	Yes
7	2	2	Yes	7	2	2	Yes
8	1	1	Yes	8	1	1	Yes
9	2	2	Yes	9	2	2	Yes
10	1	1	Yes	10	1	1	Yes
11	1	2	No	11	1	2	No
12	2	1	No	12	2	1	No
13	1	1	Yes	13	1	1	Yes
14	1	1	Yes	14	1	1	Yes
15	1	2	No	15	1	2	No
16	1	1	Yes	16	1	1	Yes
17	1	1	Yes	17	1	1	Yes
18	2	2	Yes	18	2	2	Yes
19	1	1	Yes	19	1	1	Yes
20	1	1	No	20	1	1	No
21	3	3	Yes	21	3	3	Yes
22	1	1	Yes	22	1	1	Yes
23	3	2	No	23	3	2	No
24	1	1	Yes	24	1	1	Yes
25	4	4	Yes	25	4	4	Yes
26	1	1	Yes	26	1	1	Yes
27	1	2	Yes	27	1	2	Yes
28	1	1	Yes	28	2	1	No
29	2	2	Yes	29	2	2	Yes
30	1	1	Yes	30	1	1	Yes
31	1	1	Yes	31	1	1	Yes
32	3	3	Yes	32	3	3	Yes

(continued)

Table 8 (continued)

HR (standard deviation)				SpO$_2$ (maxima)			
Sr. No.	Actual stress values	Predicted stress values	Correctly classified in its class (Yes/No)	Sr. No.	Actual stress values	Predicted stress values	Correctly classified in its class (Yes/No)
33	1	1	Yes	33	1	1	Yes
34	1	1	Yes	34	2	1	No
35	2	2	Yes	35	2	2	Yes
36	1	1	Yes	36	2	1	No
37	1	1	Yes	37	1	1	Yes
38	1	1	Yes	38	1	1	Yes
39	2	2	Yes	39	3	2	No
40	1	1	Yes	40	1	1	Yes
41	1	1	Yes	41	1	1	Yes
42	2	2	Yes	42	2	2	Yes
43	1	1	Yes	43	1	1	Yes
44	3	3	Yes	44	3	3	Yes
45	3	2	Yes	45	3	2	Yes
46	3	3	Yes	46	3	3	Yes
47	1	2	No	47	1	2	No
48	2	2	Yes	48	2	2	Yes
49	1	2	No	49	1	2	No
50	1	1	Yes	50	1	1	Yes
Correctly classified instance in percentage (%) = 84.2%				Correctly classified instance in percentage (%) = 78.2%			

6 Conclusion

Occupational stress is a significant problem faced by industrial workers. There are many issues at the workplace that cause stress, i.e. work demand, workplace environment, bad organization structure and personal-professional life imbalance, etc. Occupational stress causes various cognitive and physical loads in workers that lead to a reduction in their productivity due to injury, illness, absenteeism and violation. It is essential for an organization to measure stress amongst their workers and should adopt such policies which prevent occupational stress. Earlier, the psychological questionnaire scale-based method was preferred to gauge stress. However, the psychological questionnaire based is not effective to measure the stress level. To measure the behavioural changes, psychological scale-based (DASS 21) and physiological signals-based (SpO$_2$ and heart rate) analysis were used in this research. As per results from the DASS21 analysis, it was found that 46% of workers are in stress condition and 54% of workers are in non-stressed conditions. Discriminant analysis states that the heart rate signal performed well to measure stress with the accuracy rate of 84.2%, and on the other hand, SpO_2 78.2% accuracy was obtained. Finally, it is concluded that psychological scale-based and physiological signals-based analysis gives the improved stress detection rate. This stress detection methodology helps the organization to reduce the occupational stress and improve the work performance of their workers.

References

1. Avey JB, Luthans F, Jensen SM (2009) Psychological capital: a positive resource for combating employee stress and turnover. Hum Resour Manag 1.48(5):677–693
2. Mohan N, Ashok J (2011) Stress and depression experienced by women software professionals in Bangalore, Karnataka. Glob J Manag Bus Res 11(6):24–29
3. García-Herrero S, Mariscal MA, Gutiérrez JM, Ritzel DO (2013) Using Bayesian networks to analyze occupational stress caused by work demands: preventing stress through social support. Accid Anal Prev 57:114–123
4. Silva LS, Pinheiro TMM, Sakurai E (2007) Productive restructuring, impacts on health and mental suffering: the case of a state bank in Minhas Gerais, Brazil. Public Health Notebooks 23(12):2949–2958
5. Jinkings N (2001) Bank workers in the face of contemporary capitalist restructuring. Federal University of Santa Catarina
6. Kanthimathi S (2014) Occupational stress among the employees in the health care industry. Int Res J Bus Manag 95.
7. Vrijkotte TGM, Riese H, De Geus EJC (2001) Cardiovascular reactivity to work stress assessed by ambulatory blood pressure, heart rate, and heart rate variability. In: Progress in ambulatory assessment. Computer assisted psychological and psycho physiological methods in monitoring and field studies, pp 345–360
8. Sowa CJ, May KM, Niles SG (1994) Occupational stress within the counseling profession: implications for counselor training. Counselor Education and Supervision.
9. Sujatha J, Mishra AK (1996) Determination of the partition coefficient of 1-naphthol, an excited state acid, in DMPC membrane. J Photochem Photobiol, A Chem 101(2–3):215–219
10. Mohajan H (2012) The occupational stress and risk of it among the employees 17–34.
11. Healey JA, Picard RW (2005) Detecting stress during real-world driving tasks using physiological sensors. IEEE Trans Intell Transp Syst 6(2):156–166.
12. Zhai J, Barreto A (2006) Stress detection in computer users based on digital signal processing of noninvasive physiological variables. In: International conference of the IEEE engineering in medicine and biology society 2006. IEEE, pp. 1355–1358 (2006)
13. Fernandes CF, Mekoth N, Kumar S, George BP (2012) Organizational role stress and the function of selected organizational practices in reducing it: empirical evidence from the banking service front line in India. Int J Behav Healthc Res 3(3–4):258–272
14. Chaly PE, Anand SPJ, Reddy VCS, Nijesh JE, Srinidhi S (2014) Evaluation of occupational stress among software professionals and school teachers in Trivandrum. Int J Med Dent Sci 3(2):440–450
15. Sharawi MS, Shibli M, Sharawi MI (2008) Design and implementation of a human stress detection system: a biomechanics approach. In: 5th international symposium on mechatronics and its applications 2008. IEEE, pp 1–5
16. Chung WY, Lee YD, Jung SJ (2008) A wireless sensor network compatible wearable u-healthcare monitoring system using integrated ECG, accelerometer and SpO_2. In: 30th annual

international conference of the IEEE engineering in medicine and biology society 2008. IEEE, pp 1529–1532
17. Basha E, Kaya M (2016) Depression, anxiety and stress scale (DASS): the study of validity and reliability. Univ J Educ Res 4(12):2701–2705
18. Palanisamy K, Murugappan M, Yaacob S (2013) Multiple physiological signal-based human stress identification using non-linear classifiers. Electron Electr Eng 19(7):80–85

Exergoeconomic and Enviroeconomic Analysis of Flat Plate Collector: A Comparative Study

Prateek Negi, Ravi Kanojia, Ritvik Dobriyal, and Desh Bandhu Singh

Abstract Energy resources and their usage intimately relate to sustainable development. Solar flat plate collectors are widely used to collect solar radiation for various applications. Many hybrid systems which include flat plate collector integrated with photovoltaic/thermal panel has gone through various developments and analysis based on energy, exergy, and exergy based environment, exergy based economic analysis and shows great achievements for its applications. This presents paper briefly reviews exergoeconomic and enviroeconomic and some performance analysis of a hybrid flat plate collector to be a useful tool in assessing their thermal efficiency, economic aspect and environmental impact.

Keywords Exergoeconomic · Enviroeconomic · Flat plate collector · Photovoltaic thermal · Exergy

1 Introduction

Due to the ever-increasing demand for energy over the years and rapidly exhausting sources of energy like fossil fuels (which includes coal, petroleum, and natural gas) has put emphasis to migrate towards non-conventional sources of energy. Solar energy is considered as one of the renewable energy sources and harnessing solar energy with the help of solar thermal collectors and photovoltaic can be found useful for various applications like solar water heating, solar drying, distillation, etc. Flat plate collector is the most common among the solar thermal collectors which absorb solar radiation for heating of fluid which is then supplied to the end-user to satisfy the demand of hot water and is considered to be an alternative to conventional geysers. In order to optimise the system for improved output, various analyses need to be done. For many years thermodynamics and economic parameters are the basis for the analysis of the system [1] because the numerical method may be inadequate to optimize

---

P. Negi · R. Kanojia · R. Dobriyal · D. B. Singh (✉)
Department of Mechanical Engineering, Graphic Era Deemed to be University, Bell Road, Clement Town, Dehradun 248002, Uttarakhand, India
e-mail: dbsiit76@gmail.com

© The Author(s), under exclusive license to Springer Nature Singapore Pte Ltd. 2021
R. M. Singari et al. (eds.), *Advances in Manufacturing and Industrial Engineering*, Lecture Notes in Mechanical Engineering,
https://doi.org/10.1007/978-981-15-8542-5_28

the system due to the complexity of the models [2]. The conventional method is to calculate the unit cost on the energy basis but if it is based on the thermodynamic quantity of exergy, the output would be the parameter for which cost is determined. Hence, exergy is found to be an important parameter to determine economic analysis called exergoeconomic analysis [2]. Enviroeconomic is another important criterion for environmental and economic analyses which deals with the cost which focuses to deal with air pollution, quality of water, greenhouse, global warming, etc.

1.1 Energy and Exergy Analyses

Energy can simply be defined as "the capacity for producing an effect" or in other words "the capacity of doing work". Although it can be classified in many forms and ways broadly it is classified as:

i. Stored Energy which is contained within the system boundaries, and
ii. Energy in transition which crosses the system boundaries.

As Energy is conserved, it can neither be created nor destroyed. In other words, Energy input to the system will be equal to the energy output of the system. In order to determine how effective a particular system is, one can analyze it through its energy efficiency.

Exergy, on the other hand, is the uttermost possible work obtainable from a system undergoing a reversible process while trying to reach equilibrium with its environment.

Unlike Energy, exergy is not conserved, it obeys the second law of thermodynamics, therefore it can be said that exergy output is lesser than the exergy input. A portion of the exergy input is always destroyed. Exergy correspond to the polarity between the thermodynamic system and environment, the greater the difference between the two, the greater the amount of exergy of the system.

Exergy analysis is considered to be an important aid in order to find the causes of thermodynamics imperfection in the process [3]. With this exergy analysis, the optimization of thermodynamics processes can be achieved but the practicality of getting an improvement can only be done by economic analysis [4].

The relations between exergy and energy and exergy and environment have proved to be useful parameters in order to determine the energy analysis of flat plate collectors [5] and photovoltaic-thermal collectors [6]. Exergoeconomic parameters can be calculated on the basis of exergy gain as well as exergy loss with the help of the following formulation:

$$\text{Exergoeconomic parameter} = \frac{\text{Exergy loss}}{\text{Annual cost}} \text{"} or \text{"} \frac{\text{Exergy gain}}{\text{Annual cost}} \quad (1)$$

2 Exergoeconomic and Enviroeconomic Analyses of Flat Plate Collector

An exergo-economic analysis of shiny coated solar mixed photo-voltaic-thermal module air collector was done and compared to the photovoltaic module for energy saving [7]. The main components of a shiny coated hybrid PVT air collectors were three Photovoltaic modules each with a useful area of 0.61 m^2, the modules were of glass to Tedlar (polyvinyl fluoride polymer for insulation) type. Rectangular Poly-Vinyl-Chloride sheet air ducts were used for mounting the modules. The three modules were interconnected in series for the forced flowing air (via DC fan), i.e., the outlet of flowing air from the first module was the entrance of the second module, and so on for the third module.

The thermal modelling was done for Solar Cells, Tedlar Back Surface and Air below the Tedlar. The temperature of the air leaving the module, electrical energy dependent on temperature and useful heat gain equations were derived from the thermal modelling of the shiny coated solar mixed Photo-Voltaic-Thermal module air collectors.

Exergoeconomic analysis was done in terms of waste energy output w.r.t. capital cost of equipment and summation of exergy consumption and waste exergy output w.r.t to the capital cost of the equipment. The losses in exergy and energy were obtained by the governing equations [1] and various other analyses based on exergoeconomic [8]. The mathematical models were created based on thermal modelling and were run on MATLAB for obtaining the results. It was found that 234.7 kWh of effective electrical energy was saved by using the shiny coated solar mixed potovoltaic thermal module air collector, thus concluding that shiny coated solar mixed photovoltaic thermal module air collectors are more energy-saving and have greater potential over photovoltaic module. The exergy efficiency of hybrid PVT air collectors is to provide the conditions of achieving maximum solar efficiency at optimal inclination [9]. The result obtained from the above equation revealed that there was 12–15% efficiency in January and a slight increase in efficiency in June which was around 13–14%. When the same hybrid PVT was analysed by other researchers [10], the exergy and economic studies were investigated experimentally at four different cities in India, the highest exergy was found to be 157.22 kWh with a maximum efficiency of 14.8% which was observed in summer. However, when this system was integrated with greenhouse 12], the obtained annual exergy was found to be 12.8 kWh and the efficiency achieved was 4%.

PVT collector has been modelled numerically for which a comparative study on the exergy efficiency of PVT collector with or without glazing [11] and the results obtained showed that the collector without glass cover has better exergy efficiency when compared with collector having a glass cover. Table 1 shows the exergy efficiency of some of the solar collector systems [12].

Table 1 Solar collector energy efficiency [12]

No.	Solar collector system	Exergy efficiency (%)	References
1	The glazed PVT water collector	13.3	[13]
2	The non-glazed PVT air collector	10.75	[14]
3	The PVT (glass-to-glass) air collector	10.45	[15]
4	Water-based photo-voltaic collector with glazing	8–13	[16]
5	Greenhouse coupled photo-voltaic air collector without glazing and with earth-air heat exchanger	5.5	[17]
6	FP water collector with double glazing	3.90	[18]

2.1 Fully and Partially Covered PVT Collector

The usable thermal energy on the photovoltaic surface can be used for low-potential applications such as water and air heating. In addition, hot water or air can be used to heat up living space, solar dryers [19] greenhouse [20], and solar stills for various applications [21]. The performance of the Integrated PVT solar system without glazing was studied experimentally [22]; the results obtained showed that the system has 0.6% higher energy efficiency as compared to a conventional solar water heater. Another hybrid photovoltaic thermal system was operated on a monthly basis of Cyprus and resulted in an increase in maximum efficiency of 7.7% of the photo-voltaic solar system [23]. Hybrid PVT System without glazing which consists of PV panels and collectors cannot have the same area in order to achieve higher efficiency [24] but the system with reflectors definitely provides much more efficiency [25]. Exergy and Economic analysis is considered to be important and most valued parameters in order to achieve the maximum possible efficiency. In relation to this [5], the various relationships viz. exergy and energy, exergy and environment, and energy and sustainable development were obtained. Two configurations had been considered on hybrid PVT water collector [3]; (a) collectors which are partially covered with a semi-transparent photovoltaic module and (b) collectors that are fully covered with semi-transparent photovoltaic module. The experiments were conducted at different weather conditions of New Delhi [26] and at a temperature of constant collection. The results obtained from both the cases have shown that the electrical efficiency is comparatively much higher than the thermal efficiency. When a comparison was done on the basis of gain and loss factor, it was shown that for both the cases thermal efficiency losses were more than that of electrical efficiency due to the factor that the collector has high operating temperature. On the other side, for case (a) the gain is also high for thermal efficiency and it is due to the small collector area covered by photovoltaic module. It was concluded that for the hot production of water the suitable option is (a) and for the generation of electricity case (b) was preferred with 39.16% increase in annual exergy gain for case (a)–(b).

In an experimental work [27], which was based on exergy loss, the photovoltaic thermal water collector was experimentally and numerically analysed on the basis of the intensity of solar radiation, normal air and solar cell temperature and also the speed of wind. The results obtained experimentally were in good relationship with numerical simulation [28]. Although the exergy losses were not analyzed [26] due to the pressure drop but show improved relative values of exergy efficiency. In terms of wind speed and mass flow rate, both the results showed a similar variation of exergy efficiency as it decreases with increasing speed of wind while the exergy efficiency results obtained did not show any significant values [28], which means it has error at low radiation intensity but when compared with the results of the present study, it has exergy efficiency value from 0 to 15.5% in the range of 0–1000 W/m² solar radiation intensity.

An exergoeconomic analysis was done [29] to determine the life term cost analysis of solar distillation system in New Delhi at different climatic conditions. The analyses were done on a single slope and double slope passive solar stills. The result concluded that as the value of uniform end-of-year annual cost rises when the rate of interest increases hence the exergoeconomic parameter decreases. When comparison was done on exergoeconomic parameters, the single slope solar stills has higher value than that of double slope by 3.19% for the life term of 30 years with only 2% rate of interest. Also, both the systems were found to be feasible due to more than 100% annual productivity.

2.2 Enviro-economic Analysis of Flat Plate Collector (2016)

Enviro-economic and Exergo-economic analysis are some of the key parameters for the exergy gain or loss for any flat plate solar collector, as they indicate the impact of exergy on the environment on the basis of cost. The increase in CO_2 emissions over the last two decades has resulted in a shift from conventional to non-conventional energy resources in order to fulfil the ever-growing energy demand of the society [30]. Experiments were conducted on the partially covered photovoltaic flat plate collector for active distillation system using solar energy and results were obtained for the electrical, thermal and exergy parameters along with net exergy and thermal efficiencies on hourly basis. When compared with earlier research the proposed distillation system research was found to be more efficient. Another study [31] estimated that the emission of CO_2 per kWh of energy generated is closely 0.96 kg. The exergy-economic and enviro-economic analysis performed was based on the price of CO_2 emission [32]. As per review done by [30], 60.29$ is the approximated cost for an estimated decade on the use of hybrid photovoltaic thermal system, and this is considered as an important parameter for enviro-economic analysis. In exergo-economic analysis, the result obtained in respect to exergy gain and exergy loss, in the month of June to September maximum solar radiation is available thus exergy gain is maximum for these two months whereas in the month of October to March the exergy loss is maximum.

3 A Review of the Latest Developments in the Analysis of Solar Collector

As per the study [33], on a solar collector, the following analyses were done; energy, exergy, environmental, enviroeconomic, exergoenvironmental (environmental analysis based on exergy), exergoenviroeconomic (economic and environmental analysis based on exergy) analysis. The exergoenvironmental and Exergoenviroeconomic parameters were developed; the main conclusions based on the analysis are summarized as follows:

As the sky and glass surfaces of collector cause a huge gradient in temperature (−40 to 40 °C) which leads to major losses by radiation in energy and exergy. The losses in radiation can be further reduced by selecting a better material. The exergy efficiency (0.732%) was significantly lower than the energy efficiency (25.40%), indicating that the temperature of the environment is a contributing factor in efficiency (exergetic). The exergoenvironmental parameter results are as 4.725 $/year and the corresponding enviroeconomic parameter was 5.040 $/year. Thus the author stressed the importance of Exergoenvironmental and Exergo-enviro-economic Analysis along with Exergo-economic and Enviroeconomic analysis for assessing the carbon pricing effectively.

The performance and economic analysis of PVT collectors with or without water cooling and based on nano-fluid were done [34]. The design is made for residential purposes since better thermal properties the nanofluid helps in improving the performance of the system [35]. Various parameters have been considered in order to demonstrate the feasibility of the system. The basis of determining the exergy analysis is to find the thermal inefficiencies within the system whereas the economic analysis was done by considering various factors such as cost payback time, internal rate of return, and cost of energy produced by the system. The results concluded that the system stands suitable for residential applications for which it was designed with an improved electrical output efficiency of 8.5%, also there was thermal output efficiency of 18% by using nano-fluid rather than water-cooled system. There was a huge reduction in energy cost when compared with conventional energy system prices and it was 82% with two years of investment recovery.

An integrated multi-generation system, including a solar flat plate collector, an organic Rankine cycle (ORC), a PEM electrolyser and a single effect absorption chiller was used for electricity generation and providing cooling load using hydrogen for cooling. For obtaining the evaluation of the system two objective functions were considered:

i. First objective function was in terms of Exergy efficiency, using the input rate of exergy to the collector from the sun, exergy of the electrical system output.
ii. Second objective function was in terms of cost rate of components.

The results obtained for exergoeconomic analysis of hybrid system [36] are that the Exergy efficiency of the hybrid system comparative to other systems comes out to be as 1.83%. Cost analysis suggests that the major portion of the cost price is that of the flat plate collector. The parametric analysis suggests that system efficiency can be enhanced by 1.1–2.7% by collector mass flow rate variations. From the results of multi-objective optimization on the basis of the two objective functions formulated for the system, for the first objective function the results suggest that the exergy efficiency of the system can raise by 3.2% and from the results of the second optimization function, the rate of net cost of the system decreases up to 19.59 $/h.

In a development of a tri-generation system by using flat plate collector that can be used for heating, power generation and also for freshwater production in south Iran region and analysed the model for its feasibility [37]. The concept of doing such a study is to produce hot water and fresh water for household services can utilize it to power the residential buildings. In the end, this combined system proved to be worthy for which it was designed by producing 1300 m^3/h, 1869 kW and 300 m^3/h of hot water, power and freshwater, respectively. When exergy analyses were performed in terms of exergy efficiency and exergy destruction, solar collector was found to have very high exergy destruction, however, it has higher exergy efficiency. To achieve lower consumption of power, exergy destruction had to be reduced.

4 Conclusion

This paper provides a brief review of the literature on analysis of flat plate solar collectors on the cost-based exergy and environment to advance our understanding in this area. Exergo and enviroeconomic analysis of FPC and PVT collectors have shown itself to be a useful tool in assessing their thermal efficiency, economic aspect and environmental impact. Research also points out that hybrid systems, i.e., those having FPC integrated with PVT, have enhanced performance as compared to those comprising of FPC alone.

By making arrangements for vacuum in FPC, convective losses can be controlled which in turn will improve the exergy efficiency of the system. However, not many researchers have focused on this area yet.

Further improvements in the design of hybrid systems can be achieved by varying the packing factor of PVC panel while keeping the same watt peak. The resulting enhancement in performance can be determined by exergo and enviroeconomic analysis and further optimization techniques can be employed.

References

1. Rosen MA, Dincer I (2003) Exergoeconomic analysis of power plants operating on various fuels. Int J Appl Therm Eng 23:643–658
2. Caliskan H, Dincer I, Hepbasli A (2012) Exergoeconomic, enviroeconomic and sustainability analyses of a novel air cooler. Energy Build 55:747–756
3. Mishra RK, Tiwari GN (2013) Energy and exergy analysis of hybrid photovoltaic thermal water collector for constant collection temperature mode. Sol Energy 90:58–67

4. Rosen MA, Dincer I (2003) Exergy methods for assessing and comparing thermal storage systems. Int J Energy Res 27(4):415–430
5. Dincer I (2002) The role of exergy in energy policy making. Energy Policy 30:137–149
6. Lior N, Sarmiento-Darkin W, Al-Sharqawi Hassan S (2006) The exergy fields in transport processes: their calculation and use. Energy 31:553–578
7. Agrawal S, Tiwari GN (2012) Exergoeconomic analysis of glazed hybrid photovoltaic thermal module air collector. Sol Energy 86:2826–2838
8. Dincer I, Rosen MA (2007) Exergy: energy, environment & sustainable development. Elsevier Science and Technology
9. Joshi AS, Tiwari A (2007) Energy and exergy efficiencies of a hybrid photovoltaic–thermal (PV/T) air collector. Renew Energy 32:2223–2241
10. Raman V, Tiwari GN (2008) Life cycle cost analysis of HPVT air collector under different Indian climatic conditions. Energy Policy 36:603–611
11. Chow TT, Pei G, Fong KF, Lin Z, Chan ALS, Ji J (2009) Energy and exergy analysis of photovoltaic–thermal collector with and without glass cover. Appl Energy 86:310–316
12. Saidur R, BoroumandJazi G, Mekhlif S, Jameel M (2012) Exergy analysis of solar energy applications. Renew Sustain Energy Rev 16:350–356
13. Saitoh H, Hamada Y, Kubota H, Nakamura M, Ochifuji K, Yokoyama S (2003) Field experiments and analyses on a hybrid solar collector. Appl Therm Eng 23:2089–2105
14. Sarhaddi F, Farahat S, Ajam H, Behzadmehr A (2010) Exergetic performance assessment of a solar photovoltaic thermal (PV/T) air collector. Energy Build 42:2184–2199
15. Dubey S, Solanki SC, Tiwari A (2009) Energy and exergy analysis of PV/T air collectors connected in series. Energy Build 41:863–870
16. Bosonac M, Sorensen B, Katic I, Sorensen H, Nielsen B, Badran J (2003) Photovoltaic/ thermal solar collector and their potential in Denmark 28
17. Nayak S, Tiwari GN (2009) Theoretical performance assessment of an integrated photovoltaic and earth air heat exchanger greenhouse using energy and exergy analysis methods. Energy Build 41(8):888–896
18. Farahat S, Sarhaddi F, Ajam HC (2009) Exergeti optimization of flat plate solar collectors. Renew Energy 34:1169–1174
19. Barnwal P, Tiwari GN (2008) Design, construction and testing of hybrid photovoltaic integrated greenhouse dryer. Int J Agric Res 3(2):110–120
20. Nayak S, Tiwari GN (2008) Energy and exergy analysis of photovoltaic/thermal integrated with a solar greenhouse. Energy Build 40:2015–2021
21. Kumar S, Tiwari GN (2009) Estimation of internal heat transfer coefficients of a hybrid (PV/T) active solar still. Solar Energy 83:1656–1667
22. Huang BJ, Lin TH, Hung WC, Sun FS (2001) Performance evaluation of solar photovoltaic/thermal systems. Sol Energy 70(5):443–448
23. Kalogirou SA (2001) Use of TRYNSYS for modelling and simulation of a hybrid PV–thermal solar system for Cyprus. Renew Energy 23:247–260
24. Zakharchenko R, Licea-Jime'nez L, Pe'rez-Garci'a SA, Vorobiev P, Dehesa-Carrasco U, Pe'rez-Robels JF (2004) Photovoltaic solar panel for a hybrid PV/thermal system. Solar Energy Materials and Solar Cell 82(1–2):253–261
25. Tripanagnostopoulos Y, Nousia TH, Souliotis M, Yianoulis P (2002) Hybrid photovoltaic/thermal solar system. Sol Energy 72(3):217–234
26. Dubey S, Tiwari GN (2009) Analysis of partially covered PV/T flat plat collectors connected in series. Sol Energy 83(9):1485–1498
27. Yazdanpanahi J, Sarhaddi F, Adeli MM (2015) Experimental investigation of exergy efficiency of a solar photovoltaic thermal water collector based on exergy losses. Sol Energy 118:197–208
28. Tiwari A, Dubey S, Sandhu GS, Sodha MS, Anwar SI (2009) Exergy analysis of integrated photovoltaic thermal solar water heater under constant flow rate and constant collection temperature modes. Appl Energy 86(12):2592–2597
29. Singh DB, Tiwari GN, Al-Helal IM, Dwivedi VK, Yadav JK (2016) Effect of energy matrices on life cycle cost analysis of passive solar stills. Sol Energy 134:9–22

30. Tiwari GN, Sahota L (2016) Review on the energy and economic efficiencies of passive and active solar distillation systems. Desalination 401:151–179
31. Sovacool BK (2008) Valuing the greenhouse gas emissions from nuclear power: a critical survey. Energ Policy 36:2940–2953
32. Tiwari GN, Yadav JK, Singh DB, Al-Helal IM, Abdel-Ghany AM (2015) Exergoeconomic and enviroeconomic analyses of partially covered photovoltaic flat plate collector active solar distillation system. Desalination 367:186–196
33. Caliskan H (2017) Energy, exergy, environmental, enviroeconomic, exergoenvironmental (EXEN) and exergoenviroeconomic (EXENEC) analyses of solar collectors. Renew Sustain Energy Rev 69:488–492
34. Lari MO, Sahin AZ (2017) Design, performance and economic analysis of a nanofluid-based photovoltaic/thermal system for residential applications. Energy Convers Manag 149:467–484
35. Sarsam WS, Kazi SN, Badarudin A (2015) A review of studies on using nanofluids in flat-plate solar collectors. Sol Energy 122:1245–1265
36. Khanmohammadi S, Heidarnejad P, Javani N, Ganjehsarabi H (2017) Exergoeconomic analysis and multi-objective optimization of a solar based integrated energy system for hydrogen production 33(42):21443–21453
37. Ghorbani B, Mehrpooya M, Sadeghzadeh M (2018) Developing a tri-generation system of power, heating, and freshwater (for an industrial town) by using solar flat plate solar collectors, multi-stage desalination unit, and Kalina power generation cycle. Energy Convers Manag 165:113–126

Lead–Lag Relationship Between Spot and Futures Prices of Indian Agri Commodity Market

Raushan Kumar, Nand Kumar, Aynalem Shita, and Sanjay Kumar Pandey

Abstract The agricultural commodity market has two sections—the futures and spot market. Even though these two markets operate in a different way, they are interlinked. In this study, we focus on the lead–lag relationship between spot and futures prices of Agri Index. We have used the AGRI futures index and AGRI spot index, and have employed Augmented Dickey-Fuller test to examine a unit root in the data. To analyse the lead-lag relationship between the log of futures price and log of spot price, we employ the Johansen co-integration test, the error correction model and the Granger causality tests. The study has taken the 632 observation of daily data from April 2016 to September 2018. The results suggest that the futures market leads to the spot market.

Keywords AGRI futures index · AGRI spot index · Johansen co-integration test

1 Introduction

The agrarian sector adds to 16.1% of India's national income, and it offers occupation to 54.6% of the labour force. Rural households receive income from several sources; however, a major part of their income comes from agrarian activities. Despite the significant growth in agricultural production, the constant and perpetual growth

---

R. Kumar
Department of Economics, Zakir Husain Delhi College (Eve), University of Delhi, New Delhi, India

N. Kumar (✉)
Department of Humanities, Delhi Technological University, New Delhi, India
e-mail: nandkumar@dce.ac.in

A. Shita
Department of Economics, Debre Markos University, Debre Markos, Ethiopia

S. K. Pandey
Department of Economics, College of Commerce, Arts & Science, Patliputra University, Patna, India

in farm income remains a challenge for the policymakers. As a result, the goal of doubling farmers' income has become the centre of attention of the current government. In spite of the fact that there are numerous potential pathways[1] of increasing farmers' income, there is even now substantial opportunity of excellent price realizations by the cultivator. With the intention of ascertaining that cultivators obtain appropriate prices, there is a necessity to establish liquid, transparent and efficient spot as well as future[2] market.

India is an emerging economy. Since 1991, economic reforms in the form of liberalisation of markets and removal of state intervention are being done in India. In the changing circumstances in economy, exporters are being exposed to price volatility. Markets based mechanisms and risk management tools are needed to trim down the price uncertainty. Future market is one such tool in lessening the price risk [1]. Since the beginning of future market, the existence of price discovery associated with the spot and futures market has been of great consequence.[3]

The prices in the cash (spot) market and prices in future market are related to each other. These prices are affected by the new information whether it is economic, political or social. Since the trading cost in future market is less, the information is first reflected in future market [2]. However, this might not occur continually. It might be the case that information is first disseminated in cash market and then passed on to future market. At times, it also happens that the new information is instantaneously exhibited in futures as well as in spot market. Therefore, this study is an attempt to examine the lead–lag relationship between spot and futures prices of Agri Index.

Garbade and Silber [3] examine the price discovery in seven storable traded on the Chicago Board of Trade and find that the futures market leads the spot market. Another study [4] investigated the price leadership relationship between spot and futures prices for cotton and found that future market performs the role of price discovery. Similar results were documented for feeder cattle and live cattle [5], for live hogs [6] and for commodity sugar [7].

Brockman and Tse [8] examine the price discovery function of futures markets in four agricultural commodity markets of Canada. The study employed cointegration, error correction models and Hasbrouck's [9] model of information shared, and found that future market performs its role of price discovery in all the four agricultural commodities. Yang and Leatham [10] show that the futures price movements lead spot prices in the US wheat market.

---

[1] Income can be increased by by increasing yield, bringing efficiency in input use, changes in cropping pattern, etc.

[2] 'A futures contract is an agreement between a seller and a buyer that calls for the seller (called the short) to deliver to the buyer (called the long) a specified quantityand grade of an identified commodity, at a fixed time in the future, and at a price agreed to when the contract is first entered into' [2]. It is legal necessity that all futures contracts have to be bought and sold on future exchanges. A futures exchange is an authorised contract market under the provision of commodity exchange act.

[3] Price discovery means the making use of futures prices for getting information about spot prices.

Literature shows the importance of future trading in agricultural commodities in less developed countries [11]. Futures markets help in reducing price volatility[4] in the economy. The study also examines whether the LDCs should establish new futures exchanges or use existing exchanges and find that the use of existing (DME) exchanges are inexpensive, but creates the problem of exchange risk,[5] however, the creation of new exchanges is costly and time taking.

Kumar [1] investigate the dynamic relationship between futures and spot prices of agricultural crops. He employs semi-parametric causality test and finds the causality from futures to spot market.

The present study contributes to the existing literature by making use of more recent daily data (after setting up of e-NAM[6]) from April 2016 to September 2018. We have taken the futures and spot price data for AGRI commodities from the National Commodity and Derivative Exchange (NCDEX), Mumbai. The Johansen co-integration test, the error correction model and the Granger causality tests have been used to investigate the lead-lag relationship between spot and futures indices of AGRI commodities. The spot and futures AGRI indices are integrated of order one, so we have taken the first difference of the log of prices. The results show that there exists a unidirectional price causal relationship from future market to spot market for AGRI commodities.

The rest of the paper has been organised into four sections. In Sect. 2, we highlight the methodology. In Sect. 3, we discuss the data. In Sect. 4 we analyse the estimation results, and in Sect. 5 we conclude.

2 Methodology

To analyse the lead-lag relationship between the log of futures price and log of spot price, Johansen co-integration test, error correction model (ECM) and Granger causality test have been used. We also performed the residual diagnostic tests of no serial correlation, no heteroskedasticity and normality.

2.1 Johansen Co-integration Test

Cointegration means a long-standing equilibrium relationship between spot and futures prices that are independently nonstationary, but their linear combination is stationary. Although the two non-stationary series may drift apart in the short run but

---

[4]Market participants are facing the price risk due to increasing economic reforms in the market.
[5]There might be liquidity problem.
[6]e-NAM stands for National Agriculture Market, and is an electronic exchange for Agri commodities.

come together systematically in the long run. Co-integration between spot and future price denotes that the spot and futures prices are both non-stationary, it is, however, expected that their difference is zero. If the series are not co-integrated, there would be no long-run relationship binding the series together. It might be the case that speculators will earn huge profits via trade in future market if the co-integrating relationship between futures and spot market weakens. We employ the Johansen and Juselius [12, 13] method to examine the cointegrating relationship. This test can be written as follows:

$$\text{Let } y_t = By_{t-1} + \epsilon_t \tag{1}$$

where $y_t = (y_1, y_2, \ldots, y_m)'$ is a $m \times 1$ vector of $I(1)$ variables
$\epsilon_t = (\epsilon_1, \epsilon_2, \ldots, \epsilon_m)'$ is a $m \times 1$ vector of white noise errors
B is an $m \times m$ parameters matrix. The number of cointegrating vectors will be between 0 and $m - 1$.

We subtract y_{t-1} from both sides of the above equation, and we get

$$\Delta y_t = -(I - B)y_{t-1} + \epsilon_t$$

$$\equiv -\pi y_{t-1} + \epsilon_t$$

where I is an identity matrix of order $m \times m$. The number of cointegrating vectors is given by the rank of matrix π, which is of order $m \times m$. Rank zero implies no cointegration. If rank of matrix π is m, also implies no integration. Therefore, the rank of matrix π is r, and $0 < r < m$. We conclude that the number of co-integrating vectors is the rank of matrix π. To examine the long-run relationship, this test employs two likelihood ratios.

(a) Trace statistics:

$$\lambda_{\text{trace}}(r) = -T \sum_{i=r+1}^{m} \ln(1 - \hat{\lambda}_i)$$

where T denotes the sample size, $\hat{\lambda}_i$ is the eigenvalues (estimated) from the estimated π matrix. We test the null hypothesis of at most r cointegrating vectors against the alternative hypothesis of more than r cointegrating vectors. We get the critical values of trace statistic from Monte Carlo Simulations.

(b) Maximum eigenvalue statistic

$$\lambda_{\max}(r, r+1) = -T \ln(1 - \hat{\lambda}_{r+1})$$

We test the null hypothesis of r cointegrating vectors against the alternative hypothesis of $r + 1$ cointegrating vectors.

If the variables are cointegrated, then we can estimate an error correction model with the lagged value of the residual from a cointegrating relationship along with the other variables with lag.

2.2 Error Correction Model

Error Correction Model (ECM) represents the short run dynamics of the variables, guided by the departure from long run equilibrium.

$$\Delta y_t = \alpha_1 + \alpha_y(Y_{t-1} - \gamma X_{t-1}) + \sum_{i=1} \alpha_{11}(i)\Delta y_{t-i} + \sum_{i=1} \alpha_{12}(i)\Delta x_{t-i} + \varepsilon_{yt} \quad (2)$$

$$\Delta x_t = \alpha_2 + \alpha_x(Y_{t-1} - \gamma X_{t-1}) + \sum_{i=1} \alpha_{21}(i)\Delta y_{t-i} + \sum_{i=1} \alpha_{22}(i)\Delta x_{t-i} + \varepsilon_{xt} \quad (3)$$

where X and Y are spot and futures prices, respectively.

Equations (2 and 3) are the error correction model. In Eq. (1), $(y_{t-1} - \gamma x_{t-1})$ is the error correction term. x_t and y_t are co-integrated with γ as co-integrating coefficient. Besides, $(y_{t-1} - \gamma x_{t-1})$ is stationary and x_t and y_t are $I(1)$. Equation (1) can be interpreted as the change in the change in Y between the period $t - 1$ and t due to change in the value of independent variable X, between the time period $t - 1$ and t. α_{12} denotes the coefficient of short-run dynamics. α_y represents the speed of adjustment parameter, α_y determines the amount of last period's equilibrium error that is corrected for [14]. Similarly, we can interpret Eq. (3). Equations (2 and 3) represents the vector autoregression in first difference of log of prices, along with the error correction term $(y_{t-1} - \gamma x_{t-1})$.

2.3 Granger Causality Test

The Granger causality test examines whether previous values of the first variable X_t can assist to describe present values of a second variable Y_t, dependent on previous values of the second variable Y_t. The Granger causality test will be performed in the background of the error correction model. We do the Granger causality test by performing the joint test of the error correction term and lags of X_t.

Table 1 Data Descriptions

Statistic	Mean	Std. Dev	Skewness	Kurtosis	Observations
LAGRISP	7.711	0.163	−0.497	3.522	632
LAGRIFP	7.707	0.149	−0.072	2.596	632

Source Multi-Commodity Exchange (MCX), Mumbai

3 Data and Interpretation

We have made use of the data of futures index and spot index for Agri commodities from Multi-Commodity Exchange (MCX), Mumbai. We employ the MCXAGRI index in the present analysis. In order to compare and simplify the analysis, we have taken up a log to the data series. The data reported in the study are daily data from April 18, 2016, to September 30, 2018. The data statistics are stated in Table 1. We note that the average of the futures price is less than the mean of the spot prices (logged values). Both indices reflect negative skewness. The distribution of the series is not 3, implying that both data series do not adhere to normal distribution.

4 Estimation Results

Augmented Dickey-Fuller test [15] has been done to investigate the presence of the unit root in the spot and futures prices. Unit root results are reported in Table 2. The ADF statistic of futures price (LAGRIFP) and spot price (LAGRISP) are −2.30 and −2.74, respectively. The ADF statistic (in absolute terms) lower than the critical value −3.51(at 5% level of significance) which suggests that the null hypothesis of unit root is not rejected. Accordingly, it is established that futures and spot prices of the AGRI indices are non-stationary.

We also did a unit root test on the first difference of the log of futures and spot price of the AGRI indices. After the first differencing of the log of futures and spot

Table 2 Unit root test—augmented Dickey-Fuller (ADF) test

Series	OLL	ADF statistic	Order of integration
LAGRISP (i)	0	−2.94	LAGRISP ~ $I(1)$
LAGRISP (t)	0	−2.74	LAGRISP ~ $I(1)$
LAGRIFP (t)	4	−2.30	LAGRISP ~ $I(1)$
LAGRIFP (i)	4	−1.92	LAGRISP ~ $I(1)$
ΔLAGRISP (i)	0	−22.83	ΔLAGRISP ~ $I(0)$
ΔLAGRIFP (i)	3	−17.50	ΔLAGRIFP ~ $I(0)$

i stands for ADF test with intercept and t stands for ADF test with intercept and trend, OLL stands for optimal lag length

prices (ΔLAGRIFP & ΔLAGRISP for AGRI index), both the data series of the AGRI indices turn out to be stationary. We do not find unit root in ΔLAGRIFP & ΔLAGRISP for the reason that their ADF statistics are -17.50 and -22.83, in that order. The first difference between the log of both AGRI indices is stationary.

The present study takes the first difference of the log of indices, with the intention of making the data stationary. There is the dominance of stochastic trends in the futures and spot prices of all the four indices. However, the series become stationary after the first difference. Since the first difference of the futures and spot prices of both indices are $I(0)$, it can be stated that futures and spot prices of the AGRI indices in levels are not $I(2)$ processes.

The information about the optimum lag length in the series has been given by Schwartz criterion. The optimal lag length calculation has been executed automatically by Eviews. The LAGRIFP has the optimal lag length of 4. Also, the optimal lag lengths of first difference of future and spot prices of AGRI indices are 3 and 0, respectively.

We then test for cointegration. We examine the residual of the regression of futures (spot) price on the spot(future) price. We find that the residuals of the co-integrating regression are stationary. We then perform the Johansen test of co-integration between the futures (spot) price and the spot (futures) price. From Table 3 we find that the trace test statistic of 221.078 and the maximal eigenvalue test statistic of 78.523 are both strongly significant, rejecting the null hypothesis of no cointegration among the commodity prices. From the trace test and maximal eigenvalue test, conclude that the futures (spot) price and the spot (futures) price have long-term relationships.

The error correction model has been built from Johansen co-integration test. Table 4 presents the results of Eqs. (2) and (3) which estimates the error correction model. We can observe from Table 4 that the coefficients of the error correction terms are significant. It supports the notion that the log of future (spot) prices and log of spot (future) prices do actually co-integrate. For spot returns as dependent variable, the coefficient of error correction term for LAGRI is -0.081. This coefficient is highly significant. This signifies that if the gap between the logs of spot and future price and spot price is positive in one period, the spot price will decline in the subsequent period

Table 3 Johansen's cointegration tests for daily prices in AGRI markets

Rank	Trace test		Maximal eigen value		Conclusion
	Test value	Critical value (95%)	Test value	Critical value (95%)	
$H_0 : r = 0$	221.078 **	197.370	78.523 **	58.433	Reject H_0
$H_0 : r \leq 1$	143.581	159.529	39.806	52.362	Do not reject H_0

Notes **indicates significance at the 5% level
Number of lags—four
r—order of cointegration

Table 4 Error correction model statistics for daily returns in AGRI INDEX

Variable	Futures returns	Spot returns
Futures returns lag 1	−0.180***	0.272***
	(0.044)	(0.100)
Futures returns lag 2	−0.061	0.127
	(0.044)	(0.043)
Futures returns lag 3	−0.044	0.149
	(0.044)	(0.124)
Futures returns lag 4	0.017	0.109**
	(0.045)	(0.042)
Spot returns lag 1	0.012	−0.014
	(0.035)	(0.034)
Spot returns lag 2	0.015	0.044*
	(0.035)	(0.026)
Spot returns lag3	0.007	0.043
	(0.035)	(0.034)
Spot returns lag 4	0.056	0.086*
	(0.051)	(0.049)
α	−0.046***	−0.081***
	(0.003)	(0.006)

Notes ***, **, and * indicate significance at the 1, 5 and 10% levels
α denotes the speed of adjustment parameter
Error correction model (ECM) based on three lags of 'endogenous' variable. Standard errors are reported in parentheses

to restore the equilibrium. Similarly, for future returns as a dependent variable, the coefficient of error correction term for LAGRI, is −0.046, and is highly significant.

For spot returns as dependent variable in LAGRI index, the coefficient of future returns lag 1 is 0.272, which is positive and highly significant. This means that the lagged changes in log of future prices lead to positive changes in log of spot prices. In LAGRI index, future market leads the spot market. When we consider the coefficients of second and third lags of futures returns, these coefficients have been found to be not significant.[7] Therefore, Granger Causality test has been performed.

With futures returns as dependent variable in LAGRI index, all the coefficients of lags of spot returns have been found to be insignificant and only the coefficient of futures returns lag one has been found to be significant. This means that there is a positive autocorrelation in futures price of LAGRI index.

Diagnostic tests on residuals have been performed. These diagnostic tests indices could not satisfy the normality assumption. In case of spot returns as dependent

---

[7] Price discovery process is not clear from the error correction model.

Table 5 Granger causality test of daily returns in AGRI markets

Null hypothesis	Lags	χ^2-statistic
D (LAGRIFP) does not Granger Cause D (LAGRISP)	4	11.033***
D (LAGRISP) does not Granger Cause D (LAGRIFP)	4	1.763

Notes ***Indicate significance at the 1%
Granger causality tests based on four lags of 'endogenous' variable

variable in error correction model, there is no evidence of autocorrelation, serial correlation or heteroskedasticity.

If futures price Granger causes spot price, then futures market leads the spot market. In the same way, if spot price Granger cause futures price, then spot market leads the futures market. The results of the Granger causality test have been presented in Table 5. The test results report χ^2-Statistic. These Granger causality results show the uni-directional causality from futures to spot. This means that futures price leads to spot price.

There is a causal relationship from futures to spot due to the following reasons: Firstly, in future market all transactions occur on future exchanges where the information about the demand and supply of the commodity is collected and acted on, leading to the determination of the equilibrium prices. Secondly, lower transaction cost and greater flexibility in future market might be the reason for price discovery in future market. Without the economic burden of owning the asset, the traders can speculate on the price movements of AGRI indices. Lastly, the future markets have greater liquidity.

5 Conclusions

The present study has made use of the 632 observations of daily spot and futures price data of AGRI index, taken from the multi-commodity exchange, Mumbai (MCX). We use the Johansen co-integration test, error correction model (ECM) and Granger causality tests, to investigate the lead-lag relationship between the log of futures price and log of spot price. The evidence provided by Granger causality results suggests that there is uni-directional information flows from futures to spot market in AGRI index. This implies that there is unidirectional causality from futures to spot prices. The basic finding of the study that the futures market leads spot market, is in line with previous studies [3, 16, 17].

Futures market leads the spot market because in futures market all transactions happen in futures exchanges where the information about the demand and supply of the commodity is gathered and acted on, leading to the determination of the equilibrium prices in competitive manner. In addition, there is a low transaction cost and greater liquidity in futures market vis-à-vis spot market. Findings of the study suggest that the authorities can make rational storage decisions, depending on the signals from futures market.

References

1. Kumar R (2017) Price discovery in some primary commodity markets in India. Econ Bull 37(3):1817–1829
2. Edwards R, Ma W (1992) Futures and option. McGraw-Hill Inc., USA
3. Garbade KD, Silver WL (1983) Dominant satellite relationship between live cattle cash and futures markets. J Fut Mark 10:123–136
4. Brorsen BW, Von Bailey D, Richardson JW (1984) Investigation of price discovery and efficiency for cash and futures cotton prices. W J Agric Econ 9(1):170–176
5. Oellermann CM, Brorsen BW, Farris PL (1989) Price discovery for feeder cattle. J Fut Mark 9(2):113–121
6. Schroeder TC, Goodwin BK (1991) Cointegration tests and spatial price linkages in regional cattle markets. Am J Agr Econ 73(2):452–464
7. Zapata H, Fortenbery TR, Armstrong D (2005) Price discovery in the world sugar futures and cash markets: implications for the Dominican Republic. Staff Paper No. 469, Department of Agricultural and Applied Economics, University of Wisconsin-Madison.
8. Brockman P, Tse Y (1995) Information shares in Canadian agricultural cash and futures markets. Appl Econ Lett 2(10):335–338
9. Hasbrouck J (1995) One security, many markets: determining the contributions to price discovery. J Fin 50(4):1175–1199
10. Yang J, Leatham DJ (1999) Price discovery in wheat futures markets. J Agric Appl Econ 31(2):359–370
11. Morgan CW, Rayyner AJ, Vaillant C (1999) Agricultural futures markets in LDCs: a policy response to price volatility? J Int Dev 11(6):893–910
12. Johansen S, Juselius K (1990) Maximum likelihood estimation and inference on cointegration with applications to the demand for money. Oxford Bull Econ Stat 52(2):169–209
13. Johansen S, Juselius K (1992) Testing structural hypothesis in a multivariate cointegration analysis of PPP and the UIP for UK. J Econ 53(1–3):211–244
14. Brooks C (2014) Introductory econometrics to finance. 2nd edn. Cambridge University Press, UK.
15. Dickey DA, Fuller WA (1981) Likelihood ratio for autoregressive time series with a unit root. Econometrica 49:1057–1072
16. Moosa IA (2002) Price discovery and risk transfer in the crude oil futures market: some structural time series evidence. Blackwell Publishers.
17. Fu LQ, Qing ZJ (2006) Price discovery and volatility spillovers: evidence from Chinese spot-future markets. J Fut Mark 23:1019–1046

Learnify: An Augmented Reality-Based Application for Learning

Himanshi Sharma, Nikhil Jain, and Anamika Chauhan

Abstract Over the past few years, technology has seen a lot of advancements which are becoming an integral part of a person's life. Augmented reality (AR) is one of these technologies which has emerged out and seems to have a lot of potential. This paper tries to explore that potential of AR especially in the field of education. We devise a way to improve learning experiences by incorporating more senses so that the user can easily remember things. In addition, we try to find out how effective and adaptive this way of learning is for users so that changes can be adopted for an even better system.

Keywords Augmented reality · Education · Technology · Application · Learning

1 Introduction

1.1 Overview

Augmented reality is a field that is being worked on rigorously. Most of the times, it is confused with virtual reality. But in 'reality', both are different. In virtual reality, the user's perception of reality is completely based on virtual information, i.e. the complete reality has been changed. In augmented reality, the user is provided with additional computer-generated information (which is superimposed on the existing reality or the current reality is changed in some manner) that enhances their perception of reality [1]. For example, VR is used in computer games to provide an exciting experience to the user; whereas AR can be used to show the disassembly of a machine to help a person find the fault in an easy way.

So, we can say that augmented reality helps in bringing digital information and/or virtual objects into the physical space virtually. AR brings the digital world to life inside the view captured by any device with a camera. This is made possible by using

---

H. Sharma · N. Jain (✉) · A. Chauhan
Delhi Technological University, New Delhi 110042, India
e-mail: nikhiljainlsb@gmail.com

some triggers which bring that object to the real world inside the device's view. The triggers can be of three types based on which AR is classified as:

Marker-Based Augmented Reality
This type of AR utilizes known patterns in the scene to find out what the user is pointing his camera at. If the pattern is defined according to the application, then the corresponding virtual object is brought to the user's perception providing an augmented reality experience. For example, the application we will be presenting in this paper works on marker-based augmented reality.

Marker-Less Augmented Reality
This is exact opposite of marker-based AR. It does not require any pattern to show the virtual object. Just showing the camera or selecting the object triggers the augmented reality experience. For example, one such application which uses marker-less augmented reality involves showing virtual furniture in the user's house or other places so that the user can get an idea about how that will look in case the user is planning on buying it. Another popular application is a commonly used application amongst the youth. Snapchat works on marker-less AR.

Location-Based Augmented Reality
This type of AR uses location of the user's device as a trigger for an event which may include showing information about that place or showing a virtual object to get an augmented reality experience. For example, a game which gained a lot of popularity amongst people, Pokémon Go, used the device's location to show virtual Pokémons on the user's device which can then be interacted with and the game is continued.

These AR types are being used by a lot of tech giants including Microsoft, Facebook and Google. We will discuss some of their current technologies such as:

Microsoft HoloLens
Microsoft HoloLens is a pair of mixed reality smartglasses. It is the first head-mounted display based on the mixed reality application for Windows 10 [2].

Facebook Spark AR
Spark AR is an augmented reality application for both Mac and Windows which enables a user to create augmented reality effects for mobile cameras. It can be compared to Photoshop but with a touch of augmented reality [3].

Google ARCore
ARCore by Google is a software development kit that allows a user to build applications based on augmented reality. It uses three major technologies in integrating virtual content with the real world when seen through a user's phone's camera which are:

- Motion tracking is used to know the position of the device with respect to the world.

- Environmental understanding is used to find out the size and position of flat surfaces like a table or ground.
- Light estimation is used by the device to find out the surroundings' present lighting conditions [4].

2 Proposed Work

3 The Idea

The application involves helping users to learn easily by involving more senses of the user. The user can use any device with a camera and Internet connectivity to run this application. The user just has to use the markers provided by us. When the user points the camera of the device while using the application, the object related to that marker shows up on the screen with additional features showing what that object is and also what that object is called in the English language. As an example, if a user shows an 'A' marker to the camera, a model of an Apple, with 'Apple' written, shows up with a voice in background pronouncing 'Apple' (Fig. 1).

3.1 Implementation

This is a Web-based application which can be run on any device with a camera and supported Web browsers including Microsoft Edge, Internet Explorer, Google Chrome and Mozilla Firefox. The application was constructed with the help of some open-source augmented reality libraries.

The scene generated by camera is used as a canvas for pattern detection by the libraries. The patterns detected are then matched with the defined markers. On AR.js [5], we define specific scenes for specific markers, so when the camera identifies a marker, the application shows the 3D model on top of it. A marker is a sort of

Fig. 1 An apple showing when camera is pointed to an 'A' marker

Fig. 2 An 'A' marker

simplified QR code with a maximum identified resolution of 16 × 16 pixels. It can only contain black-and-white colours with a simple text like a single character.

Since the currently allowed resolution is not very large so if the image is too complex, the detection might not work sometimes. So, for better results, it is advised to use dissimilar images and avoid using too complex images as markers (Fig. 2).

The application is implemented using marker-based augmented reality. The markers can be constructed for different purposes depending on the application. These markers are used for pattern detection by the libraries.

4 Result

As an example, 'Apple' was shown in the previous section that was a part of the module for learning English alphabets. Those modules can be seen as different subjects in a broader aspect. Those modules are discussed in detail here:

4.1 English Alphabets

This application can be used to teach alphabets of the English language. The markers are different alphabets from A to Z. When an alphabet marker is shown to the camera, the application presents a model of the object represented by that alphabet, with its name, along with a background voice taking the name of that object (Fig. 3).

4.2 Mathematical Shapes

The most difficult thing for very young students is to visualize shapes in three-dimensional space. This application makes that task easier by showing shapes when

Fig. 3 A pumpkin on a 'P' marker

Fig. 4 A cylinder on a 'Cylinder' marker

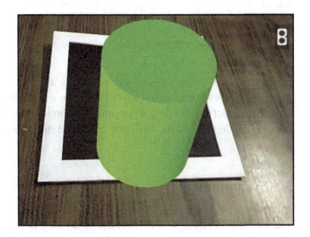

camera is pointed to a marker having the name of the shape such as cube, sphere, cone, pyramid and cylinder (Fig. 4).

4.3 Molecular Structure

Students in higher classes can use this application for visualizing the structure of various molecules by using the markers containing the molecular formula of compounds. The structure contains all the important information such as the angle between atoms, and the charge acquired by various atoms in case the bond is not covalent (Fig. 5).

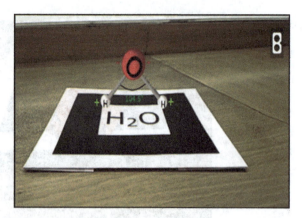

Fig. 5 Molecular structure of H_2O, with bond angle and charge acquired by atoms, on an 'H$_2$O' marker

4.4 Biology—Anatomy

This part is another very useful subject which allows users to learn anatomy without those old figures in textbooks. Just like earlier, a user when points the camera towards a marker depicting a body part or other anatomical structures. This helps the user to have an inside view of the structure in a three-dimensional form (Fig. 6).

5 Evaluation

The evaluation strategy is comprised of inviting people to use the application and then asking questions regarding their opinions about the application as well as how

Fig. 6 A human heart being shown on the screen when a 'Heart' marker is shown to the camera

Learnify: An Augmented Reality-Based Application for Learning

effective this application would be if it were to be introduced in the field of education. People from different backgrounds participated in this survey so that a varied opinion could be collected. A total of 125 entrants took part in the survey. Also, the entrants' age varied from 12 to 65 so that the opinion of different aged target audiences could be known. Out of these, 34 respondents (27.2%) were from a teaching background, another 66 respondents (52.8%) were students and the rest 25 respondents (20%) were from other professional backgrounds as shown in Fig. 7. Totally, 59 respondents (47.2%) were females and the rest 66 respondents (52.8%) were males.

On asking the entrants their opinion about including this application in the education field, most of the respondents either strongly agreed (47.2%) or agreed (47.2%) about inclusion of this application in the education system. The rest (5.6%) had a neutral opinion about this. This can be seen in Fig. 8. This showed that most of the people would be happy to see such application in the teaching spectrum as not even a single entrant disagreed to this. Another question asked to the entrants was whether including this application in education system would improve the teaching quality or not. A majority of the respondents (55.2%) strongly agreed to this and another 50 respondents (40%) simply agreed. The rest six respondents (4.8%) remained neutral about whether improvement would take place or not. This question also did not have any negative response as shown in Fig. 9.

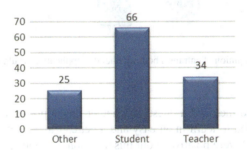

Fig. 7 Analysis of the professional background entrants

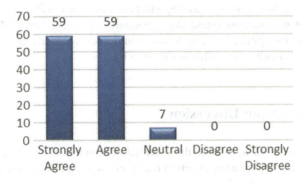

Fig. 8 Opinion of entrants about including this application in the education field

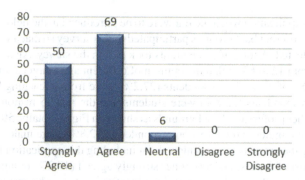

Fig. 9 Analysis of opinion of entrants about whether including this application in education system would improve the teaching quality or not

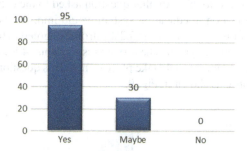

Fig. 10 Analysis of opinion of entrants about whether this application would help differently abled students or not

The survey also included a question whether this application would be helpful for differently abled to learn in an easy and a better way or not. A majority of the respondents (76%) were positive about this question, and the rest 24% respondents were not sure about this. There was not a single entrant having a negative opinion. This is shown in Fig. 10.

The last question involved asking entrants how hard using this application was. None of the entrants found the application difficult to use. Most of them (76.8%) found the application easy to use, and the rest respondents (23.2%) found it moderately difficult to use. This is shown in Fig. 11.

6 Conclusion and Discussion

In this paper, we tried to explore the potential of augmented reality in the field of education by developing an application that could be a part of the learning system. The application proved to be very promising as per the survey conducted on people from varying backgrounds and age groups. The results were very positive showing

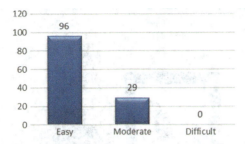

Fig. 11 Opinion about usage difficulty of the application by the entrants

that if our application based on augmented reality was introduced in the field of education, it would bring a revolution in the learning system and help a lot of people, including the instructors as well as the students and especially those with learning disabilities, since this application was found to be useful and easy to use especially by students under the age of 17 which are basically in the learning spectrum and the major target of this application.

7 Future Work

This application, although seems to have a lot of features already, can be optimized for even better learning experiences. Some of the ways in which this can be tried to achieve are:

7.1 Making the Application More Interactive

Making the application more interactive can involve a lot of things like making the models click/touch sensitive so that they perform some functions depending on the situation. As an example, consider the subject of anatomy in which biological structures were shown. Now, we can use the touches/clicks to show labels on that structure as shown (Fig. 12).

7.2 Animating the Models

In order to make the application more attractive and fun for users, animations can be added to models. These animations along with sound can be used to explain some concepts making the application a virtual teacher.

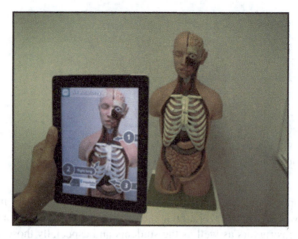

Fig. 12 Example where the model gets labelled on touching that particular part. *'The Multiple Uses of Augmented Reality in Education'* by Adriana Blum

7.3 Using a VR Headset

A VR headset such as Microsoft HoloLens or Samsung Gear VR along with this application on compatible devices so that the user can get a more immersive experience of the application (Fig. 13).

Fig. 13 An immersive experience for students using VR headsets. *'VR enriches lessons, learning experience'*

Fig. 14 A teacher from remote location being projected to students in other location. '*A New Era in Education with Holographic Telepresence Technology*'

7.4 Using Projections

Using suitable technologies which are not too costly, teaching institutions as well as individuals can also project these models on the real world making augmented reality 'almost real' (Fig. 14).

All of these features when used in conjunction can also be used to provide virtual teaching experience in real time. More specifically, teachers from remote locations can teach students without actually having to be present there just like video calling but a bit more realistic.

References

1. Introducing Virtual Environments Archived 21 April 2016 at the Wayback Machine National Center for Supercomputing Applications, University of Illinois.J. Clerk Maxwell, A Treatise on Electricity and Magnetism, 3rd ed., vol. 2. Oxford: Clarendon, 1892, pp 68–73.
2. Mcbride S (2016) With HoloLens, Microsoft aims to avoid Google's mistakes. Reuters. Retrieved May 23 2016
3. https://medium.com/%40geekydiego/10-things-you-need-to-know-about-spark-ar-baad07b2a293
4. Amadeo R (2017) Google's ARCore brings augmented reality to millions of Android devices. Ars Technica. Condé Nast. Retrieved 6 Nov 2017
5. https://github.com/jeromeetienne/AR.js/blob/master/README.md

Thermal Analysis of Friction Stir Welding for Different Tool Geometries

Umesh Kumar Singh, Avanish Kumar Dubey, and Ashutosh Pandey

Abstract Friction stir welding was developed to primarily join the aluminium alloys, but it causes thermal stresses because of thermal cycle and constraints (mechanical) during the process. Thus, reduces the life of components. Aluminium alloy AA 6061-T6 is highly demanded in transportation industries and many times requires the joining of two pieces. In the present research, the temperature and thermal stress distribution during the friction stir welding of AA 6061-T6 alloy are evaluated by using finite element method. Effect of tool shoulder diameter and cone angle have been compared with respect to maximum temperature and thermal stresses generated during the welding process. The finding shows that a tool with a 24 mm diameter and 7° cone angle has maximum temperature of 730.69 K which is greater than 6% as compared to flat shoulder with same diameter. Maximum value of thermal tensile stress is obtained for tool with flat shoulder and 24 mm diameter.

Keywords Friction stir welding · Temperature distribution · Aluminium alloy · Thermal stresses · Tool geometry

1 Introduction

Friction Stir Welding (FSW) is a solid-state joining process developed by TWI in 1991. It is derived from the old friction welding process [1]. The heat required for joining is obtained through friction between tool shoulder and workpiece (Fig.1) as well as deformation (plastic) of the material. The heat generated because of friction has a major contribution in the joining process [2]. Temperature during the welding process remains below the melting point (MP) of material. Hence, it is a very suitable welding process for materials with high thermal expansion coefficient and which gets distorted in the conventional welding process. The temperature distribution is difficult

---

U. K. Singh (✉) · A. K. Dubey · A. Pandey
Mechanical Engineering Department, Motilal Nehru National Institute of Technology Allahabad, Prayagraj, India
e-mail: umeshkumarsingh2008@gmail.com

© The Author(s), under exclusive license to Springer Nature Singapore Pte Ltd. 2021
R. M. Singari et al. (eds.), *Advances in Manufacturing and Industrial Engineering*,
Lecture Notes in Mechanical Engineering,
https://doi.org/10.1007/978-981-15-8542-5_31

to measure by experiment very near to the tool due to high plastic deformation. Numerical methods are best suited for temperature distribution analysis of FSW.

Materials such as aluminium alloys, magnesium alloys, etc. have been successfully welded by FSW in past. It is one of the green welding processes because it does not have any adverse effects on the environment. Weld formed using FSW has shown superior strength and material characteristics. Tang et al. [3] predicted temperature distribution during FSW by the experimental process. They found that maximum temperature during the welding process was 80% of the MP temperature of material for Al 6061 T-6 alloy. Gould and Feng [4] predicted temperature distribution using a point source heat flow model. They neglected heat generated due to the probe surface and considered only shoulder heat flux. Chao and Qi [5, 6] used finite element analysis to predict temperature distribution and residual stress developed during the FSW process. They were one of the first publishers who had worked on numerical simulation of FSW. The heat generated due to friction was only considered and probe heat generation (HG) was neglected in their model. Colegrove [7] found that out of the total HG, heat generated by tool probe is 20% and hence, it should not be neglected. Chao et al. [8] found that 96% of the total HG reaches to workpiece while the rest is lost through tool and surrounding. Song and Kovacevic [9] developed a thermo-mechanical model and assumed sliding conditions and neglected slipping during tool movement trough material. Chen and Kovacevic [10] produced a thermo-mechanical model as Lagrangian analysis and they also included the effects of strain hardening due to deformation (plastic) of the material to predict residual stresses developed.

AA 6061-T6 aluminium alloy is a commercially used structural material having good strength and corrosion properties. The shoulder diameter influences the maximum temperature generated during the process and maximum temperature alter the property of weld zone. Experimentally measurement of temperature at the centre of the joint is difficult due to high plastic deformation. Thermal cycle and mechanical constraints produce thermal stress during the process which reduces the working life of components. In this work, sequentially coupled thermal-structural analysis has been carried out using an analytical heat source model to evaluate the effect of tool geometry on maximum temperature and thermal stresses during the process by using ANSYS APDL software.

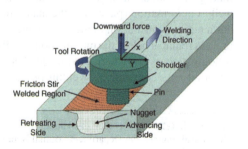

Fig. 1 Friction stir welding process [2]

2 Model Description

Only half of the model is modelled to take advantage of symmetry along the welding line. This reduces the number of elements and hence the computation time. A rectangular workpiece of dimension 200 × 50 × 6.4 mm is modelled. Model is meshed with SOLID70 elements which can be used for transient analysis. For structural analysis, SOLID180 elements are used which are equivalent to thermal elements. The workpiece material is AA 6061 T-6 an aluminium alloy whose composition is used from Mei-Ni Sua et al. [11]. Temperature-dependent material properties are used. Bilinear isotropic hardening is defined for material using Von-Mises theory. Temperature dependence curve is modelled for hardening effect incorporation using the BISO model. Temperature field data obtained in thermal analysis is used as an input file for computation of residual stresses.

2.1 Heat Generation

The expression for 3-D radially dependent HG during the FSW process per unit surface area at the tool-workpiece interface can be taken as [12, 13] (Fig. 2):

$$\frac{Q_\alpha}{A}(r) = \frac{3Q_{total}r}{2\pi\left[\left(R_{shoulder}^3 - R_{probe}^3\right)(1 + \tan\alpha) + R_{probe}^3 + 3R_{probe}^2 H_{probe}\right]} \quad (1)$$

where α is the shoulder cone angle and Q_{total} is total HG and it is given by:

$$Q_{total} = \frac{2}{3}\pi\omega\left[\delta\tau_{yield} + (1-\delta)\mu P\right]$$

Fig. 2 Tool geometry [12]

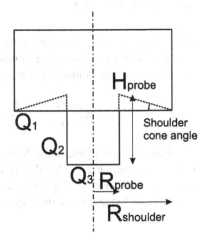

$$\left[\left(R_{shoulder}^3 - R_{probe}^3\right)(1 + \tan\alpha) + R_{probe}^3 + 3R_{probe}^2 H_{probe}\right] \quad (2)$$

where μ is friction coefficient, P is contact pressure, ω is angular velocity of tool, r is radial distance from rotational axis, $R_{shoulder}$ is shoulder radius, R_{probe} is probe radius, H_{probe} is probe height, τ_{yield} is yield strength and $\delta = \frac{\omega_{matrix}}{\omega_{tool}}$ and its value varies between 0 and 1.

Simplified form of surface heat flux expression, where the heat generated by probe is neglected, is given by "Eq. (3)":

$$\frac{Q}{A}(r) = \frac{3Q_{total} r}{2\pi R_{shoulder}^3 (1 + \tan\alpha)} \quad (3)$$

For flat shoulder "Eq. (3)" can be written as:

$$\frac{Q}{A}(r) = \frac{3Q_{total} r}{2\pi R_{shoulder}^3} \quad (4)$$

HG by pin surface is assumed to be volumetric heat flux and can be represented by "Eq. (5)"

$$\frac{Q}{V_{probe}}(r) = \frac{3Q_{total} r}{\pi (R_{shoulder}^3 + 3R_{probe}^2 H_{probe})} \quad (5)$$

In this model, shoulder heat flux and pin tip heat flux are modelled as one using surface heat flux in APDL. The heat generated by the tool pin is applied as volumetric heat flux on selected nodes using BF command in APDL. Linear surface heat flux is applied as given in the equation for surface heat flux.

2.2 Heat Transfer

The heat generated because of relative motion between workpiece and tool is transferred to workpiece following Fourier's law of heat conduction. In the cartesian coordinate system, Fourier's law can be written as:

$$k\left(\frac{\partial^2 T}{\partial x^2} + \frac{\partial^2 T}{\partial y^2} + \frac{\partial^2 T}{\partial z^2}\right) + Q_{gen} = c_p \frac{\partial T}{\partial t} \quad (6)$$

where T is the temperature distribution field and Q_{gen} is heat transferred from tool to workpiece is considered as HG in the workpiece, c_p is thermal heat capacity. The "Eq. (6)" for a moving cartesian coordinate in x-direction can be changed as control equation and given by "Eq. (7)":

$$\frac{\partial}{\partial x}\left(k_x \frac{\partial T}{\partial x}\right) + \frac{\partial}{\partial y}\left(k_y \frac{\partial T}{\partial y}\right) + \frac{\partial}{\partial z}\left(k_z \frac{\partial T}{\partial z}\right) = c_p \frac{\partial T}{\partial t}\left(\frac{\partial T}{\partial t} + U\frac{\partial T}{\partial x}\right) \quad (7)$$

2.3 Thermal Boundary Condition

Convection heat loss from all surface of workpiece except the bottom surface is applied using "Eq. (8)" given below

$$q = hA(T - T_0) \quad (8)$$

where h is the heat transfer coefficient and taken as 15 W/mK, T_0 is the ambient temperature and is assumed as 303 K. As the temperature attained during welding is not very high, hence the heat loss due to the radiation process can be neglected. For the bottom surface of the workpiece, to take account the conductive heat transfer between the bottom surface of the workpiece and backing plate, high convective heat transfer value has been taken. Heat transfer from the bottom surface of the workpiece can be defined by "Eq. (9)":

$$q = h_b A(T - T_0) \quad (9)$$

where h_b is the heat transfer coefficient at the bottom surface, h_b is taken as 300 W/m K.

2.4 Mechanical Boundary Condition

Bottom surface of the workpiece is constrained in perpendicular direction as $U_z = 0$. To take the clamping effect into account, side surface of workpiece is constrained in all direction with $U = 0$. Side along the weld line is constrained in its perpendicular direction with $U_y = 0$. End faces of workpiece are not constrained and have full degrees of freedom.

2.5 Process Parameters

All process parameter data as shown in Table 1 is taken from Soundrajan et al. [14] and validated using his model results.

Three different cases have been taken with constant radius of pin of 2.6 mm. First case has 24 mm shoulder diameter with zero cone angle, second case has 19 mm

Table 1 Process parameters used in simulation

N (rpm)	F (KN)	V (mm/s)	μ
344	12.9	2.2	0.5

shoulder diameter with zero cone angle, and third case has 24 mm shoulder diameter with 7° cone angle.

3 Results and Discussion

Temperature distribution for all three cases has nearly the same contour plots. Contour plot for the first case (24 mm shoulder diameter with zero cone angle) is shown in Fig. 3a when the tool reaches the midpoint of workpiece. At the midpoint, temperature distributions along with the nodes perpendicular to the welding direction are shown in Fig. 3b. It is observed that at the midpoint temperature is not maximum, but it increases along shoulder diameter. Temperature is maximum in the middle of the shoulder radius and after that it starts decreasing along the width of workpiece.

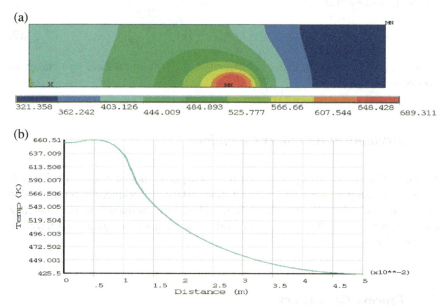

Fig. 3 Temperature distribution **a** contour plot at mid position, **b** profile along the width of workpiece

3.1 Thermal History

The maximum temperature obtained at distance 118 mm from the front end and 6 mm from the weld line is for case 3 as it can be seen from temperature history curve for all the three cases in Figs. 4a, b and 5 for case 1, case 2 and case 3, respectively. With 7° of cone angle shoulder, maximum temperature increases by 6% from 689.31 K as in case 1 (24 mm shoulder diameter with zero cone angle) to 730.69 K. With the decrease in the shoulder diameter maximum temperature is reduced to 683.15 K as in case 2. In case of shoulder with conical profile, heat input (total) is directly proportional to $R^3(1 + \tan \alpha)$ while in the case of flat shoulder, heat input (total) is directly proportional to R^3. Hence, total heat input in shoulder with conical profile is more as compared to flat shoulder. Total heat flux in case 3 (24 mm shoulder diameter with 7° cone angle) is maximum while it is minimum in case 2 (19 mm shoulder diameter with zero cone angle) due to a decrease in shoulder

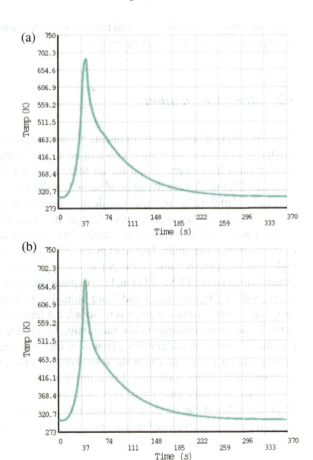

Fig. 4 Temperature history for **a** case 1, **b** case 2

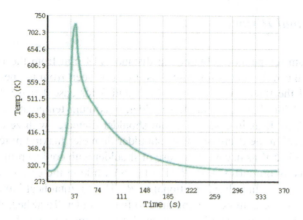

Fig. 5 Temperature history for case 3

diameter. Hence, it can be observed that the shoulder cone angle has a major impact on maximum temperature during FSW.

3.2 Stress Distribution

Stress distribution in welded components plays an important role during their working condition, as it has a major impact on fatigue failure. Longitudinal stress distribution contour plots have been plotted for all three cases when the tool is at the middle position of the workpiece. It can be observed in all the three cases that stress near the start and end of the plate is highly tensile because there is no constraint on these faces. While going from start to end, stress changes its nature from tensile to compressive. Near the middle of the workpiece, longitudinal stresses are highly compressive due to shoulder pressure applied on workpiece. Also, at the side edges stress obtained is highly compressive because of the constraint applied on side faces to take clamping effects into account. El-Sayed et al. [15] explain such type of behaviour during FSW of AA 5083-O. Figures 6a, b and 7 are plotted with longitudinal stress for case 1, case 2 and case 3, respectively. Even though the temperature obtained is maximum in third case, stress obtained is not maximum. This is due to the linear distribution of heat flux along the shoulder radius. However, maximum value of longitudinal tensile stress is obtained for case 1 near the start and end of the plate while maximum value of compressive stress is for case 2 near the shoulder.

For case 2, tensile stress near end faces cover more area due to smaller radius of shoulder. So, it can be concluded that tool with lowest diameter develops lower residual stress in workpiece but have very less effect on magnitude.

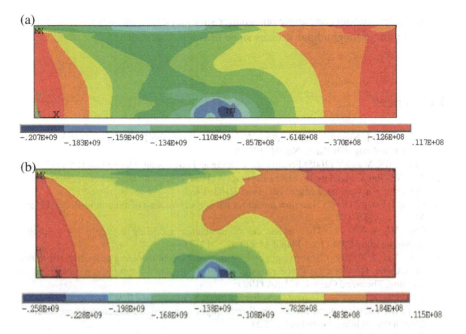

Fig. 6 Longitudinal stress for **a** case 1, **b** case 2

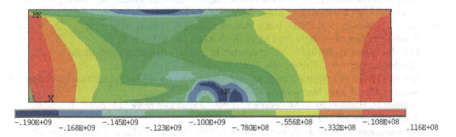

Fig. 7 Longitudinal stress for case 3

4 Conclusions

Thermal analysis for FSW of AA 6061-T6 alloy has been done for different tool geometries. Shoulder diameter and shoulder cone angle have a major role in maximum temperature obtained during FSW process. The following points can be concluded from thermal analysis and stress analysis:

1. Cone angle increases the amount of heat flux and hence the maximum temperature is obtained in case of shoulder with cone angle as compared to flat shoulder.
2. Shoulder with 7° of cone angle, increases the maximum temperature by 6% from 689.31 K as in case 1 to 730.69 K.

3. Shoulder without cone angle but smaller shoulder diameter does not cause much reduction in longitudinal stresses since heat flux is distributed in a small area.

References

1. Thomas WM, Nicholas ED, Needham JC, Murch MG, Smith PT, Dawes CJ (1991) Friction stir butt welding. GB Patent No. 9125978 8
2. Mishra RS, Mab ZY (2005) Friction stir welding and processing. Mat Sc Eng R 50(1–2):1–78
3. Tang W, Guo X, McClure JC, Murr LE, Nunes A (1998) Heat input and temperature distribution in friction stir welding. J Manuf Mat Proc 7:163–172
4. Gould J, Feng Z (1998) Heat flow model for friction stir welding of aluminum alloys. J Manuf Mat Proc 7:185–194
5. Chao YJ, Qi X (1998) Thermal and thermo-mechanical modeling of friction stir welding of aluminum alloy 6061–T6. J Manuf Mat Proc 7:215–233
6. Chao YJ, Qi X (1999) Heat transfer and thermo-mechanical modeling of friction stir joining of AA6061-T6 plates. In: Proceedings of the first international symposium on friction stir Welding, Thousand Oaks, CA, USA (1999)
7. Colegrove P, Pinter M, Graham D, Miller T (2000) 3-Dimensional flow and thermal modeling of the friction stir welding process. In: Proceedings of the section international symposium on friction stir Welding, Gothenburg, 2000
8. Chao YJ, Qi X, Tang W (2003) Heat transfer in friction stir welding—experimental and numerical studies. J Manuf Sci Eng 125(1):138–145
9. Song M, Kovacevic R (2003) Thermal modelling of friction stir welding in a moving coordinate system and its validation. Int J Mach Tools Manuf 43:605–615
10. Chen CM, Kovacevic R (2003) Finite element modelling of friction stir welding—thermal and thermo-mechanical analysis. Int J Mach Tools Manuf 43:1319–1326
11. Sua MN, Young B (2019) Material properties of normal and high strength aluminium alloys at elevated temperatures. Thin-Walled Stru 137:463–471
12. Schmidt H, Hattel J, Wert J (2003) An analytical model for the heat generation in friction stir welding. Model Simu Mat Sci Eng 12(1):143–157
13. Schmidt H, Hattel J (2005) Modelling heat flow around tool probe in friction stir welding. Sci Tech Weld Join 10(2):176–186
14. Soundararajan V, Zekovic S, Kovacevic R (2005) Thermo-mechanical model with adaptive boundary conditions for friction stir welding of Al 6061. Int J Mach Tools Manuf 45:1577–1587
15. El-Sayed MM, Shash AY, Abd-Rabou M (2018) Finite element modeling of aluminum alloy AA5083-O friction stir welding process. J Mat Proc Tech 252:13–24

Analysis of Electrolyte Flow Effects in Surface Micro-ECG

Dhruv Kant Rahi, Avanish Kumar Dubey, and Nisha Gupta

Abstract Micro-Electrochemical Grinding (ECG) is a process that produces a component with dimensions in the range of tens of nanometers to few millimeters. In this process, the material is removed by both electrochemical and abrasive action in a small interelectrode gap. Components produced by the micro-ECG process have found application in the electronics and medical industry. There have been some researches regarding the contribution of electrolytic action in the removal of material in cylindrical micro-ECG. The study of turbulence of electrolyte flow in a gap is required. This paper presents a turbulent flow modeling and analysis of surface micro-ECG to show the effect of erosion in material removal. The shear stresses generated on the workpiece surfaces are used to find the shear forces acting on the workpiece boundary. It was found that the shear stresses or shear force increase with an increase in the wheel rotation for a fixed interelectrode gap. The material removal rate by the erosion process was calculated and the theoretical values were validated previous research findings.

Keywords Micro-machining · Shear stress · MRR · CFD simulation · Shear force

1 Introduction

ECG is a compound machining process of electrochemical machining and conventional grinding. Most of the materials are removed from the workpiece through anodic dissolution (90–95%) and remaining (5–10%) material removal is done through mechanical action by conventional grinding [1–3]. Experiments and simulations have been done on the cylindrical micro-ECG about the electrolytic and abrasive action during electrolytic flow in inter-electrode gap on the material removal rate [4]. Analytical modeling of the surface micro-ECG has also been done for the study of chemical and mechanical action [5]. Analysis of erosion due to turbulence flow

---

D. K. Rahi (✉) · A. K. Dubey · N. Gupta
Mechanical Engineering Department, Motilal Nehru National Institute of Technology Allahabad, Prayagraj, India
e-mail: rahi16mnnit@gmail.com

Fig. 1 Geometry of the electrolyte domain

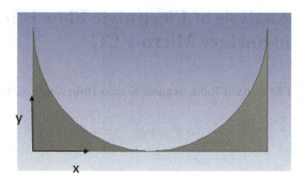

of electrolyte in inter-electrode gap is needed for the study of shear stress generation and material removal rate.

This paper presents the modeling of the surface micro-ECG for the simulation and analysis of the effects of turbulence in the interelectrode gap at various wheel rotations. It also shows the effect of erosion on the total material removal rate on validating the results with the literature review.

2 Model Approach and Geometry

FLUENT 16.0 CFD software has been used for the simulation of electrolyte in the IEG. A two-dimensional geometry as shown in Fig. 1 represents the computational domain. The diameter of the grinding wheel (ϕ70 mm) is circular and workpiece is a straight line. Wheel periphery and surface of workpiece are contacted through electrolyte.

Following assumptions have been considered:

- The flow of electrolyte is considered incompressible in nature.
- Hydrogen gas evolution effect in electrochemical action has been neglected.
- Electrolyte properties are considered unaffected from temperature.
- The wheel boundary is considered as moving boundary and is smooth in nature.
- The geometry of the IEG is carried out in 2D for the CFD analysis.
- Workpiece is given a certain length, which is assumed to be the contact length in MRR calculation.

3 Electrolyte Flow Simulation in IEG

The computational domain in IEG is meshed using ICEM application of the software. As shown in Fig. 2, the electrolyte flows in from one side and goes out from the other

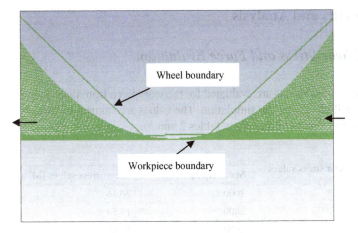

Fig. 2 Grid generation in computational domain

side. Both the side edges are separated into 70 divisions and the boundaries of wheel and workpiece are separated into 220 divisions each.

3.1 Governing Equations

- A continuity equation is given for the mass balance of the steady and incompressible electrolyte flow in the IEG:

$$\frac{\partial \bar{U}}{\partial X} + \frac{\partial \bar{V}}{\partial Y} = 0 \qquad (1)$$

- For momentum balance, a time averaged Navier–Stokes equation is given by the equation:

$$\frac{D\overline{U_j}}{Dt} = -\frac{\partial \left(\bar{P} + 2K_n/3\right)}{\partial X_j} + \frac{1}{R_e} \times \frac{\partial}{\partial X_i}\left[(1+v_{t,n})\left(\frac{\partial \overline{U_i}}{\partial X_j} + \frac{\partial \overline{U_j}}{\partial X_i}\right)\right] \qquad (2)$$

3.2 Boundary Conditions

- At $x = 0$ (electrolyte IN), $U = U_i$, $V = 0$
- At $y = 0$ (workpiece boundary), $U = 0$, $V =$ feed (m/s)
- At wheel, velocity $= W_w \times d/2$
- At $x = d$ (electrolyte OUT), $P = P_a$.

4 Results and Analysis

4.1 Shear Stress and Force Evaluation

The shear stress values are evaluated for the workpiece from the graphs obtained, as shown in Table 1 after the simulation. The values are studied and the changes in the values are also reasoned (Fig. 3; Tables 2 and 3).

Table 1 Shear stress values

Speed (rpm)	Shear stress values (MPa)
3000	20.16
2800	19.89
2600	19.06
2500	17.64
2400	16.34
2100	12.42

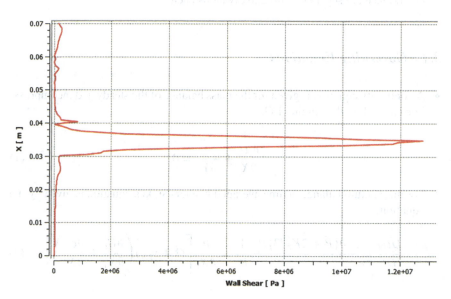

Fig. 3 Shear stress evaluation at IEG of 5 μm, flow rate 5.56×10^{-6} m/s and 2100 rpm wheel speed

Table 2 Parameters for the analysis of turbulent flow using the κ–ω model

Framework and model	Fluent 16.0, K–ω model
Flow type	Two dimensional, turbulent, incompressible
Default values	$\alpha^* = 1, \alpha_\infty = 0.52, \beta_\infty^* = 0.09, \beta_i = 0.072$ Turbulent kinetic energy Prandtl no. $= 2$ Turbulent kinetic energy dissipation rate Prandtl no. $= 2$
Electrolyte flow rate	5.56×10^{-6} mm^3/s
Operating conditions	Mass flow rate, atmospheric pressure
Boundary condition values for defined boundaries	Flow rate $= 5.56 \times 10^{-6}$ m/s Turbulence intensity $= 2\%$ Rotation of the grinding wheel $= 2100, 2400, 2500, 2600, 2800, 3000$ rpm (clockwise) Outflow Pressure at outlet Atmospheric pressure $= 1.01325$ bar Hydraulic dia $= 1.28$ mm
Discretization scheme	For momentum balance: first-order upwind Turbulent kinetic energy: first-order upwind Turbulence dissipation rate: first-order upwind
No. of iterations	500–2000 till the solution converges

Table 3 Process parameters [5, 6]

	Parameter	Details
Process	Type of process	Surface micro-ECG
	Coefficient of friction f	0.1
Grinding wheel	Abrasive	Diamond
	Bond (% volume)	Copper 12.5%
	Grain size (μm), Diameter (mm) and speed (rpm)	$d_{max} = 5, d_{min} = 1, d_{mean} = 2.5, d_g = 76;$ 70 and 3000
	Width (mm)	0.3
Workpiece	Material	Brass
Electrolyte	Type of electrolyte	Potassium Nitrate (KNO$_3$)
	Concentration	15%
	Dynamic viscosity μ (N s/m^2)	7.98×10^{-4}
	Density ρ_e (kg/m^3)	1260
	Flow rate Q_a (m^3/s), interelectrode gap (μm)	$5.56 \times 10^{-6}, 5$
	Specific conductance w (S/mm)	0.03696
	Dielectric constant	5.2

4.2 MRR Evaluation

For the calculation of total MRR, the individual MRRs of the sub-processes are summed and are given by the following equation:

$$MRR_{Total} = MRR_{ECM} + MRR_{Abrasion} + MRR_{Erosion} \qquad (3)$$

MRR_{ECM} is evaluated by:

$$MRR_{ECM} = \frac{\varepsilon CI}{z \rho F} \qquad (4)$$

where the value of current is calculated as in [4]. Equation used for calculating MRR due to erosion is represented by the following [4]

$$MRR_{Abrasion} = \int_{dcont}^{dmax} \left(dg \times \frac{\left(\left(\int_r^\theta \Delta p \times \left[(d+a) - 2\sqrt{r^2 - y^2}\right] \times t\right) + \rho g t r \left[\frac{2}{3}r - a\right]\right)^{\frac{1}{2}} \left(1 + \sqrt{\frac{Hw}{Hp}}\right)}{\left(\frac{10}{dmean} \times (\text{vol. fraction of abrasive})^{\frac{1}{3}}\right)(Hw \times Vw \times t \times \text{Prob. of active grains} \times \text{Coeff. of friction})^{\frac{1}{2}}} - \frac{dg}{2} \right) \times \frac{dl}{dt}$$

$$\times N_{Total} \times \left(\frac{1}{\sqrt{2\pi}}\right) \int_p^\infty e^{-\frac{x^2}{2}} dx \qquad (5)$$

Finally, by knowing the shear force value in the IEG due to electrolytic flow the $MRR_{Erosion}$ is evaluated and is given by

$$MRR_{Erosion} = \text{Shear Velocity} \times \text{Shear Area}(A_{shear}) \times \text{Density}(\rho_h) \qquad (6)$$

where shear velocity = Power $(p)/F_{s\_Simulation}$.
The total power of the electrolytic flow (in W), is given by:

$$P = \rho Q g H \qquad (7)$$

where $Q = V_{avg} \times A_{IEG}$.
Therefore the $MRR_{Erosion}$ is given by:

$$MRR_{erosion} = \frac{(\rho(v_{avg} \times A_{IEG}) \times g \times H)}{F_{s\_simulation}} \times A_{shear} \times \rho_h \times n \qquad (8)$$

Accordingly, the $MRR_{Erosion}$ was multiplied by a correction factor, n. This was assumed that as the IEG is so small and the wheel is rotating at a very high speed so most of the electrolyte will splash out and only 10% of the power is used by the shearing stresses (Table 4).

Evaluation of MRR and Shear force: this calculation is for 2500 rpm of wheel rotation and 6 V voltage.

Table 4 Results obtained for MRR$_{erosion}$

Speed of the grinding wheel (rpm)	Shear force (N)	MRR$_{erosion}$ mg/min
3000	187.49	0.0107
2800	184.98	0.0108
2600	177.26	0.0109
2500	164.05	0.0114
2400	151.96	0.0119
2100	115.51	0.0136

Fy = Shear stress × Area of shear (As)
= 17.64 × 0.3 × 31 = 164.052 N

$V^* = \sqrt{\frac{\tau}{\rho}} = \sqrt{\frac{17.64 \times 10^6}{1260}} = 118.32$ m/s

$f = 8\left(\frac{V^*}{V_{avg}}\right)^2$; Therefore, V_{avg} = 1058.3 m/s

MRR$_{erosion}$ = (1260 × 1058.3 × 5 × 10^{-6} × 0.3 × 10^{-3} × 9.81 × 0.508 × 31 × 10^{-6} × 0.3 × 3370 × 0.1)/164.052 = 0.0114 mg/min

MRR$_{abrasion}$ = 0.059 mg/min (for V_w = 9.163 m/s)

$\mathcal{E}$ = 0.98; C = 32.695 gm; z = 2; ρ = 8.6 gm/cm^3; F = 96,494; w = 0.03696 S/mm

Fa = tl = 0.3 × 31 × 10^{-6}; Kp = 5.2; Ve = 7.71 V

MRR$_{ECM}$ = 1.643 mg/min (for Fa = tl = 0.3 × 31 × 10^{-6})

MRR$_{total}$ = 1.7134 mg/min.

4.3 MRR Validations

The MRR$_{erosion}$ values, as shown in Fig. 4 when compared to the literature review

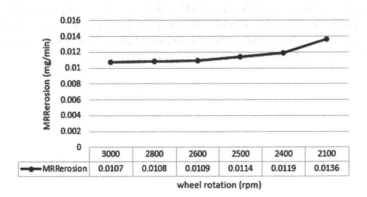

Fig. 4 Graph showing the MRR erosion values for different wheel rotation

results, showed that they have followed the trend which shows that the major contribution to the MRR$_{total}$ is of electrochemical action then abrasive action and less than 10% that of erosive action. The calculation showed that there was a considerable amount of change noticed in the MRR$_{erosion}$ values when changing the wheel speed. But not much of a change was noticed in the total MRR value. The interelectrode gap is kept constant here and the results are validated from [4–6]. The length of the workpiece is assumed to be 31 mm and this length is chosen after comparing different research papers [7–9]. This could be the reason for the difference in the values of the MRR when compared to the refs mentioned.

Results also show that the shear force value gradually increases on increasing the wheel rotation thus causes the turbulence in the interelectrode gap to increase. The obtained MRR values are found to be a little higher than that of the papers due to certain assumptions taken in the MRR calculation and also due to the assumption of workpiece length and also the actual process is done in 3D.

5 Conclusions

A considerable amount of change is noticed in the MRRerosion values while changing the wheel speed. But not much of a change was noticed in the total MRR value. It is shown that the addition of material removal due to erosion increases the total material removal rate. Results also show that the shear force value gradually increases on increasing the wheel rotation thus causes the turbulence in the interelectrode gap to increase.

6 Future Scope

An unsteady simulation needs to carry out on the electrolyte action to actually know the real changes in the Shear Stresses and MRR values. That analysis could be closer to the actual experimentation values. The only hardship that would be faced in the 3D analysis would be the time taken for the simulation. That is why it is seen that none of the research papers are there to show the 3D analysis.

References

1. Li H, Niu S, Zhang Q, Shuxing Fu, Ningsong Qu (2017) Simulation and experimental investigation of material removal in inner-jet ECG of GH4169 alloy. Chin J Aeronaut. https://doi.org/10.1038/s41598-017-03770-1
2. Ilhan RE, Sathyanarayanan G, Storer RH, Phillips RE (1992) A study of wheel wear in electrochemical surface grinding. ASME J Eng Ind 114(1):82–93
3. Kozak J, Skrabalak G (2014) Analysis of abrasive electrochemical grinding process (AECG). In: Proceedings of the world congress on engineering 2014, July 2–4, London, U.K.
4. Sapre P, Mall A, Joshi SS (2013) Analysis of electrolytic flow effects in micro electrochemical grinding. J Manuf Sci Eng 135:011012–011021
5. Gaikwad KS, Josh SS (2008) Modeling of material removal rate in micro-ECG process. J Manuf Sci Eng 130(3):034502
6. Hou ZB, Komanduri R (2003) On the mechanics of the grinding process—part I. Stochastic nature of the grinding process. Int J Mach Tools Manuf, 43:1579–1593
7. Puri AB, Banerjee S (2012) Multiple-response optimisation of ECG characteristics through response surface methodology. Int J Adv ManufTechnol 64:715–725
8. Yoshino M, Shirakashi T, Obikawa T, Usui E (1998) Electrolytic cut-off grinding machine for composite materials. J Mater Process Technol 74:131–136
9. Tehrani AF, Atkinson J (2000) Overcut in pulsed electrochemical grinding. Int J Adv Manuf Technol 214:75–86

Investigate the Effect of Design Variables of Angular Contact Ball Bearing for the Performance Requirement

Priya Tiwari and Samant Raghuwanshi

Abstract The aim of this study is to examine the effect of design variables on the performance requirements, i.e., total deformation and equivalent stress of the angular contact ball bearing. Design variables such as material properties, inner raceway curvature, and a number of balls greatly influence the deformation and stress developed on ball-bearing under different working conditions. There is a need to determine the best configuration of these design variables for given operating conditions that give minimal deformation and stress. This can be achieved by utilizing the concept of design of the experiment which will help in improving the design of angular contact ball bearing. The results obtained by mathematical calculation and ANSYS tool analysis are compared and are found to be in good correlation. Based on the results obtained the effect of these design variables on the bearing performance is studied by means of Taguchi's design of the experiment and the best design configurations can then be figured out.

Keywords Angular contact ball bearing · Inner raceway curvature · Bearing materials · Hertz contact theory · Taguchi's design of experiment · ANSYS · Total deformation

1 Introduction

Rolling bearings are the fundamental mechanical components which are exposed to high speeds and concentrated contact loads. Bearings are the highly engineered mechanical component that allows relative motion between two elements with minimal friction. It comprises of four main parts namely, inner ring, outer ring, balls or rollers, and retainer. Generally, inner ring is a moving part and outer ring is kept stationary. The retainer helps in holding the balls at a proper distance with each other. Angular contact ball bearing can take both radial and axial loads. The design of these parts has a great influence on the performance of rolling bearing.

P. Tiwari (✉) · S. Raghuwanshi
Jabalpur Engineering College, Jabalpur, Madhya Pradesh, India
e-mail: tiwari.priya033@gmail.com

Therefore, bearing variables should be analyzed in order to improve the design and performance of the bearing. This will help in improving the load-carrying capacity, increasing wear strength, reducing weight and costs associated with it, Tiwari P et al. [1].

The static structural analysis of 7008 CD/P4A angular contact ball bearing is carried out to investigate the bearing design based on a different configuration of design variables that affect the performance of ball bearing. The bearing is subjected to various operating conditions such as radial load and axial load. Hertz contact theory is used to calculate the mathematical results for deformation. The results obtained from mathematical calculation and ANSYS tool analysis for total deformation are compared. The effect of these design variables on the performance of ball bearing is studied with the help of design of experiment. Design of the experiment deals with planning analyzing and interpreting controlled variables to evaluate the factor that control the value of variables. This will reduce the number of experiments to minimal and hence consumes less time. Taguchi method in design of the experiment is adopted for analyzing the effect of the design variables of angular contact ball bearing.

2 Mathematical Model

Hertz contact theory assumes an elliptical shape contact area developed when a point contact occurs between two elastic solids subjected to load W [2], Khonsari et al. [3]. Then,

$$\text{The curvature sum,} \sum \rho = \rho_{11} + \rho_{21} + \rho_{12} + \rho_{22}$$

For inner raceway,

$$\rho_{11} = \frac{1}{r_{11}} = \frac{2}{d_b};$$

$$\rho_{12} = \frac{1}{r_{12}} = \frac{2}{d_b};$$

$$\rho_{21} = \frac{1}{r_{21}} = \frac{2}{d_b}\left(\frac{\gamma}{1-\gamma}\right);$$

$$\rho_{22} = -\frac{1}{r_{22}} = -\frac{1}{f_0 d_b};$$

where,

γ $\frac{d_b(\cos\alpha_0)}{d_m}$
f_0 curvature ratio.
d_m Pitch radius.
α_0 Contact angle.

Then, equivalent curvature radius is

$$R_x = \rho_{11} + \rho_{21}$$
$$R_y = \rho_{12} + \rho_{22}$$

Ellipticity ratio is

$$K = 1.0339 \left(\frac{R_y}{R_x}\right)^{0.636}$$

The first complete elliptic integral is,

$$F = 1.0003 + \frac{0.5968}{\left(\frac{R_y}{R_x}\right)}$$

The second complete elliptic integral is

$$E = 1.5277 + 0.6023 \ln\left(\frac{R_y}{R_x}\right)$$

For combined load, load on each bearing ball is given by

$$Q = \frac{F_r}{Z J_r(\varepsilon) \cos \alpha} = \frac{F_a}{Z J_a(\varepsilon) \sin \alpha}$$

where Z denotes Number of balls and $J_r(\varepsilon)$, $J_a(\varepsilon)$ depends on ratio $\frac{F_r}{F_a} \tan \alpha$.
The equivalent elastic modulus,

$$\frac{2}{E'} = \frac{1-\mu_1^2}{E_1} + \frac{1-\mu_2^2}{E_2}$$

Then, deformation,

$$\delta = \delta^* \left(\frac{3Q}{\sum \rho E'}\right)^{\frac{2}{3}} \left(\frac{\sum \rho}{2}\right)$$

where,

$$\delta^* = \left(\frac{2F}{\pi}\right) \left(\frac{\pi}{2EK^2}\right)^{\frac{1}{3}}$$

Table 1 Bearing design variables

Bearing material	Inner raceway curvature	Number of balls
Stainless steel	0.52d	12
Aluminum Oxide	0.53d	15
Zirconium Oxide	0.51d	18

3 Design Variables Selection

The design variables to be considered for the analysis are inner raceway curvature, number of balls, and material as described in Table 1.

4 Design of Single Row Angular Contact Ball Bearing

See Fig. 1.

Three-dimensional modeling of 7008 CD/P4A angular contact ball bearing is done with the help of CATIA V5 software using the following standard dimensions (Table 2).

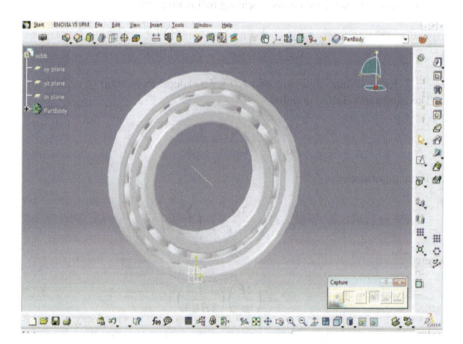

Fig. 1 Three-dimensional model of 7008A ball bearing

Table 2 Ball bearing specifications

Attributes	Values
Inner diameter (d)	40 mm
Outer diameter (D)	68 mm
Pitch circle diameter (D_m)	54 mm
Initial contact angle (α)	15°
Basic dynamic load rating (C)	16.8 kN
Basic static load rating (C_0)	11 kN
Mass	0.19 kg
Attainable speed	20,000 r/min
Outer raceway curvature	0.53d
Ball diameter (d)	7.938 mm

Table 3 Material properties

Variables	Density (kg/m³)	Elastic modulus (GPa)	Poisson's ratio	Tensile strength (MPa)
Stainless steel	7750	193	0.31	465
Aluminum oxide	3000	215	0.21	69
Zirconium oxide	5000	100	0.22	115

5 Material Properties

The following are the material properties for different materials used in this project (Table 3).

6 Design Variable Matrix

The following design variable matrix is obtained using Taguchi's method of design of experiment in MINITAB software. Based on this design variable matrix, nine three dimensional design models that contain different configurations of design variables are prepared using standard dimensions with the help of CATIA V5 software (Table 4).

Table 4 Design variable matrix

S. No.	Material	Inner raceway curvature	Number of balls
1.	Stainless steel	0.52d	12
2.	Stainless steel	0.53d	15
3.	Stainless steel	0.51d	18
4.	Aluminum oxide	0.52d	15
5.	Aluminum oxide	0.53d	18
6.	Aluminum oxide	0.51d	12
7.	Zirconium oxide	0.52d	18
8.	Zirconium oxide	0.53d	12
9.	Zirconium oxide	0.51d	15

7 Static Structural Analysis

7.1 Meshing and Boundary Conditions

With the help of ANSYS 18 software, the elements of angular contact ball bearing are finely divided which increases the accuracy of the analysis. The meshing size used is 5 mm. The boundary conditions are applied by fixing the outer ring of the ball bearing and, by applying a radial load of magnitude 4000 N in x-direction of inner race and axial load of magnitude 2000 N is applied in the y-direction of inner race.

7.2 Inputs and Results

See Tables 5 and 6.

8 Effect of Design Variables

With the help of MINITAB software, response table and graph are obtained for total deformation and equivalent stress using ANSYS results such that smaller value for total deformation and equivalent stress is considered to analyze the effect of design variables on the performance of bearing. Based on the signal-to-noise ratio graph, the best design configuration is found out. The S-N ration graph is basically the log function of the desired output that helps in optimization and data analysis of the results. The following results are obtained.

Table 5 Static structural analysis results

Configuration	Total deformation	Equivalent stress
SS 0.52d 12		
SS 0.53d 15		
SS 0.51d 18		

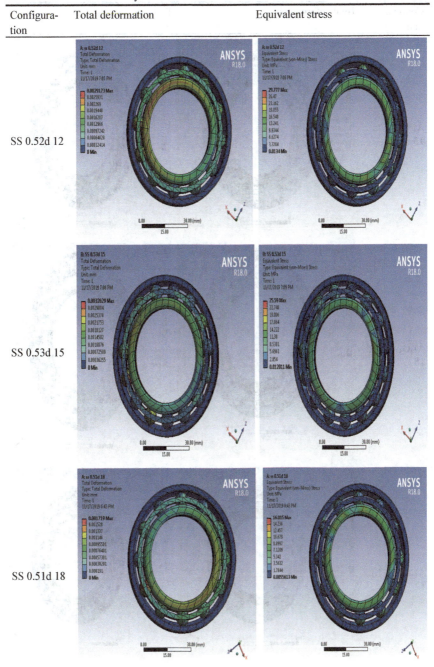

(continued)

Table 5 (continued)

Configuration	Total deformation	Equivalent stress

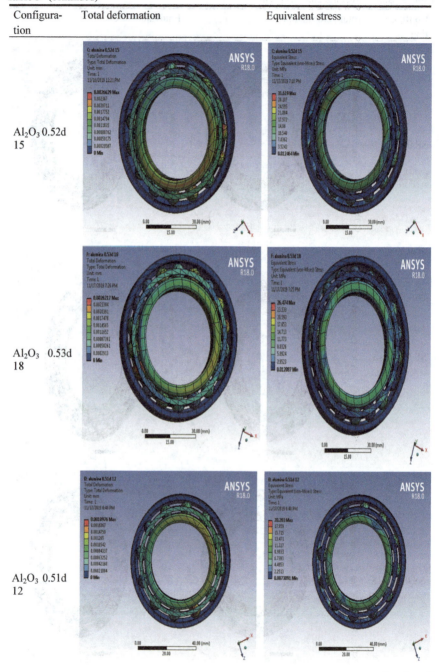

Al$_2$O$_3$ 0.52d 15

Al$_2$O$_3$ 0.53d 18

Al$_2$O$_3$ 0.51d 12

(continued)

Table 5 (continued)

Configuration	Total deformation	Equivalent stress
ZrO$_2$ 0.52d 18		
ZrO$_2$ 0.53d 12		
ZrO$_2$ 0.51d 15		

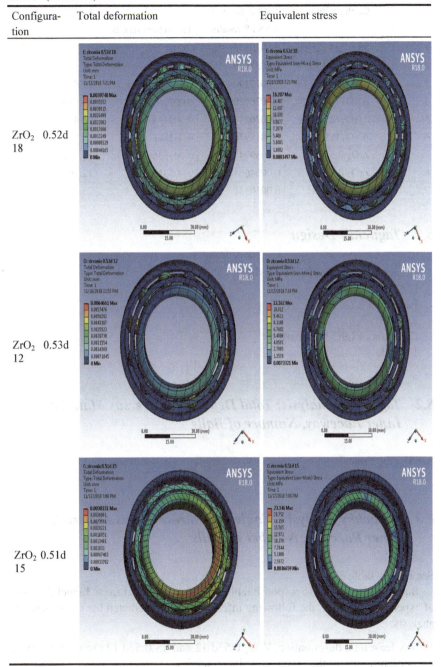

Table 6 Theoretical and static structural ANSYS results

S. No.	Configuration	Total deformation		Equivalent stress
		ANSYS results	Theoretical results	
1.	SS 0.52d 12	0.0029173	0.002723	29.777
2.	SS 0.53d 15	0.0032629	0.002862	25.590
3.	SS 0.51d 18	0.0017190	0.001492	16.015
4.	Al_2O_3 0.52d 15	0.0026629	0.002268	31.619
5.	Al_2O_3 0.53d 18	0.0026217	0.002444	26.474
6.	Al_2O_3 0.51d 12	0.0018976	0.001888	20.203
7.	ZrO_2 0.52d 18	0.0039748	0.003336	16.207
8.	ZrO_2 0.53d 12	0.0064661	0.005319	12.162
9.	ZrO_2 0.51d 15	0.0030331	0.002702	23.346

8.1 Taguchi's Design

Taguchi Orthogonal Array Design
L9(3**3)
Factors: 3
Runs: 9
Columns of L9(3**4) Array
1 2 3.

8.2 Taguchi's Analysis: Total Deformation Versus Material, Inner Raceway, Number of Balls

See Fig. 2; Table 7.

8.3 Taguchi's Analysis: Equivalent Stress Versus Material, Inner Raceway, Number of Balls

See Fig. 3; Table 8.

Based on the S-N ratio response table and graph obtained from Taguchi's design of experiment method, the following inference can be predicted for static structural analysis:

1. For least total deformation, ZrO_2 0.53d 12 and SS 0.52d 12 are the best design configuration for the given operating condition.
2. SS 0.52d 12 and Al_2O_3 0.52d 15 are the models that give minimum value of equivalent stress.

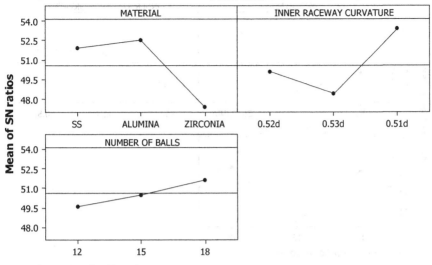

Fig. 2 S-N ratio graph for total deformation

Table 7 Response table of total deformation for signal-to-noise ratios (smaller is better)

Level	Material	Inner raceway curvature	Number of balls
1	51.91	50.07	49.64
2	52.52	48.38	50.53
3	47.39	53.36	51.65
Delta	5.13	4.98	2.00
Rank	1	2	3

9 Conclusion

The mathematical and ANSYS results are compared and it is found that outcomes of both results are in good correlation. The effect of design variables for the given boundary conditions can be studied using Taguchi's design of experiment method. Therefore, it can be concluded that the best configuration that gives minimal total deformation and equivalent stress is as follows (Table 9).

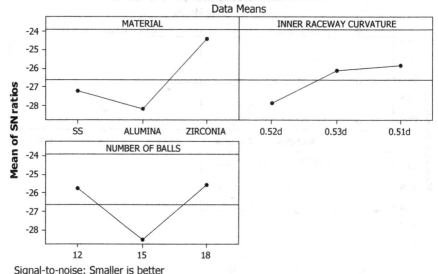

Fig. 3 S-N ratio graph for equivalent stress

Table 8 Response table of equivalent stress for signal-to-noise ratios (smaller is better)

Level	Material	Inner raceway curvature	Number of balls
1	−27.24	−27.89	−25.76
2	−28.19	−26.11	−28.51
3	−24.42	−25.85	−25.58
Delta	3.77	2.04	2.93
Rank	1	3	2

Table 9 Best model configuration

Material	Inner raceway curvature	Number of balls
Stainless steel	0.52d	12

References

1. Tiwari P, Raghuwanshi S (2019) Investigation of design parameters that affect the performance of angular contact ball bearing: a Review. Int Res J Eng Technol 06(08):167–169
2. Wang F, Jing M, Fan H, Wei Y, Zhao Y, Liu H (2016) Investigation on contact angle of ball bearings. Proc Inst Mech Eng, Part K, J Multi-body Dyn 231(1):230–251
3. Khonsari, MM, Booser ER (2017) Applied tribology: bearing design and lubrication, 3rd edn. Wiley Publication
4. Boyer HE (1986) Atlas of Fatigue curves. American Society for Metals
5. Bhushan B (2013) Introduction to tribology, 2nd edn. Wiley Publication

Effect of Flow of Fluid Mass Per Unit Time on Life Cycle Conversion Efficiency of Double Slope Solar Desalination Unit Coupled with N Identical Evacuated Tubular Collectors

Desh Bandhu Singh, Navneet Kumar, Anuj Raturi, Gagan Bansal, Akhileshwar Nirala, and Neeraj Sengar

Abstract In this research paper, life cycle conversion efficiency (LCCE) of double slope solar desalination unit (DSSDU) augmented with N alike evacuated tubular collectors (ETCs) have been discussed. An archetypal day of May for New Delhi climatic situation has been considered for the computation. The computation has been done using a programming code written in MATLAB. The input data for the evaluation of LCCE has been taken from IMD, Pune, India. It has been concluded that values of LCCE get reduced as the value of fluid mass flowing through tubes on a unit time basis is enhanced for the selected value of number of ETCs.

Keywords Mass of fluid flow/unit time · Double slope solar desalination unit (DSSDU) · ETC

D. B. Singh (✉) · A. Raturi · G. Bansal · N. Sengar
Department of Mechanical Engineering, Graphic Era Deemed to be University, Bell Road, Clement Town, Dehradun 248002, Uttarakhand, India
e-mail: Deshbandhusingh.me@geu.ac.in; dbsiit76@geu.ac.in

A. Raturi
e-mail: anujraturi@geu.ac.in

G. Bansal
e-mail: gaganbansal@geu.ac.in

N. Sengar
e-mail: neerajsengar111@geu.ac.in

N. Kumar · A. Nirala
Galgotias College of Engineering and Technology, Plot No. 1, Knowledge Park II, Greater Noida, Uttar Pradesh 201306, India
e-mail: navneet.kumar@galgotiacollege.edu

A. Nirala
e-mail: akhileshwar.nirala@galgotiacollege.edu

© The Author(s), under exclusive license to Springer Nature Singapore Pte Ltd. 2021
R. M. Singari et al. (eds.), *Advances in Manufacturing and Industrial Engineering*, Lecture Notes in Mechanical Engineering, https://doi.org/10.1007/978-981-15-8542-5_34

1 Introduction

The human beings on the planet earth are having the problem of freshwater availability for drinking and other uses and hence the purification of polluted water is essential because the use of polluted water by human beings is responsible for lots of diseases in the body and sometimes, it is so severe that the death of human being takes place. The human being can access a very low percentage of water (less than 1%). So, the contemporary situation demands a self-sustainable water purification system that can provide competitive output. The conventional system of water purification unit being used is not self-sustainable because electrical power is needed as input for the working and this type of water purification unit creates various pollutants which are detrimental for the environment. So, the use of solar energy for running the water purification unit is a good alternative for mitigating the contemporary problem of fresh water scarcity.

There are two modes namely passive and active in which the water purification unit based on solar energy commonly known as a solar still can work. The passive mode of solar-based water purification units are self-sustainable and it can successfully provide fresh water even at distant locations where availability of sunlight is in abundance. However, these types of solar-based water purification unit suffer from the issue of low fresh water output (one to three lires pr day). It can be overcome by allowing to operate the solar-based water purification unit in an active mode which needs the attachment of an auxiliary unit to basin so that extra heat can be supplied for enhancing the output. This type of solar-based water purification unit was introduced in 1983 [1, 2]; the reported system consisted of the attachment of single flat plate collector to the basin. After investigation, It was reported that the fresh water output on daily basis got enhanced as compared to solar-based water purification unit of the same basin area acting in passive mode. Afterwards, many solar-based water purification units have been reported throughout the world.

The experimental investigation of basin type solar-based water purification unit augmented with evacuated tubular collectors was done by Sampathkumar et al. [3] and it was reported that that the fresh water output from that system got enhanced by 129% than the similar set-up working in passive mode due to the supply of heat from evacuated tubular collectors to the basin. The solar-based water purification unit augmented with evacuated tubes and working in natural mode was investigated by Singh et al. [4] and they reported the efficiency based on exergy in the range of 0.15–8%. This work was extended by Kumar et al. [5] in which a pump was inserted between tube and basin and reported higher fresh water production over the natural circulation mode operated system because of the enhanced circulation in the system reported by Kumar et al. [5].

The expression for heat gain by N alike ETCs in series was reported by Mishra et al. [6]. It was further extended by Singh et al. [7–9] in which N identical ETCs were attached to basin type solar-based water purification units. The expressions for different unknown variables were reported followed by the comparative investigation taking energy, exergy and cost as the basis. Issa and Chang [10] investigated the basin type solar-based water purification unit augmented with mixed-mode connected ETCs. It was reported that the output of the proposed system was better than the similar unit in a passive mode of operation. The comparative energy metrics were further studied by Singh and Al-Helal [11] and Singh [12].

The current literature study suggests that the evaluation of LCCE of solar energy based water purification unit by incorporating the dissimilarity of the mass of fluid flow per unit time has not been done by any researchers throughout the world. The work of Singh and Al-Helal [11] has been extended in the investigation reported in this paper. The investigation reported in this paper is different in the sense that LCCE has been computed by incorporating the dissimilarity of mass of fluid flow per unit time; however, Singh and Al-Helal [11] reported the energy metrics computation at the considered values of mass of fluid flow per unit time and number of ETCs and results obtained was compared with results of similar set-ups on taking the same parameter as the basis of computation.

2 Working of DSSDU Augmented with N Series Connected ETCs

The solar energy based double slope water purification unit augmented with N alike ETCs has been represented by Fig. 1. The principle of working of the proposed system is similar to greenhouse effect. The detailed specification of the proposed system is given in Table 1. N alike ETCs in series have been attached to the basin of solar energy based double slope water purification unit for providing energy in the form of heat so that fresh water production can be enhanced. The direct input energy in the form of solar energy impinged at outer surface of the top cover is

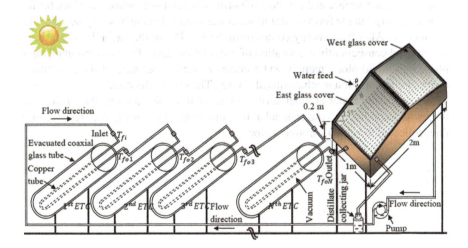

Fig. 1 The solar energy based double slope water purification unit augmented with N alike ETCs

Table 1 Specification in detail for the solar energy based double slope water purification unit augmented with N alike ETCs

DSSS

Component	Specification	Component	Specification
Length	2 m	Width	1 m
Inclination of glass cover	15°	Lower side height	0.2 m
Body material	GRP	Supporting stand material	GI
Cover material	Glass	Orientation	East-West
Thickness of glass cover	0.004 m	Thickness of insulation	0.1 m
K_g	0.816 W/m K	K_i	0.166 W/m K

ETC

Component	Specification	Component	Specification
Type and no. of collectors	ETC, N	α_p	0.8
DC motor rating	12 V, 24 W	F'	0.968
Tube (copper) inside radius	0.0125 m	τ_g	0.95
Tube (copper) thickness	0.0005 m	$K_g (\text{Wm}^{-1}\text{K}^{-1})$	1.09
Outer glass tube outside radius	0.024 m	Angle of ETC with horizontal	30°
Inside glass tube Inner radius	0.0165 m	Length of each copper tube	2.0 m
Outer/inner glass tube thickness	0.002 m		

transmitted to water kept in basin after occurring two other phenomena namely reflection and absorption. A major portion (95%) is transmitted to water and a minor portion (5%) is contributed for reflection as well as absorption. Further, solar radiation from water mass is transmitted to basin liner after absorbing. The solar radiation reaching the blackened surface gets absorbed almost totally. Hence, the temperature of the blackened surface at the bottom of purification unit gets enhanced which further provides energy in the form of heat to water mass placed in the basin by heat transfer mechanism. Also, heat energy comes from the N ETCs to the basin. In this fashion, heating and consequently evaporation of water takes place. The conversion of vapor phase into liquid phase namely water occurs at the inside plane of the condensing cover by condensation mechanism (filmwise). The water after condensation of vapor obtained as fresh water at inner surface of cover flows down along the surface due to gravitational force to the channel at the smaller height of wall. The fresh water is then collected to beaker through a pipe.

3 Equations Used for Analysis

Main equations used for computation of LCCE for the proposed system by incorporating dissimilarity of fluid mass flowing through tubes on the basis of unit time are the following:

$$T_{foN} = \frac{(AF_R(\alpha\tau))_1}{\dot{m}_f C_f} \frac{\left(1 - K_k^N\right)}{(1 - K_k)} I(t) + \frac{(AF_R U_L)_1}{\dot{m}_f C_f} \frac{\left(1 - K_k^N\right)}{(1 - K_k)} T_a + K_k^N T_{fi} \quad (1)$$

where, T_{foN} is the temperature at outer end of last collector, $I(t)$ is the solar radiation at collectors, $\dot{m}_f$ is mass of fluid mass flowing through tubes on the basis of unit time, C_f is heat capacity per unit mass, T_a is environmental temperature, T_{fi} is fluid temperature at inlet of first ETC. Here, $T_{fi} = T_{wo}$ in which T_{wo} is the water temperature at the start.

The expressions for water temperature inside still (T_w) and temperature at the inside surface of condensing cover are as follows:

$$T_w = \frac{\bar{f}(t)}{a}\left(1 - e^{-at}\right) + T_{w0} e^{-at} \quad (2)$$

$$T_{giE} = \frac{A_1 + A_2 T_w}{P} \quad (3)$$

$$T_{giW} = \frac{B_1 + B_2 T_w}{P} \quad (4)$$

where $\bar{f}(t)$ gives average value of $f(t)$ for time interval 0 to t. The computation of hourly fresh water production ($\dot{m}_{ew}$) by proposed system after evaluating of T_w, T_{giE} and T_{giW} from Eqs. (2)–(4) respectively can be done as follows:

$$\dot{m}_{ew} = \frac{h_{ewgE} \frac{A_b}{2} (T_w - T_{giE}) + h_{ewgW} \frac{A_b}{2} (T_w - T_{giW})}{L} 3600 \quad (5)$$

Here, L is latent heat of vaporization. The methodology for developing Eqs. (1)–(4) can be seen in Singh and Tiwari [8]. Also, values all unknown terms in these equations are present in Singh and Tiwari [8]. After the computation of fresh water from solar-based double slope active water purification unit on hourly basis using Eq. (5), fresh water output from solar based double slope active water purification unit on daily basis can be evaluated as a summation of fresh water output from solar-based double slope active water purification unit for 24 h followed by the evaluation of fresh water output from solar-based double slope active water purification unit on basis of one year (M_{ew}). The fresh water output from solar-based double slope active water purification unit on basis of one year can be done as multiplication of fresh water output from solar-based double slope active water purification unit on daily basis and number of clear days for a year. The energy output from solar-based double slope active water purification unit on the basis of one year can be computed in the following way:

$$E_{out} = (M_{ew} \times L) \tag{6}$$

The computation of LCCE [11] can be carried out as follows:

Life cycle conversion efficiency

$$= \frac{(\text{Energy output from system on the basis of 1 year} \times \text{system life span}) - (\text{Embodied energy})}{(\text{Solar energy on the basis of 1 year}) \times (\text{system life span})} \tag{7}$$

The explanation and detailed discussion of Eq. (7) can be seen in Singh and Tiwari [11].

4 Procedure for Computation of LCCE for the Proposed System

The solar radiation on horizontal plane and the environmental temperature have been taken from IMD located at Pune in India. Its values on inclined planes have been evaluated using the formula given by Liu and Jordan. The computational programming has been done in MATLAB for the evaluation of radiation on inclined planes. Then computation of T_w, T_{giE} and T_{giW} for solar-based double slope active water purification unit have been done using Eqs. (2)–(4) respectively followed by the calculation of fresh water output from solar based double slope active water purification unit from Eq. (5). Energy output from solar-based double slope active water purification unit on the basis of 1 year has been calculated using Eq. (6). The value of total energy required for the manufacturing of solar energy based double slope active water purification unit (embodied energy) and annual solar energy for solar-based double slope active water purification unit have been taken from Singh and Tiwari [11]. At last, LCCE has been calculated using Eq. (7) for various values of fluid mass flowing through tubes on the basis of unit time.

5 Results and Discussion

MATLAB has been used for evaluating LCCE of the solar energy based double slope water purification unit by writing programming code. All input data and main equations have been fed to MATLAB. After execution of programming code in MATLAB, outputs have been obtained and they have been presented as Figs. 2, 3, 4 and 5. The dissimilarity of hourly distillate from the solar-based double slope water purification unit at various values of fluid mass flowing through tubes on unit time

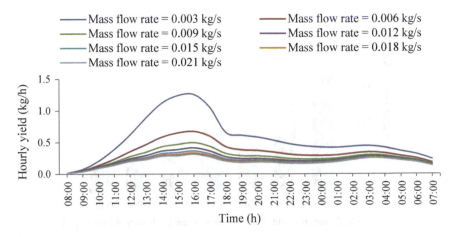

Fig. 2 Dissimilarity of fresh water production on hourly basis with fluid mass flowing through tubes on unit time basis at $N = 4$

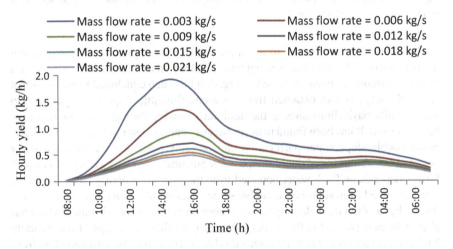

Fig. 3 Dissimilarity of fresh water production on hourly basis with fluid mass flowing through tubes on unit time basis at $N = 8$

basis considering the value of N (number of collectors) as 4. It is clear from Fig. 2 that the production of distillate from the solar-based double slope water purification unit diminishes as fluid mass flowing through tubes on the basis of unit time is raised. It has been found to occur because fluid mass in tubes of collector has less time to gain heat energy with the enhanced fluid mass flowing through tubes on the basis of unit time and ultimately less amount of energy in the form of heat is added to basin. Further, the same type of dissimilarity has been found at the selected value of collectors as 8. This dissimilarity has been presented in Fig. 3.

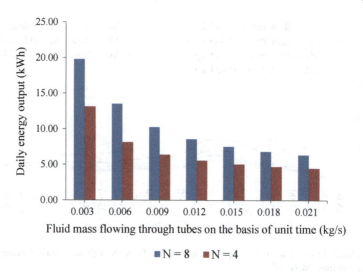

Fig. 4 Dissimilarity of fresh water production on daily basis with fluid mass flowing through tubes on unit time basis at selected values of N

The energy output obtained from solar-based double slope water purification unit on daily basis with fluid mass flowing through tubes on unit time basis taking values of N as constant has been presented as Fig. 4. It has been concluded from Fig. 4 that value of energy output obtained from solar-based double slope water purification unit on daily basis diminishes as the fluid mass flowing through tubes on unit time basis is raised. It has been found to occur because energy output obtained from solar-based double slope water purification unit on daily basis depends on fresh water obtained from solar-based double slope water purification unit on a daily basis which further depends on fresh water obtained from the system on hourly basis. The fresh water obtained from solar-based double slope water purification unit on hourly basis diminishes as fluid mass flowing through tubes on unit time basis is raised which has already been explained in the earlier paragraph. Further, same type of dissimilarity has been found to occur at other selected value N. However, the corresponding fresh water production from the proposed solar-based double slope water purification unit at a particular value of fluid mass flowing through tubes on unit time basis is more if value of N is higher and it has been found to occur because the energy in the form of heat added to the basin is more due to increased area of collecting heat energy at an enhanced value of N.

The computation of LCCE for solar-based double slope active water purification unit for various values of fluid mass flowing through tubes on unit time basis at considered value of N has been presented as Table 2. The life span of solar-based double slope active water purification unit has been taken as 50. Some other values can also be taken. However, life span of the proposed system has been considered as 50 in this research work. It has been concluded from Table 2 that LCCE of solar-based double slope active water purification unit diminishes as fluid mass flowing

Table 2 Computation of LCCE of solar-based double slope active water purification unit for various values of fluid mass flowing through tubes on unit time basis at $N = 8$

Flow of fluid mass/unit time	Daily energy	No. of clear days	Annual energy	Embodied energy [11]	Annual solar energy [11]	Life of system	LCCE
(kg/s)	(kWh)		(kWh)	(kWh)	(kWh)	(year)	(Fraction)
0.003	19.80	47	930.43	2364.19	227,641.38	50	0.00388
0.006	13.51	47	635.07	2364.19	227,641.38	50	0.00258
0.009	10.26	47	482.07	2364.19	227,641.38	50	0.00191
0.012	8.55	47	401.81	2364.19	227,641.38	50	0.00156
0.015	7.52	47	353.37	2364.19	227,641.38	50	0.00103
0.018	6.83	47	321.12	2364.19	227,641.38	50	0.00120
0.021	6.34	47	298.16	2364.19	227,641.38	50	0.00110

through tubes on unit time basis is raised at the considered value of N. It has been found to occur because energy output from solar-based double slope active water purification unit diminishes as fluid mass flowing through tubes on unit time basis is raised as evident from Fig. 4.

6 Conclusions

The computation of LCCE of solar based double slope active water purification unit has been done for various values of fluid mass flowing through tubes on unit time basis using programming code written in MATLAB. Based on the present work, the following conclusions have been drawn:

i. The value of LCCE of solar based double slope active water purification unit diminishes as fluid mass flowing through tubes on unit time basis is raised.
ii. The same kind of dissimilarity of LCCE for solar based double slope active water purification unit has been found for other considered values of N.

References

1. Zaki GM, Dali TE, Shafie HEl (1983) Improved performance of solar still. In: Proceedings of the first Arab international solar energy conference, Kuwait, pp 331–335
2. Rai SN, Tiwari GN (1983) Single basin solar still coupled with flat plate collector. Energy Convers Manage 23(3):145–149
3. Sampathkumar K, Arjunan TV, Senthilkumar P (2013) The experimental investigation of a solar still coupled with an evacuated tube collector. Energy Soui, Part A: Recovery, Utilization, Environ Eff 35(3):261–270

4. Singh RV, Kumar S, Hasan MM, Khan ME, Tiwari GN (2013) Performance of a solar still integrated with evacuated tube collector in natural mode. Desalination 318:25–33
5. Kumar S, Dubey A, TiwariG N (2014) A solar still augmented with an evacuated tube collector in forced mode. Desalination 347:15–24
6. Mishra RK, Garg V, Tiwari GN (2015) Thermal modeling and development of characteristic equations of evacuated tubular collector (ETC). Sol Energy 116:165–176
7. Singh DB, Dwivedi VK, Tiwari GN, Kumar N (2017) Analytical characteristic equation of N identical evacuated tubular collectors integrated single slope solar still. Desalination and Water Treatment, Taylor and Francis 88:41–51
8. Singh DB, Tiwari GN (2017) Analytical characteristic equation of N identical evacuated tubular collectors integrated double slope solar still. J Solar Energy Eng: Includ Wind Energy Build Energy Conserv, ASME, 135(5):051003(1–11)
9. Singh DB, Tiwari GN (2017) Energy, exergy and cost analyses of N identical evacuated tubular collectors integrated basin type solar stills: a comparative study. Sol Energy 155:829–846
10. Issa RJ, Chang B (2017) Performance study on evacuated tubular collector coupled solar stillin west texas climate. Int J Green Energy. https://doi.org/10.1080/15435075.2017.1328422
11. Singh DB, Al-Helal IM (2018a) Energy metrics analysis of N identical evacuated tubular collectors integrated double slope solar still. Desalination 432:10–22
12. Singh DB, Al-Helal IM (2018b) Energy metrics analysis of N identical evacuated tubular collectors integrated single slope solar still. Energy 148.546–560

Micro-milling Processes: A Review

Kriti Sahai, Audhesh Narayan, and Vinod Yadava

Abstract The micro-milling is a micromachining process which can be characterized by the interaction of sharp rotating microtool and workpiece. This paper presents a brief review of various conventional and modern micro-milling techniques. The aim of this study is to introduce various micro-milling techniques so far developed such as CNC-based conventional micro-milling technique, advance micro-milling processes (AMMPs) and hybrid micro-milling processes (HMMPs) such as micro-milling electrochemical spark machining (MM-ECSM) which shows a massive possibility for production of miniaturized three-dimensional structures on hard and brittle materials. The author's research focuses on MM-ECSM laying emphasis on its future growth in R&D.

Keywords Micro-milling · Advanced machining processes (AMPs) · Hybrid machining processes (HMPs)

1 Introduction

The term "micro" means very "small" in order of magnitude of 1…999 μm which indicates that too small can also be machined by normal machining. In that sense, "Micromachining" can be ranged in between 1 and 50 μm. The milling process, when defined in microdomain, shows many similarities to conventional milling. The specialty of this process is in terms of "miniaturization" and "reducing size." The micro-milling is a micromachining process which can be characterized by the interaction of sharp rotating microtool and workpiece. In this fabrication technique, the tool moves along the defined path for material removal from the workpiece in form microchips, debris or vaporized material. Thus, 2D or 3D profiles in the form

---

K. Sahai (✉) · A. Narayan
Mechanical Engineering Department, MNNIT Allahabad, Prayagraj, India
e-mail: kritisahai@mnnit.in

V. Yadava
Mechanical Engineering Department, NIT Hamirpur, Hamirpur, H.P., India

of microgroove, microchannels or micro-wall are created. The interaction may be thermal, mechanical or chemical. The micro-milling has emerged as the most vital aspect of the present-day industrial advancement. The need and demand for small, reliable, consumer and industrial products have pushed the current limit of research interest toward miniaturization.

Out of the existing technologies up till now and their limited application and economic utilization only for mass production pave way for replacement of conventional material removal processes into the microdomain. Micro-milling offers the most variable machining possibility and a very promising alternative. Though it may not be able to produce miniaturized components as lithography does, the connection between the microdomain and macrodomain is important for making a component. There exists enough room for micro-milling in applications such as in defense and commercial sector (e.g., asymmetric high-precision molds, masks for deep X-ray lithography) that requires high strength along with complicated profiles that microelectromechanical systems (MEMS) manufacturing cannot produce with the existing technologies. Micro-milling has capabilities to fill so far existing limitations of MEMS technologies. For instance, aerospace industries require miniaturized products that can sustain harsh environmental conditions and have a lightweight and high load-bearing capacity along with complex geometry.

Technologies such as computer numerical control (CNC)-based machines, laser beam, ultrasonic, electrodischarge, electrochemical, and lithographic machining in combination with micro-milling are used for the production of ultra-precision miniaturized products and can also achieve high accuracy. The advantage of micro-milling is that it has few material restrictions and the potential to fabricate complex three-dimensional features. Electrochemical spark machining, when combined with conventional micro-milling further, adds up an advantage to this process. The most complicated is the micro-milling of 3D miniaturized molds, chemical microreactors, and fluidic parts. Fig. 1a shows a diagrammatic representation of the cutting operation of the macro- and micro-milling process [1].

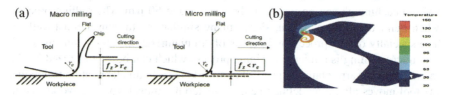

Fig. 1 **a** Diagrammatic representation of cutting operation in the macro and micro-milling process [2]. **b** Figure showing simulation by finite element of micro-milling of AISI 4340 steel [3]

2 Various Micro-milling Techniques

The micro-milling process has three major classification as: (a) conventional micro-milling processes, (b) advanced micro-milling processes and (c) hybrid micro-milling processes. The sections below give a brief idea and classification of the processes.

2.1 Conventional Micro-milling Processes

Conventional micro-milling techniques are conducted on conventional CNC-automated milling machines for removal of material where tool–workpiece interaction is mechanical. This process offers no restriction on the type of material to be used. The only limitation is with microtool which undergoes high tool wear and fabrication of complex structures with a high aspect ratio; hard and brittle material like glass is challenging by mechanical micromachining. This makes the process unproductive.

CNC-Based Micro-Milling Process
According to the literature review so far, micro-end mill can be used for machining of three-dimensional structures and milling of straight grooves. However, the CNC-based micro-milling process is also considered for the manufacturing of different micro-structures. Ardila et al. [4] presented initial results based on experiments conducted on micro-milling. Part geometry based on fundamental micro-features commonly used on molds for microfluidic devices was developed using CAD/CAM software. Finally, the procedures "design," "tool path generation," "process setup," "machining" and "part inspection" were documented and analyzed with regard to their time consumption.

Özel et al. [3] conducted preliminary experiments and modeling on micro-milling of AL 2024-T6 aluminum and AISI 4340 steel. Micro-end mill with 0.635 mm diameter was used in the experiments at spindle speeds up to 80,000 rpm, and finite element model of micro-milling was used to predict chip formation and temperature (Fig. 1b) without considering dynamics. Bang et al. [5] introduce a 5-axis machine that provides higher ease of use and relatively greater precision. By moving the workpiece in $X-Y$ direction and rotating the A-axis, the machine may also be operated as lathe, as shown in Fig. 2b.

2.2 Advanced Micro-milling Processes (AMMPs)

With the emergence of microelectromechanical systems (MEMS), various micromachining techniques were developed to enhance the use of newer and harder materials. All types of methods are not suitable for all types of machining problems.

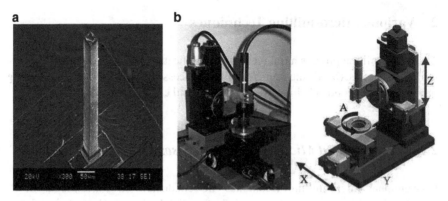

Fig. 2 a Micro-column with rectangle section machined by 5-axis micro-milling machine tool on workpiece material: brass [5]. **b** Constructed 5-axis micro-milling machine tool [5]

So, newer methods used such as micro-ultrasonic machining (M-USM), micro-laser beam machining (M-LBM), micro-electrodischarge machining (M-EDM), focused ion beam machining (FIB), and micro-electrochemical machining (M-ECM) are selected for micro-milling depending on desirable geometry required, material and tooling and various other conditions.

Micro-Ultrasonic Machining (M-USM)

Das et al. [6] in his chapter states achieving stiff tolerances and dimensions in terms of requisite surface quality and texture by micro-ultrasonic machining (M-USM) on hard and brittle materials. Other advanced micro-milling processes like micro-EDM, micro-ECM, and micro-LBM, etc., find difficult to machine materials which are hard and brittle in proper shape and size, when it comes to surface impressions, complex shaped holes, machine square, and irregular features. Egashira et al. [7, 8] mentioned that micro-tool development and fixing on the USMM system is very complicated. Jain et al. [1] have made a complete study of three-dimensional contour machining with diamond tools and found that it is a very accurate method of machining intricate shapes in ceramics as are usually milled in metals (Fig. 3a).

Micro-Electric Discharge Machining (M-EDM)

Zhaoqi et al. [9] carried investigation on M-EDM and M-ECM integrated with milling for 3D micro-structure (Fig. 3c). In this work, shaping and machining were done with micro-EDM process, finishing the machined surface with micro-ECM process. The process can be exactly controlled and is carried out in different dielectric mediums in sequence on the same machine tool along with the same electrode. Since the EDM process generates surface defects and there is recast layer formation, the results show that material is removed completely, and the surface quality and mechanical property of the workpiece are improved. The machining precision and shape accuracy are much better in this case. It proves that this combined milling method is possible. This is highly recommended in the field of 3D metallic micro-structure milling.

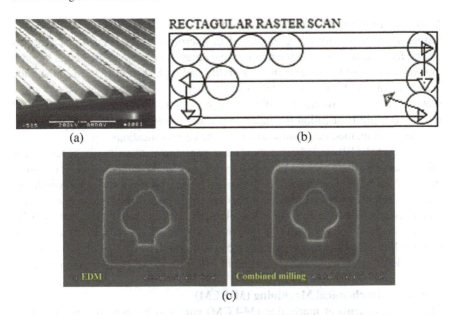

Fig. 3 **a** Micro-milled surface generated using the PCD tool and complex micro-milled structure [1]. **b** Scanning of the rectangular pattern in M-FIB [1]. **c** SEM of micro-petal machined by micro-EDM combined with micro-milling [9]

Table 1 shows various variants of M-EDM, capabilities, and manufacturing applications.

Table 1 Micro-electric discharge machining (M-EDM) variants, capabilities, and their application [1]

Variants	Capabilities (features geometry, size, and surface roughness)	Used in manufacturing
Micro-EDM die sinking	3-D; width = 20 μm (min), aspect ratio = 20; Ra ~0.50 μm	Replication molds, embossing tools
Micro-EDM drilling	2-D; depth of hole = 40 μm (min), aspect ratio = 25.50; Ra ~0.30 μm	Injection nozzle
Micro-EDM milling	3-D, width = electrode dia μm (min); Ra ~0.20 μm	Micro-injection molds. Embossing or coining tools
Micro-EDM grinding	2.5-D, width = 50 μm (min), aspect ratio = 15; Ra ~0.80 μm	Fluidic structure, embossing or coining tools
Micro-wire EDM	2.5-D; depth of hole = 30 μm (min), h(max) = 5 mm; Ra ~0.07 μm	Opto-electronic components, stamping tools
Micro-wire ED grinding	Rotational symmetry, depth of ole = 20 μm (min), aspect ratio = 30; Ra ~0.015 μm	Pin electrodes, rolling tools

Micro-Laser Beam Machining (M-LBM)
Lasers are increasingly being employed in micromachining because their beams can be focused precisely on the microscopic areas. Lasers are employed in engravings or micro-milling because of their ability and high flexibility to machine wide variety of materials such as metals, plastics, ceramics, semiconductors, and other materials that are conventionally difficult to process such as diamonds, graphite, and glass. Laser miniaturization is applied in a wide variety of industries of photonics, instrumentation, medical, semiconductors, and communications.

Biswas et al. [10] carried experiments using one-factor-at-a-time for generation of microchannel utilizing laser micro-milling process on flat zirconia (ZrO_2) ceramics using pulsed fiber laser system. Performance measures of microchannel width and surface roughness were considered in this experimentation and studied microscopic images of machined surface. For FIB milling, the Ga^+ ion beam can either digitally scanned in rectangular pattern (Fig. 3b) over the region to be milled or line scanned over the line. Also, the surface damages can be minimized with lower beam current in nano-level machining, and simulation of FIB milling is equally important area.

Micro-Electrochemical Machining (M-ECM)
Micro-electrochemical machining (M-ECM) can also be used for machining of micro-slots and 3D complex configurations. It has applications in many sectors such as automobile, defense, and medical. Micro-milling ECM is used in the manufacturing of titanium and cobalt alloys based on artificial joints, valve parts, microtool, and microgrooves for fluid film bearing, without any distortion.

2.3 Hybrid Micro-milling Processes (HMMPs)

It is the machining process in which there is an integration of two or more machining processes called as hybrid micro-milling process (HMMPs). HMMPs include the micro-electrochemical spark machining process (M-ECSM) and several other similar variants.

Micro-Electrochemical Spark Machining Process (M-ECSM)
Micro-electrochemical spark machining process (M-ECSM) is a versatile method and is a union of M-ECM and M-EDM. This process uses a simple shaped micro-sized tool used to produce 3D microcavity by adapting a similar tool movement strategy as conventional milling. Microstructure components made of various engineering materials can be utilized in a number of applications [microreactors and micro-total analysis systems (μ-TAS)], micro-structuring, as well as two-dimensional microchannel fabrication in which overlap of step milling depth method or layer-by-layer machining method is used.

Bhattacharya et al. [11, 12] carried out an investigation on the M-ECSM process on glass for microchannel cutting using the genetic algorithm, and parametric influence

Micro-milling Processes: A Review

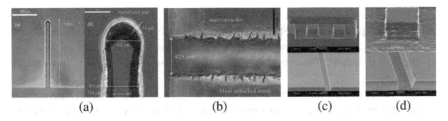

Fig. 4 Optical image of microchannel obtained through MM-ECSM after the 5th pass at 60 V: **a** Through microchannel having 5 mm channel length. **b** Overcut at the channel end and the heat-affected zone [14]. **c** Optical images showing microcracks formed in the heat affected region in the microchannels obtained in the 20th pass [14]. **d** SEM image of micro grooves and column machined on quartz [13]

on MRR, overcut (OC), and heat-affected zone (HAZ) was studied and achieved 332.067 μm micro slot.

Nguyen et al. [13] proved in his paper that micro-electrochemical spark machining The process has great potential in drilling and milling of microholes and microstructures, respectively, through proper optimization of process parameters. Thus, he investigates on machining conditions for micro-milling on quartz sample to improve ECSM accuracy and efficiency.

Mishra et al. [14] also fabricated microchannels by multipaass electrochemical discharge-based micromilling techniques, with channel depth varying from 400 to 1100 μm on a glass substrate with a tool feed rate of 50 μm/s such deep microchannels with high feed rate has not been reported in the literature as shown in Fig. 4a–c.

3 Conclusions

1. Micromilling is the most extensible, versatile, and lucrative machining process for the production of miniaturized components especially for machining real 3D geometries. The experiments are only in their initial phase yet.
2. Mechanical micro-milling in general is not used at the level of micromachining due to considerable difficulties in obtaining conditions defined microtools causing tool breakages.
3. Advanced machining processes for milling on various materials are widely used, especially, FIB which can be used in combination for milling as well for generation and instant generation of SEM images of machined samples. The only requirement is that the samples are at least air dry before they can be tested in the FIB. The redeposition of the evaporated material during milling is the major drawback in this process.
4. Another widely used method is M-LBM but that to have a limitation of having a low material removal rate (MRR) and unable to machine on a sample having more thickness. Also, the efficiency of the process is low.

5. Miniaturized dies, molds, micro-trenches, microreactor, micro-gear, 3D micromachined part, and micro-projection array show the most important application. The market demands show a great opportunity for micro-devices as well as in medicinal sectors, biological-based natural research, home care, and medical devices.
6. Currently, the research trends are focused on fabrications of 2D geometries instead of 3D geometries in micro-milling ECSM because of lack of knowledge of various controlling process parameters and scientific research. Hence, we need to know a relative study of all micromilling process so that a comparative study may provide us sources to choose appropriate process, materials, and parameters to perform micromilling as per manufacturing and an industrial requirement on the commercial scale.

Acknowledgements Authors would like to thank UPCST for providing financial assistance on the project entitled "Development and Performance Study of Milling Electro-Chemical Spark Micro-Machining process."

References

1. Jain VK Introduction to micromachining, 2nd edn. Narosa Publications.
2. Mallick B, Sarkar BR, Doloi B, Bhattacharya B (2014) Multi criteria optimization of electrochemical discharge micro-machining process during micro-channel generation on glass. Appl Mech Mater 592–594:525–529
3. Özel T, Liu X, Dhanorker A Modelling and simulation of micro-milling process. Manufacturing Automation and Research Laboratory, Dept. of Industrial & Systems Eng. Rutgers University, Piscataway, New Jersey, U.S.A. & Microsolution Inc., Chicago, IL 60612, USA.
4. Virginia Vázquez Lepe E (2014) Contribution to micro-milling process parameters selection for process planning operations. universitat de Girona. Doctoral Thesis
5. Das S, Doloi B, Bhattacharya B (2017) Chapter 2, Recent advancement on ultrasonic micromachining (USMM) process. Non-traditional micromachining processes, Material Forming, Machining and Tribology, Springer International Publishing AG
6. Ul Rehman G, Jaffery SHI, Ali MKL, Khan A, Butt SI (2018) Analysis of Burr formation in low speed micro-milling of titanium alloy (Ti–6Al–4V). Mech Sci 9:231–243
7. Rahman M, Wong YS (2014) Advanced machining technologies, comprehensive material processing (2014).
8. Egashira K, Masuzawa T, Fujino M, Sun XQ (1997) A of USM to micromachining by on the machine tool fabrication. Int J Electr Mach 2:31–36
9. Bang Y-B, Lee Seungryul Oh K (2005) 5-axis micro-milling machine for machining micro parts Int J AdvManuf Tec 25:888–894
10. Egashira K, Masuzawa T (1991) Micro ultrasonic machining by the application of workpiece vibration. CIRP Ann-Manuf 48(1):131–134
11. Mallick B, Sarkar BR, Doloi B, Bhattacharyya B (2018) Analysis of the effect of ECDM process parameter during micromachining of glass using genetic algorithm. J Mech Eng Sci 12(3):3942–3960
12. Ardila LKR, Del Conte EG, Picarelli TC, de Oliveria Perroni FA, Abackerli AJ, Schutzer K et al (2015) Micro-milling process for manufacturing of microfluidics moulds. In: 23 ABCM international congress of mechanical engineering, Rio de Janeiro, RJ, Brazil, 6–11 Dec

13. Nguyen K-H, Lee PA, Kim BH (2015) Experimental investigation of ECDM for fabricating micro structures of quartz. Int J of Precision Eng Manuf 16(1):5–12
14. Mishra DK, Arab J, Magar Y, Dixit P (2019) High aspect ratio glass micromachining by multi-pass electro-chemical discharge based micromilling technique. ECS J Solid-State Sci Tech, pp 322–331

Strategic Enhancement of Operating Efficiency and Reliability of Process Steam Boilers System in Industry

Debashis Pramanik and Dinesh Kumar Singh

Abstract This paper describes the economy of process steam boilers system with its strategic improvement, evaluating efficiency (including losses, formulating maintenance), studying feasibility, ensuring energy conservation/higher reliability and productivity based on "Optimization Study of Operating Thermal Efficiency of Steam Boilers" in Haryana. The ways are identified for the elimination of needless time, materials, energy, money and resources wastage with focus on efficiency and reliability. The practical concept at the intersection of business and engineering is the job of an industrial engineer. The operation analysis of these boilers showed regular deviations due to multiple reasons. The efficiency represents the cost-effective maintenance, coupled with devices calibration, tracking and computerized maintenance management system (CMMS) capabilities—all targeting/ensuring reliability, safety and energy performance. The result oriented actions made served as a starting point from where the plants would have started moving toward improved efficiency and reliability of boilers systems. Long-term energy efficiency and reliability require close attention to a range of factors. Inadequate maintenance is a major cause of wastage and deteriorating reliability at plant. In order to analyze, highlight the size of deviations, improvement and adopting up-gradations to documenting in report and to this paper compilation is done. The objective of this project findings-based research work was to analyze for boilers system efficiency and reliability improvement and enhance the systems performance for timely development for better planning/maintenance result and plant-specific productivity enhancement. It provides a real-time solution without which it cannot be fully accomplished by existing poorly performing utilities/facilities at site and units not maintaining its level best and processes managing upgraded systems.

D. Pramanik (✉) · D. K. Singh
Mechanical Engineering Department, Netaji Subhas University of Technology, Sector 3, Dwarka, New Delhi 110078, India
e-mail: debashisteri@gmail.com

D. K. Singh
e-mail: dks662002@yahoo.com

© The Author(s), under exclusive license to Springer Nature Singapore Pte Ltd. 2021
R. M. Singari et al. (eds.), *Advances in Manufacturing and Industrial Engineering*,
Lecture Notes in Mechanical Engineering,
https://doi.org/10.1007/978-981-15-8542-5_36

Keywords Boiler · Steam · Efficiency · Optimization · Energy · Maintenance · Deviation · Upgradation · Wastage · Reliability

1 Introduction

This paper presents the comprehensive findings of the "Optimization Study of Operating Thermal Efficiency of Steam Boilers in Haryana", carried out by a Research Group. The measuring of operational efficiency increases overall boiler lifespan, and adopting energy and resources conservation improves productivity, profitability and sustainability. The study was aimed at evaluating the operating thermal efficiency (including calculation of heat losses and formulating waste stream utilization/recovery from clue/technology and techniques improvement and systematic maintenance suggestions on inefficiently working areas/sub-systems) of fifteen representative boilers installed at different industrial units in the state of Haryana, studying techno-economics and retrofitting/implementing energy conservation (including categorical maintenance) strategies/schemes. There were about 1000 boilers of varied sizes and types operating in small- and medium-scale industries in the state, and it was strongly felt that most of these boilers were operating at less than optimum efficiency levels due to multiple reasons (including major factors deterioration, inadequate implementation/maintenance and reliability strategies against regular wear and tear). Realizing the scope that exists in improving the operating efficiency of these boilers, State Council for Science and Technology (SCST) entrusted Research Group with the study to come out with the set of short-, medium- and long-term energy conservation measures/ECMs (include ensuring of adequate cost-effective mix of preventive, predictive and reliability-centered/running maintenance technologies, coupled with equipment/instrument calibration and tracking) that can be adopted to improve the long-term end-use energy efficiency [1]. The ways were identified by the professionals for the elimination of needless time, materials, energy, money and resources wastage with the final focus on operating efficiency and reliability. The practical concept at the intersection of business and engineering is the perfect job of an industrial engineer.

For each of the recommended measure, the cost/benefit analysis was also provided giving such details as the investment required for implementing the energy conservation measure, payback periods, etc. Project objectives, selection of industry, methodology followed, major ECMs identified and devices maintenance/calibration proposed for different categories of boilers and associated systems are detailed in the following sections.

2 Design and Objectives of the Study

The main objectives of the study were listed below:

- Selection of representative boilers in the state of Haryana operating on different fuel feed;
- Observation on operating parameters comparison and deviation from designing values;
- Determination of thermal efficiency of the boilers;
- Analysis of various losses and identification of short- and medium-term ECMs for improving the efficiency of boilers;
- Identification of long-term measures and study the feasibility of retrofitting/replacing the existing boilers with fluidized bed combustion (FBC) boilers;
- Techno-economic feasibility of implementing the recommended ECMs and
- Preparation of a set of guidelines for better operation, maintenance (O&M) and more reliability of the boilers system.

The overall goal of the project report and guideline preparation was to provide technical information and best practice guidance with regards to six types of boilers and associated equipment installation, O&M (including reliability enhancement) and instruments calibration [2]. The plants provided the necessary data to evaluate the technology's performance and allowed for accurate comparison with these boilers and other available upgraded technologies/techniques.

3 Details of the Process Boilers Studied

A total of 15 process boilers operating in the following industrial units were covered under the study.

1. Associated Distilleries Limited, Hisar.
2. Ballarpur Industries Limited, Yamuna Nagar.
3. Bhiwani Vanaspati Limited, Bhiwani.
4. Coir Foam India Private Limited, Faridabad.
5. East India Cotton Manufacturing Company Limited, Faridabad.
6. Hilton Rubbers Limited, Rai.
7. Mercury Rubber Mills, Sonepat.
8. National Dairy Research Institute, Karnal.
9. Northland Rubbers Limited, Sonepat.
10. Rollatainers Limited, Kundli.
11. Rubber Reclaims Company of India Limited, Bahalgarh.
12. Saria Industries, Sirsa.
13. Technological Institute of Textile and Science, Bhiwani.

While selecting the units, care was taken to ensure that the selected boilers represent a wide range, both in terms of the boiler type and fuel feed so that the outcome of the study can be applied to the entire cross section of the boilers operating in the state of Haryana. The final selection included both water and fire-tube boiler operating

Table 1 Different categories of boilers studied

Type & Number	Type of boiler	Fuel used	Number of boilers
Type 1	Lancashire	Coal	4
Type 2	Stoker fired, water-tube	Coal	4
Type 3	Manual fired, water-tube	Coal	2
Type 4	Step grate, locomotive	Bagasse/rice husk	2
Type 5	Step grate, water-tube	Bagasse	1
Type 6	Package, fire-tube	Oil	2

on different fuels like coal, oil, bagasse and rice husk. The different categories of boilers were evaluated for the study listed in Table 1.

4 Methodology

The thermal efficiency of boiler can be evaluated by both direct and indirect methods [3]. In the **direct method,** the thermal efficiency is evaluated by taking the ratio of the heat content of the steam generated to the heat content of the fuel input to the boiler [4]. This method does not provide an energy balance of the boiler system and is used only as a spot check and for confirming the efficiency value as computed by the indirect method. In the **indirect method** of efficiency evaluation, a detailed energy balances carried out. Different heat loss fractions are computed separately, and the thermal efficiency is obtained by subtracting the total loss fraction from 100. The different losses considered and evaluated for the study are

- Dry flue gas loss;
- Heat loss due to CO in flue gas;
- Structural heat loss;
- Heat loss due to unburnt carbon in bottom ash;
- Heat loss due to unburnt carbon in fly ash;
- Heat loss due to moisture in fuel;
- Heat loss due to hydrogen in fuel;
- Heat loss due to moisture in air;
- Heat loss due to sensible heat in bottom ash and
- Heat loss due to blowdown.

The different parameters were measured for computing the efficiency (and various losses) of the boiler includes [5]:

1. Exit flue gas temperature from the boiler;
2. Analysis of exit flue gas for O_2 and CO;
3. Total dissolved solids (TDS) in blowdown, feed water, raw water and condensate;
4. Ambient dry bulb temperature;

5. Structural heat losses from the boiler surfaces and
6. Temperature of ash.

Apart from these, the fuel consumption, steam flow rate, boiler operating conditions, boiler specifications, etc., were collected from the individual plant. In addition to boiler efficiency evaluation, other boiler sub-systems such as condensate recovery, heat losses from condensate tank and performance of steam traps were also studied, and recommendations to improve the overall system efficiency by making suitable modifications in these systems were also provided.

5 Test Results

The boiler specifications, measurements taken, details of efficiency calculation, etc., for each individual unit were given separately in the detailed report. The operation analysis of these boilers at different plants/sites showed regular operating deviations of parameters in relation to their design values. In order to analyze, highlight the size of deviations, improvement and adopting upgradations from sites to documenting in project report to this paper, compilation is done. Here, the consolidated test results for all the 15 boilers are shown in Table 2.

Table 2 Operating efficiencies of boilers at different industrial units

S. No	Name of the unit	Type of boiler	Thermal efficiency (%)
1	Associated Distilleries Ltd.	Step grate, locomotive	54.33
2	Ballarpur Industries Ltd.	Stoker fired; Stoker fired; Stoker fired	75.18; 73.70; 77.75
3	Bhiwani Vanaspati Ltd.	Fixed-grate, water tube	47.60
4	Coir Foam India Pvt. Ltd.	Lancashire	44.79
5	East India Cotton Mfg. Co. Ltd.	Step grate, water tube	55.61
6	Hilton Rubbers Ltd.	Lancashire	43.08
7	Mercury Rubber Mills	Lancashire	43.89
8	National Dairy Research Institute	Package	73.70
9	Northland Rubbers Ltd.	Lancashire	45.97
10	Rollatainers Ltd.	Stoker fired	61.76
11	Rubber Reclaims Co. India Ltd.	Package	76.41
12	Saria Industries	Step grate, locomotive	47.57
13	Technological Institute of Textile and Science	Fixed-grate, water tube	63.14

A summary table showing the range of efficiency values and major heat loss figures for each of the six types of boiler studied (please refer to Table 1 for boiler types) is given in Table 3.

It was observed that most of these boilers were operating at less than optimum levels due to multiple reasons (including major factors/dry flue gas, latent heat of steam in the flue gas, combustion loss or the loss of unburned fuel, and radiation and convection losses deterioration, inadequate maintenance and reliability against regular wear and tear).

Table 3 Range of performance for different boiler types

Parameter	Type 1	Type 2	Type 3	Type 4	Type 5	Type 6
Capacity (tph)	1.0–1.5	6–35	3–4	0.8–6.0	8	3
Efficiency (%)	44–46	61–78	47–63	47–54	55	74–76
Flue gas loss (%)	22.92–28.67	5.39–15.27	15.9–20.9	13.84–27.12	17.79	12.22 13.76
Structural heat loss (%)	071–1.38	0.63–1.5	0.7–2.0	3.81–4.69	1.00	0.89–1.26
Unburnt carbon loss in bottom ash (%)	15.05–21	7.38–14.25	11.8–17.34	0.69–5.77	0.69	N.A.
Unburnt carbon loss in fly ash (%)	N.A.	0.75–0.91	N.A.	N.A.	N.A.	N.A.
Heat loss due to hydrogen in fuel (%)	4.15–4.34	2.90–4.27	4.01–4.30	7.28–8.61	7.61	6.42–6.86
Heat loss due to moisture in fuel (%)	0.72–0.75	0.76–0.81	0.69–0.75	1.80–13.56	14.18	0.06
Heat loss due to moisture in air (%)	2.23–2.70	0.98–1.71	1.7–1.76	2.45–3.07	1.64	1.12–1.20
Heat loss due to CO in flue gas (%)	0.01–0.18	0.01–0.32	0.01–0.04	0.05–0.52	0.13	N.A.
Heat loss due to sensible heat in ash (%)	0.62–0.73	0.37–0.54	0.64–0.66	0.04–0.29	0.40	N.A.
Heat loss due to blowdown (%)	0.62–2.89	0.49–2.85	1.27–4.65	0.81–3.05	1.32	2.43–3.65

6 Measures to Improve Boiler Efficiency

Boiler operating efficiency is called to some extent of total thermal energy which can be recovered from the fuel. Based on the analysis of the test results, recommendations were made for each type of boiler for improving the operating thermal efficiency. These measures were classified under short-, medium- and long-term measures depending on the magnitude of investment and the associated payback period [6]. Short-term measures involve marginal or no investment at all, and the payback period will normally be less than 6 months. For medium-term investment, the payback period would be from 1 to 2 years. Long-term measures involve large investments were demanding major retrofitting/modification work. Various ECMs recommended along with the investment, and payback period is given separately in the following sections for each of the six types of boilers and the sub-systems covered under the study.

Working process boilers with O&M skill utilization, improved categorical maintenance schedule/action and ensuring optimum efficiency is the mandatory requirement in process industry today. The present O&M opportunity in boilers and boiler systems highlights the need for a functional O&M organizational structure also and actions as a necessary starting point for an advance level practical O&M program. Therefore, the plants/industries are required to do boilers performance levels comparison, set priority and invest in maintenance work, procurement of adequate controls/instrumentation, equipment/components and implementation of energy conservation measures/schemes (ECMs) at site. The costly downtimes and maintenance could be minimized by accurate and reliable measuring and monitoring solutions. The impact of good maintenance on boiler would not only improve performance but also extend the reliability and life of the boiler. The proper sizing of a boiler was also a key issue when considering a replacement [7, 8]. If a boiler is undersized or oversized, then replacing it with a proper sized boiler is a sensible alternative. However, a combination of the performance characteristics, systems structural integrity and energy performance of the boiler was also evaluated before replacement decision was made.

6.1 Type 1 Boilers: Lancashire Boilers

Type 1 boilers were of lancashire type. The fuel fired was coal. The capacity of the boilers tested ranges from 1 to 1.5 tph. The efficiency range of the boilers was 44–46%. The major losses were the flue gas loss which accounts for 23–29 percent and the unburnt carbon loss which is as high as 15–21%. Four boilers were studied under type 1. A generalized summary of recommended measures to improve boiler efficiency is presented below.

6.1.1 Medium-Term Measures

1. Excess air Optimization
   The optimum excess air level for a coal fired boiler of lancashire type was 40%. The excess air levels observed in these bowlers were 182–208%. Reducing and monitoring the excess air to the optimum level help in minimizing the flue gas losses, thus improved the thermal efficiency of the boiler. This required installing a CO_2/O_2 recorder and an automatic control system for excess air [9]. The investment requirement was estimated to be Rs. 8 lakhs. The payback period was in the 7–12 months range.
2. Installation of economizer
   The temperature of few gases leaving the boiler was ranged from 260 to 320 °C. The heat available in the flue gas could be utilized to preheat the feed water [10]. The investment requirement for installing an economizer in the flue gas circuit to harness the heat available in flue gas was estimated Rs. 1.5–2.5 lakhs. The payback period was estimated to be 4–36 months.
3. Improvement of feed water quality and optimization of blowdown frequency
   The TDS contents of feed water were found to be very high. This increased the blowdown frequency and hence increased blowdown losses in the boiler. By incorporating suitable water treatment system, the quality of feed water to the boiler can be improved which directly reduce the blowdown losses [11]. The investment required for such a water treatment plant was estimated to be about Rs. 5.0 lakhs, and the payback period was about 18 months.

6.1.2 Long-Term Measures

Replacement with FBC boilers:

As a long-term measure, it was recommended to replace these boilers with the FBC boilers [12]. The design efficiency of FBC boilers was about 80%. The investment required was estimated to be about in the range of Rs. 22–30 lakhs, and the payback period was 15–21 months.

6.2 Type 2 Boilers: Stoker Fired and Water Tube

Type 2 boilers were of stoker fired, water tube type. The fuel fired there was coal. The capacity of these boilers studied ranges from 6 to 36 tph. The efficiency range under this category was 61–78%. The flue gas loss in this case varied 5–15%. The unburnt carbon loss was 7–14%. Four boilers were studied under Type 2.

6.2.1 Short-Term Measures

Reduction of unburnt carbon in bottom ash:

The unburnt carbon content in bottom ash varied 18–21%. The corresponding heat loss was 7–14%. The factors which contribute to unburnt carbon loss were inadequate air-fuel ratio, improper firing and coal size [13]. The losses could be reduced by

1. proper firing;
2. ensuring even grate loading;
3. better air distribution and
4. minimizing coal dust fed to the boiler.

The unburnt carbon in this case could be reduced to 18% which would improve the efficiency of the boiler by 3–4%.

6.2.2 Medium-Term Measures

1. Excess air optimization
   The optimum excess level for a coal fired boiler of stoker fired and water tube type was 40%. The level of excess air observed in these boilers ranged 50–134%. Reducing and monitoring the excess air to the optimum level would help in minimizing the flue gas losses, thus improving the thermal efficiency of the boiler. As already pointed out, this required installing a CO_2/O_2 recorder and an automatic system to control excess air. The investment requirement was estimated to be about Rs. 8 lakhs. The feedback period was 4–5 months.
2. Installation of economizer
   The flue gas leaving the boiler was at a temperature of 135–225 °C. The flue gas heat could be utilized to preheat the feed water [14]. The investment requirement for installing economizer in the flue gas circuit to recover the heat available in flue gas was about Rs. 4.0 lakhs. The feedback period was 4–5 months.

6.2.3 Long-Term Measures

Retrofitting with FBC boilers:

The efficiency of the high capacity boilers at maximum load under type 2 was found to be about 73–78%. For lower load (capacity required), it was less than 65%. Retrofitting the lower capacity boilers with FBC boilers would have improved the efficiency to 80% [15, 16]. The investment required would be around Rs. 65 lakhs, and the payback period was 14–15 months.

6.3 Type 3 Boilers: Manual Fired and Water Tube Type

Type 3 boilers are of manual fired, water tube type. The fuel fired here is coal. The capacity of the boilers ranges from 3 to 4 tph. The efficiency range under this category was 47–63%. The flue gas loss in this case was varying 21–26%. The unburnt carbon loss was 12–17%. Two boilers were studied under type 3.

6.3.1 Medium-Term Measures

1. Excess air optimization
   The optimum excess air level for coal fired boiler of manual fired and water tube type is 40%. The level of excess air noticed in these boilers was 125–163%. The investment required to control the excess air within reasonable limits was about Rs. 8 lakhs. The payback period was 2–4 months.
2. Installation of economizer
   The flue gas was leaving the boiler at temperature of 210–305 °C. Investment required for installing and economizer in the flue gas circuit would be about Rs. 2.0 lakhs. The payback period was 2–3 months.
3. Improvement of feed water quality and optimization of blowdown frequency
   The TDS contents of feed water were found to be very high. This increased the blowdown frequency and hence increased blowdown losses in the boiler. Because of the scaling inside the tubes, uneven heating of tubes takes place. This leads to frequent tube failures, affecting the normal functioning of the plant and leading to overall energy inefficiency. By incorporating suitable water treatment system, the quality of feedwater to the boiler could be improved which directly reduces the blowdown losses [17, 18]. Because of improved water quality, the incidence of frequent tube failure would have also been avoided resulting in increased reliability and availability of the boiler. The investment required for such a water treatment plant would be about Rs. 4 lakhs, and the payback period was about 4–5 months.

6.3.2 Long-Term Measures

Retrofitting/replacing with FBC boilers:

The thermal efficiency of these boilers was found to vary 47–63%. Retrofitting/replacing these boilers with FBC boilers would have improved the efficiency to 80%. The investment required would be Rs. 85 lakhs, and the payback period was 11–15 months.

6.4 Type 4 Boilers: Step Grate and Fire Tube

Type 4 boilers are of step grate, fire tube (locomotive) type. The fuel fired here was either bagasse or rice husk. The capacity of type 4 boilers considered during the study was 0.8–6 tph. The efficiency range under this category was 47–54 percent. Out of this, the flue gas loss alone accounts for 14–28%. Two boilers were studied under type 4.

6.4.1 Medium-Term Measures

Excess air optimization:
   The optimum excess air level for bagasse/rice husk fired boiler of fire tube type is 60%. The level of excess air observed in these boilers was 163–210%. The investment requirement for controlling excess air would be about Rs. 8 lakhs, and the payback period was 20–26 months.

6.4.2 Long-Term Measures

Replacing with FBC boilers:
   Replacing these step grate locomotive type boilers with FBC boilers would have improved the efficiency to 80%. The investment required would be Rs. 25 lakhs, and the payback period is 40–41 months.

6.5 Type 5 Boilers: Step Grate and Water Tube

Type 5 boilers are of step grate, water tube type. The fuel fired here was bagasse. Only one boiler of type 5 was studied. The boiler capacity was 8 tph, and the efficiency of boiler found to be 55%. The flue gas loss was 18 percent.

6.5.1 Medium-Term Measures

Excess air optimization:
   The optimum excess air level for bagasse fired boiler of water tube type is 60%. The level of excess air was found to be more than 150%. Reducing and monitoring the excess air to the optimum level would help in minimizing the flue gas losses, thus improving the thermal efficiency of the boiler. This required installing a CO_2/O_2 recorder and an automatic control system for excess air. The investment requirement would be about Rs. 8 lakhs, and the payback period was 2–3 months.

6.5.2 Long-Term Measures

Retrofitting with FBC boiler:
    The thermal efficiency of the boiler was around 55%. Retrofitting with FBC boiler would have improved the efficiency to 80% [19, 20]. The investment required would be Rs. 70 lakhs. The payback period was 11–12 months.

6.6 Type 6 Boilers: Packaged and Fire Tube Type

Type 6 was of package type, fire-tube boilers. The fuel fired was oil. The capacity of the boilers was 3 tph. The efficiency range under this category is 74–76%. The flue gas loss accounts for 12–14%. Two boilers were studied under type 6.

6.6.1 Medium-Term Measures

1. Excess air optimization
   The optimum excess air level for an oil fired boiler of package type is 5%. The excess air levels observed in these boilers were 39–56%. Installation of control system to monitor excess air would have caused about Rs. 8 lakhs. The payback period was 12–34 months.
2. Installation of economizer/air preheater
   The flue gas leaves the boiler at a temperature of 262–320 °C. This heat available in the flue gas could be utilized to preheat the feed water. Economizers are heat exchangers. Most consist of bundles of straight finned tubes installed in a rectangular enclosure. The investment required for installing an economizer in the flue gas circuit would be about Rs. 2 lakhs, and the payback period was 3–15 months.

   Key Points:

- An economizer allows a boiler to respond more quickly to rapid load demands.
- Economizers recover waste heat, improving boiler efficiency by up to 10%.
- Efficiency depends on the stack temperature and the amount of hot or makeup water required.

7 General Measures

Apart from the measures listed above, there were some general finding with regard to the boiler sub-systems primarily related to the steam traps and the condensate recovery schemes [21]. Some major observations and recommended measures to improve the above which would have led to an increase in overall boiler system operating efficiency are listed below.

7.1 Steam Traps

Irregularities were observed with the functioning of the steam traps. The defects identified were (i) leaking steam traps (ii) steam traps not working at all or (iii) oversized steam traps. It was recommended to replace the faulty steam traps to prevent steam leakages, which would have also result in proper functioning of the process equipments.

7.2 Condensate Recovery

In most of the cases, the extent of recovery of process condensate was very poor. In many plants, the condensate was not recovered at all. In some other plants, even if condensate was recovered partly, the condensate recovery system is not properly designed so that by the time the condensate was mixed with the feed water, it loses most of its sensible heat. It was strongly recommended to improve the condensate recovery system which could lead to substantial energy/monetary savings [22, 23]. The plant-specific economies of improving the condensate recovery were provided in the detailed report.

A consolidated summary of major recommendations for different categories of boilers is listed in Table 4 along with the investment required and the annual savings.

The focus of this project findings-based research work was to highlight the efficiency improvement of boilers and associated systems for its efficient operation by minimizing all energy losses (including recovery/improvement suggestions on clue). The actions and recommendations made in report served as a starting point from where the plants would have started moving toward improved operational efficiency of boilers and boilers systems/ facilities in a systematic manner. The operational efficiency represents the life cycle, cost-effective mix of preventive, predictive and reliability-centered/running maintenance technologies, coupled with equipment/instrument calibration, tracking and computerized maintenance management system (CMMS) capabilities—all targeting reliability, safety, operating personnel comfort and system efficiency [24, 25]. CMMS is a type of management tool/software that performs function in support of management and tracking of O&M activities. These systems automate most of the logistical functions performed by maintenance staff and management.

Table 4 Major recommendations, investment and savings for different boiler types

Type	Recommendation	Investment (Rs lakhs)	Annual savings (Rs lakhs)
Type 1	Improvement of feedwater quality and optimization of blowdown frequency	5.0	3.0
	Excess air optimization	8.0	8–14
	Installation of economizer	1.5–2.5	1–4.0
	Condensate recovery	2.5	2.2
	Replacing with FBC boiler	22–30	15–25
Type 2	Excess air optimization	8.0	20–150
	Installation of economizer/air preheater	4.0	10
	Retrofitting with FBC boiler	65	50
Type 3	Improvement of feedwater quality and optimization of blowdown frequency	4.0	12
	Excess air optimization	8.0	22–27
	Installation of economizer	2.0	13
	Installation of air preheater	2.8	2.5
	Retrofitting/replacing with FBC boiler	35–85	40–45
Type 4	Excess air optimization	8.0	3.8–4.5
	Replacing with FBC boiler	25	8.0
Type 5	Condensate recovery	2.0	4.3
	Excess air optimization	8.0	32
	Retrofitting with FBC boiler	70	90
Type 6	Excess air optimization	7–8	2.8–8.0
	Installation of economizer/air preheater	2.0	1.5–7.0

8 Other Steps Identified to Motivate Energy Conservation and Prolonged Reliability

8.1 Proper Instrumentation

As noted during the study, most of the industries in small- and medium-scale sector did not have proper instrumentation to monitor the performance of their boilers and steam consuming equipments. Only when monitoring of the boiler parameters were done on a continuous basis and compared regularly with the standard/design values, action could be taken in a proper direction to reduce the losses and improve the efficiency levels in boiler and related equipment/components/systems. List of equipments necessary to monitor and regulate the performance of the boiler and its related equipments is given below:

- Flue gas monitor (CO_2/O_2);
- Steam flow meter;
- Flow meters to measure feed water and makeup water flow rates;
- Contact type temperature indicators for surfaces;
- Immersion type temperature indicators for condensate and feed water and
- Portable conductivity meters for monitoring TDS levels of raw water, feed water and blowdown.

It was observed during site visits that in most of the plants/industries even if the instruments were installed, they were showing erroneous readings. The instruments had to be recalibrated periodically for meaningful analysis.

8.2 Efficiency, Reliability and Maintenance

The newly retrofitted/replaced or installed modern equipment would contain a very advance/effective instrumentation for controlling the four key operating parameters (temperature, time, turbulence and oxygen). Simultaneous with operating control, these installed instruments would provide the indications that become helpful to operators for predicting the type and time of required maintenance.

During day to day operation, the operator prepares logbook with data, information collected from installed/portable instrumentation at frequent sufficient intervals to detect problems in advance. This early detection is always critical for operating efficiently and controlling energy and monetary wastages. To understand this control sequence, it becomes necessary to look at some of the common types of equipment, devices and their makeup.

The furnace and steam-generating process boiler are to be made up of an operating and peak level setting, installations, fuel supply/handling system, combustion control system, boiler for transferring heat to the water, steam generation/storing header and sub-header, air and ash-handling (for solid fuel firing) equipment and associated components that can include condensing pumps, feed water pumps, deaerators, water softeners and soot blowers.

The effective O&M is one of the most cost-effective methods for ensuring energy performance/efficiency, reliability and safety. Inadequate maintenance of energy using systems is a major cause of energy wastage and deteriorating reliability/safety level at individual plant. Energy losses from steam, water and air leaks, bare/damaged/uninsulated lines, maladjusted/uncalibrated or inoperable controls and other losses from poor maintenance are often significant/considerable.

8.3 Training/Education

It was felt that one of the essential components in the energy conservation and maintenance/reliability programme lies in the training/education of the boiler operators. The programme should have covered different aspects of energy conservation and maintenance like better housekeeping, operational adjustments and making the operators aware of how the various boiler operating parameters affect energy consumption trend/pattern. As a part of the present study, the list of tips was prepared and submitted to sponsor for providing guidelines on proper boiler operational, maintenance and reliability-centric practices. The guidelines also listed out dos and don'ts separately for the managers and for the boiler operators.

8.4 Incentive Schemes

A large percentage of boilers operating in the Haryana state were in small-/medium-scale sectors and faced financial constraints in implementing the recommended energy conservation measures. Additionally, the lack of awareness with regard to the possible benefits of implementing the ECMs and maintenance practices might act as a non-encouraging factor. Some kind of financial support (either in terms of part grant or soft loans) might have helped the units in undertaking the energy conservation programme.

The Petroleum Conservation Research Association (PCRA) also had a Boiler Modernization Scheme; but the scheme was restricted to the replacement of inefficient oil fired boilers by efficient coal/oil fired ones.

9 Conclusion

The objective of this project findings-based research work is to do extensive monitoring/analyzing operating data of boilers and associated systems at different load, comparative study to highlight/prioritize the areas to improve efficiency of boilers and associated systems and reliability enhancement based on modification/retrofitting/maintaining for its efficient operation by minimizing all energy losses (including recovery/improvement suggestions on clue) and overall system reliability enhancement. The main part of this research is comprised of analyzing the effect of various parameters on performance/efficiency such as excess air, fuel moisture, air humidity, fuel and air temperature, temperature of combustion gases and calorific value of the fuel. Based on the obtained results, it is possible to analyze and make recommendations for optimizing process steam boilers and associated systems in industry using loss calculating indirect method. The significant improvement in operating thermal efficiency of the lancashire boilers, stoker fired water-tube boilers, manual fired water-tube boilers, step grate locomotive boilers, step grate water-tube boilers, package fire-tube boilers in overall systems was being realized/in progress at site over the computed existing efficiency level 44–46%, 61–78%, 47–63%, 47–54%, 55% and 74–76%, respectively.

The key benefit of this project undertaking was to update field-based knowledge, enhance the systems performance/reliability (from maintenance against regular wear and tear, design modification approach and previous relevant research work), right time action/implementation and leadership skill that can be developed by interacting equipment/components manufacturing industry people, quotations procurement, compatibility evaluating work for better planning/maintenance results, plant-specific implementation and associated systems.

Out of many industries detailed study, the following conclusions have been drawn from the study carried out for the different types of boilers.

1. Except for stoker fired boilers of higher capacities all the other boilers tested were found to be operating at lower than achievable efficiency levels.
2. There was a substantial scope for energy savings by implementing the energy conservation measures and maintenance actions/practices recommended through the study.
3. The major ECMs identified were those related to improvement in combustion efficiency, condensate recovery, use of waste heat recovery systems to utilize the heat available in flue gas and newer technology options like use of fluidized bed combustion boilers in small sizes.
4. Close monitoring of the system performance by installing suitable instrumentation was a pre-requisite for any energy conservation programme. Therefore, it was strongly recommended to provide adequate instrumentation to keep a close check on boiler operating efficiency.
5. Many of the units visited were in small-/medium-scale sectors and might have faced financial constraints in implementing the ECMs. Some sort of financial assistance would have been necessary to encourage the units to take up the implementation programme.

Good maintenance practices can generate substantial energy savings, upgrade reliability/safety and emerge as a critical resource in the entire plant. Moreover, improvements to plants facility maintenance schedule can often be accomplished immediately and at a relatively low cost. It provides a real-time solution for the capabilities of existing systems/plants without which (including best practices, approaches/schemes/funding priorities) it cannot be fully accomplished by existing poorly performing utilities/facilities at site and units not maintaining its level best and managing upgraded processes/systems.

For smaller steam requirements/generations, the plant personnel may select/install high efficiency packaged boilers, where loads fluctuate widely, or where frequent start-ups and shutdowns are necessary. They may prefer to install more than one smaller packaged units, in place of one large furnace and boiler.

The industries are required to set priority and invest in procurement of adequate instrumentation and implementation of energy conservation and maintenance schemes at site. Hence, there is a need to adopt practices for plant operators team practical learning also for upgradation of knowledge, skill, accuracy and speed in bringing an improvement in the overall system performance and reliability. The study was sponsored by State Council of Science and Technology 10 years back and was aimed at evaluating the operating thermal efficiency of fifteen representative boilers installed at different industrial units in the state of Haryana.

References

1. Chattopadhyay P Boiler operation engineering. Tata McGraw Hill Publication
2. National Productivity Council (Bureau of Energy Efficiency), Document on "Energy Efficiency in Thermal Utilities"—for preparation of National Certification Examination for Energy Managers & Energy Auditors
3. Kothandarman CP, Khajuria PR, Arora C, Domkundwar S (2000) A course in thermodynamics & heat engines. Dhanpat Rai & Co.
4. Ballncy PL (1991) Thermal engineering
5. The Energy Research Institute, New Delhi (2007) Handbook on energy audits and management
6. Murphy W, McCkay G (1995) Energy management. Butterworth Heinemann
7. Reference manual for Revomax–850, Thermax R&D
8. Central Fuel research Institute (CFRI) (Nagpur Unit), "Reference Document"
9. Gupta MK (2012) Power plant engineering. PHI Learning Private Limited, New Delhi
10. Parthiban KK Boiler consultant, venus energy audit system, "Design of fluidised bed combustion boilers for multifuels"
11. Bureau of Energy Efficiency, New Delhi (2008) 'Energy performance assessment of boilers' document for "energy auditors examination"
12. NTPC "Best Practices in O&M Area" (2018)
13. 'Integrated Energy Policy' Document by Planning Commission (2006)
14. US Dept. of Energy Efficiency and Renewables (2007) Document on 'Improving Process Heating System Performance: A Sourcebook for Industry'
15. Deshpande VS, Tambe PP Performance evaluation of an oil fired boiler—a case study in dairy industry
16. Khurmi RS (1995) Steam Table. S. Chand & Co. Ltd.
17. ACEEE Report, Lawrence Berkeley National Laboratory, USA and American Council for an Energy-Efficient Economy, Document on "Emerging Energy-Efficient Industrial Technologies" (2000)
18. Ernest Orlando Lawrence Berkeley National Laboratory, USA, Industrial Energy Audit Guidebook: Guidelines for Conducting an Energy Audit in Industrial Facilities (2010)
19. Market Assessment of Industrial Size Coal-fired Boilers in China (2001)
20. 'Energy Savings in Industry' Report by UNEP (2007)
21. 'Building Operational Excellence in the Processing Industry' STTP
22. Prepared under DCCI-CIPE/ ERRA Project (An Affiliate of the US Chamber of Commerce, Washington D.C., USA), "Economic Policy Paper on Assessing Appropriate Technology for SMEs" (2003)
23. Cost & coal management of a CFBC boiler in a power station (2019). Int J Eng Adv Technol
24. Bramwell FH (2006) Mechanical engineering in the chemical industry. In: Proceedings of the institution of mechanical engineers
25. Dahlan NY, Shaari MS, Putra MS, Mohd Shokri SM, Mohammad H (2013) Energy and environmental audit for Bangunan Menara Seri Wilayah and Bangunan Kastam, Putrajaya: analysis and recommendations. IEEE conference on clean energy and technology (CEAT) 2013

A Step Towards Responsive Healthcare Supply Chain Management: An Overview

Shashank Srivastava, Dixit Garg, and Ashish Agarwal

Abstract The Healthcare sector is one of the high revenue generators in a developing country. The purpose of healthcare industry is to serve the human being by providing the high quality of patient care and patient safety at minimum cost. This objective can be achieved by making a responsive supply chain. To obtain the aforesaid purpose, it is essential to assess the current situation of healthcare industry. It is very difficult to achieve responsive supply chain as it required the collaborative effort of every member of the chain. This paper is an attempt to determine the key issues, barriers and enablers of responsive supply chain management. This paper also shows the challenges faced by healthcare industry to become responsive. Responsiveness in healthcare supply chain is required the more than one hundred percent involvement of every member. Expected performance outcomes of responsive supply chain management are also discussed in this paper.

Keywords Healthcare supply chain management · Responsiveness · Patient care · Patient safety

1 Introduction

To gain a competitive advantage in today's scenario, organizations are trying to make their supply chain more flexible and responsive because the level of product variety and performance expectations of the customers has been increased rapidly. All industries like electronics, fashions are making their strategies which can increase

S. Srivastava (✉)
GLA University, Mathura 281406, India
e-mail: shashank.srivastava@gla.ac.in

D. Garg
National Institute of Technology Kurukshetra, Kurukshetra, India

A. Agarwal
School of Engineering & Technology, Indira Gandhi National Open University, New Delhi, Delhi, India

their responsiveness by fulfilling the customer's requirements by providing a high quality of product variety with reduced lead time [2, 3]. Automobile industries, fashion industries personal computer making industries optimize their operations and tries to develop the customer-driven products [4]. Service industries like healthcare though try to use the concept of supply chain management but still did not achieve fruitful results. Healthcare industry is a very important industry as the services offer here are directly or indirectly relates to the lives of human beings. Therefore it is important to understand the patient's requirements and proper care is required. The cost burden has been increased on the patients because here patient's safety is the prime concern. In order to gain competitive advantage, it is essential that lead time should be minimum otherwise patients will suffer. Therefore the concept of responsiveness has been emerging and gain attention to the researchers since last decades. Patient's service and patient's safety become the key concept in the recent health care supply chain management concept. Supply chain responsiveness is the mother of all the core concepts which were used by the researcher's in order to be competitive in the market. Lean manufacturing, agile manufacturing, six sigma all concepts were used in operations strategies and having the same objective of providing the required level of satisfaction to the customers with minimum possible cost burden [5]. Managing the healthcare supply chain is a difficult task as it requires continuous effort and support from all members of the chain [6]. Goods/medicines, surgical or clinical instruments funds, and information simultaneously flow in the chain. If responsiveness is achieved in the chain that means it ensures the availability of the required level of products or service at the right time to the right person at the right place at minimum cost burden. Responsiveness in supply chain ensures the improvement in overall productivity of the chain [7]. Responsiveness in the manufacturing industries is interchanged by flexibility in the operations earlier on. But as the concept of used agile manufacturing and lean manufacturing emerges it develop the concept of responsiveness in new dimensions as discussed by Kumar et al. [1]. Responsiveness in the supply chain can be achieved if coordination and cooperation are provided by all members. The objectives of the different stakeholders should be aligned in one direction only.

1.1 Healthcare Supply Chain Management

Healthcare supply chain management has been defined by different researchers in various ways. It is defined as a process in which the information, money, and services or goods have been flows between the supplier to the end-users. End-user is sometimes the patients or some time the clinics or healthcare. Its sole purpose is sometimes to improve the clinical outcomes or sometimes to improve the patient's safety. Healthcare or clinics are the places where the services have been provided to the patients. Healthcare supply chain management can be defined as the process of integration of material flow and information flow within healthcare. Health care supply chain management can be broadly classified in the two categories: Internal Supply Chain

Management (commonly known as Healthcare Supply chain management) deals with the traditional functions of all the supply chain activities related to healthcare. External Supply Management generally includes the distributors and manufacturers also in supply chain process. External supply chain management combined customer relation management, technical knowledge management, and fund flow management along with the materials and information flow management.

Several research studies suggest that both external and internal integration is required to fulfill the objectives of supply chain management. Healthcare supply chain management can be considered as the process of flow of medical products, equipment (surgical and clinical), and service from producer to the patient. It can be improved significantly by managing the proper flow of information throughout the whole supply chain. Beier [8] mentions that the performance on internal supply chain management depends not only the pharmaceuticals goods, medical instruments, and health-related accessories but also closely related to the movement of patients within healthcare (Fig. 1).

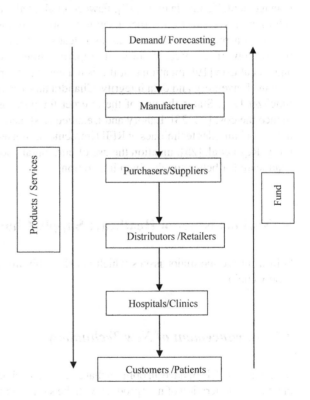

Fig. 1 General structure of health care supply chain management (modified from Gabler [9])

2 Literature Review

Healthcare supply chain management is still a main point of research as it directly involves human life. Landry and Philippe [10] mention that most of the researches focus on external supply chain integration as compared to internal supply chain management. They tried for better management of inventory while applying vendor management techniques, faster response time, and electronically management of data. Rappold et al. [11] concluded that healthcare supply chain can better be managed by proper planning and coordination and management of materials. Bowersox et al. [12] mention that by accurate forecasting, proper management of inventory, timely delivery of materials, and by proper and accurate information sharing can widely improve the internal integration and healthcare supply chain management. Some of the authors like Volland et al. [13], Bhakoo et al. [14], states the importance of inventory management on internal integration. According to Schneller [15], Burns and Lee [16] state the central purchasing policy is effective for efficient materials management. Tummala et al. [17], Fawcett et al. [18] emphasize the critical role of information systems for internal integration. Some authors mention that information system is a problem for healthcare as it leads to duplicacy of data, under utilization of flow of resources if not used properly while other researchers like Tutuncu and Kucukusta [19], mention that if it is managed properly results well planned and managed inventory and data integrity. Chandra and Kachhal [20] and Schneller and Smeltzer [21]. States the use of the internet for purchasing purposes significantly reduce the cost [22, 23], Landry and Beaulieu [24], and Volland et al. [13] mention the use of barcodes techniques or RFID implementation techniques for inventory visibility. Rego et al. [25], mention the use of horizontal cooperation between neighbor healthcare for better supply chain integration.

3 Various Issues of Heathcare Supply Chain Management

Following are the major issues which considerably affects the management of the supply chain.

3.1 Advancement of New Technology

Jayaraman et al. [26] have an opinion that due to rapid change of technology advancement in the generation of new product with the same composition and new and better method of diagnosis and treatment also increases the SCM practices in the healthcare as the members do not have good practices of procurement, inventory management system. Also, it increases the new way of delivery and distribution of medicines.

3.2 Communication (Lack of Information Flow)

Breen and Crawford [27], Harland and Caldwell [28] mention that Information sharing and the use of e-business are directly correlated to the co-ordination and combination of operational processes like procurement, inventory visibility, etc. Therefore a lot of researcher focuses to find the role e-services in healthcare supply chain management.

3.3 Cooperation

Ineffective cooperation among the supply chain partner in healthcare sector leads to an increase in the waiting time, duplication of work/test of the patient, unnecessary delay in the medicine/equipment supply.

3.4 Coordination (Integration Issue)

Shukla et al. [29] the objective of increasing patient care at minimum cost generally depends on how well the different processes are integrated. The basic approach to achieve the high performance of SCM can be achieved by integrating the operational processes in healthcare.

3.5 Non Availability of Essential Medicines, Modern Equipment

Non-availability of materials like medicines, surgical or clinical equipment is also a problem while implementing the SCM practices in health sector.

3.6 Shortfall of Trained Human Resources

The other issues are the lack of motivation, interest, and knowledge of SCM principle increases the complexity while implementing the SCM practices.

4 Challenges Faced by Healthcare Supply Chain Management

In recent times, challenges faced by the health care sector are as follows:

4.1 Accurate diagnosis,
4.2 Healthcare redesign,
4.3 Workforce planning and scheduling,
4.4 Streamlining of patient flow,
4.5 Performance management, and
4.6 Chaurasia et al. [30], lack of coordination between the streams of supply chain members.
4.7 Prakash et al. [31] the objective of supply chain member is to provide the services as per the requirements of the patients but due to conflicting objectives of the members, it is very difficult to achieve.
4.8 Agarwal et al. [32] mention another big challenge faced by healthcare supply chain management is that the members did not have enough belief on other members of the chain. Because of this they did not cooperate among others and also did not share the information regarding the process.
4.9 Shukla et al. [33] shows technology has been changed n every field and the emergence of new technology generated new methods of treatment and diagnosis of the patient. This situation is good on one side as it increases the quality of patient care. But on the other end, it generates an outdated inventory of medicines.
4.10 Agarwal et al. [34] mention healthcare supply chain management involves the various decision-makers as funds involved are by various stakeholders. Due to this and their conflicting objectives, it is very difficult to handle and manage the activities of chain.
4.11 Physicians are providing medicines by their own knowledge and symptoms of patients and there it is very much difficult to standardize the medicines for a particular disease. But it becomes a challenge for an inventory manager to stock every medicine which is required by the physician/doctors.

5 Enablers of Healthcare Supply Chain Management

There are certain variables that reduce the impact of barriers of healthcare supply chain management and are the main causes to achieve integration and collaboration of healthcare supply chain operations known as barriers. These enablers to achieve responsiveness in the healthcare supply chain identified in literature review are shown in Table 1.

Table 1 Enablers of healthcare supply chain management

S. N.	Authors (Year)	Enabler	Remark
1	Lawson and Potter [35], Lane et al. [36], Shankar et al. [37]	Transferability of knowledge	For sustainability of supply chain process, it is essential that knowledge should be exchange among the members
2	Mishra and Bhaskar [38], Song and Di Benedetto [39], Filieri and Alguezaui [40], Kogut and Zander [41]	Updated knowledge	Technology has been changed at a faster rate and if it is not updated then it will become impossible to provide the required services to the customer
3	Kayakutlu and Büyüközkan [42], Huang and Lin [43]	Top management support	Without the top management support and commitment, it is impossible to execute the supply chain process and derive the fruitful results
4	Wu [44], Sodhi [45]	Strategic planning	It is the foundation of supply chain. If planning is not done in the correct way then it is difficult to deliver the requisite services to the customer
5	McLaughlin and Marwick [46], Lin and Lee [47], Christopher and Gaudenzi [48]	Motivation of Employee	Without motivated employees, supply chain performance can be improved
6	Oke e al. [49], Paulraj and Chen [50], He et al. [51]	Strategic relationship	This is possible just because of the top management commitment and more than a hundred percent involvement of the employee
7	Nagati and Rebolledo [52], Yang et al. [53]	Employee training	It is essential to improve the operational efficiency of the chain and ultimately improve the performance of the chain
8	Harrington [54], Al-Karaghouli [55], Paulraj et al. [56]	Continuous communication	Without proper and sufficient communication results poor impact on responsiveness in the chain
9	Al-Mutawah et al. [57], Ke et al. [58], Samuel et al. [59]	IT infrastructure	It is essential to have a good base by which information can be flow among the members as it results a positive impact on supply chain performance

6 Barriers of Healthcare Supply Chain Management

There are certain variables that adversely affect healthcare supply chain management and are the main causes of difficulties in the integration and collaboration of healthcare supply chain operations known as barriers. These barriers to achieve responsiveness in healthcare supply chain identified in the literature review are shown in Table 2.

Table 2 Barriers of healthcare supply chain management

S. No.	Authors (year)	Healthcare supply chain management barriers
1	Jayaraman et al. [26]	Advancement of New Technology
2	Breen and Crawford [27], Harland and Caldwell [28], Agarwal et al. [60]	Communication (lack of information flow)
3	Nelson et al. [61]	Support from the Strategic Management Level
		Fragmented Behavior of Members of HSCM
		Technical Understanding of HSCM
4	Fawcett and Magnan [62], Lauer [63]	Less understanding and knowledge about SCM
5	Lauer [63]	Lack of proper standardized and accepted nomenclature for health care products,
6	Nachtmann and Pohl [64], Shukla et al. [65]	Lack Of Proper Communication/Information Flow Between The Different Actors of HSCM
		Lack of top management commitment and support
		Accurate diagnosis, Healthcare redesign
		Workforce planning and scheduling, Streamlining of patient flow
7	Smith et al. [66]	Lack of Supply Chain Planning Data
8	Sadi and Aldubaisi [67], Kumar et al. [68]	Lack of Motivation and Employee Involvement
10	Schneller and Smeltzer [21], Shukla et al. [69]	Communication, integration, information gathering and processing

7 Benefits of Responsive Healthcare Supply Chain Management

The following benefits can be expected by responsive healthcare supply chain managements:

7.1 Streamlined workflow
7.2 Control over inventory
7.3 Improved synergy between finance and material department
7.4 Improvement in vendor management
7.5 Empowered procurement staff
7.6 Improvement in sourcing, pricing, contract management, inventory management, etc.

8 Results and Interpretation

The different actors of SCM are generally acted in a more disjointed manner and provide a deferred response to the requests of the other members. The actions of all the members in the SCM are more personal profit centred to a certain extent than exact requirement of supply chain profit-oriented. Supply chain management involves the different actors like Physician/Doctor, Staff members, Nurses, Top management, patients, and all of them have their individual objectives which are collaboratively conflict among others which represent a problem to adopt the best practices of SCM. For example, patient wants the high-quality care at the minimum cost while the owner wants the high profitability of the healthcare which required a higher cost. As in SCM, more than one member is involved therefore none of them generally take the responsibility, as a result, some time stock out occurs or some time huge inventory is piled up. Often there is the disintegration of responsibility and supremacy between the ministry of health, the healthcare, and health staff at the district level which will reduce the practices of SCM.

9 Conclusion and Scope for Future Research

All the organizations (maybe manufacturing or service industry), trying to adopt new technologies or restructuring their whole system to adopt the various techniques of supply chain management. Because of the advancement of new technologies, changes in customer's preferences health care industries also changes their processes to remain competitive in the market. Health care industry is the most complex and fragmented industry due to involvement of many stakeholders like customers' preferences, manufacturers preferences as well as suppliers choices. Thus the requirement of adoption of principles of supply chain management is increased significantly. The effectively adoption of supply chain principles requires the effort from all the stakeholders from top to bottom. It also required proper coordination and cooperation both from suppliers and customers in order to gain competitive advantages. Basic principles of supply chain management as related to the health care industry have been proposed in this study. Different activities that are generally used in the supply chain management of the manufacturing sector have been described in the perspective of health care industry. By doing so the managers of health care industry will be more aware of the supply chain strategies.

From the literature review, it has been observed that the researchers have now become more aware of the SCM principles and tries to improve the processes but still, there are many inhibitors that are needed to be conquered. Researchers in this field did not focus on overall supply chain management till now. For example, some of them focus on logistics problems or some on inventory management. But in order to get the full advantages, they must have to consider all areas at a time. Therefore it is necessary to understand first the basic concept of supply chain management principles only then improvement can be made. It can be concluded that still there is a requirement of more efficient and customer-oriented supply chain in health sector to serve the human in better ways. This study presents the current status of a role supply chain in health care sector and found that future research is required in different fields like an integration of different processes, logistics, information flow, capacity management, and inventory management. Only resolving all these problems of supply chain, health care sector will provide fruitful results.

References

1. Kumar R, Banavara R, Agarwal A, Sharma MK (2016) Lean management—a step towards sustainable green supply chain. Competitiveness Rev Int Bus J 26(3)
2. Ireland RD, Webb JW (2007) A multi-theoretic perspective on trust and power in strategic supply chains. J Oper Manage 25(2):482–497
3. Sharma A, Garg D, Agarwal A (2012) Quality management IN supply chains: the literature review. Int J Qual Res 6(3)
4. Chand M, Raj T, Shankar R, Agarwal A (2017) Select the best supply chain by risk analysis for Indian industries environment using MCDM approaches. Benchmarking Int J 24(5)
5. Chaurasia B, Garg D, Agarwal A (2017) Lean Six Sigma application in healthcare of patients. Int J Intell Enterprise 4(3)
6. Chaurasia B, Garg D, Agarwal A (2019) Lean Six Sigma approach: a strategy to enhance performance of first through time and scrap reduction in an automotive industry. Int J Bus Excell 17(1)
7. Agarwal A, Shankar R, Tiwari MK (2007) Modeling agility of supply chain. Ind Market Manag 36
8. Beier FJ (1995) The management of the supply chain for hospital pharmacies: a focus on inventory management practices. J Bus Logistics 16(2):153
9. Gabler, Sunil C, Meindl P (2007) Supply chain management. Strategy, planning & operation. In: Das summa summarum des management, pp 265–275
10. Landry S, Philippe R (2006) How logistics can service healthcare. In Supply Chain Forum: Int J 5(2):24–30

11. Rappold J, Van Roo B, Di Martinelly C, Riane F (2011) An inventory optimization model to support operating room schedules. In supply chain forum: Int J 12(1):56–69
12. Bowersox DJ, Closs DJ, Cooper MB (2010) Supply chain logistics management. 3rd Ed., New York: McGraw-Hill Higher Education
13. Volland J, Fügener A, Brunner JO (2017) A column generation approach for the integrated shift and task scheduling problem of logistics assistants in hospitals. Eur J Oper Res 260(1):316–334
14. Bhakoo V, Singh P, Sohal A (2012) Collaborative management of inventory in Australian hospital supply chains: practices and issues. Supply Chain Manage: Int J 17(2):217–230
15. Schneller ES (2009) The value of group purchasing-2009: meeting the needs for strategic savings. Health Care Sec Adv Inc 1–26
16. Burns LR, Lee JA (2008) Hospital purchasing alliances: utilization, services, and performance. Health Care Manage Rev 33(3):203–215
17. Tummala VMR, Tobias S (2008) Best practices for the implementation of supply chain management initiatives. Int J Logistics Syst Manage 4(4):391–410
18. Fawcett SE, Magnan GM, McCarter MW (2008) Benefits, barriers, and bridges to effective supply chain management. Supply Chain Manage: Int J
19. Tutuncu O, Kucukusta D (2008) The role of supply chain management integration in quality management system for hospitals. Int J Manage Perspect 1(1):31–39
20. Chandra C, Kachhal SK (2004) Managing health care supply chain: trends, issues, and solutions from a logistics perspective. In: proceedings of the sixteenth annual society of health systems management engineering forum, February pp 20–21
21. Schneller ES, Smeltzer LR (2006) Building supply chain leadership and resources for the future. Health Care Purchasing News 30(9):72–73
22. Mathew J, John J, Kumar S (2013) New trends in healthcare supply chain. In: annals of POMS conference proceedings; Denver pp 1–10
23. Nabelsi V, Gagnon S (2017) Information technology strategy for a patient-oriented, lean, and agile integration of hospital pharmacy and medical equipment supply chains. Int J Prod Res 55(14):3929–3945
24. Landry S, Beaulieu M (2013) The challenges of hospital supply chain management, from central stores to nursing units. In: handbook of healthcare operations management pp 465–482
25. Rego N, Claro J, de Sousa JP (2014) A hybrid approach for integrated healthcare cooperative purchasing and supply chain configuration. Health Care Manage Sci 17(4):303–320
26. Jayaraman R, Taha K, Park KS, Lee J (2014) Impacts and role of group purchasing organization in healthcare supply chain. In: IIE annual conference. Proceedings (p 3842). Institute of Industrial and Systems Engineers (IISE)
27. Breen L, Crawford H (2005) Improving the pharmaceutical supply chain. Int J Qual & Reliab Manage 22(6):572–590
28. Harland CM, Caldwell ND, Powell P, Zheng J (2007) Barriers to supply chain information integration: SMEs adrift of eLands. J Oper Manage 25(6):1234–1254
29. Shukla RK, Garg D, Agarwal A (2018) Modelling supply chain coordination for performance improvement using analytical network process-based approach. Int J Bus Perform Supply Chain Model 8(2)
30. Chaurasia B, Garg D, Agarwal A (2016) Framework to improve performance through implementing leansix sigma strategies to oil exporting countries during recession or depression. Int J Prod Perform Manag 65(3)
31. Prakash R, Singhal S, Agrawal A (2018) An integrated fuzzy-based multi-criteria decision making approach for selection of effective manufacturing system: a case study of an Indian manufacturing company. Benchmarking Int J
32. Agarwal A, Shankar R (2003) On-line trust building in e-enabled supply chain. Supply Chain Manag Int J 8(4):324–334.
33. Shukla RK, Garg D, Agarwal A (2016) Modelling supply chain coordination in fuzzy environment. Int J Bus Perform Supply Chain Model 8(2)
34. Agarwal A, Shankar R (2005) Modeling supply chain performance variables. Asian Acad Manag J 10(2):47–68

35. Lawson B, Potter A (2012) Determinants of knowledge transfer in inter-firm new product development projects. Int J Oper Prod Manage
36. Lane PJ, Koka BR, Pathak S (2006) The reification of absorptive capacity: acritical review and rejuvenation of the construct. Academy of Manage Rev 31(4):833–863
37. Shankar R, Narain R, Agarwal A (2003) An interpretive structural modeling of knowledge management in engineering industries. J Adv Manag Res 1(1), 28–40
38. Mishra B, Bhaskar AU (2011) Knowledge management process in two learning organisations. J Knowl Manage 15(2):344–359
39. Song M, Di Benedetto CA (2008) Supplier's involvement and success of radical new product development in new ventures. J Oper Manage 26(1):1–22
40. Filieri R, Alguezaui S (2012) Extending the enterprise for improved innovation. J Bus Strategy 25(1):105–136
41. Kogut B, Zander U (1992) Knowledge of the firm, combinative capabilities, and the replication of technology. Organ Sci 3(3):383–397
42. Kayakutlu G, Büyüközkan G (2010) Effective supply value chain based on competence success. Supply Chain Manage: Int J 15(2):129–138
43. Huang CC, Lin SH (2010) Sharing knowledge in a supply chain using the semantic web. Expert Syst Appl 37(4):3145–3161
44. Wu C (2008) Knowledge creation in a supply chain. Supply Chain Manage: Int J 13(3):241–250
45. Sodhi MS (2003) How to do strategic supply-chain planning. Sloan Manage Rev 45(1):69–75
46. Marwick AD (2001) Knowledge management technology. IBM Syst J 40(4):814–830
47. Lin HF, Lee G (2005) Impact of organizational learning and knowledge management factors on e-business adoption. Manage Decis 43(2):171–188
48. Christopher M, Gaudenzi B (2009) Exploiting knowledge across networks through reputation management. Ind Mark Manage 38(2):191–197
49. Oke A, Prajogo DI, Jayaram J (2013) Strengthening the innovation chain: the role of internal innovation climate and strategic relationships with supply chain partners. J Supply Chain Manage 49(4):43–58
50. Paulraj A, Chen IJ (2007) Environmental uncertainty and strategic supply management: a resource dependence perspective and performance implications. J Supply Chain Manage 43(3):29–42
51. He Q, Ghobadian A, Gallear D (2013) Knowledge acquisition in supply chain partnerships: the role of power. Int J Prod Econ 141(2):605–618
52. Nagati H, Rebolledo C (2013) Supplier development efforts: the suppliers' point of view. Ind Mark Manage 42(2):180–188
53. Yang CS, Lu CS, Haider JJ, Marlow PB (2013) The effect of green supply chain management on green performance and firm competitiveness in the context of container shipping in Taiwan. Transp Res Part E: Logistics Transp Rev 55:55–73
54. Harrington T (2001) How multi-tech RF systems can provide complete tracking. Frontline Solutions 2(9):51–51
55. Al-Karaghouli W, Ghoneim A, Sharif A, Dwivedi YK (2013) The effect of knowledge management in enhancing the procurement process in the UK healthcare supply chain. Inf Syst Manage 30(1):35–49
56. Paulraj A, Lado AA, Chen IJ (2008) Inter-organizational communication as a relational competency: antecedents and performance outcomes in collaborative buyer–supplier relationships. J Oper Manage 26(1):45–64
57. Al-Mutawah K, Lee V, Cheung Y (2009) A new multi-agent system framework for tacit knowledge management in manufacturing supply chains. J Intell Manuf 20(5):593
58. Ke Y, Wang S, Chan AP, Lam PT (2010) Preferred risk allocation in China's public–private partnership (PPP) projects. Int J Proj Manage 28(5):482–492
59. Samuel KE, Goury ML, Gunasekaran A, Spalanzani A (2011) Knowledge management in supply chain: an empirical study from France. J Strategic Inf Syst 20(3):283–306
60. Agarwal A, Shankar R, Mandal P (2006) Effectiveness of information systems in supply chain performance: a system dynamics study. Int J Inf Syst Change Manag 1(3)

61. Nelson ML, Gray LA, Friedlander ML, Ladany N, Walker JA (2001) Toward relationship-centered supervision
62. Fawcett SE, Magnan GM (2001) Achieving world-class supply chain alignment: benefits, barriers, and bridges. Tempe, AZ: center for advanced purchasing studies.
63. Lauer C (2004) Excellence in supply chain management. Mod Healthc 34(50):29–32
64. Nachtmann H, Pohl EA (2009) The state of healthcare logistics: cost and quality improvement opportunities. Center for Innovation in Healthcare Logistics, University of Arkansas
65. Shukla RK, Garg D, Agarwal A (2012) Modeling barriers in supply chain coordination. Int J Manag Sci Eng Manag 7(1):69–80
66. Smith BK, Nachtmann H, Pohl EA (2011) Quality measurement in the healthcare supply chain. Qual Manage J 18(4):50–60
67. Sadi MA, Al-Dubaisi AH (2008) Barriers to organizational creativity. J Manage Dev 27(6):574–599
68. Kumar RB, Sharma MK, Agarwal A (2015) An experimental investigation of lean management in aviation. J Manuf Technol Manag 26(2), 231–260
69. Shukla RK, Garg D, Agarwal A (2014) An integrated approach of Fuzzy AHP and Fuzzy TOPSIS in modeling supply chain coordination. Prod Manuf Res 2(1):415–437

Designing of Fractional Order Controller Using SQP Algorithm for Industrial Scale Polymerization Reactor

D. Naithani, M. Chaturvedi, P. K. Juneja, and V. Joshi

Abstract The present analysis focuses on designing and implementation of fractional order-based PID controller designed using SQP algorithm and compared to a PID controller regarding time response characteristics. The objective is to minimize settling time and overshoot for the MIMO process including parameters to be controlled and pairing of controlled and manipulated variables. The paper also highlights performing of optimization using various performance indices. Another issue that is discussed in the paper is decentralized control.

Keywords Decentralized control · FOPID · PID · SQP

1 Introduction

Chemical industries basically use PID controllers for controlling any process or plant. PID controllers are most extensively used in industries since decades due to their ease of operation and better performance. But with the changing time, the use of PID controller is upgraded by fractional order instead of integer order PID controllers as these are more reliable and accurate. It has two extra degrees of freedom than integer order PID controllers. PID controllers have a number of tuning techniques, whereas in addition to tuning techniques, FOPID controllers can be tuned using algorithms with the variations of performance indices for much optimal values of controller parameters [1].

The original version of this chapter was revised: The author name "Vivek Joshi" has been changed to "V. Joshi". The correction to this chapter is available at https://doi.org/10.1007/978-981-15-8542-5_108

D. Naithani
Graphic Era Hill University, Dehradun, India

M. Chaturvedi (✉) · P. K. Juneja · V. Joshi
Graphic Era (Deemed To Be University), Dehradun, India
e-mail: mayankchaturvedi.geit@gmail.com

© The Author(s), under exclusive license to Springer Nature Singapore Pte Ltd. 2021, corrected publication 2021 R. M. Singari et al. (eds.), *Advances in Manufacturing and Industrial Engineering*, Lecture Notes in Mechanical Engineering, https://doi.org/10.1007/978-981-15-8542-5_38

Practically, a control system includes the dynamic system of fractional order that is to be controlled by the fractional order controller. Most common type of controllers is the fractional order controllers, because the real-time plant model has dynamic characteristics [2]. In present analysis, performance of a fractional order controller is compared to integer order controller designed for polymerization reactor that is a MIMO process.

Combination of a control system design for chemical plants is always an attractive topic for a control system engineer. But these combinations of one or more control schemes can raise operation problems. Purpose of decentralization is to concern about the information structure that is immanent in the solution of the given problem [3]. The decentralization is the online information about the systems interconnected states. It helps in an independent implementation of the states.

For eliminating weak coupling of subsystems, decentralized control is used. In this control, a MIMO system is divided into SISO systems. Using RGA, ERGA, CFA and RFA matrix, an effective transfer function is calculated, and SISO systems are obtained and controlled individually [4].

Sequential quadratic programming (SQP) is one of the reliable and powerful algorithmic for constrained nonlinear optimization problems [5, 6]. It is a simplification of Newton's method for unconstrained optimization [7].

2 Problem Formulation

In the present work, an industrial scale polymerization reactor is considered which is a 2×2 MIMO process, and using decentralization, two subsystems are designed for individual control of the system [8]. The two systems obtained are SISO systems for which the controllers are tuned and designed using SQP algorithm, and then, the integer order and fractional order controller are compared. The tuning technique applied is Cohen–Coon for PID controllers [9], and for fractional order controllers, the use of SQP algorithm is additional for calculating fractional parameters of the controllers. Performance indices are also varied for optimal results for the systems.

A 2×2 MIMO system is taken for investigation, and by calculating RGA, it is clear that 1–1 and 2–2 recommendations are feasible, and then, CFA, ERGA and RFA are calculated for obtaining an effective transfer function which is two SISO systems that are to be controlled using PID and FOPID controllers are compared for the systems using algorithms and varying performance indices for FOPID controllers.

The transfer function of polymerization reactor is [10]:

$$G(s) = \begin{bmatrix} \frac{22.89e^{-0.2s}}{4.572s+1} & \frac{-11.64e^{-0.4s}}{1.807s+1} \\ \frac{4.689e^{-0.2s}}{2.174s+1} & \frac{5.80e^{-0.4s}}{1.801s+1} \end{bmatrix} \quad (1)$$

For obtaining decentralized SISO systems from a centralized MIMO system, RGA, ERGA, CFA and RFA are computed and shown below:

$$\Lambda = \begin{bmatrix} 0.7087 & 0.2913 \\ 0.2913 & 0.7087 \end{bmatrix} \quad (2)$$

The critical frequency array is computed by determining the cross over frequency for the transfer functions describing the process:

$$\Omega = \begin{bmatrix} 8.05 & 4.18 \\ 7.85 & 4.30 \end{bmatrix} \quad (3)$$

The ERGA is computed as shown below:

$$E = \begin{bmatrix} 184.38 & -48.75 \\ 36.82 & 24.96 \end{bmatrix} \quad (4)$$

Thus,

$$\emptyset = E * [E]^T \quad (5)$$

$$\emptyset = \begin{bmatrix} 0.7193 & 0.2807 \\ 0.2807 & 0.7193 \end{bmatrix} \quad (6)$$

The relative frequency array can be obtained by performing the Hadamard division of ERGA by RGA. The RFA was found to be

$$\Gamma = \begin{bmatrix} 1.0151 & 0.9633 \\ 0.9633 & 1.0151 \end{bmatrix} \quad (7)$$

Both ERGA and RGA indicate diagonal pairing according to ETF method, and the equivalent process for the two loops is calculated as

$$G_1(s) = \frac{32.30 e^{0.2s}}{4.572 s + 1} \quad (8)$$

$$G_2(s) = \frac{8.18 e^{0.4s}}{1.801 s + 1} \quad (9)$$

Obtaining two SISO systems, FOPID and PID controller are designed using Cohen–Coon tuning technique, and FOPID is also designed using SQP algorithm. The time response characteristics and performance indices for both FOPID and PID are calculated and compared.

3 Results and Discussion

The transient response of $G_1(s)$ for minimizing ISE performance index, is shown in Fig. 1 that shows the comparison of PID and FOPID for SQP algorithm. Figure 2 shows comparison of PID and FOPID for SQP algorithm and minimizing IAE performance index, and Fig. 3 is the comparison of PID and FOPID controller for SQP algorithm and ITSE performance index. Figure 4 is the comparison of PID and FOPID for SQP algorithm, when the performance index was taken is ITAE. Table 1 shows the comparison of time response characteristics for PID and FOPID controllers varying performance indices for and keeping algorithm SQP for FOPID controllers. For PID controllers, Cohen–Coon tuning technique is used to calculate the time response characteristics.

Table 2 shows the comparison of performance indices for PID and FOPID controllers. For obtained $G_2(s)$ transfer function, the comparison of time response characteristics is shown in Fig. 5 shows the comparison of PID and FOPID for minimizing ISE performance index by SQP algorithm. The rise time and settling time for FOPID are greater than PID, but overshoot is less of FOPID than PID.

Figure 6 shows comparison of PID and FOPID for minimizing IAE performance index by SQP algorithm. Figure 7 is the comparison of PID and FOPID controller for minimizing ITAE by SQP algorithm, Fig. 8 is the comparison of PID and FOPID for minimizing ITSE by SQP algorithm, and which apperently shoes that settling time and overshoot are better for FOPID controller.

Table 3 shows the comparison of time response characteristics for PID and FOPID controllers for various performance indices by SQP algorithm. Table 4 shows the comparison of performance indices for PID and FOPID controllers which shows

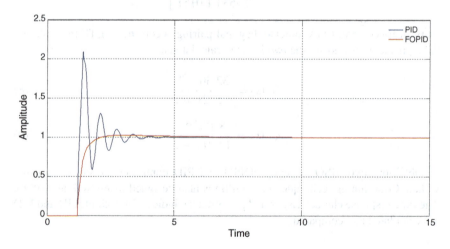

Fig. 1 Comparison of PID and FOPID for minimizing ISE performance index by SQP algorithms using Cohen–Coon tuning technique

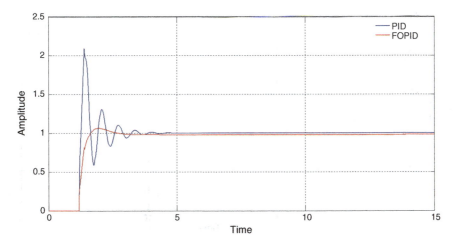

Fig. 2 Comparison of PID and FOPID for minimizing IAE performance index by SQP algorithms using Cohen–Coon tuning technique

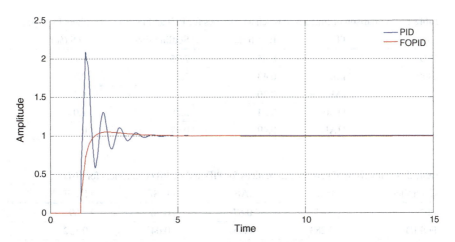

Fig. 3 Comparison of PID and FOPID for minimizing ITSE performance index by SQP algorithm using Cohen–Coon tuning technique

that, the performance indices are much better in case of FOPID controllers than PID controllers.

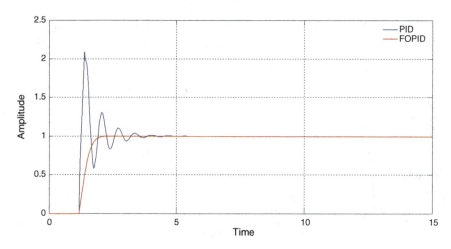

Fig. 4 Comparison of PID and FOPID for minimizing ITAE performance index by SQP algorithm using Cohen–Coon tuning technique

Table 1 Comparison of time response characteristics for PID and FOPID controllers for $G_1(s)$

Controller	PI	Rise time	Settling time	O.S (%)
PID	–	0.05	2.75	105
FOPID	ISE	0.43	3.08	2.5
	IAE	0.30	–	6.2
	ITSE	0.34	2.28	1.04
	ITAE	0.49	0.97	0

Table 2 Comparison of performance indices for PID and FOPID controllers for $G_1(s)$

Controller	ISE	IAE	ITSE	ITAE
PID	0.44	0.67	0.60	1.09
FOPID	0.283	0.457	0.043	0.122

4 Conclusion

In the present analysis, a 2×2 MIMO system is selected. By using the decentralized approach, two SISO systems are recommended to control for which PID and FOPID controllers are designed using Cohen-Coon tuning technique, SQP algorithm and various performance indices. The comparative analysis of the controller performance parameters, i.e., time response characteristics and performance indices, for the controllers designed for industrial scale polymerization reactor clearly exhibits that for both $G_1(s)$ and $G_2(s)$ transfer function, ITAE performance index for FOPID give optimal results for transient response characteristics. It can also be concluded

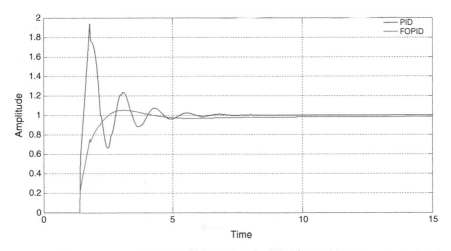

Fig. 5 Comparison of PID and FOPID for minimizing ISE performance index by SQP algorithm using Cohen–Coon tuning technique

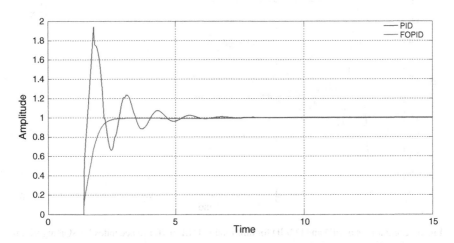

Fig. 6 Comparison of PID and FOPID for minimizing IAE performance index by SQP algorithm using Cohen–Coon tuning technique

that all the performance indices are better in case of FOPID controllers, and for the overall process, FOPID controller gives optimum response in comparison with PID controller.

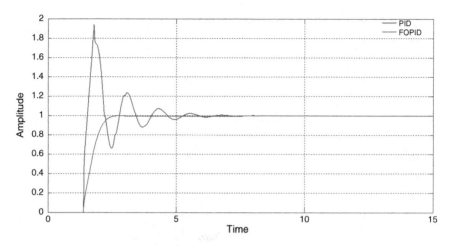

Fig. 7 Comparison of PID and FOPID for minimizing ITAE performance index by SQP algorithm using Cohen–Coon tuning technique

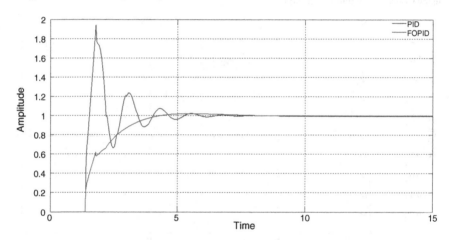

Fig. 8 Comparison of PID and FOPID for minimizing ITSE performance index by SQP algorithm using Cohen–Coon tuning technique

Table 3 Comparison of time response characteristics for PID and FOPID controllers for $G_2(s)$

Controller	PI	Rise time	Settling time	O.S (%)
PID	–	0.116	4.65	92.1
FOPID	ISE	0.78	9.74	5.1
	IAE	0.70	1.53	0
	ITSE	1.80	5.3	2.35
	ITAE	0.70	1.41	0

Table 4 Comparison of performance indices for PID and FOPID controllers for $G_2(s)$

Controller	ISE	IAE	ITSE	ITAE
PID	0.68	1.09	1.06	2.25
FOPID	0.541	0.75	0.29	0.31

References

1. Naithani D, Chaturvedi M, Juneja PK (2016) Controller performance analysis for a delayed process based on integral error performance indices. Orient J Chem 32(3):1671–1674
2. Luo Y, Chen Y (2009) Fractional order [proportional derivative] controller for a class of fractional order systems. Automatica 45(10):2446–2450
3. Bakule L (2008) Decentralized control: an overview. Ann Rev Control 32(1):87–98
4. Stanković SS, Stipanović DM, Šiljak DD (2007) Decentralized dynamic output feedback for robust stabilization of a class of nonlinear interconnected systems. Automatica 43(5):861–867
5. Boggs PT, Tolle JW (2008) Sequential quadratic programming. Actanumerica 4:1–51
6. Fesanghary M, Mahdavi M, Minary-Jolandan M, Alizadeh Y (2008) Hybridizing harmony search algorithm with sequential quadratic programming for engineering optimization problems. Comput Methods Appl Mech Eng 197(33–40):3080–3091
7. Kampouropoulos K, Andrade F, Sala E, Romeral L (2014) Optimal control of energy hub systems by use of SQP algorithm and energy prediction. In: IECON 2014—40th annual conference of the IEEE Industrial Electronics Society. IEEE, USA, pp 221–227
8. Lucia S, Finkler T, Engell S (2013) Multi-stage nonlinear model predictive control applied to a semi-batch polymerization reactor under uncertainty. J Process Control 23(9):1306–1319
9. Tavakoli S, Tavakoli M (2003) Optimal tuning of PID controllers for first order plus time delay models using dimensional analysis. In: 2003 4th international conference on control and automation proceedings. IEEE, pp 942–946
10. Kosek J, Grof Z, Štěpánek F, Marek M (2001) Dynamics of particle growth and overheating in gas-phase polymerization reactors. Chem Eng Sci 56(13):3951–3977

Additive Manufacturing in Supply Chain Management: A Systematic Review

Archana Devi, Kaliyan Mathiyazhagan, and Harish Kumar

Abstract Additive manufacturing (AM) in supply chain management (SCM) is one of the most emerging research topics nowadays, due to increased competition among manufacturing industries. AM techanology has capabilities to revolutionize manufacturing industry due to its huge advantages over traditional manufacturing (customized design, complex free and flexibility). AM technology provides small quantities of customized, personalized and geometrically complex products relatively at low cost as compared to traditional manufacturing. AM technology also provides customized shape, digital interaction with customers, direct manufacturing that give benefits in terms of lower cost, reduced supply chain complexity and lead times, etc. AM technology has been already adopted in fields of dental, biomedical, fashion and apparel due to its hyper flexibilities in supply chain. This review explores future scope, opportunities and challenges associated with supply chain management in adoption of AM technologies and also highlights the research gap by researchers from research background. AM technology is developing since 1980s and undergone huge transformation, but the integration of AM in supply chain is not explored so much. Therefore, this topic has good research scope. Systematic literature review has been done to find impacts (challenges, risks opportunities, advantages, etc.) of AM technologies on supply chain as well as future scope of AM in SCM. This review paper can serve as basis for future research related to this topic and add value to existing literature.

Keywords Additive manufacturing · Supply chain management · Digital interaction · Challenges and opportunities

---

A. Devi (✉)
National Institute of Technology, New Delhi 110040, India
e-mail: arvi662013@gmail.com

K. Mathiyazhagan · H. Kumar
Amity University, 201303 Greater Noida, India

© The Author(s), under exclusive license to Springer Nature Singapore Pte Ltd. 2021
R. M. Singari et al. (eds.), *Advances in Manufacturing and Industrial Engineering*,
Lecture Notes in Mechanical Engineering,
https://doi.org/10.1007/978-981-15-8542-5_39

1 Introduction of Additive Manufacturing

Additive manufacturing is also known as E-Manufacturing that creates a physical object from digital design, through layer by layer depositions. AM technologies that are widely used are electron beam melting (EBM), selective laser sintering (SLS), digital light processing (DLP), stereolithography (SLA), laminated object manufacture (LOM), fused deposition modeling (FDM), Zcorp and selective laser melting (SLM). The detailed explanation of all these technologies is given by Gibson [1]. Generally, three types of materials are used in AM technology, and those are (1) metals (2) polymers and (3) ceramics [2]. AM technologies have capability to reduce waste of material, use of resources (fuels, water, energy and non-renewable resources) and carbon emissions. This also does not require harmful chemicals such as cutting fluids, cooling fluids and chips. Therefore, this technology is more sustainable than traditional manufacturing.

AM technology provides small quantities of customized, personalized and geometrically complex products relatively low costs as compared to traditional manufacturing. 3D printers are used in many sectors such automotive replacement parts, dental crowns, artificial limbs, aviation industry, clothing and even in foodstuff. This has lasting implications on multiple industries such as production research, business development and design [3]. Main advantage of AM technologies is that it can remove extra part of components which is very useful in case of aircraft, fighter jets and medicals instruments as well as in artificial human being parts, where mass customization and design customization are very important. About 16.4% of the total revenue of AM market was collected from medical applications [4]. Part count of fuel nozzle of GE aviation has reduced from 18 to 1, while life cycle increased by 5 times and weight is decreased by 25% [5].

2 Literature Review

2.1 Background of Additive Manufacturing

Additive manufacturing is also called "rapid manufacturing", "digital manufacturing", "direct manufacturing" and "generative manufacturing" [6–9]. A decision-making framework or methodological framework is developed by Ch. Achillas for selection of effective production strategy for focused factory by using AM [10].

AM technologies have capability to reduce use of resources (fuel, non-renewable resources, water and energy) and carbon emissions for sustainable development in manufacturing [11]. AM is developing from 1980s [7, 11], but recently, it is being more popular due to its some advantages (low cost, mass and design customization, free from complexity, etc.) over traditional manufacturing [12, 13]. According to Wohlers analysis, global markets size of AM had increased from $3 billion in 2013 to $13 billion in 2018 and given the estimation to surpass $21 billion by the 2020

[14]. AM is a hyper flexible technology which provides a specific set of opportunities and challenges for developing supply chain management [15].

AM technologies have huge capability to revolutionize the way products are produced [16–19]. Simon Ford has done study on special issues of technological forecasting and social change on adoption of AM [20].

AM methods over traditional manufacturing methods are given bellow [21].

1. There is no need of tooling.
2. AM is economical and feasible for small productions.
3. Products can be optimized for specific function.
4. Product design can be changed easily in faster way
5. More economical for producing customized, personalized and complex products.
6. It has simper supply chain with sorter lead time and low inventories.
7. AM allows to produce lighter, stronger and more durable part.
8. It adds more precise features and complexity to products without increasing cast.
9. AM has potential to customize the design at any stage of production.
10. It is mold-free manufacturing process that reduces maintenance and mold tooling costs.

Other than above benefits, there is the possibility to reduce material waste up to 90% according to a report by Markillie in AM [22].

2.2 Review of AM in SCM

Some research on challenges associated SCM on adoption of AM is done, but more research on this topic is needed. Katrin Oettmeier has analyzed two case studies on hearing aid industry to find impact of AM on supply chain business process and management components [23]. To analyzed impacts on supply network, a computational study has been done by A. Barz, by dividing supply network into two parts [24]. Impact of 3D printing on supply chain has become disruptive technology of this decade. This technology has already has popular among manufacturing industries, and Mohr (2015) has explored how we will enter into homes, schools and private consumer sector. To identify impacts of AM on SCM, a relation is established between 3D printing and supply chain by Mohr as well as interviews are also conducted with supply chain managers from different manufacturing industries [25]. AM technologies have huge potential to revolutionize business model in terms of the ways of products are produced and delivered to customer [26]. Analysis of SCM on the adoption of AM has done by Zanoni through the review of quantitative and qualitative cases that is observed from literature and by doing empirical observations on real industries and also explored impacts (advantages, impediments, etc.) of AM on each phases of supply chain [27]. Peng Liu has done case study on aircraft spare parts industry to evaluate the various impacts of AM on aircraft spare parts supply chain, and these impacts are reduced shipping costs, safety inventory and delivery lead time of aircraft spare parts in the supply chain [28]. Impacts on SCM on adoption of AM studied with help of case study on medical industry. AM technology has capability to change structural composition, types of business processes among supply chain components as well as structural dimension [29]. Henk Zijm has done analysis of advance level impacts of AM on supply chain, summarized basics of AM technology and also study spare part delivery of aerospace industry [30]. 3D printing has potential to affect our personal and professional lives with revolution and replacement of existing manufacturing technology as well as emerged as one of the most innovative technologies to impact logistic industry and global supply chain. Sebastian Mohr and Omera Khan have examined the area of supply chain that effected by 3D printing technology and evaluate key questions for future research [31].

2.3 Review of AM in Supply Chain Through Simulation Model

Very less research works are done on supply chain through simulation model. Therefore, there is need to explore this field. Two different supply networks are considered for healthcare industry, which are traditional supply chain network (TSCN) and 3D printing supply chain network (3DPSCN) as well as a simulation model has been developed by Ozceylan to find potential impact of 3D printing on supply chain for producing orthopedic insole [32].

2.4 Review Results

Review of this paper divided into three sections (1) Review of AM (2) Review of AM in SCM (3) Review of AM in SCM through simulation model. From above review, there are different research gaps have been founded. And these are (1) AM in supply chain is still immature technology (2) AM in supply chain still is not used for large production (3) AM technology is still not total replacement of traditional production (4) There is lock of optimization of AM parameter (5) AM in SCM is still not adopted by all manufacturing industries (6) There is missing of economic scale of AM (Table 1).

Table 1 List of challenges associated with SCM in adoption of AM

Challenges	Short description	References
AM technologies limitations	Lower surface finish and lower level of precision as compared to some other manufacturing technologies. Restrictions of size, strength, object detail and high materials cost, all are another challenges for AM	Berman [16], Tuck et al. [19], Strong et al. [33], Gao et al. [34], Janssen et al. [35]
Material limitation of AM	AM technologies are restricted to some specific materials	Berman [16], Zanoni et al. [27], Strong et al. [33], Janssen et al. [35]
Immature technology	There is need of more research and more skilled AM operator and require improvement in AM performance	Berman [16], Petrick and Simpson (2013), Petrovic et al. [36]
Large-scale production limitations	AM technologies provide lower speed as well become costly for large-scale productions	Berman [16], Florentin et al. [37], Tuck et al. [19], Scott [38]
Lack of know-how	Lack of skill in optimizing process parameters and operating of AM machines	Zanoni et al. [27], Florentin et al. [37]
Undefined guidelines/regulations	There are no proper availability of guidelines and standards for design, qualification of materials and finished parts, cyber security of the system and proprietary rights	Zanoni et al. [27], Berman [16], Petrick et al. (2013)
High capital expenditure	Acquisition cost of AM machines and its procurements become costly for small and medium-sized companies	Zanoni et al. [27], Mohr et al. [25], Strong et al. [33]
Sustainability of AM	There is no proper research on sustainability of AM technology	Ford et al. [20]
Systematic study of AM in SCM	Systematic analysis of impacts of AM technology on supply chain is still missing	Katrin Oettmeier
Limitation for metal	Use of metal in AM is one of the primary barrier and challenge on adoption of AM	Petrick and Simpson (2013)
Higher operating cost	AM needs more maintenance and more care that increases operating cost	Strong et al. [33]
Major post-processing steps	Need some treatments to get improved material properties and process variability that increases lead time and cost	Zijm et al. [30]

(continued)

Table 1 (continued)

Challenges	Short description	References
Digital nature of AM	It can create problem in the protection of intellectual property rights and liability and cyber security	Zijm et al. [30], Gao et al. [34]

3 Potential Impacts of AM on SCM

Systematic analysis of potential impacts of AM on supply chain management is divided in different stages, and these stages are design, production and usage.

3.1 Design

Many industries like aerospace, medical sector and defense sectors need high performance and more precise products with low resources and low cost that can only achieve through appropriate design of components with [31, 39].

Product customization: Nowadays, competition among manufacturing firms is increasing rapidly, and to keep them in competition, customization can play an important role. Customization of products provides increased profit margins, customer's satisfaction, products with high added value, customer's involvement in product design and development. AM provides online platform which leads to easy changes in design and easy availability of CAD file that means file can be downloaded from anywhere and changes is done according to customer requirement [31, 35, 39, 40].

Complex design: AM technologies have freedom of design, and these can be produced any type of complex geometries without using any special tools and without adding any complexity cost. Cores and support need for hollow products, specials tools needed for complex geometries, and undercuts needed for easy machining that results in increased lead time and cost. On other hand, AM technologies directly use CAD models in manufacturing machines that remove some primary activities such as tooling, molding that leads to reduce lead time cost and production complexity [34, 35].

3.2 Production and Supply Chain

This stage discusses about manufacturing and distribution of production. This stage first discusses about how additive manufacturing makes supply chain from centralization to decentralization and then explains how resources can efficiently use with help of AM, and in last, it explains about effect of AM on inventory management and logistic in supply chain [39].

Manufacturing centralization to decentralization: AM technology provides high digitalization that means CAD file can be downloaded from anywhere and production is done near to customer which result in distributed supply chain. This can reduce global transportation with help of digital file transfer instead of physical flow in supply chain. Distributed supply network can reach disconnected market and remote location [3, 35] and also provides reduction in warehousing, packaging and transportation [35, 39–41].

Higher resource efficiency: AM technologies produce components with help of layer by layer depositions of materials that produce very few wastes and waste materials can be used for further production of other parts. AM produces high resource efficiency that provides various advantages. And these advantages are preserved natural resources, reduced global warming and low carbon emissions that provide sustainable SCM for production [39].

Inventory management and logistic: AM only needs CAD file and AM machine which result in simple logistic. AM technology needs less material and provides distributed SCM that lead to reduced inventory cost, reduced stock levels, material handling, cost and lead times. This technology can convert physical stock keeping units in to virtual CAD file as well as has capability to meet uncertain demand. AM technology can also produce consolidated parts. Therefore, this does not need longer downstream inventory [40, 42].

Reduced Assembly: AM technology produces consolidated part which results to reduce flow number of parts, and many assembly steps in production phase that means many steps in assembly are converted into single steps. Therefore, it provides convenient material handling, easier to control and reduced process complexity, lead time and cost of assembly [34, 35].

3.3 Use

Last stage provides information about how product used by customer, how will products effect environment and how can products recycled and reused [39].

Energy efficiency: Impacts of AM technologies on environment and resources consumed by AM are the important factor for AM sustainability [39]. AM technology uses less material, less consumption of energy and provides higher resources efficiency which result in less wasted, preserved natural resources and fewer emissions that provide sustainability to AM. Complex and customized products with high strength and lightweight can be produced easily and involve less steps with AM technology that lead less energy and material consumption.

Service and maintenance: AM has potential to redesign products with the help of given technology such as repair, refurbishment and remanufacture, and this increases products life, preserves natural resources and reduces production lead time and resources damping steps that make AM sustainable to environment [5, 43].

Material utilization and recycling: AM technologies produce products with the help of layer by layer material deposition; therefore, no need unwanted material that leads to higher utilization of material and resource efficiency. AM technologies produce a few material's waste, yet this waste material can be reused in next cycle, or in newer production, these wastes can be used 5–10 times. And these all lead to sustainability of AM to the environment.

4 Conclusions and Future Scope

This paper is discussed about various impacts of AM on SCM and challenges associated with this technology in systematic way. Impacts of AM associated with design phase are customization, complex shape that lead to provide reduced lead time, free-design and more customer's involvement. It also provides distributed manufacturing, reduced inventory stock, simple assembly and higher resource efficiency that lead to low transportation cost, less material handling, reduced lead time and a few material. AM has capability to utilize row materials up to 90 to 95% and can reuse waste material in next cycle of production as well need less energy as compared to traditional manufacturing that lead to preserved natural resource, less emission that make AM more sustainable to environment. And this technology also has potential to redesign product such as repair, refurbishment and remanufacture as well as provide product in market in faster way. This paper also discussed about barriers and challenges associated with AM in SCM, and these are listed out from the literature review. AM is still limited to some specific materials, and use of metal is primary barrier for this technology. Surface finish, strength, size of bed of AM machines and speed are others challenges for this technology. There is lack of skilled operators, customer's awareness and proper knowledge of AM machine as well as there is proper availability of regulation and standard for material, design and cyber security. Cost of acquisition, repairing and maintenance and procurement is high for small- and medium-sized industries.

AM is one of the emerging manufacturing technologies, and huge transformation is done. But AM in SCM of the most recent technology and there is need to explore it. Future scope of this technology is (1) There is need of systematic study of sustainability of AM technology (2) There is need of proper integration of AM with SCM to find various impacts of AM on each and every step of SCM (3) Systematic study of challenges of AM in SCM from the literature (4) There is no systematic study of protection of online transfer of 3D CAD file from one location to another location.

References

1. Gibson L et al (2015) Additive manufacturing technologies: 3D printing, rapid prototyping, and direct digital manufacturing, 2nd edn. 978-1-4939-2112-6
2. AM Platform, Additive manufacturing: strategic research agenda (2014)
3. Cohen D, Sargeant M, Somers K (2014) 3-D printing takes shape. McKinsey Quarterly, January 2014
4. Wohlers T, Gornet T (2014) History of additive manufacturing. Wohlers Report 2014
5. Knofius N (2016) Selecting parts for additive manufacturing in service logistics
6. Ebert (2009) Direct inkjet printing of dental prostheses made of zirconia. J Dental Res 88(7):673–676
7. Khajavi SH, Partanen J, Holmström J (2014) Additive manufacturing in the spare parts supply chain. Comput Ind 65(1):50–63
8. Hopkinson N, Dickens P (2003) Analysis of rapid manufacturing—using layer manufacturing processes for production. Proc Inst Mech Eng Part C: J Mech Eng Sci
9. Vinodh S, Sundararaj G, Devadasan SR, Kuttalingam D, Rajanayagam D (2009) Agility through rapid prototyping technology in a manufacturing environment using a 3D printer. J Manuf Technol Manag 20(7):1023–1041
10. Cozemei C et al (2012) Additive manufacturing flickering at the beginning of existence
11. Rogers H, Baricz N, Pawar KS (2016) 3D printing services: classification, supply chain implications and research agenda. Int J Phys Distrib Logist Manag 46(10):886–907
12. Gosselin C et al (2016) Large-scale 3D printing of ultra-high performance concrete—a new processing route for architects and builders. Mater Design 100:102–109. https://doi.org/10.1016/j.matdes.2016.03.097
13. Rayna T, Striukova L (2016) From rapid prototyping to home fabrication: how 3D printing is changing business model innovation. Technol Forecast Soc Chang 102:214–224
14. Wohlers (2012) Additive manufacturing and 3-D printing state of the industry. Annual Worldwide Progress Report, 2012 edn. Wohlers Associates, Colorado, USA
15. Ponfoort et al. (2014) Successfull business models for 3D printing: seizing opportunities with a game changing technology. Berenschot, Utrecht
16. Berman B (2012) 3P printing: new industrial revolution. Bus Horiz 55:155–162
17. Huang et al (2013) Additive manufacturing and its societal impact: a literature review. Int J Adv Manuf Technol 67:1191–1203
18. Gibson I, Rosen DW, Stucker B (2010) Additive manufacturing technologies: rapid prototyping to direct digital manufacturing. ISBN: 978-1-4419-1119-3. e-ISBN: 978-1-4419-1120-9
19. Tuck CJ, Hague RJM, Burns ND (2007) Rapid manufacturing: impact on supply chain methodologies and practice. Int J Serv Oper Manag 3(1):1–22
20. Ford S, Mortara L, Minshall T (2016) The emergence of additive manufacturing: introduction to the special issue of technological forecasting and social change
21. Holmström J, Gutowski T (2017) Additive manufacturing in operations and supply chain management: no sustainability benefit or virtuous knock-on opportunities? J Ind Ecol 21(S1):S21–S24
22. Markillie P (2012) A third industrial revolution
23. Oettmeier K et al (2016) How additive manufacturing impacts supply chain business processes and management components
24. Barz A et al (2016) A study on the effects of additive manufacturing on the structure of supply networks
25. Mohr S et al (2015) 3D printing and supply chains of the future
26. Bogers M et al (2016) Additive manufacturing for consumer-centric business models: implications for supply chains in consumer goods manufacturing
27. Zanoni S et al (2019) Supply chain implications of additive manufacturing: a holistic synopsis through a collection of case studies
28. Liu P et al (2013) The impact of additive manufacturing in the aircraft spare parts supply chain: supply chain operation reference (SCOR) model based analysis

29. Oettmeier K, "Impact of additive manufacturing technology adoption on supply chain network structures—an exploratory case study analysis. Chair of Logistics Management, University of St .Gallen, Switzerland
30. Zijm H, Knofius N, van der Heijden M (2019) Additive manufacturing and its impact on the supply chain. Department of Industrial Engineering and Business Information Systems, University of Twente, Enschede, Overijssel, The Netherlands
31. Mohr S, Khan O (2015) 3D printing and its disruptive impacts on supply chains of the future
32. Ozcelylan E et al (2017) Impacts of additive manufacturing on supply chain flow: a simulation approach in healthcare industry
33. Strong D, Kay M, Conner B, Wakefield T, Manogharan G (2018) Hybrid manufacturing – integrating traditional manufacturers with additive manufacturing (AM) supply chain. Add Manuf 21:159–173
34. Gao W, Zhang Y, Ramanujan D, Ramani K, Chen Y, Williams CB, Zavattieri PD (2015) The status, challenges, and future of additive manufacturing in engineering. Comput Aided Des 69:65–89
35. Janssen GR, Blankers IJ, Moolenburgh EA, Posthumus AL (2014) TNO: the impact of 3-D printing on supply chain management
36. Petrovic V et.al (2011) Additive layered manufacturing: sectors of industrial application shown through case studies
37. Florentin C et.al (2012) Additive manufacturing flickering at the beginning of existence. Procedia Econ Financ 3:457–462.
38. Scott (2012) Additive manufacturing: status and opportunities
39. Zanon S et al (2019) Supply chain implications of additive manufacturing: a holistic synopsis through a collection of case studies
40. Durach CF, Kurpjuweit S, Wagner SM (2017) The impact of additive manufacturing on supply chains (2017).
41. Garrett B (2014) 3D printing: new economic paradigms and strategic shifts. Global Policy 5(1):70–75
42. Scott A, Harrison TP (2015) additive manufacturing in an end-to-end supply chain setting. 3D Print Addit Manuf 2(2):65–77
43. Wits WW, García JRR, Becker JMJ (2016) How additive manufacturing enables more sustainable end-user maintenance, repair and overhaul (MRO) strategies. Procedia CIRP 40:693–698
44. Achillas Ch et al (2014) A methodological framework for the inclusion of modern additive manufacturing into the production portfolio of a focused factory

A Step Towards Next-Generation Mobile Communication: 5G Cellular Mobile Communication

Ayush Kumar Agrawal and Manisha Bharti

Abstract The mobile cellphone network in this generation has become one of the most effective networking innovations of the last several years. The introduction of the latest phones and devices in the past few decades has reflected in a massive increase in network traffic. Various new systems have already appeared to stay on top of wireless technology with the prevalence of more intelligent platforms that communicate with data centers and with each other via high-speed internet-connected devices. In the next generation, the data rate required will be very much more than that of the current data rate which we are currently using for our mobile communication. Fifth-generation mobile communication is still a concept with many ideas of the type of technology that can be used in it. The author will enlighten and discuss several technologies that can be used in 5th generation wireless communication in this paper.

Keywords 5G communication · Latest trend · High-speed internet · Wireless connectivity · Mobile communication

1 Introduction

Rising International data traffic had already powered the density requirements for presently mobilized third generation as well as fourth-generation mobile communication. Today, efficient analysis of 5G mobile communication systems is advancing in several areas. It is estimated that fifth-generation technology will be in use by 2021. In evaluating a broad range of Western research work, current literature, and fifth-generation papers from major players in mobile communication, this paper makes sense in the context of numerous fifth-generation activities [1].

The goal was to better explain what fifth-generation mobile communication is all about and how various fifth-generation projects strive to get there. The fifth-generation mobile system covers the advancement of different standards of mobile communication technological innovation. Advancement involves the advancement

---

A. K. Agrawal (✉) · M. Bharti
National Institute of Technology Delhi, New Delhi 110040, India
e-mail: ayush6295@gmail.com

© The Author(s), under exclusive license to Springer Nature Singapore Pte Ltd. 2021
R. M. Singari et al. (eds.), *Advances in Manufacturing and Industrial Engineering*,
Lecture Notes in Mechanical Engineering,
https://doi.org/10.1007/978-981-15-8542-5_40

of technology and social engagement. Many scientists have started working in this area to build up enhanced standards and regulations within the knowledge of the technology-society activity [2]. The actor-network concept in this respect is among the key concepts.

For illustration, utilizing a decreased wireless spectrum, researchers and academics are experimenting with new methods to power-efficient as well as price-effective fifth-generation mobile communication. Within the next decade, they are also going to launch fifth-generation technology [3]. In China, every era's creation is extremely competitive through first-generation to fifth-generation. Second-generation mobile communication uses TDMA, which provides digital communication and low-speed data connectivity. Third-generation mobile communication offers high-frequency voice and data connectivity utilizing, for example, the CDMA. Fourth-generation uses the OFDMA to increase data speeds of 150 Mbps to 1.5 Gbps which provides different mobile telecommunications mobile services. Fifth-generation technical analysis is based on framework schematic design, mm-wave, air application, light MAC (Media Access Control), RRM (Radio Resource Management), multi-cell shared storing, huge multi-input (MIMO) transceivers and dissemination, smart devices, Machine-to-Machine (M2M) connectivity, etc. [4].

2 Backend Explanation of 5G

Universal technological specifications were established through the first generation to the fifth generation on the basis of both technological and social activity. Its transition from the first generation to the fourth generation has advancements that bust interoperability. Fifth-generation requires high spectrum bandwidth including multiple cell towers and networks as a revolutionary innovation. The introduction to fifth generation-based networking networks such as those of the Internet of Things (IoT) would also greatly impact our daily lives. Fourth-generation communication widely used is alluded to as the wireless technology network of the fourth generation, which primarily involves TD-LTE (Long-Term Evolution, LTE) as well as FDD-LTE systems. TD, as well as FDD modulation, is separate from the air system, which can be regarded as another sort of technology [5]. In terms of functionality, fourth-generation infrastructure, which combines third-generation and wireless LAN, allows equivalent to 110 Megabits per second of data transmission or connection speed. In addition, it can be expanded and easily implemented. Furthermore, which has several weaknesses that can occur because of many circumstances. The major drawback is that it has many frequency ranges [6] (Fig. 1).

The next generation of technical standards is the fifth generation (5G) mobile communication network. 5G also addressed most of those technological shortcomings of fourth-generation systems, enhancing the quality of service, time delay, I/O frequency, energy consumption, and network performance. Fifth-generation wireless technologies are 15–150 times better over fourth-generation, achieving a peak data transmission speed of 11 Gbps compared with 110 Mbps of fourth-generation.

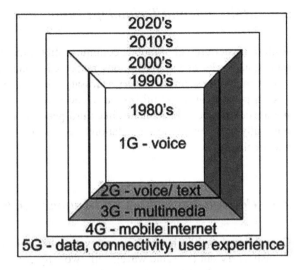

Fig. 1 Dates and the type of technology evolved for wireless mobile communication

And hence the average lag between the transmitter and the phone was decreased approximately 6–12 times compared to fourth-generation. The cellular data for each unit area of fifth-generation is one thousand times greater than the fourth generation in terms of broadband service.

Fifth-generation communication development does have two schemes. The first is to line by line upgrade the infrastructure, that involves building the infrastructure to enhance data traffic and performance related to already deployed fourth-generation LTE systems. This system used supranational systems such as TDD & FDD, improved link, 3D-MIMO, improved CoMP, LTE-Hi with a tiny base, etc. One strategy is developing entirely new infrastructure architectures and consumer devices to create the whole new mobile communication system including the following major innovations [7–11].

2.1 Large-Frequency Bandwidth Usage

Many of the feature bandwidth of wireless transmission is less than 4 GHz, which allows more consumers with a limited range. Nevertheless, the range with a frequency above 4 GHz was not completely used. Using transmission bandwidth over 4 GHz would help mitigate the lack of available space in the network. Large-frequency spectrum (70 GHz, for example) has higher anti-interference properties, an adequate spectrum with interchangeable network capacity, smaller size facilities, and high output antenna. Furthermore, the working situation for higher frequency broadband should be considered, which is sometimes used in cooperation with these other wireless techniques.

2.2 Multi-antenna Technology Innovation

With the faster growth of mobile communications, there is an increase in supply for mobile networks, and scarce options are present. Thus, increasing the quality of use of the bandwidth seems to be very necessary. This is an innovative way of improving system performance and utilization of the network. It is currently enforced to all mobile communication elements, like third-generation, LTE, as well as LTE-A. The unusually high number of base stations ensures both data communication accuracy and areal density.

2.3 Cofrequency Cotime Full-Duplex (CCFD)

Conventional mobile networking innovation has constraints; it could not simultaneously carry out the same wavelength two-way connection. It leads to substantial energy loss and CCFD software using the same uploading and update rate tools for two-way interaction concurrently. This uses the internet twice technically. But CCFD also experiences a minor problem of extreme self-interference; thus, eliminating interference is the main problem. Furthermore, frequency interaction with individual cells with full-duplex is indeed a concern.

3 Architecture of Cellular Mobile Communication

The cellular mobile communication architecture is as similar as from the 1G communication but the connections between them change with respect to the current generation of mobile communication which is being used. This architecture is widely used for mobile communication and for voice and data services as shown above. As we widely use this and our important data is also involved in it, hence it needs to be highly secured as our data (voice or mail) should not be misused at any instance of the transmission [12–15].

The security requirements are very much necessary in cellular mobile communication, as it consists of many personal and private data.

Which can be further divided into security threats and security services which are further classified as:

3.1 Security Threats

As the threat is defined as some damage or any hostile action to the data where the original signal is being disturbed by an external source. Such of the security threat in mobile communication is Radio Eavesdropping and MS impersonation.

Radio Interface Eavesdropping Eavesdropping is defined as unauthorized interception at real-time of some private communication, which may be a phone call, FAX, text message, or video transmission.

The term eavesdrop came from one of the practices which is actually standing at the eaves of building premises, for listening to talks inside of the premises.

Confidentiality of Data To secure the data from the access of any other third party. Which is further discussed in this chapter.

Anonymity of User Anonymity of user means that the real author of a message is not shown/displayed. Anonymity can be implemented to make it impossible or very difficult to find the true author of a message/sender.

Impersonation of MS (masquerade) In this kind of attack to the communication system the attacker pretends to be authenticated or so-called real user who do have all the access but in real scenario it is the attacker, who is misusing the data which is being sent.

Masking can be attempted by using stolen login credentials and passwords, looking for security vulnerabilities in programs, or bypassing the authentication mechanism. The attempt may come from an organization, for example, an employee or an external user via a public network connection.

3.2 Security Services

The services which are being used for which the data which is being transmitted is secured and the original data can easily be received at the receiver side without any loss of the information and also the misuse of the data where no external source can get your original data.

Identity Protection of Subscriber Identity of the subscriber is the sole important thing, which is one of the most important things which should and must be protected so that any external source cannot get the real-time details of the sender or the receiver who is involved in the communication process.

Authentication of Subscriber Verification or authentication is necessary for mobile communication as the access of the subscriber services can be secured and no other can be able to access the data of any user.

Confidentiality of Data It means the protection of the important data or maybe any kind of the data so that it is inaccessible by any of the third party and your data is

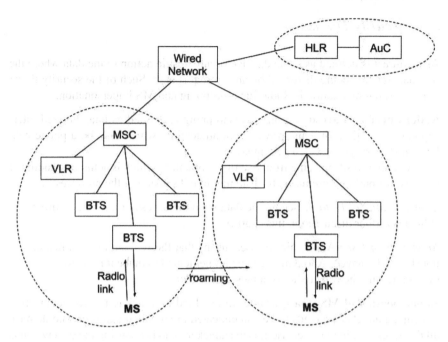

Fig. 2 Cellular mobile communication architecture

highly protective. Which can also be said as protecting the data from someone not to access it who can gain access (Fig. 2).

3.3 Legends of Fig. 2

HLR—Home location register It is the main backbone component of mobile communication as all the MSCs are connected to only one HLR(which is basically present in Hyderabad).

The main work of HLR is to have all the permanent database of all the subscribers whosoever is using cellular mobile communication, which may be of CDMA, TDMA, or GSM.

For example- "ABC" is a boy whose address in his id is "XYZ". So the location of "ABC" is saved as "XYZ" in the HLR even if he changes the city, BTS, VLR, etc. his permanent location will be saved in HLR and which is irrespective of his current location.

AuC—Authentication center The main work of the authentication centre is to verify the location form HLR whensoever is required. Basically, it is used to authenticate the user when asked by MSCs or VLRs.

MSC—Mobile switching center It is present in each telecom circle, depending upon the telecom service provider.

Suppose, "X" is a company which has 28 MSCs in INDIA. Generally, in local language, we call it a telecom circle, like:

- MP—CG circle**
- Bihar—Jharkhand circle**
- Punjab circle
- Gujarat circle.

**(as before 2000 these are only one state, hence the circle is not been separated).

VLR—Visitor location register The work of this is as same as the HLR but this contains the subscribers' temporary location.

For example- "ABC" is a boy whose address in his id is "XYZ" but his current location is "MNO" from last few days. So the location of "ABC" is saved as "XYZ" in the HLR even if he changes the city, but in VLR his location will be saved as "MNO" irrespective of his permanent location in HLR.

BTS—Base transceiver station It is the interface between the MS and MSCs, whose work is to transfer the data/signal from MS or subscriber to the service provider and also from that of the service provider to the subscriber. (It is the structure which is also called as "Tower" commonly).

MS (Mobile Subscriber)—ME (Mobile Equipment) + SIM (Subscriber Identity Module) Mobile subscriber consists of Mobile equipment and Subscriber identity module, where Mobile equipment is the electronic device which helps SIM to connect to BTS. And the Subscriber Identity module is card which keeps the subscriber data which consists of the user phone number, location of a user, contacts, user identity, text messages, and so on...

4 Fifth-Generation Communication System Evolution and Future

Discussion on fifth-generation is widely divided into two systems of thinking: a network-led perspective that views fifth-generation as a reorganization of second-generation, third-generation, fourth-generation, Wireless-Fidelity, and other technologies that provide far wider coverage and efficiency at all times; and a second perspective propelled by a massive shift in internet speed and intensity elimination in edge-to-end lag. Such terms are always examined around each other, furthermore, leading to an increase in demands that are often conflicting [16–18].

Fourth-generation or other networks can allow a few of the criteria recognized for fifth-generation. Sub-1 ms delay and >1 Gbps downlink frequency are the technical requirements that require a true generational shift, and also only services that allow at least one of those could be called fifth-generation usage scenarios according to both concepts. It will be challenging to deliver 1 ms latency over a large-scale network, and this condition may be relaxed. If that were to occur, many of the proposed future fifth-generation networks might no longer be necessary and fifth-generation third example would be far less apparent. This paper looks at some of the barriers to achieve 1 ms latency that needs to be addressed.

4.1 Since Different Scenarios Require Different Approaches to Technology and the Following Issues Need to Be Addressed in Order to Meet the Needs of the User

Smooth transition from 4 to 5 G needs more private enterprises to be identified as a development platform for the spectrum usage analysis It is necessary to support the growth of small and medium-sized private enterprises, particularly in spectrum efficiency research, for sustainable and innovative innovation. It is important to plan a smooth transition from 4 to 5G. Market demand for increases in data services will not be slowed down because the 5 G infrastructure is not yet mature. Operators must, therefore, continue to improve the current 4 G networks and take full advantage of the seamless future migration.

Spectrum performance should now be improved at least three times as much as 4.5 G, and similar technology will be applied directly to provide a smoother transition from 4 to 5G.

Standardization requires cooperation in a global context At the same time, standardization work will concentrate on both international and local dimensions. It must stress on eMBB and IoT within the context of the Union and 3GPP. 3GPP and ITU could only propose standards through the development of compliant, reliable, and efficient technologies. The benefits will be outweighed by premature overseas agendas.

5G standardization requires cooperation and results sharing. The sharing program of experimental results is important and the forum for collaboration should be strengthened.

User-oriented technology requires high performance and efficiency of the spectrum As the key finding of a change from policy approach to consumer orientation is exposed, service providers, content providers, and vendors may benefit from scientifically gathering, researching, and analyzing client expectations for a more fulfilling end-user experience at the induction point.

Improving bandwidth and frequency utilization of 5G information, in particular, is an important issue that needs to be addressed in 5G environments in the large rising network.

Fifth Generation technology is the technology of the upcoming telecommunication era. Mobile networks are intended to be a minimum of 30 times faster than fourth-generation mobile communication. Fifth-generation tech could use a range of frequencies, namely millimeter wave (mmWave), that can also transmit quite huge

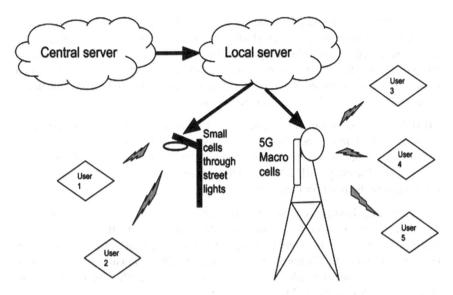

Fig. 3 Possible 5G architecture

quantities of data over a couple of miles. The disadvantage of the bigger frequencies is that they are very easily hampered by the buildings, forests as well as other lays, and sometimes even the atmosphere.

The upcoming fifth-generation technologies might come from different manufacturers and will consist of devices aimed at providing extremely quick download rates and low latency. Industries like Ericsson, Samsung, AT&T, and Verizon are testing new breakthroughs in data processing, chips, as well as transceiver devices which will allow the upcoming generation of cell phone networking today–ahead of the expected fifth generation, will hit the market next year. Even though the demand for fifth-generation tech is focused, we see a range of innovations evolving while essential to the implementation of fifth-generation system (Fig. 3).

4.2 Innovations During 5G Communication

MM wave Signals are delivered at speeds ranging 30–300 GHz relative with channels under 6 GHz of 4 G Mobile. The upcoming fifth-generation networks are going to attempt to relay massive amounts of data—and yet only certain frames at a moment. Even though the fifth-generation protocol would provide the biggest merits over these larger frequencies, it'll also work in small frequencies and also unauthorized frequencies already used by WiFi, without causing clashes with current WiFi devices. For such a purpose, in addition to common cellular towers, fifth-generation networks will be using small cells.

Small Cells These are compact BTS that are small-powered and may be positioned in towns. In order to create a heavy, multidimensional network, providers will configure several small cells. The minimal-profile antennas of small cells allow themselves inoffensive, yet the sheer volume makes it difficult for them to put up in semi-urban areas. End-users should continue to see universal fifth-generation antennas as fifth-generation technology matures, within their homes.

MIMO Fifth-generation tech requires cell towers to accommodate several more antennas than fourth-generation cell towers. With MIMO, there are numerous antennas for either the source (transmitter) and the destination (receiver), enhancing speed and efficiency. MIMO thus adds opportunities for interference, which contributes to the need for beamforming.

Beamforming Beamforming is a fifth-generation tech that provides end-users with the most flexible data delivery route. Greater-frequency antennas require smaller transmitting signals to be steered. This customer-specific beamforming enables both horizontally and vertically transmissions. The path of the beam may shift many times a fraction of a second. Beamforming can help to make more effective use of the range around them by massive MIMO architectures.

Full duplex A way to ultimately increase the overall speed of communication technology is full-duplex communication. By using a fifth-generation full-duplex system on a single network, the transmission of information into and out of the cell tower requires just one channel, instead of two. A possible downside of full-duplex is that it can infringe with the waveform.

5 Conclusion

The intent of this study would be to start moving towards explaining what' Fifth-generation' really means throughout the technical context, through decreasing fifth-generation to its root (such as recognizing what it may not be); enlarging on a few use possible circumstances fifth-generation could allow; and describing imaginable network topology and business opportunities consequences for providers. It can only be done by positioning the fifth-generation debate between wireless network technology and those currently under development in a wider context. The present state of the fifth-generation test is constrained to innovation assessments and similar technologies, which include Cofrequency Cotime Full-Duplex (CCFD), Device-to-Device (D2D) technology, all-spectrum access, massive MIMO, and extremely-dense network. Innovation networks, including companies and research centers supported by the government, as well as universities, are collaborating in the standardization process on the related research for energy. The 5 G development could only grow healthily by satisfying the interests of each actor with actual involvement.

References

1. Zhang JC, Ariyavisitakul S, Tao M (2012) LTE-advanced and 4G wireless communications. IEEE Commun Mag 50(2):102–103
2. 3GPP (2010) TR 36.912, Feasibility study for further advancements for E-UTRA (LTE-advanced)
3. Hirosaki MB (1981) An orthogonally multiplexed QAM system using the discrete Fourier transform. IEEE Trans Commun 29(7):982–989
4. ITU-R WP-5D (2015) Document 5D/TEMP/625-E, IMTVision—framework and overall objectives of the future development of IMT for 2020 and beyond
5. Bogucka H, Kryszkiewicz P, Kliks A (2015) Dynamic spectrum aggregation for future 5G communications. IEEE Commun Mag 53(5):35–43
6. 3GPP (2014) TR 36.873, Study on 3D channel model for LTE
7. 3GPP (2015) TR 36.897, Study on elevation beamforming/full-dimension (FD) MIMO for LTE
8. 3GPP (2015) TR 36.889, Study on licensed-assisted access using LTE
9. 3GPP (2013) TR 36.888, Study on provision of low-cost machine-type communications (MTC) user equipments(UEs) based on LTE
10. ITU-R (2015) WP-5D, Att. 2.12 to 5D/1042, ITU-R Working party 5D structure and work plan
11. Chang R (1966) High-speed multichannel data transmission with bandlimited orthogonal signals. Bell Syst Tech J 45:1775–1796
12. Saltzberg B (1967) Performance of an efficient parallel data transmission system. IEEE Trans Commun Tech 15(6):805–811
13. METIS (2015), ICT-317669-METIS/D8.4, METIS final project report
14. Farhang-Boroujeny B (2011) OFDM versus filter bank multicarrier. IEEE Signal Process Mag 28(3):92–112
15. Premnath S, Wasden D, Kasera S, Patwari N, Farhang-Boroujeny B (2013) Beyond OFDM: best-effort dynamic spectrum access using filterbank multicarrier. IEEE/ACM Trans Network 21(3):869–882
16. 3GPPTR 36.819 (2011) Coordinated multi-point operation for LTE physical layer aspects
17. Pi Z, Khan F (2011) An introduction to millimeter-wave mobile broadband systems. IEEE Commun Mag 49(6):101–107
18. Kim C, Kim T, Seol JY (2013) Multi-beam transmission diversity with hybrid beamforming for MIMO-OFDM systems. In Proceedings of the IEEE GLOBECOM'13 Workshop, pp 61–65

Efficacy and Challenges of Carbon Trading in India: A Comparative Analysis

Naveen Rai and Meha Joshi

Abstract As per June 2018 World Bank's report on South Asian Development, climate change is going to cost 2.8% of India's GDP and lower the standard of living of nearly half of its population by 2050. Green financing in general and carbon trading in particular have emerged as important tools to incentivize industries to move toward a lower carbon footprint. Emissions trading, as laid out in article 17 of Kyoto Protocol, allows buying and selling of carbon credits. The carbon market thus created offers an ocean of opportunities to developing countries. To the developed and developing countries alike, it also opens a window of possibilities to meet the Sustainable Development Goals (SDGs) by 2030. The objective of this paper is to assess India's development in the field of carbon trading and to compare it against European Union Emissions Trading Scheme (EU ETS). It also assesses major systemic issues that impede India's progress toward achieving its full potential.

Keywords Carbon credits · Environmental finance · Carbon trading

1 Introduction

Carbon financing is a way to incentivize companies to cause less pollution [1]. There is a permissible limit of carbon emission[1] a company can do as per Kyoto Protocol. It is measured in terms of carbon credits (1 carbon credit is equal to 1 tonne of CO_2 emission). There is an amount of money associated with carbon credit. Just like stock exchanges trade in stocks, their prices go up and down, Climate Exchanges

---

[1]Note: In this paper, the term carbon has been invariably used to represent all Green House Gases (GHGs), viz. carbon di-oxide, nitrous oxide, hydro-fluoro-carbons, per-fluoro-carbons, and sulfur hexafluoride as given in Annex A of Kyoto Protocol.

N. Rai · M. Joshi (✉)
Delhi School of Management, Delhi Technological University, New Delhi, India
e-mail: meha.joshi83@gmail.com

N. Rai
e-mail: naveenrai.dce@gmail.com

or Mercantile Exchanges (for example, NYMCX) trade in carbon credits. Now, the companies that cause lesser pollution (generally by using Clean Development Mechanism (CDM) or other technological interventions), than what they are permitted to, are rewarded with equivalents (known as carbon Emission Reduction (CER) units or simply put carbon credits). Along with carbon trading, other two important tools to inhibit carbon emissions (as per Kyoto Protocol) are Joint Implementation (JI) and Clean Development Mechanism (CDM) [2]. While in Joint Implementation, an Annex I country is allowed to invest in another Annex I country [3] as an alternative to domestically reducing emissions, in CDM, the investment can take place in any country having commitment of reducing emissions under Kyoto Protocol [4]. This paper primarily focuses on carbon Trading and touches upon CDM with respect to Indian carbon market. The rationale for choosing Indian carbon market is the fact that it is in its nascent stage. For the same reason, comparative analysis has been done with European emission market, which happens to be one of the largest in the world [5].

2 Emissions Versus Economic Growth Trade-Off: A Perennial Issue

The common but differentiated goals, as enshrined in the seven principles of Rio declaration of 1992 and further adopted by Kyoto Protocol [6], realize that developing and developed countries have different capabilities and hence differing responsibilities toward bridling their carbon footprint. The developmental requirements of developing and developed countries are different. The industries of the former must grow at a faster rate than those of the former to meet their economic aspirations. This gives rise to a paradox where the carbon emissions must be contained despite the inevitable industrial growth [7].

3 Need for Political Will and Market Forces

The success of carbon trading schemes hinges upon various factors. Political will to enact stronger national legislations, aimed at curbing industrial and non-industrial pollution and simultaneously incentivizing the stakeholders to reduce their carbon footprint, for one, is a key determinant in evaluating the efficacy of carbon trading schemes. Tighter the regulations, higher would be the prices of carbon credits [8] [9]. The reverse is also true. Other determinants are the external market conditions and the lucrativeness of carbon credits viz-a-viz other financial instruments. If the aforementioned conditions are unfavorable, the prices may fluctuate and plummet, disincentivizing the investors.

4 Data Set

Analysis has been done based on the data sourced from EU ETS Dashboard, which extracts data from European Union Transaction Log (EU TL) [10]. Data from more than 11,000 stationary industrial installations as well as 500 aircraft operators flying between EEA's airports has been considered for the purpose of this study. The regression analysis has been performed on ETS price data from 2014 to 2019.

5 Opportunities in Emissions Trading

European Union has capitalized upon the opportunity presented by the Emissions Trading Scheme. In particular, Germany, which happens to cause the greatest emission among the EU countries, has turned it around into a success story by registering for highest Certified Emissions Reductions (CERs) [11]. In the past five years, there has been an upsurge in the prices of carbon credit, thereby enabling European countries to reap maximum financial benefit of the same [12].

6 Calculations

The future price of carbon credits can be calculated by running a simple regression. Results of the regression analysis are tabulated in Tables 1, 2, and 3. Upon regressing price data against independent variable year, the coefficient as 3.343 and intercept as −6731.941. The standard linear equation takes a form of **Y = 3.343X − 6731.941**.

Table 1 Result of regression analysis

	Values (rounded off)
Multiple R	0.77
R square	0.59
Adjusted R square	0.59
Standard error	4.74
Observations	308

Table 2 Result of regression analysis continued

	df	SS	MS	F	Significance F
Regression	1	9846.18	9846.18	437.67	5.843E−61
Residual	306	6883.96	21.26		
Total	307	16,730.14			

Table 3 Result of regression analysis continued

	Coefficient	Std. error	t-Stat	P-value	Lower 95%	Upper 95%
Intercept	−6731.94	322.29	−20.88	7.79E−61	−7366.14	−6097.74
X variable	3.34	0.15	20.92	5.84E−61	3.02	3.65

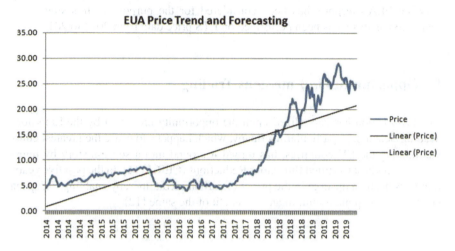

Fig. 1 EUA price trend and forecasting based on 2014–2019 EU ETS TL data

For independent variable **X = 2025** (value of year), we get price of EUA as **EUR 37.634** (Fig. 1).

7 Results

The result is not astonishing considering the upsurge of prices due to stronger policy measures taken by EU. Keeping policy matters constant or favorable, we can expect to see a stellar growth in the emissions trading market. This provides a lucrative opportunity to not only the countries but also to the organizations and other individual investors to make fortunes along with achieving the shared goal of pollution abatement. One limitation of the calculation is that only five-year prices have been considered (2014–2019) to make the projection. The rationale for considering only five-year prices is to remove the extreme price fluctuations in period before 2014. Prior to 2014, the EU carbon trading market was only in the infancy, and hence, a high degree to price fluctuation can be observed.

8 Role of CDM in Carbon Abatement

CDM or Clean Development Mechanism, as defined in article 12 of Kyoto Protocol, allows industries to reduce their carbon footprint by application of technology, thereby providing the host country an opportunity to avail CER units (Certified Emissions Reduction) which can be further traded in one of the several Emissions Trading Schemes. This presents a cost-effective proposition to the Annex I countries (industrialized nations) [3] enabling them to purchase carbon credits from the non-Annex I countries (developing nations). It also facilitates technology transfers [13] and collaborations toward the common objective of environmental sustainability and financial viability (Figs. 2, 3 and 4).

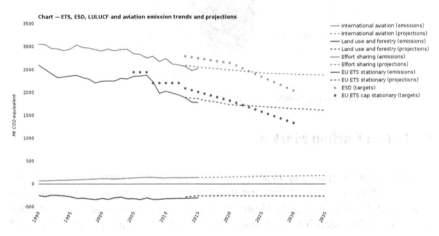

Fig. 2 Recent ETS, ESD, LULUCF, and aviation emission trends

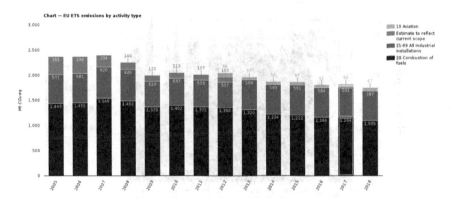

Fig. 3 Trends in EU ETS GHG emissions from 1990 and projections until 2035

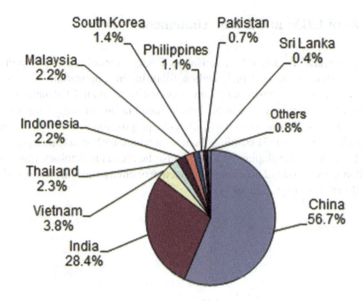

Fig. 4 Share of CDM projects among Asian countries

9 Indian Carbon Market

Off late, India has sprung up in action to have a piece of the carbon pie. Data points to the fact that along with China and Brazil, India is a global leader in terms of its share in CDM projects (Fig. 5). However, most of the projects are small and medium sized. Several projects have kick started in the state of Gujarat [14]. In June 2019,

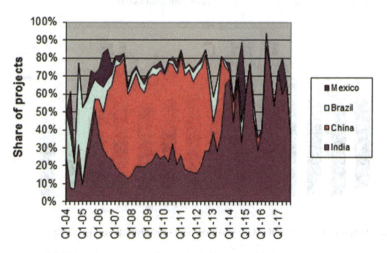

Fig. 5 Share of CDM projects among Mexico Brazil China and India. *Source* UNEP website

Gujarat unveiled its first Emissions Trading Scheme which allows industries to buy and sell carbon credits. 350 industries around Surat (Gujarat) have installed real-time emissions monitoring and tracking systems. Delhi Metro Rail Corporation (DMRC) has become the world's first metro rail organization to register its carbon credits with UNFCCC [15]. The policy climate of India is also conducive to creating an enabling environment for proliferation of carbon emission reduction market. The National Clean Air Program (NCAP) proposed in 2018–2019 envisages reducing air pollution by 30% by the year 2024.

10 India Versus European Union

Development of the field of carbon market is minuscule compared to that of European markets. European market alone commands a lion share of $175 billion market [16]. While European Union has a unified Emissions Trading Scheme since 2005, India is just foraying in this field with some states coming up with their emissions trading programs.

11 Challenges and Threats

There is a growing demand among the developed countries to disallow the exchange of carbon credits from developing countries. Such a move may give India and other developing countries a body blow to their emission reduction programs. Millions of CER units are in pipeline among various CDM projects in India alone. Another threat is the low level of carbon pricing. The financial incentives received by the industries must by far exceed the cost incurred in developing new technology required to cut the carbon emissions. The current low price of carbon credits defeats that purpose. Another challenge that the carbon emission schemes are marred with is the exaggerated reporting in carbon credits by the industries in order to avail greater incentives [17].

12 Conclusion

Carbon trading provides a great way to accomplish dual objective of achieving pollution abatement and receiving financial incentives while doing so. The government regulations have to be well directed such that it discourages the polluting industries and at the same time incentivizes the industries that are working toward neutralizing their carbon footprint. India shows the political will and the industrial backing required to become a global leader in the carbon trading market. Hence, the outlook of carbon trading market in India is quite positive.

13 Suggestions

The central government of India should come up with a unified action plan to develop a centralized carbon Emissions Trading Scheme on the lines of EU ETS. Some sort of floor price should be proposed to insulate the market from the vagaries of pricing fluctuations. Stronger audit mechanism must be kept in place to ensure that the companies do not inflate their emission reductions. Indian industry and the business community needs to be sensitized more about the grave challenges that confront the environment at large and their confidence needs to be reinforced in green technologies in general and carbon trading in particular.

References

1. Disch D, Rai K, Maheshwari S (2010) Carbon finance: a guide for sustainable energy enterprises and NGOs. Global Village Energy Partnership 46
2. Dechezleprêtre A, Glachant M, Ménière Y (2009) Technology transfer by CDM projects: a comparison of Brazil, China, India and Mexico. Energy Policy 37(2):703–711
3. UNFCCC website, https://unfccc.int/process/parties-non-party-stakeholders/parties-convention-and-observer-states. Last accessed 5 Dec 2019
4. Haites E, Duan M, Seres S (2006) Technology transfer by CDM projects. Climate Policy 6(3):327–344
5. Reuters website, https://www.reuters.com/article/us-global-carbontrading-report/value-of-global-co2-markets-hit-record-144-billion-euros-in-2018-report-idUSKCN1PA27H. Last accessed 5 Dec 2019
6. UN website, https://sustainabledevelopment.un.org/getWSDoc.php?id=4086. Last accessed 5 Dec 2019
7. Aghion P et al (2016) Carbon taxes, path dependency, and directed technical change: evidence from the auto industry. J Polit Econ 124(1):1–51
8. Xu X, Xiaoping Xu, He P (2016) Joint production and pricing decisions for multiple products with cap-and-trade and carbon tax regulations. J Clean Prod 112:4093–4106
9. Financial Times website, https://www.ft.com/content/d1d9fcf4-a7c0-11e9-984c-fac8325aaa04. Last accessed 5 Dec 2019
10. European Energy Exchange website, https://www.eex.com/en/market-data/environmental-markets/auction-market/european-emission-allowances-auction. Last accessed 5 Dec 2019
11. Rogge KS, Schneider M, Hoffmann VH (2011) The innovation impact of the EU Emission Trading System—findings of company case studies in the German power sector. Ecol Econ 70(3):513–523
12. Carbon Price Viewer by Sandbag, https://sandbag.org.uk/Carbon-price-viewer/. Last accessed 5 Dec 2019
13. UNEP Partner website, https://www.cdmpipeline.org/cdm-projects-region.htm2. Last accessed 5 Dec 2019
14. Live Mint website, https://www.livemint.com/science/news/india-launches-emissions-trading-programme-to-reduce-air-pollution-1559799447842.html. Last accessed 5 Dec 2019
15. DMRC website, https://www.delhimetrorail.com/. Last accessed 5 Dec 2019
16. European Commission 2019 Report, https://ec.europa.eu. Last accessed 5 Dec 2019
17. Martin P, Walters R (2013) Fraud risk and the visibility of carbon. Int J Crime Justice Soc Democracy 2(2):27–42

Micro-structural Investigation of Embedded Cam Tri-flute Tool Pin During Friction Stir Welding

Nadeem Fayaz Lone, Arbaz Ashraf, Md Masroor Alam, Azad Mustafa, Amanullah Mahmood, Muskan Siraj, Homi Hussain, and Dhruv Bajaj

Abstract Leading organizations in the aerospace sector have adopted extensive application of the friction stir welding (FSW) process. Since its inception, material flow and micro-structural changes occurring in the base metal during FSW have been a subject of prime interest among the researchers. One of the most effective ways, that reveals these mechanisms is the examination of plan view or coronal section of the tool pin embedded into the workpiece. Such investigations provide insight into the chronological order of events that occur on the leading side of the tool which are otherwise very difficult to observe. Therefore, assisted breaking of tool pin was performed in the course of FSW on 15.4 mm thick aerospace-grade aluminium alloy in this study. Different areas of the coronal section around the pin were studies via optical microscopy. A narrow circular void was observed in the leading retreating side of the pin which is created by the rotating edge of the flute. Notably, aberrant variation in grain size occurs on the flute edge at leading advancing side and between the flutes at trailing retreating side of the tool pin.

Keywords Friction stir welding · Embedded pin · Micro-structure

1 Introduction

FSW was invented and patented by TWI in 1991 [1] and was consequently used primarily for the welding of high strength aluminium alloys. Conventional fusion welding techniques like MIG, TIG and other arc welding processes are not suitable for such alloys and also result in distortion, high residual stresses, severe metallurgical defects like porosity, inclusions and formation of the brittle dendrite structure which reduce the joint efficiency [2–5], whereas FSW, a solid-state welding process, does not involve the melting of base materials but involve heating upto a temperature

---

N. F. Lone · A. Ashraf (✉) · M. M. Alam · A. Mustafa · A. Mahmood · M. Siraj · H. Hussain · D. Bajaj
Department of Mechanical Engineering, Jamia Millia Islamia (A Central University), New Delhi 110025, India
e-mail: arbazashraf.a@gmail.com

below melting point to 'soften' the material. FSW uses a combination of frictional heating (generated by the friction between the tool shoulder and material surface) and plastic deformation to weld various similar and dissimilar materials. FSW involves (1) plunging of a rotating tool with a probe into the abutting surfaces of base metals (2) traversing of the rotating tool along the joint line and (3) retraction of the tool at the completion of the weld. The tool is usually given a small tilt angle of about 2° such that the leading edge of the shoulder is above the plane of workpiece. This creates a pressure difference between the trailing half and the leading half of the tool, with higher forging pressure on the trailing side. The simultaneous rotation and traversing of the tool generate plastic deformation in such a manner that the material gets extruded from the advancing side (AS) and gets forged in the retreating side. The extensive strain undergone by the metal during this process also results in fragmentation of oxide layer from the abutting surface, and a weld is formed. Three distinct zones, namely (1) the nugget zone or stir zone (SZ) identified by the formation of ultra-refined equi-axed grains due to the continuous dynamic recrystallization, (2) the thermo-mechanically affected zone (TMAZ) identified by induced plastic deformation of the grains and the (3) heat-affected zone (HAZ), are formed [6–8].

Considerable studies on material flow during FSW have been performed to have a better understanding of the process [9–11]. For instance, Siedel and Reynolds [11] used marker insert technique and found that most markers deposited at the same location with respect to the transverse direction. The most commonly analysed transverse cross section of the weld provides little information about the material flow prevalent during the joining consolidation. However, if the FSW process is halted midway and the tool is kept 'frozen' or embedded in the materials, very useful insight can be obtained by examining the plan view or coronal cross section of the region surrounding the tool. This is also known as the 'stop action' technique and has previously been used by few researchers to understand the material flow [12–15]. Chen and Cui [12] used this technique and showed how the material flows and deposits in the form of shear layers in the wake of the weld formation. Development of grain structure during FSW was explained by Prangnell and Heason [13] using this technique. Nunes et al. [14] demonstrated the concept of plastic deformation via a shear surface similar to the metal cutting shear plane using the mid-thickness coronal section. Moreover, Chen and Cui [16] also calculated strain and strain rate prevailing ahead of the pin by facilitating the breaking of pin and studying the coronal section. Such studies are precious for the clear understanding of the interaction mechanisms between the tool pin surface and flowing material. However, still very few studies are present on this subject and such investigations conducted with the commonly used tri-flute pin profile are even rarer. Thus, in the current study, plan view or horizontal section of the embedded cam tri-flute pin was studied with reference to macro- and micro-structural features. The granular structure around the peaks and valleys of the tri-flute pin has been addressed in detail.

2 Experimental Procedure

FSW was performed on 15.4 mm thick plates of armour grade AA2519 alloy in butt joint configuration using a vertical milling machine. The machine was adapted for FSW process using an in-house tool adopter and work fixture. High-speed steel (HSS) was used as tool material. A tapered cam tri-flute pin profile was employed for the welding process. The outer diameter of pin at the root and bottom was 7 and 10 mm, respectively. The welding was performed at a tool rotational speed, traverse speed and shoulder diameter of 450 rpm, 31.5 mm/min and 26 mm, respectively. The breaking of the pin was facilitated to obtain an embedded frozen pin in the base metal. After the experimentation, the frozen portion of the workpiece was sliced from the horizontal section (coronal) using wire electric discharge machining (WEDM). After standard polishing procedure, the specimen was etched with 2 mL HF, 4 mL HNO_3 and 94 mL H_2O for observation under the optical microscope.

3 Result and Discussion

3.1 Macro-structure

The macro-structure of coronal plan section with embedded pin is shown in Fig. 1. The three valleys between the flutes of frozen cross section of pin can be clearly observed in this figure. For convenience, polar angle, θ, is used to refer to different locations in this cross-sectional view. In the vicinity of the pin, base metal can be seen to be re-crystallized due to severe plastic deformation (SPD) in the rotational stir zone (RSZ). The SPD results in very fine equi-axed grains in the SZ in the wake of the weld. A very sharp linear boundary can clearly be observed between SZ and advancing side TMAZ (AS-TMAZ). On the other hand, material on the RS is dragged into the wake of the weld, forming a hazy interface between the SZ and retreating side TMAZ (RS-TMAZ). This occurs because of the induced material flow in the RS of the weld due to smaller degree of plastic deformation on RS as compared to the AS. Due to the same reason, the grains in the AS-TMAZ can be seen to be elongated along the welding direction. The U-shaped marks in the SZ are actually the horizontal section view of the popular 'onion rings' which are formed due to layer-by-layer deposition of the base metal.

Interestingly, a long circular arc-shaped cavity can be seen from $90° < \theta < 180°$ on the leading side of the pin in Fig. 1. The formation of this void on the leading RS can be explained in the following manner: on the RS leading periphery of the pin, tangential velocity provided by the tool is such that the base metal is driven against the tool's traversing direction, thus promoting the transport of the material into the wake of the weld. Generally, for a circular pin cross section, the leading RS is simultaneously filled up by (1) the material coming from the AS along with the rotating tool and (2) from the relatively less deformed material in the front. However,

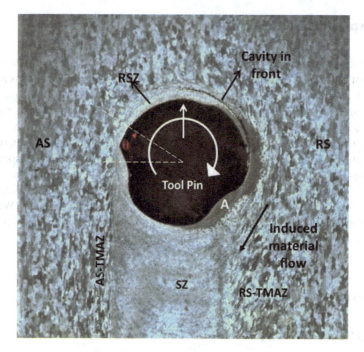

Fig. 1 Coronal section of macro-structure of the embedded pin

in the current scenario, the flute of the pin presently existing at $\theta \sim 200°$ in Fig. 1 cuts the material ahead like a 'knife' while engulfing this material in a valley on the pin periphery. Due to the much higher magnitude of tangential velocity as compared to the traverse velocity, it takes some time to fill up this material deficiency at the leading RS. This deficiency is observable as a cavity in this region.

3.2 Micro-structure

Optical microscopy was employed to reveal the granular features on the coronal section with the embedded pin. Figure 2 depicts the micro-structures of horizontal plan view at different values of polar angle θ. In the wake of the weld, at $\theta = 270°$, coarsening of grains can be observed with increasing distance from the pin, as shown in Fig. 2a. In front of the tool, at $\theta = 90°$, a similar increase in grain size with increasing distance from the pin surface can be observed in Fig. 2d. This primarily occurs due to greater plastic deformation of the material in the vicinity of pin. However, at the same angle of $\theta = 270°$, a cluster of relatively coarse grains in between relatively finer grains can also be observed in Fig. 2b, taken at a greater distance from the pin. An alternative pattern of increasing and decreasing grain size is also evident in Fig. 2c at $\theta = 300°$. Such variation in grain size arises because

Micro-structural Investigation of Embedded Cam ...

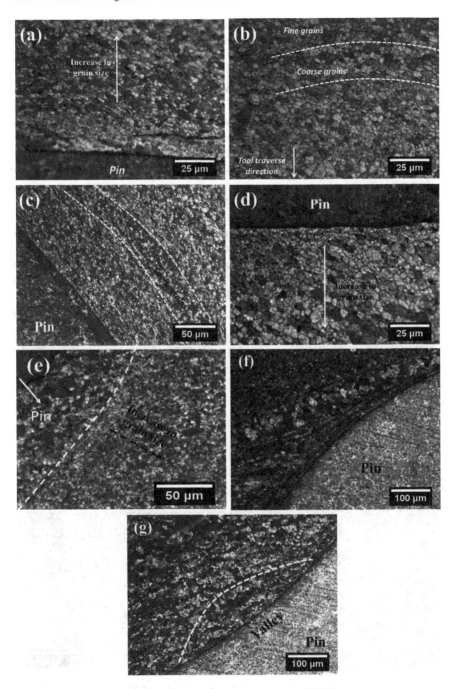

Fig. 2 Micro-structures of coronal section of embedded pin specimen at different values of θ: **a** $\theta = 270°$, **b** $\theta = 270°$ at a greater radial distance, **c** $\theta = 300°$, **d** $\theta = 90°$, **e** $\theta = 60°$, **f** $\theta = 240°$, **g** $\theta = 210°$

of the difference in magnitude of strain and strain rate undergone by the deforming material at different polar angles, encountering distinct sections of the three flutes of the pin.

Further, at $\theta = 60°$ which lies in the leading AS regime, existence of a band of much coarser grains adjacent to the pin is unexpected in Fig. 2e. Possibly, the rotating pin picks up a small amount of material at $\theta \sim 0°$ which has undergone relatively lesser amount of plastic deformation, thus pertaining to larger grain sizes. Importantly, this observation occurs at the peak of the flute, which exists at greater radial distance from the tool centre. The tip of the flute has greater proximity to the material existing farther from the centre at AS-TMAZ, thus supporting the above inference.

Figure 2f, g depicts the micro-structural images at location A marked in Fig. 1 which lies in $180° < \theta < 270°$. As evident from Sect. 3.1, the flute of the pin spades the material ahead of the tool and transports it what is also known as pulsating action. Again, the material which is transported through the pulsating action from the vicinity of RS-TMAZ undergoes lesser amount of plastic strain, resulting in larger grains. As a result of this intricate material flow, the region between the two flutes consists of a very large range of grain sizes, as evident from Fig. 2f.

The circular arc-shaped cavity at the leading RS has been shown in Fig. 3a, b. Fine

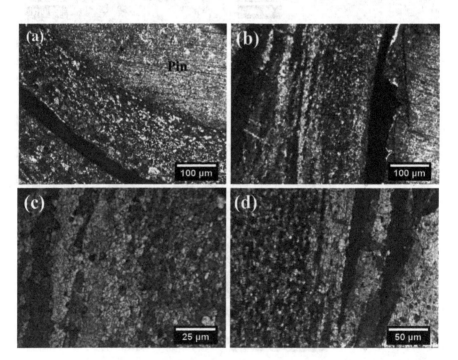

Fig. 3 Micro-structural images of **a** Arc-shaped cavity on leading RS. **b** Interface between RSZ and RS-TMAZ at 100X magnification. **c** Interface between RSZ and RS-TMAZ at 400X magnification. **d** Interface between RSZ and AS-TMAZ at 200X magnification

equi-axed grains can be observed on either side of the void in Fig. 3a. The interface between the RSZ and RS-TMAZ can be clearly seen in Fig. 3b. This region has been shown at a higher magnification by Fig. 3c. In this image, larger grains can be seen to be transforming into much smaller, equi-axed grains in the RS-TMAZ (left). This could be due to more static nature of recrystallization in the TMAZ as compared to the dynamic recrystallization in the RSZ. Similar micro-structural feature was also observed in the AS-TMAZ, as depicted by Fig. 3d.

4 Conclusion

Based upon the macro- and micro-structural investigation around the frozen pin during FSW of AA2519 alloy, following insightful conclusions can be drawn:

1. An arc-shaped void exists at the leading RS of the rotating tool. The occurrence of this void can be attributed to the cutting action of the edge of flute in the pin, which behaves as shovel, engulfing the material in the trailing side of the weld.
2. On the leading AS, a band of relatively coarser grains existing adjacent to the pin suggests the sticking of material from AS-TMAZ on the edge of the flute.
3. A large range of grain size is prevalent between the two flutes in trailing RS of the pin. This primarily occurs due to shovel or knife action of the edge of the flute while moving from leading RS to trailing RS.
4. A general increase in grain size on moving away from the pin surface occurs due to more severe plastic deformation of the base metal in closer proximity of the pin.

References

1. Bitondo C, Prisco U, Squillace A, Giorleo G, Buonadonna P, Dionoro G, Campanile G (2010) Friction stir welding of AA2198-T3 butt joints for aeronautical applications. Int J Mater Form 3(1):1079–1082
2. Rhodes CG, Mahoney MW, Bingel WH, Spurling RA, Bampton CC (1997) Effects of friction stir welding on microstructure of 7075 aluminum. Scripta Mater 36(1):69–75
3. Mishra RS, Ma ZY (2005) Friction stir welding and processing. Mater Sci Eng R Rep 50(1–2):1–78
4. Zhang Z, Li W, Wang F, Li J (2016) Sample geometry and size effects on tensile properties of friction stir welded AA2024 joints. Mater Lett 162:94–96
5. Zhang ZH, Li WY, Feng Y, Li JL, Chao YJ (2015) Global anisotropic response of friction stir welded 2024 aluminum sheets. Acta Mater 92:117–125
6. Ahmed MMZ, Ataya S, Seleman MES, Ammar HR, Ahmed E (2017) Friction stir welding of similar and dissimilar AA7075 and AA5083. J Mater Process Technol 242:77–91
7. Yan Z, Liu X, Fang H (2016) Effect of sheet configuration on microstructure and mechanical behaviors of dissimilar Al–Mg–Si/Al–Zn–Mg aluminum alloys friction stir welding joints. J Mater Sci Technol 32(12):1378–1385

8. Ilangovan M, Boopathy SR, Balasubramanian V (2015) Microstructure and tensile properties of friction stir welded dissimilar AA6061–AA5086 aluminium alloy joints. Trans Nonferrous Met Soc China 25(4):1080–1090
9. Long T, Tang W, Reynolds AP (2007) Process response parameter relationships in aluminium alloy friction stir welds. Sci Technol Weld Joining 12(4):311–317
10. Schmidt HNB, Dickerson TL, Hattel JH (2006) Material flow in butt friction stir welds in AA2024-T3. Acta Mater 54(4):1199–1209
11. Seidel TU, Reynolds AP (2001) Visualization of the material flow in AA2195 friction-stir welds using a marker insert technique. Metall Mater Trans A 32(11):2879–2884
12. Chen ZW, Cui S (2005) Tool-workpiece interaction and shear layer flow during friction stir welding of aluminium alloys. Trans Nonferrous Met Soc China 17(s1A):s258–s261
13. Prangnell PB, Heason CP (2005) Grain structure formation during friction stir welding observed by the 'stop action technique.' Acta Mater 53(11):3179–3192
14. Nunes Jr AC (2001) Friction stir welding at MSFC: Kinematics
15. Fonda RW, Bingert JF, Colligan KJ (2004) Development of grain structure during friction stir welding. Scripta Mater 51(3):243–248
16. Chen ZW, Cui S (2009) Strain and strain rate during friction stir welding/processing of Al-7Si-0.3 Mg alloy. IOP Conf Ser Mater Sci Eng 4(1):012026

Design and FEM Analysis of Connecting Rod of Different Materials

Sanjay Kumar, Vipin Verma, and Neelesh Gupta

Abstract The connecting rod is used in the internal combustion engine (IC engine) to convert the reciprocating motion of the piston to rotary motion of the crankshaft. This research article is used to investigate connecting rod's material which will be stronger as well as lighter. Three different materials such as carbon steel 40C8, aluminium 7075-T651 and Al 2014-7651 have been analysed analytically and by using finite element software ABAQUS. Various plots have been drawn which shows von Mises stress, maximum and minimum principal stress and displacement at critical points. For regular passenger vehicles, carbon steels are better option since they have to bear different traffic situations, and for high-performance vehicles like race cars where low weight and high strength is required, aluminium alloys are preferred.

Keywords Connecting rod · Carbon steel 40C8 · AL-7075-T651 · AL 2014-T651 · Finite element analysis · ABAQUS

1 Introduction

A connecting rod is one of the most significant parts in the engine that get together going about as a connection between crankshaft and piston. The cross-section of the connecting rod might be taken as rectangular, circular, H-section type and I-section; however, because of the buckling effect I-section type is the most reasonable recommendation for passenger cars since engines are low powered and thus to resist forces in the plane of motion only. The H-section type connecting rod is utilized in high rpm engines since it is progressively impervious to high twisting burdens [1]. The connecting rod is intended to withstand cyclic stacking (rehashed tension and compression) conditions. The connecting rod is one of the reversals of the slider-crank system keeping cylinder chamber fixed.

S. Kumar (✉) · V. Verma · N. Gupta
Delhi Technological University, New Delhi 110042, India
e-mail: sanjaydce2008@gmail.com

© The Author(s), under exclusive license to Springer Nature Singapore Pte Ltd. 2021
R. M. Singari et al. (eds.), *Advances in Manufacturing and Industrial Engineering*, Lecture Notes in Mechanical Engineering,
https://doi.org/10.1007/978-981-15-8542-5_43

Fig. 1 A connecting rod [2]

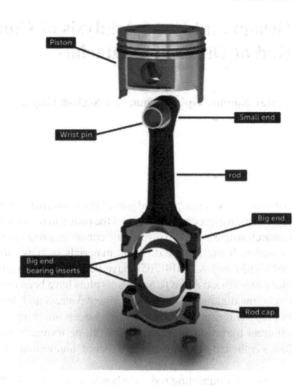

The connecting rod executes complex (rotational + translational) movement consequently moving vitality from cylinder to crankshaft and changes over the linear movement of the piston to rotating movement of the crankshaft. The principal point of this exploration is to decide the best appropriate material for connecting rod by examining von Mises stresses, most extreme and least chief anxieties and disfigurement for various materials utilized in the assembling of the connecting rod. In this exploration, a CAD model is set up in SOLIDWORKS 2018 and loads and boundary conditions are applied to the model for the examination which is performed on programming ABAQUS CAE 2019 (Fig. 1).

2 Objective

A lot of experiments are performed to explore the efficiency and life of connecting rod by introducing design changes and material changes such as replacing steel with aluminium and titanium alloys in order to reduce dimensions like weight and increase performance without affecting its load carrying and transmission characteristics [3]. This inspires us to put on material changes in the study that we knew from our curriculum and note the factors that are affected. This study's main target is to examine the resistance and strength bearing of various materials used in connecting rod during engine operation. The step-by-step objectives are as follows:

1. The connecting rod is imitated and modelled in SOLIDWORKS.
2. To survey the failure of connecting rod according to strength criteria.
3. Using ABAQUS CAE, determining von Mises stresses, maximum and minimum principal stresses and deflections under applied operating conditions.
4. Comparison of stresses and deformations induced in different materials of connecting rods.

The strength and distortion diagnostics of connecting rod have been accessed utilizing FEM methods to determine load at different nodes.

3 Materials Used in Connecting Rod

Considering strength, weight, dynamic load life, resistance to deformation, cost, availability, etc., various materials are being studied and experimented which gives different grades of steel alloyed with different materials, aluminium alloys and others [4]. Steel connecting rods are the most widely used types of connecting rods because of high strength and long fatigue life. Many materials are being examined because steel is denser, strong, easily available, requiring more power and tension to rotating assembly. Variation in heat treatment processes can produce extremely different results with ultimate tensile strength and yield strength.

Different types of materials to be discussed in this research include:

a. Aluminium 2014-T651 [5]
b. Aluminium 7075-T651
c. Carbon steel 40C8.

4 Design Specifications

The detailed parameters that are used for design of connecting rod are shown in Table 1 (Figs. 2 and 3).

5 Determination of Forces Acting on Connecting Rod

See Fig. 4.
Connecting rod is subjected to following forces:

1. Due to gas pressure,
2. Due to inertia of masses,
3. Due to friction acting in opposite direction of motion.

Table 1 Design specifications for connecting rod

S. No	Specification	Magnitude
1	Speed of IC engine (RPM)	2000
2	Bore diameter (d)	85 mm
3	Length of connecting rod (l)	350 mm
4	Mass of reciprocating parts (m)	1.5 kg
5	Factor of safety (FOS)	5
6	Wall pressure for piston rings	0.135 Pa
7	No. of rings (i)	3
8	Coefficient of friction (μ)	0.045
9	Explosion pressure (P_{gas})	3 Mpa
10	Piston pin diameter	30 mm
11	Crankpin diameter	36 mm
12	Axial thickness of piston rings (h)	600 mm

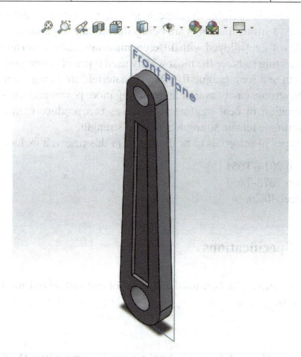

Fig. 2 SOLIDWORKS CAD used in simulation

Fig. 3 Dimensions (in mm)

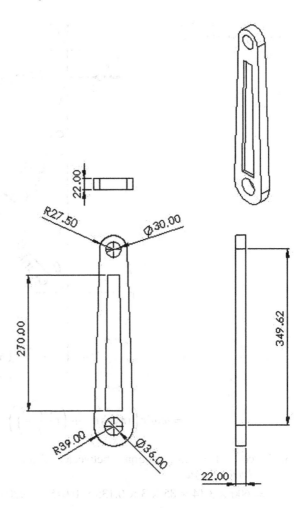

6 Design Calculation and Analysis

1. Forces due to gas pressure:

$$= \pi d^2 \left(\frac{P^{gas}}{4} \right)$$
$$= 17023.5 \text{ N}$$

2. Forces due to inertia of reciprocating masses:
   Taking $\theta = 0$, since inertia force is maximum at the top dead centre position of the piston.

Fig. 4 Different forces are shown on connecting rod [6]

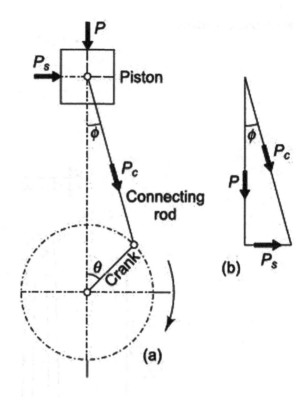

$$= m\omega^2 r \left((r\cos\theta) + \left(r\frac{\cos\theta}{l} \right) \right) = 5527 \text{ N}$$

3. Forces due to friction between piston and wall of cylinder:
$$= h\pi di(Pw)\mu$$
$$= 600 \times 3.14 \times 85 \times 3 \times 0.135 \times 0.045 = 2920.03 \text{ N}$$

Thus [7], resultant maximum force on connecting rod at critical position of dead centre is (Figs. 5, 6, 7 and 8),

$$\begin{aligned} F &= Fg + Fi - Ff \\ &= 17023.5 + 5527 - 2920.03 \\ &= 19630.47 \text{ N} \end{aligned}$$

Design and FEM Analysis of Connecting Rod of Different Materials 499

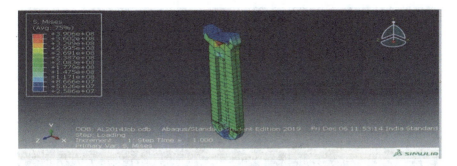

Fig. 5 Von Mises stresses (Pa)

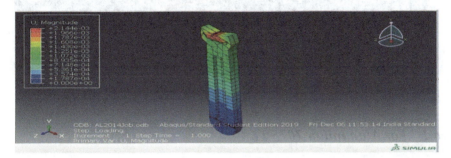

Fig. 6 Deformation (m)

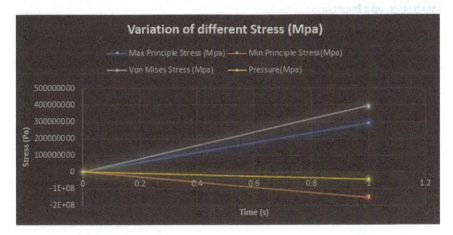

Fig. 7 Variation of stresses (in Pa) with time (s)

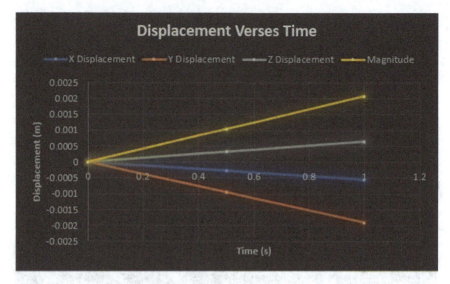

Fig. 8 Variation of displacement (in m) with time (s)

6.1 For Aluminium 7075 T6

Type of element used in analysis: C3D8R (an eight-node linear brick, reduced integration and hourglass).

1. Number of elements: 382
2. Number of nodes: 699
3. Material properties [8] used in the analysis:

    (a) Density: 2810 kg/m^3
    (b) Young's modulus: 71.7 Gpa
    (c) Poisson ratio: 0.33.

4. Boundary conditions:

    (a) Bottom half of the big end of connecting rod is fixed.
    (b) Load is applied on the bottom half of the small end in downward direction (−Y direction) (Figs. 9, 10, 11 and 12).

6.2 For Carbon Steel 40C8

Type of element used in analysis: C3D8R (an eight-node linear brick, reduced integration and hourglass).

5. Number of elements: 382
6. Number of nodes: 699

Design and FEM Analysis of Connecting Rod of Different Materials 501

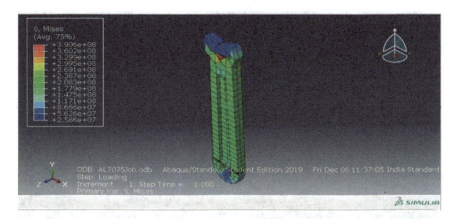

Fig. 9 Von Mises stress (Pa)

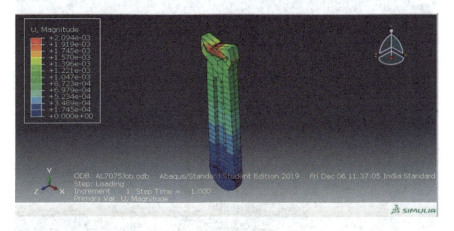

Fig. 10 Deformation (in m)

7. Material properties [9] used in the analysis:

    (a) Density: 7850 kg/m^3
    (b) Young's modulus: 210 Gpa
    (c) Poisson ratio: 0.3.

8. Boundary conditions:

    (a) Bottom half of the big end of connecting rod is fixed.
    (b) Load is applied on the bottom half of the small end in downward direction (−Y direction) (Figs. 13, 14, 15 and 16).

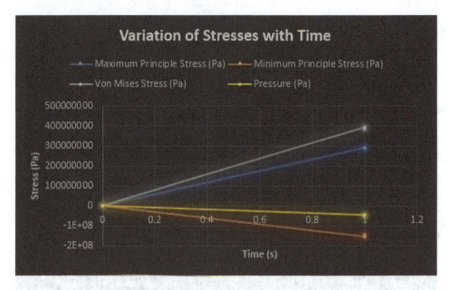

Fig. 11 Variation of stress at maximum stress node at small end (Pa) with time (s)

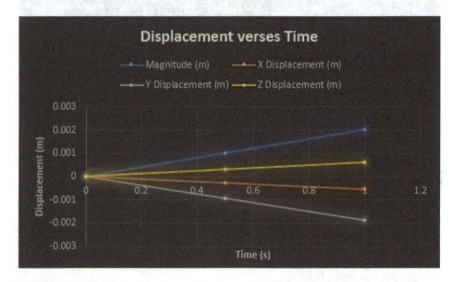

Fig. 12 Variation of displacement at small end node (m) with time (s)

7 Results and Discussion

Figure 17 shows that von Mises stress is similar for all materials since it is calculated using:

Fig. 13 Von Mises stress (Pa)

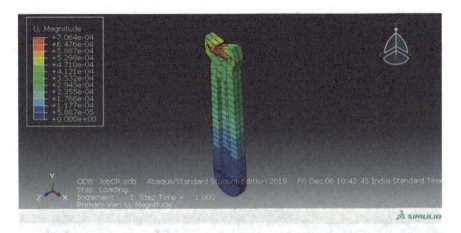

Fig. 14 Deformation (m)

$$\frac{(\sigma_1 - \sigma_2)^2}{2} + \frac{(\sigma_2 - \sigma_3)^2}{2} + \frac{(\sigma_3 - \sigma_1)^2}{2} \leq (\sigma)^2$$

where σ_1, σ_2 and σ_3 are principal stresses along with three different planes and are same for different materials mentioned in the study (Table 2).

Figures 18 and 19 show that maximum and minimum principal stresses are almost same for different materials mentioned in the study as it is calculated as

$$\left(\frac{\sigma x + \sigma y}{2}\right) \pm \left(\left(\frac{\sigma x - \sigma y}{2}\right)^2 + (\tau xy)\right)^2$$

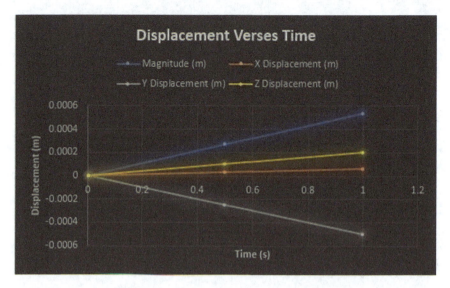

Fig. 15 Variation of displacement of small end node (m) with time (s)

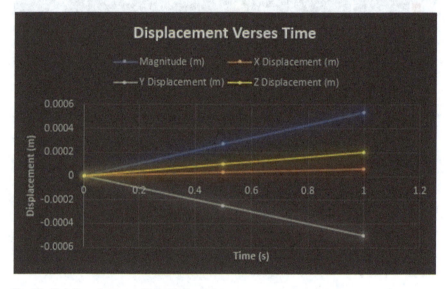

Fig. 16 Variation of stresses (Pa) with time (s)

where stresses in x, y and z axes are same.

Stress values are calculated as load applied to the effective area which is similar for the connecting rod for different materials; hence, stress value is similar for aluminium and steel (Fig. 20).

Since induced maximum stress is same and $\Delta l \propto \frac{1}{E}$.

Design and FEM Analysis of Connecting Rod of Different Materials

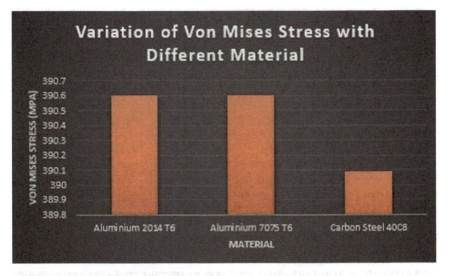

Fig. 17 Variation of von Mises stress with different materials

Table 2 Results obtained through simulations

Material	Von Mises stress (Mpa)	Maximum principal stress (Mpa)	Minimum principal stress (Mpa)	Displacement (in mm)
Aluminium 2014 T6	390.6	300.2	−2.341	2.1
Aluminium 7075 T6	390.6	300.2	−2.341	2.09
Carbon Steel (40C8)	390.09	300.2	−2.132	0.706

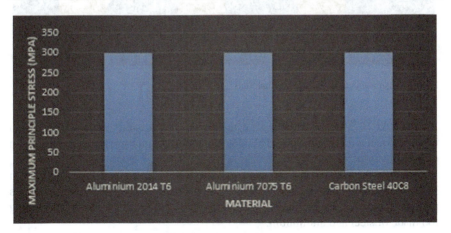

Fig. 18 Variation of maximum principal stress with different materials

Fig. 19 Variation of minimum principal stress with different materials

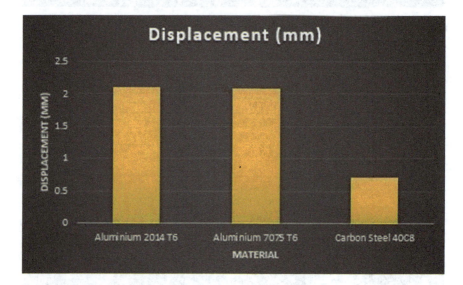

Fig. 20 Displacement with different materials

(since $\Delta l = \frac{PL}{AE}$ in axial loading, i.e. tension or compression), therefore deflection in case of steel is minimum since its Young's modulus is much greater than aluminium (Table 3).

Figure 21 shows that connecting rod made of steel is heavier than that of aluminium since the volume calculated from software is same for the design and density of steel is roughly three times that of aluminium; hence, there is a huge difference in connecting rod made of steel and aluminium.

Design and FEM Analysis of Connecting Rod of Different Materials 507

Table 3 Variation of mass with different materials as per simulation

Materials	Volume (m^3)	Mass (kg)
Aluminium 2014 T6	4.25041×10^{-4}	1.19012
Aluminium 7075 T6	4.25041×10^{-4}	1.19437
Carbon steel (40C8)	4.25041×10^{-4}	3.33657

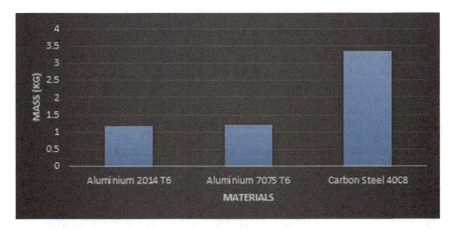

Fig. 21 Variation of mass with different materials

8 Conclusions

From the analysis performed using ABACUS CAE software and comparing the results for different materials used in manufacturing of connecting rod, following conclusions were made:

1. Values obtained for von Mises stress are same for all materials, i.e. about 390.6 MPa.
2. Value of deflection is maximum for AL-2014, i.e. 2.1 mm and least for 40C8 carbon steel, i.e. 0.706 mm.
3. The weight of carbon steel is roughly three times that of aluminium.
4. The value of maximum principal and minimum principal stresses, i.e. 300.2 MPa, −2.34 MPa, respectively, is similar for different materials discussed above.
5. Also steel can sustain greater no. of stress cycles when compared to aluminium when subjected to compression.
6. Thus, above results lead to a conclusion that steel is chosen material for connecting link as it is having greater resistance to load deflection and higher strength though it adds weight to it.

Acknowledgements We would like to express our gratitude to our supervisor Sh. Sanjay Kumar, Assistant Professor, Delhi Technological University, Delhi, for providing their valuable feedback, guidance, comments and suggestions throughout the course of the project.

References

1. Vegi LK, Gopal V (2013) Design analysis of connecting rod using forged steel
2. https://en.wikipedia.org/wiki/Connecting_rod
3. Haider AA, Kumar A, Chowdhury A, Khan M, Suresh P (2018) Design and structural analysis of connecting rod
4. Singh P, Pramanik D, Singh RV (2015) Fatigue and structural analysis of connecting rod's material due to (C.I) Using FEA
5. https://asm.matweb.com/search/SpecificMaterial.asp?bassnum=MA2014T6
6. Bhandari VB (2010) Design of machine elements, 3rd edn. McGraw-Hill. ISBN 13:978-0-07068179-8
7. Khurmi RS, Gupta JK (2005) Machine design, 14th edn. S. Chand & Co. Ltd. ISBN - 9788121925242
8. https://asm.matweb.com/search/SpecificMaterial.asp?bassnum=MA7075T6
9. Bhandari VB (2013) Design data handbook. McGraw-Hill. ISBN-13:978-93-5134-284-7

Numerical Study on Heat Affected Zone and Material Removal Rate of Shape Memory Alloy in Wire Electric Discharge Machining

Deepak Kumar Gupta, Avanish Kumar Dubey, and Alok Kumar Mishra

Abstract This work provides a numerical analysis of the heat affected zone and material removal rate for shape memory alloys (SMAs) through wire electric discharge machining (WEDM). A finite element method computer program was created using MATLAB. SMAs are widely used in many areas like aerospace, biomedical and automobile due to their unique properties, such as superelasticity/pseudoelasticity, shape memory effect, high specific resistance, high abrasion resistance and many other enriched mechanical and physical properties. SMAs machining is an important concept in the field of SMAs applications. However, conventional shape memory alloy machining is extremely difficult. To overcome difficulties during conventional machining, WEDM can be used successfully to machine SMAs. In WEDM, about 70% of the cost of machining is associated with the wire used in machining. A shape memory alloy nickel-titanium alloy has been taken as work material for numerical simulation. The responses heat affected zone and material removal rate curve are plotted with respect to input process parameters pulse on time and discharge current.

Keywords Shape memory alloy · FEM · HAZ · MRR · MATLAB

1 Introduction

WEDM has emerging enormously for over 30 years. To self-regulate the geometry of the parts to be machined using the WEDM, Dulebohn launched the optical line follower system [1]. This process popularity has risen quickly after 1975 in order to comprehend the process and its capabilities [2]. Far 1970s, the CNC scheme was came into picture in wire electrical discharge machining, which brought about a

---

D. K. Gupta (✉) · A. K. Dubey · A. K. Mishra
Mechanical Engineering Department, Motilal Nehru National Institute of Technology Allahabad, Prayagraj, India
e-mail: rme1604@mnnit.ac.in

significant machining process development. It's wide capacity enabled it to cover the manufacturing, aviation, automotive sectors and nearly all fields of conductive material machining. Wire electrical discharge machining offers the finest option for the manufacture of conductive high strength and conductive engineering ceramics with complex forms and profiles. WEDM is an electrical discharge method with a continually moving conductive wire. Tool electrode made of copper, brass, molybdenum or 0.05–0.30 mm diameter tungsten that can achieve a very tiny corner radius. The wire is maintained in tension with a tensioning (mechanical) device that reduces the tendency to produce incorrect shape of components. Peak current (I_P), voltage (V), pulse on time (T_{ON}), pulse off time (T_{OFF}), flushing pressure (FP), wire speed (WS), wire feed rate (WF), wire tension (WT), and dielectric fluid etc. are the characteristics of machine, and the characteristics of workpiece and tool electrode materials are thermal conductivity (K), electrical resistivity, specific heat (C_P), melting temperature (T_m), the thickness of the workpiece, and size and composition of detritus in dielectric liquid and these characteristics can be regarded as the primary system physical parameters [1]. SMAs are 'smart materials' due to their special characteristics such as pseudoelasticity, shape memory effects, biocompatibility, fortitudinous and lightweight [3, 4]. In biomedical, industrial and aerospace applications, nickel-titanium SMAs are used prominently. Machining of such alloy by traditional way is terrifically difficult, leading to a degradation in part quality, and consumption of energy and resources. It needs that novel techniques to be explored and developed to machine this material. WEDM, which is an advanced technique, has been tried to machine nickel-titanium SMAs, and better outcomes are recorded under most part performance and improved workability [4–7]. Sivaprakasam et al. [8] used the biologically inspired ANN and quantitative RSM to model and analyze the system input features of the WEDM method on Al2024 T351 material. They found that the importance of all three input machining parameters along with their interaction effects was also discovered to be crucial in obtaining the necessary surface roughness (SR) and extraction MRR. Prasad et al. [9] made a mathematical model between surface roughness of the workpiece and WEDM cutting parameters that have been created using multiple regression analysis methods. Predicted model values with on hidden layer were found to be very close to the experimental results in the ANN. Manjaiah et al. [10, 11] conducted a detailed study of various grades of NiTi material on WEDM. Increasing T_{ON} increases MRR and also, increase in SR with increase in pulse off time were found. The reason behind this is high discharge energy and spark intensity at longer T_{ON} and appropriate flushing pressure in the machining area because of longer pulse off time. Srivastava et al. [12] used SiC strengthened composite of Al2024 to assess the impacts of three distinct levels of each parameters I_p, T_{ON} and percentage reinforcement of surface quality and MRR. In EDM with the rise in I_P, the speed of the MRR increases, as well as on rising T_{ON}, MRR also increases. Nayak et al. [13] designed and undertaken an experimental inquiry to optimize multiple machining parameters during taper machining of all WEDM Inconel 718 deep cryo-treated. There is a shortage of work on NiTi's WEDM, and further research is needed. This void is covered by the present work, and the purpose of the study includes a numerical analysis of WEDM of shape memory alloy of heat affected zone and material removal rate. A computer program for the finite element method has been developed using MATLAB. For numerical simulation, NiTi shape memory alloy has been taken as working material. The HAZ and the MRR curve are plotted with respect to the time and discharge current as an input parameter.

2 Methodology

2.1 FEM Approach of Heat Affected Zone (HAZ)

To evaluate the distribution of temperature within the computational domain range, the Galerkin method of finite element was applied. For applying Galerkin's technique, the following equations were used.

$$[K]e = \int_{De} [B]^{eT}[B]^e dD_e \tag{1}$$

$$[K]^b = \int_{Bh} h\{N\}^b\{N\}^{bT} dB_h \tag{2}$$

$$[C]^e = \int_{De} \rho C p\{N\}^e\{N\}^{eT} dDe \tag{3}$$

$$\{fc\}^b = \int_{B_h} T_0 h\{N\}^b dB_h \tag{4}$$

$$\{f_q\}^b = \int_{B_q} \{N\}b\, q_w dB_q \tag{5}$$

Here, $[B]^e$ is the nodal interpolation function derivatives matrix for the typical area element. B_h is the convective borders, and B_q is the input heat flux limit. The method of Gauss quadrature is used to assess basic matrices and vectors. Assembled equation in the domain of workpiece is given by:

$$[GC]_{nnm \times nnm} \times \{T\}_{nnm \times 1} + [GK]_{nnm \times nnm} \times \{T\}_{nnm \times 1} = \{GF\}_{nnm \times nnm} \tag{6}$$

The objective of FEM model is to identify the temperature variation at different points on the workpiece.

Governing equation:

$$-\frac{d}{dx}\left(kA\frac{dT}{dx}\right) + hP(T - T_\infty) = q_w + AQ_g \tag{7}$$

Boundary conditions

$$T = T_0 \quad \text{at } x = 0 \text{ on } B_T$$
$$-KA\frac{dT}{dx} = q_w \quad x < R_p$$
$$-KA\frac{dT}{dx} = hA(T - T_\infty) \text{ at } x > R_p \tag{8}$$

where T_0 is the initial temperature or melting point of a workpiece, R_p is the spark radius, k is the thermal conductivity of a material, B_T is the boundary at starting where heat is supplied, and q_w is heat flux which is given by

$$q_w = \frac{13.4878 \times 10^5 \, f_c U I^{0.14} \exp(4.5(t/T_{\text{on}}))^{0.88}}{T_{\text{on}}^{0.44}} \tag{9}$$

T_{ON} Pulse on time
t Duration of machining
U Voltage supplied
I Discharge current (A).

2.2 FEM Approach for MRR

Calculation of cavity quantity at the end of each discharge is very crucial. Many scientists have made some efforts, and the suggested model of Yeo [14] predicts the crater cavity (CC) shape to be hemispheric. Joshi et al. [15] predicted crater cavity shape as shallow bowl-shaped and calculated CC volume by separating into cylindrical disks. The methods of image processing are used to calculate the quantity of crater cavity [16]. It has been shown that the quantity of the crater acquired using the technique suggested is equal to the quantity by assuming for the circular crater of parabolic geometry [15]. The theoretical amount of craters is calculated by the following equation:

$$V_{\text{th}} = (2/3)\pi(R_c)^3 \tag{10}$$

where R_c is crater radius which is pretended as equal to the spark radius. Experimentally measuring, the R_p is very difficult. Researchers have suggested various methods in the literature. A semi-empirical equation suggested that provides the R_p as a function of I and T_{ON} is more realistic to calculate the radius in the electric discharge machining method compared to other methods [16].

Where R_p is the spark radius.

$$R_p = (2.04 \times 10^{-3}) I^{0.43} T_{\text{on}}^{0.44} \tag{11}$$

MRR is evaluated by using the following formula:

$$\text{MRR} \, (\text{mm}^3/\text{min}) = (60 \times V_{\text{th}})/(T_{\text{on}} + T_{\text{off}}) \tag{12}$$

3 Computational Analysis

A workpiece of NiTi alloy (Table 1) of rectangular cuboidal shape is taken for analysis having specifications:

Length (L) = 150 mm.
Cross-sectional area (A) = 100 mm².

Figure 1 shows the WEDM process's thermophysical model of the workpiece three-dimensional axis symmetry and Table 2 shows process parameters. In this model, the heat flux is implemented with its boundary condition in terms of Gaussian heat distribution. In this method, the spark discharge is enforced only to the upper surface B_1, and due to this spark, energy transfer to the workpiece by the distribution of Gaussian heat flow (Eq. 7) to the R_p. The boundary regions B_2 and B_3 are sufficiently far from the R_p, so heat is not transmitted to that region. The B_5 boundary is an axis of surface symmetry, so there is no gain or loss of heat in that area. The heat flux at the boundary B_5 is presumed to be zero.

Assumptions for MRR

Table 1 Properties of nickel-titanium SMAs

Properties	Value
Mass density	6.45 G/cc
UTS	754–960 MPa
Young modulus	75.0 GPa
Modulus of rigidity	28.8 GPa
K	0.18 W/cm k
T_m	1240–1310 °C

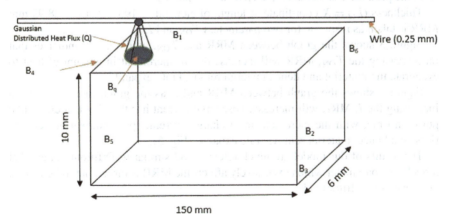

Fig. 1 Geometrical model for WEDM

Table 2 Parameters for MRR

T_{ON} (μs)	T_{off} (μs)	I (A)
5.6	1	2.34
7.5	1.3	2.85
13	2.4	3.67
18	2.4	5.3
24	2.4	8.5
32	2.4	10
42	3.2	12.8
56	3.2	20
100	4.2	25

1. Spark radius depends on the I and T_{ON}.
2. The crater geometry is taken as hemispherical.
3. The R_c is same as R_p for simplicity.
4. The tool electrode moves in a straight line.
5. The efficiency of flushing is taken as 100%.
6. The MRR is taken purely by melting and vaporization.

4 Results and Discussion

The dependency of HAZ on process parameters can be shown with the following plot (Fig. 2).

The HAZ thickness will be up to the recrystallization. And as mentioned above, the recrystallization temperature (T_r) of NiTi SMAs is approximately 430 °C. So from the plot, we can calculate the HAZ layer thickness.

Thickness (L_t) = X coordinate × length of element = 193 × 0.15 = 28.95 mm
MRR is taken as response for two parameters T_{ON} and I.

Figure 3 shows the graph between MRR and T_{ON}, as the graph indicates that on increasing the T_{ON}, MRR will increase due to increment in amount of heat to evaporate the material and hence material removal rate expand.

Figure 4 shows the graph between MRR and I, as the graph indicates that on increasing the I, MRR will increase. Discharge current has the identical effect like pulse on time, with the increment in discharge current, spark and surface energy rises, and hence, material removal rate expands (Fig. 5).

The results of our model are much relevant and similar to DiBitonto et al. [17] model. The parameter which excessively affects the MRR is turned out to be T_{ON} as also given by DiBitonto [17].

Numerical Study on Heat Affected Zone and Material Removal Rate ...

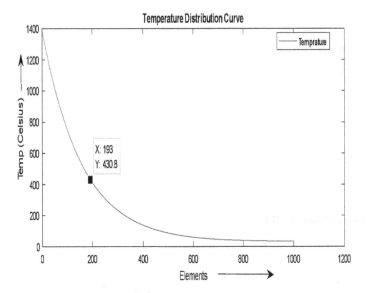

Fig. 2 Temperature distribution at elements

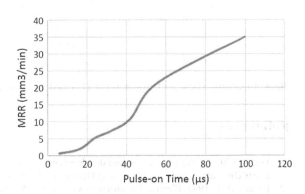

Fig. 3 Effect of the T_{ON} on MRR

5 Conclusion

The effects of WEDM parameters on HAZ and MRR have been carried out. It can be concluded that our model is closer to the experimental value and offers very satisfactory results. The T_{ON} is the most important parameter that affects MRR. At high pulse on time, high energy is discharged resulting in a high MRR. The trend of the discharge current was similar to that of the T_{ON}, as the pulse must be maintained in time of the average level of the discharge current for better performance.

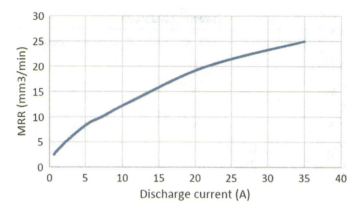

Fig. 4 Effect of the I on MRR

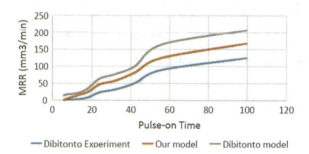

Fig. 5 Comparison between DiBitonto et al. [17] model and our model

References

1. Jameson EC (2001) Description and development of electrical discharge machining (EDM). Electr Discharge Mach Soc Manuf Eng Dearbern, Michigan 16
2. Benedict GF (1987) Electrical discharge machining (EDM). In: Non-traditional manufacturing processes. Marcel Dekker, Inc, New York & Basel, pp 231–232
3. Miller T, Gupta K, Laubscher RF (2018) An experimental study on MQL assisted high speed machining of NiTi shape memory alloy. In: Proceedings of 16th international conference on manufacturing research, Skovde (Sweden), Advances in Manufacturing Technology XXXII, IOS Press, pp 80–85
4. Manjaiah M, Laubscher RF, Narendranath S, Basavarajappa S, Gaitonde VN (2016) Evaluation of wire electro discharge machining Characteristics of Ti50Ni50-xCux shape memory alloys. J Mater Res 31:1801–1808
5. Sharma N, Raj T, Jangra KK (2017) Parameter optimization and experimental study on wire electrical discharge machining of porous $Ni_{40}Ti_{60}$ alloy. Proc I Mech E Part B: J Eng Manuf 231:956–970
6. Sharma N, Gupta K, Paulo Davim J (2019) On Wire spark erosion machining induced surface integrity of $Ni_{55.8}Ti$ shape memory alloys accepted. Arch Civil Mech Eng 19(3):680–693
7. Raymond M., Sharma N., Gupta K., Davim JP.: Modeling and optimization of Wire-EDM parameters for machining of Ni55.8Ti shape memory alloy using hybrid approach of Taguchi and NSGA-II. Int J Adv Manuf Technol. https://doi.org/10.1007/s00170-019-03287-z

8. Sivaprakasam P, Hariharan P, Gowri S (2014) Modelling and analysis of micro-WEDM process of titanium alloy (Ti–6Al–4V) using response surface approach. Int J Eng Sci Technol 17:1–9
9. Prasad VK, Rajyalakshmi G, Venkata Ramaiah P (2013) Simulation and modeling of performance characteristic of Wire Cut EDM on Inconel825 using multiple Regression and ANN. In: International conference on mathematical computer engineering-ICMCE 147
10. Manjaiah M, Narendranath S, Basavarajappa S, Gaitonde VN (2015) Effect of electrode material in wire electro discharge machining characteristics of $Ti_{50}Ni_{50}$-xCux shape memory alloy. Precis Eng 41:68–77
11. Jani JM, Leary M, Subic A, Gibson MA (2014) A review of shape memory alloy research, applications and opportunities. Mater Des 56:1078–1113
12. Srivastava A, Dixit AR, Tiwari S (2014) Experimental investigation of wire EDM process parameters on aluminium metalmatrix composite Al2024/SiC. Int J Adv Res Innov 2:511–515
13. Nayak BB, Mahapatra SS (2016) Optimization of WEDM process parameters using deep cryo-treated Inconel 718 as work material. Int J Eng Sci Technol Int J 19:161–170
14. Yeo SH, Kurnia W, Tan PC (2008) Critical assessment and numerical comparision of electrothermal models in EDM. J Mater Process Technol 203:241–251
15. Joshi SN, Pande S (2010) Thermo-physical modeling of die-sinking EDM process. J Manuf Process 12:45–56
16. Marafona J, Chousal JAG (2006) A finite element model of EDM based on the Joule effect. Int J Mach Tools Manuf 46:595–602
17. DiBitonto DD, Eubank PT, Patel MR, Barrufet MA (1989) Theoretical models of the electrical discharge machining process. I. A simple cathode erosion model. J Appl Phys 66:4095–4103

A Hybrid Multi-criteria Decision-Making Approach for Selection of Sustainable Dielectric Fluid for Electric Discharge Machining Process

Md Nadeem Alam, Zahid A. Khan, and Arshad Noor Siddiquee

Abstract Electric discharge machining (EDM) process is a highly precise and fast-growing non-conventional machining process that is employed for the fabrication of high hardness materials including metal matrix composites (MMCs), alloy of different metals, and ceramic materials. The selection of adequate coolant and lubricant during the EDM process is an important issue, which influences the EDM efficiency by reducing the high heat generated during machining. This paper presents the application of a hybrid multi-criteria decision-making (MCDM) technique for the selection of suitable sustainable dielectric fluid for EDM process. The criteria for selection of best dielectric fluid are considered as the properties of a dielectric fluid required for better machining during EDM process. Standard deviation method is used to calculate the weights of the criteria. Further, six sustainable dielectric fluid used in EDM process are ranked using PIV method. It has been found in the study that the most important property of the dielectric fluid required for better machining in EDM process is flash point. Further, it has also been found that among the considered sustainable dielectric fluids, sunflower oil is found to be the best dielectric fluid for better performance during EDM process.

Keywords EDM process · Sustainable dielectric fluid · Standard deviation method · Proximity index value method

1 Introduction

EDM process is an advanced thermo-electric process, most widely employed for machining of complex and hardened materials for different biomedical and industrial applications. Removal of material mainly accomplishes due to the thermal energy of the sparks that lead to generating extremely high temperatures between 10,000 and 12,000 °C [1, 2]. The opening between tool and workpiece is termed as a spark gap that provides the minimum distance to maintain the maximum electric field. A

M. N. Alam (✉) · Z. A. Khan · A. N. Siddiquee
Department of Mechanical Engineering, Jamia Millia Islamia, New Delhi 110025, India
e-mail: mdnadeemalam7@gmail.com

© The Author(s), under exclusive license to Springer Nature Singapore Pte Ltd. 2021
R. M. Singari et al. (eds.), *Advances in Manufacturing and Industrial Engineering*,
Lecture Notes in Mechanical Engineering,
https://doi.org/10.1007/978-981-15-8542-5_45

constant spark gap is maintained by the servo control unit, which is detected through the average voltage across this gap [3, 4]. Dielectric fluid (DF) acts as a coolant and flushes rubbish and detritus from the machining region and reduces overheating during the EDM process. Therefore, applying an adequate DF during the machining process is an important issue not only for machining performance but also for the health and environmental anxieties. As the EDM process has several special characteristics for machining of hard to machine materials, it also witnesses various impediments such as sustainability. That includes consumption of high specific energy, emission of hazardous gases, operator health, production of toxic waste, and sludge as foremost issues [5]. The selection of suitable DF during EDM process is indeed a complex problem, as it influenced by several properties of the fluid. Researchers have successfully employed several MCDM methods and other mathematical techniques to analyze different MCDM problems, for example, Pattnaik et al. (2015) employed a fuzzy TOPSIS multi-optimization technique to optimize several cutting parameters during machining of stainless steel [6]. The objective of Jagadish and Ray (2016) was to examine the MCDM approach, for the green EDM process by employing an amalgamation of GRA associated with principal component analysis (PCA) method [7]. A non-dominated sorting genetic algorithm-II (NSGA-II) was employed by Mandal et al. (2007) to explore EDM process [8].

Some of the abovementioned multi-optimization methods comprise quite complex computational procedures. Recently, Mufazzal and Muzakkir developed a novel MCDM method, based on PIV technique. It has been verified that this technique minimizes the rank reversal problem and enhances the robustness and reliability of the results during decision making [9]. To solve an MCDM problem, criteria weights are essential to calculate, and for this purpose, several techniques are available such as standard deviation (SD), analytic hierarchy process (AHP), entropy, and principal component analysis (PCA). However, for the determination of criteria weights standard deviation (SD) method seems to be the simplest one.

2 Methodology

The methodology used in this work for the selection and ranking of sustainable DF has been shown in Fig. 1.

This paper made an effort to explore the flexibility, simplicity, utility, and efficacy of the SD-based PIV method. The objective is to select the best suitable sustainable DF from six considered sustainable fluids for EDM process. The required properties of a DF, i.e., density (ρ), viscosity (μ), thermal conductivity (k), specific heat (C), flash point (FP), breakdown voltage (BV), and dielectric constant (DC) for better machining during EDM process, are used as the decision criteria. On the basis of these properties, six sustainable dielectric fluids, viz. waste vegetable oil (WVO), bio-dielectric fluid (BDF), sunflower oil (SF), canola oil (CO), jatropha oil (JO), and waste palm oil blended with kerosene (WPO), are ranked.

A Hybrid Multi-Criteria Decision-Making Approach ...

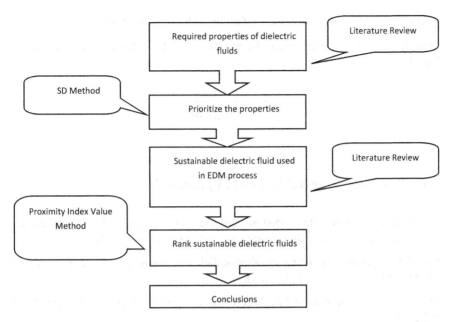

Fig. 1 Methodology involved in the selection of sustainable dielectric fluid for EDM process

2.1 Standard Deviation (SD) Method

SD method is used to determine criteria weights. The steps involved in SD method are as follows:

Step 1: Standardization: The different values of the criteria are defined in different ranges. Therefore, they need to be converted into a comparable range. In SD method, Eq. (1) is used to convert the different range values into a comparable range.

$$X'_{lm} = \frac{X_{lm} - \min_{1<m<j} X_{lm}}{\max_{1<m<j} X_{ij} - \min_{1<m<j} X_{lm}} \quad (1)$$

where min X_{lm} and max X_{lm} represent the minimum and maximum values of the criterion (m), respectively. $l = 1, 2, \ldots i$ and $m = 1, 2, \ldots, j$.

Step 2: Determine standard deviation (SD): The standard deviation of each alternative is calculated using Eq. (2)

$$SD'_m = \sqrt{\frac{1}{j} \sum_{m=1}^{j} (X' - \overline{X'}_m)^2} \quad (2)$$

where $\overline{X'}_m$ refers to the mean value of all the mth normalized criterion.

Step 3: Calculate weights (w_m): the weight of all the criteria is calculated using Eq. (3).

$$w_m = \frac{SD'_m}{\sum_{m=1}^{j} SD'_m} \tag{3}$$

2.2 Proximity Indexed Value (PIV) Method

PIV method is a newly developed method used for ranking in MCDM techniques. The simple stepwise procedure involved in the PIV method is as follows:

Step 1: Categories the existing sustainable dielectric fluids F_l ($l = 1, 2, 3, \ldots, i$) and DF properties P_m ($m = 1, 2, 3, \ldots, j$).

Step 2: Arrange F_l in rows and P_m in columns resulting in a decision matrix as shown in Eq. (4)

$$Z = [Z_{lm}]_{i \times j} = \begin{bmatrix} Z_{11} & Z_{12} & \ldots & Z_{1j} & \ldots & Z_{1m} \\ Z_{21} & Z_{22} & \ldots & \ldots & \ldots & Z_{2m} \\ \ldots & \ldots & \ldots & \ldots & \ldots & \ldots \\ Z_{i1} & \ldots & \ldots & Z_{ij} & \ldots & Z_{im} \\ \ldots & \ldots & \ldots & \ldots & \ldots & \ldots \\ Z_{l1} & \ldots & \ldots & Z_{mj} & \ldots & Z_{lm} \end{bmatrix} \tag{4}$$

where Z_{lm} refers to the value of mth sustainable DF property for lth DF.

Step 3: Normalized Z using Eq. (5)

$$N_{lm} = \frac{Z_l}{\sqrt{\sum_{l=1}^{i} Z_{lm}^2}} \tag{5}$$

Step 4: Computation of the weighted normalized decision matrix (WNDM) using Eq. (6)

$$V_{lm} = w_m \times N_{lm} \tag{6}$$

where w_j represents the weight of the jth criterion.

Step 5: Evaluation of the weighted proximity index (WPI), u_i by employing Eq. (7)

$$u_i = \begin{cases} V_{\max} - V_i; & \text{for beneficial attributes} \\ V_i - V_{\min}; & \text{for cost attributes} \end{cases} \quad (7)$$

Step 6: Calculation of the overall proximity value, d_i using Eq. (8)

$$d_l = \sum_{l=1}^{i} u_l \quad (8)$$

Step 7: Alternatives are ranked on the basis of d_i values. Least value of d_i for an alternative signifies a minimum deviation from the finest value and ranked first. Similarly, ranking order is followed by alternatives with increasing value of d_i.

3 Illustrative Examples

In this section, some examples of EDM process are presented to select suitable sustainable DF during machining. Many researchers tried to stabilize and optimize performance characteristics such as surface quality, MRR, and TWR, but very few works have been carried out toward the sustainability issue. The following are some examples of sustainable DF used during the machining process: Srinivas et al. (2017) considered the conventional EDM process as a hazardous process due to the emission of toxic gases and debris mixed DF. It has been explored that sustainable DF enhances EDM performance characteristics and limits the environmental impact [10]. Bhatia et al. (2019) examined the impact of various dielectric fluids on different EDM responses including MRR, Ra, TWR, and surface hardness (SH) during machining of Al_2O_3 and Si_3N_4 ceramic [11]. According to Hegab et al. (2019), flood cooling provides a fruitful solution to moderate the high heat generated during the machining [12]. Chakraborty et al. (2014) concluded that low viscosity DF significantly influences the efficiency and machining cycle time [13]. Janak et al. (2016) analyzed the working feasibility of waste vegetable oil (WVO) as a sustainable DF for EDM process to minimize the adverse environmental effects and poor operational safety [14]. Singaravel et al. (2019) compared the influence of several vegetable oils on Ra during EDM of Ti-6Al-4V [15]. According to Valaki et al. (2016), poor biodegradability and produced wastes from hydrocarbons of the conventional DF lead to produced toxic gases and other harmful wastes. Further, concluded that proposed palm oil-based bio-DF minimizes harmful effects. However, it increases by 38% MRR and almost the same Ra and surface hardness [16]. Moreover, Kumar and Krishnaiah (2016) also presented some fruitful research to the selection of suitable sustainable DF [17].

4 Result and Discussion

This work helps in the selection of sustainable DF for better performance during EDM process. The properties of DF that are required for better machining during EDM process are displayed in Table 1.

In order to select a suitable sustainable DF, six dielectric fluids that are found to be used by different researchers in EDM process are considered. The properties of the selected sustainable dielectric fluids are used to evaluate and rank them. The values of the properties of the considered dielectric fluids are displayed in Table 2.

In order to determine the weights of all the seven criteria, SD method is employed. The criteria values as shown in Table 2 are standardized using Eq. (1). The standard deviation of the standardized values is calculated using Eq. (2), and finally by using Eq. (3), weights of the criteria are obtained. The weights so obtained are as follows $W_\rho = 0.1239$, $W_\mu = 0.1380$, $W_k = 0.1354$, $W_C = 0.1455$, $W_{FP} = 0.1642$, $W_{BV} = 0.1415$, and $W_{DC} = 0.1515$, respectively. It has been found from the weights so obtained that the most important property among the considered properties of DF is FP followed by DC followed by C followed by BV followed by μ followed by k followed by ρ.

Table 1 Required properties of the dielectric fluids (DF)

Symbol	Required property	Benefit obtained in machining
ρ	High density	Gives better flushing effect
μ	Low viscosity	Gives better cooling capacity
k	High thermal conductivity	Gives enhanced better cooling of electrode and work material
C	High specific heat	Gives improved cooling of electrode and work material
FP	High flash point	Gives good fire prevention
BV	High breakdown voltage	Gives minimum arcing
DC	Low dielectric constant	Gives minimum energy loss

Valaki et al. 2019 [5], Janak et al. 2016 [14], and Singaravel et al. [15]

Table 2 Properties of considered DF

Dielectric fluid	ρ	μ	k	C	FP	BV	DC
WVO	0.8932	9.5500	0.2	1.670	171	26.0	2.86
BDF	0.8787	4.7200	0.157	1.900	158	35.8	2.53
SF	0.8790	5.2000	0.152	1.833	330	60.0	3.00
CO	0.9200	5.0650	0.168	1.910	330	50.0	3.20
JO	0.8700	6.5836	0.157	1.900	240	35.8	2.53
WPO	0.8450	6.5600	0.168	1.900	130	22.0	3.40

In order to rank the different identified sustainable cutting fluids, PIV method has been implemented. Table 2 acts as the decision matrix for PIV method. The very first step in PIV method is to normalize the decision matrix; therefore, Eq. (1) has been applied to normalize the decision matrix. The next step of PIV method is to multiply the weights of each criterion with corresponding normalized values of the different DF. So the weights obtained using SD methods are used along with the normalized decision matrix to acquire weighted normalized values, and Table 3 shows the weighted normalized values thus obtained.

Weighted proximity index (u_i) which shows the dispersion from the best alternative is calculated using Eq. (7). Finally, the overall proximity value (d_i) for each sustainable DF is calculated using Eq. (8). Ranking in PIV method is based on d_i values. Hence, the DF whose d_i value is minimum is considered to be closest to the best solution and is ranked first. The ranking of the DF so obtained along with the d_i values is shown in Table 4.

The rank so obtained demonstrated that the SF occupies the first rank followed by CO, JO, BDF, WVO, and WPO, respectively. Further, Table 4 also reveals that among the selected sustainable dielectric fluids, sunflower oil is found the best suitable sustainable DF for EDM process.

Table 3 Weighted normalized decision matrix

DF	ρ	μ	k	C	FP	BV	DC
WVO	0.0513	0.0829	0.0659	0.0535	0.0477	0.0371	0.0602
BDF	0.0504	0.0410	0.0517	0.0609	0.0441	0.0511	0.0533
SF	0.0504	0.0452	0.0501	0.0587	0.0921	0.0857	0.0631
CO	0.0528	0.0440	0.0554	0.0612	0.0921	0.0714	0.0674
JO	0.0499	0.0572	0.0517	0.0609	0.0669	0.0511	0.0533
WPO	0.0485	0.0570	0.0554	0.0609	0.0363	0.0314	0.0716

Table 4 Weighted proximity value and rank

DF	ρ	μ	k	C	FP	BV	DC	d_i	Rank
WVO	0.0015	0.0419	0.0000	0.0077	0.0444	0.0485	0.0069	0.1510	5
BDF	0.0024	0.0000	0.0142	0.0003	0.0480	0.0346	0.0000	0.0994	4
SF	0.0024	0.0042	0.0158	0.0025	0.0000	0.0000	0.0099	0.0347	1
CO	0.0000	0.0030	0.0105	0.0000	0.0000	0.0143	0.0141	0.0419	2
JO	0.0029	0.0162	0.0142	0.0003	0.0251	0.0346	0.0000	0.0932	3
WPO	0.0043	0.0160	0.0105	0.0003	0.0558	0.0543	0.0183	0.1595	6

5 Conclusion

The use of appropriate DF during EDM process plays a significantly important role during machining. However, in the current era, the hazardous effects of dielectric fluid need to be considered while selecting the DF. This work is intended to help in the selection of suitable DF for EDM process. The standard deviation (SD) based PIV method is employed to select a suitable sustainable DF among the available fluids. The conclusion drawn from the results obtained in this work is as follows:

Among the considered properties, it has been found that the most important property of DF is flash point and the least important property is density. However, it was also observed that the variation in the weights of the properties is not large, which shows that each of the considered property has a significant influence on the performance of the EDM process. Among the considered sustainable fluids, sunflower oil is ranked first. This shows that for better performance in EDM process, sunflower oil is the best sustainable DF.

References

1. Kalpakjian S, Steven RS (2013) Manufacturing engineering and technology, 4th edn. Pearson Education, 7th floor, Knowledge Boulevard, A-8(A), Sector-62, Noida (U. P.), India
2. Das MK, Kumar K, Barman TK, Sahoo P (2014) Application of artificial bee colony algorithm for optimization of MRR and surface roughness in EDM of EN31 tool steel. In: 3rd international conference on materials processing and characterisation 2014, pp 741–751. Procedia Materials Science 6
3. Ghosh A, Mallik AK (2002) Manufacturing Science, 1st edn. East-West Press Private Limited, pp 283–287. ISBN 81-85095-85-X
4. Abbas NM, Solomon DG, Bahari MF (2007) A review on current research trends in electric discharge machining (EDM). Int J Mach Tool Manuf 47:1214–1228
5. Valaki JB, Rathod PP, Sidpara AM (2019) Sustainability issues in electric discharge machining. In: Innovations in manufacturing for sustainability, materials forming, machining and tribology. Springer Nature Switzerland AG
6. Pattaik SK, Priyadarshini M, Mahapatra KD, Mishra D, Panda S (2015) Multi objective optimization of EDM process parameters using FUZZY TOPSIS method. In: IEEE sponsored 2nd international conference on innovation in information, embedded and communication systems
7. Jagadish RA (2016) Optimization of process parameters of green electrical discharge machining using principal component analysis (PCA). Int J Adv Manuf Technol 87:1299–1311
8. Mandal D, Pal SK, Saha P (2007) Modeling of electrical discharge machining process using back propagation neural network and multi-objective optimization using non-dominating sorting genetic algorithm-II. J Mater Process Technol 186:154–162
9. Mufazzal S, Muzakkir SM (2018) A new multi-criterion decision making (MCDM) method based on proximity indexed value for minimizing rank reversals. Comput Ind Eng 119:427–438
10. Srinivas VV, Ramannujam R, Rajyalakshmi G (2018) A review of research scope on sustainable and eco-friendly electrical Discharge machining (E-EDM). Mater Today Proc 5:12525–12533
11. Bhatia K, Singla A, Sharma A, Sengar SS, Selokar A (2019) A Review on dielectric fluids and machining of Si_3N_4 and Al_2O_3 composites via EDM. In: Advances in industrial and production engineering, Lecture Notes in Mechanical Engineering. Springer Nature Singapore Pte Ltd.
12. Hegab H, Kishawy HA, Darras B (2019) Sustainable cooling and lubrication strategies in machining processes: a comparative study. Elsevier, Procedia Manuf 33:786–793

13. Chakraborty S, Dey V, Ghosh SK (2014) A review on the use of dielectric fluids and their effects in electrical discharge machining characteristics. Elsevier, Precision Eng
14. Valaki JB, Rathod PP (2016) Assessment of operational feasibility of waste vegetable oil based bio-dielectric fluid for sustainable electric discharge machining (EDM). Int J Adv Manuf Technol 8:1509–1518
15. Singaravel B, Shekar KC, Reddy GG, Prased SD (2019) Experimental investigation of vegetable oil as dielectric fluid in Electric discharge machining of Ti-6Al-4V. Ain Shams Eng J
16. Valaki JB, Rathod PP, Khatri BC, Vaghela JR (2016) Investigations on palm oil based biodielectric fluid for sustainable electric discharge machining. In: International conference on advances in materials and manufacturing
17. Kumar AH, Krishnaiah G (2016) Optimization of process parameters and dielectric fluids on machining EN 31 by using Topsis. Int J Eng Res Appl 6:13–18

Preference Selection Index Approach as MADM Method for Ranking of FMS Flexibility

Vineet Jain, Mohd. Iqbal, and Ashok Kumar Madan

Abstract The idea of this study is to find the impact of FMS flexibilities because it is one of the key factors of the performance analysis of manufacturing system. "Preference selection index" (PSI) as a decision-making technique is used to detect the best flexibility from among flexibilities without deciding the weight of the attributes. PSI is authenticated in this work by differentiating the outcome of this method with the available results of different MADM approaches like AHP, TOPSIS, modified TOPSIS, improved PROMETHEE and VIKOR. The result of PSI approach shows that the topmost flexibility is production flexibility whenever related to the production with the new part configuration in FMS. This investigation research has accomplished that the PSI method is suitable for the selections of alternatives.

Keywords PSI · FMS · Flexibility · MADM

1 Introduction

Manufacturing companies are focusing on flexible manufacturing system (FMS) to improve the competitive advantage, inflexible customer demands, reduce direct labor cost, save indirect labor cost and enhance productivity as increased in customer service and on-time delivery. Stecke [1] defined that "FMS consists innumerable programmable and computerized machine tools connected by an automatic material handling system like robots and automatic guided vehicles (AGVs) and automatic

---

V. Jain (✉) · Mohd. Iqbal
Department of Mechanical Engineering, Mewat Engineering College, District Nuh, Palla, Haryana 122107, India
e-mail: vjdj2004@gmail.com

Mohd. Iqbal
e-mail: mohmadiqbal_86@yahoo.com

Mohd. Iqbal · A. K. Madan
Department of Mechanical Engineering, Delhi Technological University, New Delhi 110042, India
e-mail: ashokmadan79@gmail.com

© The Author(s), under exclusive license to Springer Nature Singapore Pte Ltd. 2021
R. M. Singari et al. (eds.), *Advances in Manufacturing and Industrial Engineering*, Lecture Notes in Mechanical Engineering, https://doi.org/10.1007/978-981-15-8542-5_46

storage and retrieval system (AS/RS) that can process simultaneously medium-sized volumes of the different parts". Rao [2] presented combined MADM methods like TOPSIS and AHP for ranking of FMSs. Raj, Shankar [3] applied AHP methodology for the ranking of manufacturing system. Jain and Raj [4] stated that "flexibility is one of the critical dimensions of enhancing the competitiveness of organizations". Jain and Raj [5] analyzed that "flexibility in manufacturing has been identified as one of the key factors to improve the performance of FMS". Jain and Raj [6] also discussed that flexibility is a significant factor of FMS productivity. Jain and Raj [4]used AHP, TOPSIS and improved PROMETHEE MADM methods for FMS flexibilities by different decision-making method and accomplished that "production flexibility is the topmost flexibility in FMS". VIKOR modified TOPSIS for the flexibility evaluations (Jain and Raj [7, 8]). Jain and Soni [9], Jain and Ajmera [10] discussed the performance factor by fuzzy TISM, AHP, CMBA and ELECTRE methodology.

In this research, fifteen flexibilities and variables from literature are considered as fifteen flexibilities (machine flexibility, routing flexibility, process flexibility, product flexibility, volume flexibility, material handling flexibility, operation flexibility, expansion flexibility, production flexibility, program flexibility, market flexibility, response flexibility, product mix flexibility, size flexibility and range flexibility) and fifteen variables (ability to manufacture a variety of products, capacity to handle new products, flexibility in production, flexible fixturing, combination of operation, automation, use of automated material handling devices, increase machine utilization, use of the reconfigurable machine tool, manufacturing lead time and setup time reduction, speed of response, reduced WIP inventories, reduction in material flow, quality consciousness and reduction in scrap) which effect the flexibility of FMS [4, 7, 8, 11–16].

The main concern of this research is to execute a novel approach as preference selection index (PSI) for ranking of flexibility based on variables which effect the flexibility of FMS. The PSI method suggests the effective alternative among the different alternatives without considering any subjective or relative importance between attributes [17]. In this paper, an overview of preference selection index approach is under in Sect. 2. In Sect. 3, analysis of ranking of flexibilities by preference selection index approach is discussed. Discussion and conclusion are discussed in Sect. 4.

2 PSI Methodology

PSI methodology was proposed by Maniya and Bhatt [18] as a MADM method. In this approach, relative importance between attributes is not necessary. Even, there is no requirement of defining the weights of attributes to solve the problems. In the previous studies, a number of MADM techniques are discussed as "graph theory and matrix approach (GTMA)", "analytic hierarchy process (AHP)", "analytic network process (ANP)", "technique for order preferences by similarity to ideal solution (TOPSIS)", "modified TOPSIS", "improved preference ranking organization method for enrichment evaluation method (PROMETHEE)", "compromise ranking method (VIKOR)", etc. These techniques look bit complex when numbers of variables are more in the problem [18]. While in the PSI method, calculations are very simple and results are found with minimum time as compared to other methods, and no weights of attributes are necessary for the calculations. According to Attri and Grover [19], it may be applied to any number of attributes.

PSI methods are used in different field to found the best choice. The literature has been reviewed from the perspective of this methodology.

Jain [20] analyzed the FMS performance factors by MOORA and PSI. Chauhan and Singh [21] applied preference selection index (PSI) methodology to find the optimal design parameters inside the duct. Singh and Patnaik [22] applied PSI for the ranking of the friction materials. Attri and Dev [23] used for selection of cutting fluids. Almomani and Aladeemy [24] determined the best setup technique based on AHP, TOPSIS and PSI methods. Maniya and Bhatt [25] applied for electrical energy equipment. Khorshidi and Hassani [26] did comparative analysis for selection of materials. Maniya and Bhatt [27] applied for the layout design. Vahdani and Zandieh [28] used for alternative fuel for buses. Maniya and Bhatt [17] solved for the FMS selection. Sawant and Mohite [29] used for automated guided vehicle selection. Joseph and Sridharan [30] applied PSI method in FMS for the ranking of scheduling rules. Maniya and Bhatt [18] used for the materials.

The following are the steps involved in the overview of the PSI approach [18–20]:

Step 1: To define the objective
Firstly, find out all alternatives, i.e., flexibilities, and there selection variables related to the application.

Step 2: To construct the decision matrix (D_{MXN})
After defining the objective, construct the decision matrix, i.e., the package of all information related to each alternative and attributes. In the decision matrix, where M is the "alternatives" which shows row and N is the "attributes" which shows column, which is expressed as the A_i alternative, i.e., A_i ($i = 1, 2, 3, ..., M$) and for attribute B_j ($j = 1, 2, 3, ..., N$). If the data is not quantitative mean qualitative, then convert it into qualitative with the help of fuzzy sets. The decision matrix is shown by Eq. (1).

$$D_{MXN} = \begin{array}{c} \text{Attributes} \\ A_1 \\ A_2 \\ A_3 \\ - \\ - \\ A_M \end{array} \begin{bmatrix} B_1 & B_2 & B_3 & - & - & B_N \\ d_{11} & d_{12} & d_{13} & - & - & d_{1N} \\ d_{21} & d_{22} & d_{23} & - & - & d_{2N} \\ d_{31} & d_{32} & d_{33} & - & - & d_{3N} \\ - & - & - & - & - & - \\ - & - & - & - & - & - \\ d_{M1} & d_{M2} & d_{M3} & - & - & d_{MN} \end{bmatrix} \quad (1)$$

Chen and Hwang [31] indicated "an approach to solve more than ten alternatives and they proposed first converts linguistic terms into fuzzy numbers and then the fuzzy numbers into crisp scores" [4].

Step 3: To normalize the attribute data (N_{ij})
In this decision-making approach, attribute value should be dimensionless. In this part, normalization takes place. The obtained values called as normalized values in terms of binary form, i.e., 0 and 1. In PSI methodology, normalization is done as given below

$$N_{ij} = \frac{d_{ij}}{d_j^{max}}; \quad \text{"(if } i\text{th attribute is beneficial)"} \tag{2}$$

$$N_{ij} = \frac{d_j^{min}}{d_{ij}}; \quad \text{"(if } j\text{th attribute is non-beneficial)"} \tag{3}$$

Step 4: To determine the "mean value of normalized attribute" data (N_{mean})
It is determined as per equation:

$$N_{mean} = \frac{1}{N} \sum_{i=1}^{N} N_{ij}$$

(where N_{mean} is the mean value of normalized attribute data) (4)

Step 5: To reckon the "preference variation value" (Ω_j)
It is reckoned as per equation:

$$\Omega_j = \sum_{i=1}^{N} \left[N_{ij} - N_{mean} \right]^2 \tag{5}$$

Step 6: To evaluate the deviation in "preference value" (Φ_j)
It is evaluated as per equation:

$$\Phi_j = \left[1 - \Omega_j \right] \tag{6}$$

Step 7: To obtain the "overall preference value" (Ψ_j)
It is obtained as per equation:

$$\Psi_j = \frac{\Phi_j}{\sum_{j=1}^{N} \Phi_j} \tag{7}$$

There is one condition to check, i.e., the "overall preference value" should be one and shown in Eq. 8.

$$\sum_{j=1}^{N} \Psi_j = 1 \tag{8}$$

Step 8: To quantify the "preference selection index" (PSI$_i$)
Now, it is quantified as per equation:

$$\text{PSI}_i = \sum_{j=1}^{N} \left(d_{ij} \times \Psi_j \right) \tag{9}$$

Step 9: To rank the alternatives
Each alternative is ranked either "ascending or descending" order according to PSI values. Highest PSI value alternative is ranked one, i.e., best alternative, and rest is so on.

3 Ranking of Flexibility by PSI

In this part, PSI methodology is applied for the ranking of FMS flexibility as given below.

Step 1: As per the objective, rank the flexibilities of FMS, fifteen flexibilities as alternatives and fifteen attributes are taken to evaluate the flexibilities.
Step 2: The values of attribute are in qualitative. So, fuzzy sets are applied to transform the linguistic data into crisp value, and it is shown as a decision matrix in Table 1.
Step 3: The normalization of attribute data is done as per Eq. 2.
Step 4: The "normalized mean value of each attribute" is determined by Eq. 4. It is depicted in Table 2.
Step 5: Each attributes' preference variation value is reckoned by Eq. 5. It is depicted in Table 3.
Step 6: The preference value deviation is evaluated by using Eq. 6. It is depicted in Table 4.
Step 7: The overall preference value is calculated by using Eq. 7. It is depicted in Table 5.
Step 8: By using Eq. 8, quantification of each alternative as the preference selection index (PSI$_i$) is depicted in Table 6.
Step 9: Now, alternatives are sorted as per preference selection index in descending order and shown in Table 6. From Table 6, according to PSI values production flexibility (9) is the top one rank.

4 Discussion and Conclusion

This PSI methodology is easy to understand in comparison with other methods. There is no requirement of weights of attributes because it uses the concept of statistics.

Table 1 Quantitative data for decision matrix

Alternatives (flexibilities)	Attributes														
	1	2	3	4	5	6	7	8	9	10	11	12	13	14	15
1	0.865	0.665	0.665	0.5	0.59	0.5	0.41	0.59	0.665	0.665	0.59	0.335	0.255	0.5	0.41
2	0.41	0.41	0.665	0.5	0.255	0.5	0.59	0.59	0.41	0.41	0.665	0.59	0.5	0.41	0.5
3	0.665	0.5	0.59	0.59	0.5	0.5	0.41	0.59	0.5	0.5	0.59	0.5	0.5	0.41	0.5
4	0.745	0.865	0.665	0.59	0.41	0.5	0.41	0.59	0.59	0.5	0.665	0.41	0.41	0.5	0.41
5	0.41	0.41	0.41	0.5	0.5	0.59	0.59	0.5	0.5	0.665	0.5	0.41	0.41	0.41	0.41
6	0.255	0.255	0.41	0.41	0.5	0.59	0.745	0.41	0.41	0.59	0.59	0.5	0.41	0.335	0.255
7	0.335	0.255	0.41	0.5	0.41	0.5	0.41	0.41	0.59	0.5	0.41	0.335	0.335	0.255	0.255
8	0.41	0.335	0.665	0.5	0.5	0.41	0.5	0.665	0.745	0.745	0.5	0.335	0.255	0.41	0.255
9	0.665	0.59	0.59	0.59	0.5	0.665	0.59	0.665	0.865	0.41	0.41	0.255	0.335	0.5	0.135
10	0.255	0.255	0.335	0.255	0.135	0.5	0.59	0.335	0.41	0.335	0.59	0.255	0.255	0.135	0.135
11	0.5	0.59	0.5	0.335	0.255	0.665	0.135	0.255	0.59	0.255	0.5	0.255	0.135	0.5	0.135
12	0.5	0.59	0.665	0.59	0.335	0.745	0.59	0.41	0.5	0.335	0.5	0.255	0.41	0.665	0.5
13	0.59	0.5	0.665	0.5	0.5	0.59	0.5	0.5	0.5	0.59	0.5	0.59	0.41	0.5	0.5
14	0.665	0.59	0.5	0.5	0.5	0.5	0.59	0.5	0.59	0.665	0.5	0.41	0.335	0.335	0.255
15	0.5	0.5	0.59	0.5	0.41	0.5	0.59	0.41	0.5	0.5	0.59	0.335	0.255	0.255	0.135

Table 2 Mean value of normalized data

N_{mean}	1	2	3	4	5	6	7	8	9	10	11	12	13	14	15
	0.599	0.563	0.835	0.832	0.712	0.739	0.685	0.744	0.645	0.686	0.812	0.652	0.695	0.614	0.639

Table 3 Preference variation value

Ω_j	1	2	3	4	5	6	7	8	9	10	11	12	13	14	15
	0.601	0.553	0.438	0.363	0.623	0.196	0.506	0.466	0.304	0.517	0.197	0.546	0.584	0.544	1.231

Table 4 Deviation in preference value

Φ_j	1	2	3	4	5	6	7	8	9	10	11	12	13	14	15
	0.399	0.447	0.562	0.637	0.377	0.804	0.494	0.534	0.696	0.483	0.803	0.454	0.416	0.456	−0.231

Table 5 Overall preference value

Ψ_j	1	2	3	4	5	6	7	8	9	10	11	12	13	14	15
	0.054	0.061	0.077	0.087	0.051	0.110	0.067	0.073	0.095	0.066	0.110	0.062	0.057	0.062	−0.031

Table 6 Preference selection index

PSI$_i$	1	2	3	4	5	6	7	8	9	10	11	12	13	14	15
	0.563	0.506	0.528	0.571	0.496	0.476	0.424	0.517	0.571	0.363	0.422	0.523	0.534	0.522	0.484
Ranking	3	9	5	2	10	12	13	8	1	15	14	6	4	7	11

Secondly, no extra parameters are required in the calculation. The "computational time" of the PSI method in comparison with other MADM methods is less.

The main aim of this work is to concentrate on the ordering of fifteen FMS flexibility. In this study, ranking of flexibilities is found by a PSI approach, i.e., MADM method.

Jain and Raj [4] determined the ordering of flexibilities in flexible manufacturing system formed by approaches as AHP methodology, TOPSIS approach and improved PROMETHEE method are 9-4-1-12-13-3-14-5-8-2-15-6-11-7-10. As per ordering, top ranking is the production flexibility, i.e., no. 9, and last is program flexibility, i.e., no. 10, in flexible manufacturing system. By PSI method, got the ranking is 9-4-1-13-3-12-14-8-2-5-15-6-7-11-10. Ranking of flexibilities by different MADM is shown in Fig. 1.

To check the inconsistency with other MADM method correlation is found out by Spearman's rank among the PSI approach and the other methods. The correlation coefficient of Spearman's rank among the PSI approach and the other is shown in Fig. 2.

Finally concluded that PSI methodology can be used productively by the researcher or industrial persons for finding in different areas such as "material selection, product and process design, plant facility location, plant facility layout and material handling system selection".

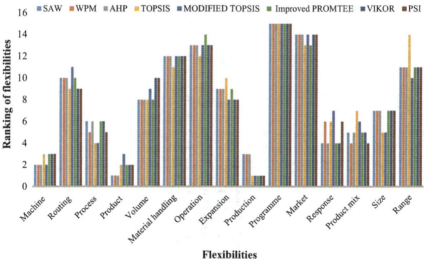

Fig. 1 Ranking of flexibilities

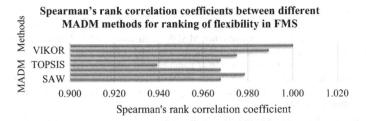

Fig. 2 Spearman's rank correlation coefficients between different MADM methods for ranking of flexibility in FMS

References

1. Stecke KE (1983) Formulation and solution of nonlinear integer production planning problems for flexible manufacturing systems. Manage Sci 29(3):273–288
2. Rao RV (2008) Evaluating flexible manufacturing systems using a combined multiple attribute decision making method. Int J Prod Res 46(7):1975–1989
3. Raj T et al (2008) An AHP approach for the selection of Advanced Manufacturing System: a case study. Int J Manuf Res 3(4):471–498
4. Jain V, Raj T (2013a) Ranking of flexibility in flexible manufacturing system by using a combined multiple attribute decision making method. Global J Flexible Syst Manag 14(3):125–141
5. Jain V, Raj T (2016) Modeling and analysis of FMS performance variables by ISM, SEM and GTMA approach. Int J Prod Econ 171(1):84–96
6. Jain V, Raj T (2014a) Modelling and analysis of FMS productivity variables by ISM, SEM and GTMA approach. Front Mech Eng 9(3):218–232
7. Jain V, Raj T (2014b) Evaluation of flexibility in FMS by VIKOR methodology. Int J Ind Syst Eng 18(4):483–498
8. Jain V, Raj T (2015a) A hybrid approach using ISM and modified TOPSIS for the evaluation of flexibility in FMS. Int J Ind Syst Eng 19(3):389–406
9. Jain V, Soni VK (2019) Modeling and analysis of FMS performance variables by fuzzy TISM. J Model Manag 14(1):2–30
10. Jain V, Ajmera P (2019a) Evaluation of performance factors of FMS by combined decision making methods as AHP, CMBA and ELECTRE methodology. Manag Sci Lett 9(4):519–534
11. Jain V, Raj T (2015b) Modeling and analysis of FMS flexibility factors by TISM and fuzzy MICMAC. Int J Syst Assur Eng Manag 6(3):350–371
12. Jain V, Raj T (2013b) Evaluating the variables affecting flexibility in FMS by exploratory and confirmatory factor analysis. Global J Flexible Syst Manag 14(4):181–193
13. Jain V, Raj T (2013c) Evaluation of flexibility in FMS using SAW and WPM. Decision Sci Lett 2(4):223–230
14. Jain V, Raj T (2015c) Evaluating the intensity of variables affecting flexibility in FMS by graph theory and matrix approach. Int J Ind Syst Eng 19(2):137–154
15. Jain V, Raj T (2018) Identification of performance variables which affect the FMS: a state-of-the-art review. Int J Process Manag Benchmark 8(4):470–489
16. Jain V, Ajmera P (2019b) Application of MADM methods as MOORA and WEDBA for ranking of FMS flexibility. Int J Data Netw Sci 3(2):119–136
17. Maniya K, Bhatt M (2011a) The selection of flexible manufacturing system using preference selection index method. Int J Ind Syst Eng 9(3):330–349
18. Maniya K, Bhatt M (2010) A selection of material using a novel type decision-making method: preference selection index method. Mater Des 31(4):1785–1789

19. Attri R, Grover S (2015) Application of preference selection index method for decision making over the design stage of production system life cycle. J King Saud Univ-Eng Sci 27(2):207–216
20. Jain V (2018) Application of combined MADM methods as MOORA and PSI for ranking of FMS performance factors. Benchmark Int J 25(6):1903–1920
21. Chauhan R et al (2016) Optimization of parameters in solar thermal collector provided with impinging air jets based upon preference selection index method. Renew Energy 99:118–126
22. Singh T et al (2015) Optimization of tribo-performance of brake friction materials: effect of nano filler. Wear 324:10–16
23. Attri R et al (2014) Selection of cutting-fluids using a novel, decision-making method: preference selection index method. Int J Inf Decision Sci 6(4):393–410
24. Almomani MA et al (2013) A proposed approach for setup time reduction through integrating conventional SMED method with multiple criteria decision-making techniques. Comput Ind Eng 66(2):461–469
25. Maniya K, Bhatt M (2013) A selection of optimal electrical energy equipment using integrated multi criteria decision making methodology. Int J Energy Optim Eng (IJEOE) 2(1):101–116
26. Khorshidi R, Hassani A (2013) Comparative analysis between TOPSIS and PSI methods of materials selection to achieve a desirable combination of strength and workability in Al/SiC composite. Mater Des 52:999–1010
27. Maniya K, Bhatt M (2011b) An alternative multiple attribute decision making methodology for solving optimal facility layout design selection problems. Comput Ind Eng 61(3):542–549
28. Vahdani B, Zandieh M, Tavakkoli-Moghaddam R (2011) Two novel FMCDM methods for alternative-fuel buses selection. Appl Math Model 35(3):1396–1412
29. Sawant V, Mohite s, Patil R (2011) A decision-making methodology for automated guided vehicle selection problem using a preference selection index method. In: Technology systems and management. Springer, pp 176–181
30. Joseph O, Sridharan R (2011) Ranking of scheduling rule combinations in a flexible manufacturing system using preference selection index method. Int J Adv Oper Manag 3(2):201–216
31. Chen SJ, Hwang CL (1992) Fuzzy multiple attribute decision making methods. In: Lecture notes in economics and mathematical systems. Springer Berlin, Heidelberg, pp 289–486

Impact of Additive Manufacturing in Value Creation, Methods, Applications and Challenges

Rishabh Teharia, Gulshan Kaur, Md Jamil Akhtar, and Ranganath M. Singari

Abstract Today, additive manufacturing (AM) reveals changes in complete value creation, strategic system and processes. Mass customization with minimization of waste and ability to manufacture complex structure makes this process superior over others. Freedoms of design and rapid prototyping are some benefits. Some of the methods like stereolithography (SLA), fused deposition modeling (FDM), laser engineered net shaping (LENS) and powder bed fusion were discussed. Also this paper examined the progressive use of AM in bioprinting, biomedical, aviation and working alongside fundamental preparing difficulties like void development anisotropic conduct. To this end, we distinguish the connection between used potential and difficulties and esteem creation process.

Keywords Additive manufacturing · Stereolithography · Fused deposition modeling · Laser engineered net shaping · Bioprinting · Anisotropy

1 Introduction

The production of added substances is a system used to create structures all-inclusive complicated geometries from the 3D representation [1]. AM is a procedure for creating a 3D object of the desired shape from a 3D replica or further sources of electronic information along which additional substance forms in progressive layers of material are established in the presence of PC [2]. The procedure involves printing progressive layers of materials that are framed on top of each other. Charles Hull had an unmistakable fascination with the strong imaging process known as stereolithography and stereolithography (STL) recording. AM can 3D print small quantities of modified items at generally less expense which is explicitly valuable in biomedical, where exceptional items altered by patient are normally required. WinSun printed a

---

R. Teharia (✉) · G. Kaur · M. J. Akhtar · R. M. Singari
Department of Mechanical, Production and Industrial Engineering, Delhi Technological University, New Delhi 110042, India
e-mail: rishabhteharia@gmail.com

© The Author(s), under exclusive license to Springer Nature Singapore Pte Ltd. 2021
R. M. Singari et al. (eds.), *Advances in Manufacturing and Industrial Engineering*, Lecture Notes in Mechanical Engineering,
https://doi.org/10.1007/978-981-15-8542-5_47

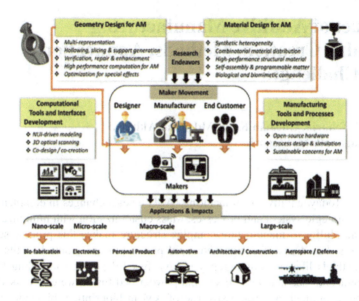

Fig. 1 A geometry-material-machine-process roadmap for AM and maker movement [10]

mass gathering of generally modest houses in China ($4800 per unit) in less than a day [3].

AM is used for intelligent and adaptable creation frameworks [4]. AM can possibly modify the entire structure of the supply chain [5, 6], and it is important to consider the changes throughout the value creation framework, in particular to recognize the consequences for the customer or the transport site [7, 8].

The advantage of innovation in the production of added substances is that of continuing to develop through the search for efforts, adopted to interpret and exclude the requirements which hinder the utilization of innovation [1]. Switching from one facility to another is easy with included intangible costs and without exceptional preparation requirements. The accuracy of the molded parts is subject to the precision of the strategy used and the size of the print, for example, sintering [9] (Fig. 1).

2 Main Methods

See Fig. 2.

Impact of Additive Manufacturing in Value Creation ...

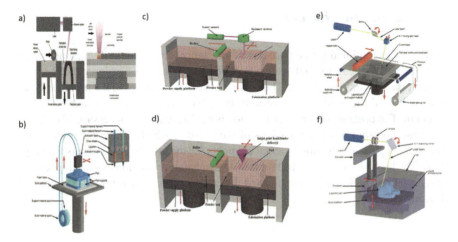

Fig. 2 Different methods of additive manufacturing. **a** Selective laser sintering; **b** Fused deposing modeling; **c** Powder bed fusion; **d** Inkjet printing; **e** LOM; **f** Stereolithography [1,11]

2.1 Selective Laser Sintering

Selective laser sintering (SLS) is one of the 3D printing strategies that use a high strength laser axis to sinter powder material (usually metal) [1]. The attention of the laser on approaches in space is characterized by a 3D model to bind the material and create a strong structure. SLS uses an equivalent idea, but in SLM, the material is completely liquefied with respect to sintering, which allows various properties (structure of the gem, porosity) [11].

2.2 Fused Deposition Melting

In FDM technique, thermoplastic polymer fiber is utilized to 3D-printed layers. The fiber is warmed at the spout to arrive at a semi-fluid condition and afterward expelled on the plat-structure or over recently printed layers. The thermo pliancy permits the fibers to combine while printing, afterward to cement at room temperature (RT) subsequent to printing. Thickness of layers, width, direction of fibers and air bubbles (in alternate layer or individual layers) are fundamental preparing factors which influence the mechanical characteristics of printed component [1, 12]. Mechanical shortcoming is brought about by mostly bury layer twisting [1, 13]. Confined to thermoplastic materials used and poor surface are fundamental downsides this method [12].

2.3 Powder Bed Fusion

Layers of exceptionally fine particles of substances are scattered which are firmly pressed onto a stage and used in the powder bed combination. The laser clamp is used to melt multiple layers of dust. The resulting dust layers move to the previous layers and melt until the preprinted portion is assembled. Excess dust is evacuated under vacuum. The coating, sintering or invasion is performed when necessary. The dispersion and pressing of the dimensions of the powder, which decide the thickness of the molded part, are the fundamental elements for the effectiveness of this technique [14]. The laser should be utilized for powders of less liquefaction or sintering temperatures, while fluid cover must be used differently. Low cost is one most important factor.

2.4 Inkjet Printing (IJP) and Contour Crafting (CC)

The inkjet printing used for the added substance produces only for the production of porcelain. This method is utilized for printing of intricate, guided clay structures like platforms used in tissue engineering. This technique consist of stable suspension of clay, for example zirconium oxide (ZrO_2) powder in water (H_2O) [1, 15], is pumped and deposited in form of droplets via injection nozzle on the substrate. This technique is fast and efficient, including adaptability to plan and print intricate shapes. Two basic kinds of clay impressions are wax-based and fluid suspensions. The creation of shapes has been prototyped to be used for development on the moon [16].

2.5 Stereolithography (SLA)

Stereolithography was created in 1986 and is one of the most consolidated strategies for the generation of added substances [17]. It uses intense light (or electron column) to start a chain reaction to a film of tar or monomer matrix, mainly acrylic or epoxy based which is UV based and can quickly change into polymer chains after radicalization [1]. This diffusion of the molecules in the monomers can be utilized to print lit polymeric organizations [18] or mud monomers derived from polymers, for example, silicon oxide carbide. SLA prints excellent parts with fine lenses up to 10 μm [19].

2.6 Laminated Object Manufacturing

The laminated object manufacturing (LOM) depends on the cutting of layers and overlapping of films or movements of material. The progressive sheets are definitively cut with a laser or a mechanical molder, and then they are reinforced together (structure, therefore union) or vice versa (union, therefore structure). The structure then-bond technique is especially valuable for warm holding of earthenware production and metallic materials, which additionally encourages the development of interior highlights by expelling abundance materials before holding. The overabundance materials in the wake of cutting are left for the help and after finishing of the procedure can be expelled and reused [20]. Various materials are used, for example, polymeric composites, production of terracotta, paper and tapes with metal filling.

2.7 Binderjetting

In the folio stream, a fluid polymer is specifically stored on a powder bed. The thick drop of polymer invades the outside of the powder, bringing about a crude agglomerated powder. The coating is caused by the dispersion of the powder, as occurs in the combination systems of the powder bed. The molded parts are made of glued powder and therefore requisite invasion while subsequent handling for adequate quality. This technique was first learned at MIT and promoted by Z Corporation and ExOne. Any powdered substance that can be effectively dispersed and impregnated with fly closure can be managed by this innovation [10]. Specialists used this innovation to develop a variety of metallic, artistic, molten sand and polymers materials [10] (Table 1).

3 Value Creation System

VCS is a series of activities that create value for customers by economic agents, using tangible and intangible supplies [7, 21]. In addition to the characteristics of the materials and the quality, more details should be taken in consideration for the VCS (e.g., process stability, productivity, automation, post-processing, flexibility and decentralization of production, product life cycle, etc.) and the degree to which added substance advancements are made to accomplish more noteworthy potential. Besides, this more extensive viewpoint permits the advancement of AMT innovation from financial models [7].

The use of AM technologies provides added (potential) value in various fields (test areas). However, the necessary changes to the VCS can put them at a disadvantage or undergo changes to the process or product, with consequent additional

Table 1 classification of AM process by ASTM international [10]

Categories	Technologies	Printed 'INK'	Power source	Strengths/downsides
Material extrusion	Fused deposition modeling (FDM)	Thermoplastics, Ceramic slurries,	Thermal energy	• Inexpensive extrusion machine
	Contour crafting	Metal pastes		• Multi-material printing • Limited part resolution • Poor surface finish
Powder bed fusion	Selective laser sintering (SLS)	Polyamides/polymer	High-powered laser beam	• High accuracy and details • Fully dense parts
	Direct metal laser sintering (DMLS)	Atomized metal powder (17-4 PH stainless steel, cobalt chromium, titanium Ti6At-4V), ceramic powder	Electron beam	• High specific strength and stiffness • Powder handling and recycling
	Selective laser melting (SLM)			• Support and anchor structure • Fully dense parts
	Electron beam belting (EBM)			• High specific strength and stiffness

(continued)

Table 1 (continued)

Categories	Technologies	Printed 'INK'	Power source	Strengths/downsides
Vat Photopolymerization	Stereolithography (SLA)	Photopolymer, Ceramics (alumina, zirconia, PZT)	Ultraviolet laser	• High building speed • Good part resolution • Overcuring, scanned line shape • High cost for supplies and materials
Material jetting	Polyjet/inkjet printing	Photopolymer, wax	Thermal energy, photo curing	• Multi-material printing • High surface finish • Low-strength material
Binder jetting	Indirect inkjet printing (Binder 3DP)	Polymer powder (plaster, resin), ceramic powder, metal powder	Thermal energy	• Full-color objects printing • Require infiltration during postprocessing • Wide material selection • High porosities on finished parts
Sheet lamination	Laminated object manufacturing (LOM)	Plastic film, metallic sheet, ceramic tape	Laser beam	• High surface finish • Low material, machine, process cost • Decubing issues
Directed energy deposition	Laser engineered net shaping (LENS) Electronic beam welding (EBW)	Molten metal	Laser beam	• Repair of damaged worn parts • Functionally graded material printing • Require post-processing machine

Table 2 Impact to VCS by AM [7]

Field of investigation	Key potential/key challenges	Impacts to the VCS
Product	Product quality	− Limited strength and resistance to heat etc.
	Performance enhancing geometries	+ Durable product due to optimized geometry
		+ Increased comfort because of a flexible and adaptable product
	Manufacturing driven design	− Lack of formal standards
Technology	Material availability	− Limitations on materials
Process	Production process	+ Reduced amount of components due to integration of assembly
Value chain	Digitalization of value chain	+ Reduced excess production due to production co demand
		+ Short delivery time due to the lapse of lead time

costs (challenges). The use of AM technologies provides added (potential) value in various fields (test areas). However, the necessary changes to the VCS can put them at a disadvantage or undergo changes to the process or product, with consequent additional costs (challenges) (Table 2).

4 Trending Application

4.1 *Biomaterials*

An ongoing Wohler's report [22] estimates that the industry that produces added substances will increase from $ 6.1 billion in 2016 to $ 21 billion continuously in 2020. Biomedical sector today speaks of 11% of absolute share of the sector in general and will be one of the drivers for the advancement as well as development of AM. Biomedical applications have extraordinary needs that incorporate high complexity, customization and explicit patient needs, free of charge and more. Biofabrication includes the age of tissues and organs with the help of bioprinting, bioassembly and development. Bioprinting with biological imprinting is incorporated with laser-activated direct exchange (LiFT).

4.2 Aerospace

Flying and space industries disregard the breaking points of geometric multifaceted nature. Because of the uncommon highlights of AM, they are perfect for the shuttle. These incorporate the capacity to create complex shapes. Metallic and non-metallic parts, (for example, metametals) can be created or improved by AM methods, for example, stereolithography and FDM. "Thales Alenia Space" in collaboration with "Norsk Titanium AS", have significantly reduced the purchase/flight ratio and decreased lead time by six months. The sections created for the Airbus A350 XWB are 30% lighter and reduce the generation time by about 75% [23].

4.3 Building

Wohler's report [22] indicates that structural usage only 3% of the entire AM application. 3D printing can also be used in the development sector in areas with requirements, such as geometric complexity and exposed structures. In this sense, his duty is due to his ability to create with great precision and open different plan options. Khoshnevis [16] created the Form Advancement (CC) innovation for mechanized development of foundations and structures and space research.

5 Challenges

5.1 Void Formation

The porosity produced by AM can be extremely high which affects the mechanical performance of last application because of reduced surface retention between the printed layers [1, 19]. The size of the vacuum layout is highly dependent on the 3D printing strategy and literature. The increased porousness of the printed segments always not create problem and can be used incorrectly in operations, where under control porousness is taken as favorable condition for AM, for example, the structure of permeable scaffolding in the fabrics. The expanded thickness of the solid layers after a time between successive layers produces a better retention between the layers and a less empty development during the creation of the added solid.

5.2 Anisotropic Microstructure and Mechanical Properties

Anisotropic behavior is an important test for AM. In AM, printing is done in progressive layer form, and the microstructure of the material in individual layer contrasts with the boundaries of intermediate layer. The anisotropic behavior produces an alternative mechanical behavior of the 3D-printed part in vertical tension or weight comparative with the flat course. The changes in morphology and surface result in greater stiffness and flexibility in the transverse form (developmental support) than the longitudinal course of the 3D-printed titanium (Ti) compound using SLM system [24]. Materials such as amalgams, ceramics and polymers show this type of anisotropic behavior [25, 26], which can be useful in some applications.

5.3 Divergence from Design to Execution

Computer-aided design programming is the essential device for planning 3D printable parts. Due to AM limitations, the printed part may contain errors that were not expected in the component. To limit the contrast between structure and development, it is important to preplan and identify the ideal direction of piece, cut the segment into successful layers which support materials that help the expansion of the resulting layers and are effectively expelled from the press.

5.4 Layer-By-Layer Appearance

In various types of applications like structures, toys and flight, the flat appearance is preferred over the layer. Physical handling substances or techniques, such as sintering, can reduce this error [27]; however, installation times and costs increase.

6 Conclusions

3D printing has capacity to print complex structures, has opportunity of configuration and tweaked items with negligible material waste.

As far as procedure powder bed techniques, for example, specific laser sintering (SLS) and specific laser melting (SLM) forms give improved mechanical characteristics and great nature of printed component then melded statement displaying (FDM). FDM is the regular 3D printing procedure as a result of its minimal effort, effortlessness of process and rapid preparing. Direct vitality testimony utilizes laser or electron shaft vitality source to dissolve powder metals in an equivalent manner as in FDM process. Shape creating is utilized to print bigger structure, for example, building. Stereolithography (SLA) is one of the most established AM processes primarily utilized for polymers that can print leaves behind fine goals, yet it is increasingly slow strategy with constrained materials utilized. LOM is based on layer-by-layer cutting and lamination of sheets. Ultrasonic AM is proficient to development of metal structure at low temperature.

AM technology is replacing traditional methods of manufacturing. This is helpful in setting distributed manufacturing of spare parts, and also AM has a huge scope supply chain management system as it can change entire system. Also it is utilized for prototyping but recent research and development of biomaterials for bio-fabrications of tissues and organs using bioprinting. AM has various potential in creating VCS. The impact was investigated in different fields like product design, technology, process and value chain. It has found that due to lack of materials and product quality, AM is still struggling with conventional methods. But it helps in digitization of value chain increasing the rate of production, and also it not only increases the process but also inventory cost decreases.

Aeronautic trade has created altered part with higher solidarity to weight proportion; however, it faces difficulties, for example, significant expense and conflicting nature of 3D-printed parts and constrained materials. The advantages of AM is design freedom from manufacturing constraints. Anisotropic conduct of microstructure is another test of AM which brings about various mechanical conduct under vertical strain or pressure contrasted with that of the even heading. For structuring 3D printable parts, computer-aided design programming is essential device, and there is uniqueness from plan to execution. In spite of being a progressive technique for altered items, proper utilization of 3D printing needs greater advancement to wipe out these disadvantages.

References

1. Ngo TD et al (2018) Additive manufacturing (3D printing): a review of materials, methods, applications and challenges. Compos B Eng 143:172–196
2. Lee D et al (2006) 3D microfabrication of photosensitive resin reinforced with ceramic nanoparticles using LCD microstereolithography. J Laser Micro/Nanoeng 1:142–148
3. Wu P, Wang J, Wang X (2016) A critical review of the use of 3-D printing in the construction industry. Auto Constr 68:21–31
4. Seidenberg U, Ansari F (2017) Qualitätsmanagement in der additiven Fertigung: Herausforderungen und Handlungsempfehlungen. 3D-Printing: Recht, Wirtschaft und Technik des industriellen 3D-Drucks. CH Beck 159–214
5. Steinwender A, Mayrhofer W, SIHN W (2013) The 4th party production provider: enabeling additive manufacturing in industrial environments. management of technology-step to sustainable production, 2013
6. Hofmann E, Oettmeier K (2016) 3D-Druck: wie additive Fertigungsverfahren die Wirtschaft und deren Supply Chains revolutionieren. Zeitschrift Führung + Organisation 85(2):84–90
7. Kritzinger W et al (2018) Impacts of additive manufacturing in value creation system. Procedia CIRP 72:1518–1523
8. Bourell DL, Leu MC, Rosen DW (2009) Roadmap for additive manufacturing: identifying the future of freeform processing. The University of Texas at Austin, Austin, TX, pp 11–15
9. Vaezi M, Seitz H, Yang S (2013) A review on 3D micro-additive manufacturing technologies. Int J Adv Manuf Technol 67(5–8):1721–1754
10. Gao W et al (2015) The status, challenges, and future of additive manufacturing in engineering. Comput Aided Des 69:65–89
11. Gokhare VG, Raut D, Shinde D (2017) A review paper on 3D-Printing aspects and various processes used in the 3D-Printing. Int J Eng Res Technol 6(6):953–958

12. Mohamed OA, Masood SH, Bhowmik JL (2015) Optimization of fused deposition modeling process parameters: a review of current research and future prospects. Adv Manuf 3(1):42–53
13. Sood AK, Ohdar RK, Mahapatra SS (2010) Parametric appraisal of mechanical property of fused deposition modelling processed parts. Mater Des 31(1):287–295
14. Utela B et al (2008) A review of process development steps for new material systems in three dimensional printing (3DP). J Manuf Proc 10(2):96–104
15. Dou R et al (2011) Ink-Jet printing of zirconia: coffee staining and line stability. J Am Ceram Soc 94(11):3787–3792
16. Khoshnevis B (2004) Automated construction by contour crafting—related robotics and information technologies. Auto Constr 13(1):5–19
17. Melchels FP, Feijen J, Grijpma DW (2010) A review on stereolithography and its applications in biomedical engineering. Biomaterials 31(24):6121–6130
18. Travitzky N et al (2014) Additive manufacturing of ceramic-based materials. Adv Eng Mater 16(6):729–754
19. Wang X et al (2017) 3D printing of polymer matrix composites: a review and prospective. Compos B Eng 110:442–458
20. Edgar J, Tint S (2015) Additive manufacturing technologies: 3D printing, rapid prototyping, and direct digital manufacturing. Johnson Matthey Technol Rev 59(3):193–198
21. Richter S, Wischmann S (2016) Additive Fertigungsmethoden–Entwicklungsstand, Marktperspektiven für den industriellen Einsatz und IKT-spezifische Herausforderungen bei Forschung und Entwicklung. Berlin, www.vdivde-it.de/publikationen/studien/additive-fertigungsmethoden/at_download/pdf (7.6. 2016)
22. Campbell I et al (2017) Wohlers report 2017 3D printing and additive manufacturing state of the industry: annual worldwide progress report. 2017: Wohlers Associates
23. A world first: additively manufactured titanium components now onboard the Airbus A350 XWB. Available from: https://www.etmm-online.com/a-world-first-additively-manufactured-titanium-components-now-onboard-the-airbus-a350-xwb-a-486310/
24. Carroll BE, Palmer TA, Beese AM (2015) Anisotropic tensile behavior of Ti–6Al–4 V components fabricated with directed energy deposition additive manufacturing. Acta Mater 87:309–320
25. Cooke W et al (2011) Anisotropy, homogeneity and ageing in an SLS polymer. Rapid Prototyping J 17(4):269–279
26. Guessasma S et al (2016) Anisotropic damage inferred to 3D printed polymers using fused deposition modelling and subject to severe compression. Eur Polymer J 85:324–340
27. Oropallo W, Piegl LA (2016) Ten challenges in 3D printing. Eng Comput 32(1):135–148

3D Printing: A Review of Material, Properties and Application

Gulshan Kaur, Rishabh Teharia, Md Jamil Akhtar, and Ranganath M. Singari

Abstract 3D printing technique is employed for making broad range of complex geometrical structure of distinct sizes from micro- to macro-scale. It provides customized products, reduction in waste of expensive material and fast prototyping so that 3D printing can utilize from prototype to product. A comprehensive review of the material used in single and multimaterial 3D printing was carried out. The material which includes polymers, composites, concrete, biomaterials, ceramics, and metal alloys was discussed, including properties of these materials. In addition, the paper discussed the application of 3D printing in aerospace, architectural, toy fabrication, and other industrial application, expanding in schools, home, library, and research institutes laboratories. This paper discussed an overview of 3D printable material and properties including applications—250 words.

Keywords 3D printing · Multimaterial · Biomaterial · Aerospace · Architecture

1 Introduction

In the present scenario, 3D printing has high potential applications in all fields such as aerospace, building industries, and biomaterials with wide range of materials like metal alloys, ceramics, thermoplastic polymers and composites; pure metals and biomaterial with single and multimaterial printing can be done with lower thickness. It converts models generated by designing software to the physical objects by using layer-by-layer techniques [1].

The primary points of interest of 3D printings are minimization in squander, opportunity of structure, and mechanization. Tweaked goods are a measure for marks because of their large generation costs; however, 3D printing furnishes small quantity of recondition items with generally less outlay. This advantage is uncommonly used in the medicinal sectors; by this, special patient redid items are effectively accessible,

---

G. Kaur (✉) · R. Teharia · M. J. Akhtar · R. M. Singari
Department of Mechanical, Production & Industrial Engineering, Delhi Technological University, New Delhi 110042, India
e-mail: Gulshan.kaur.9494@gmail.com

© The Author(s), under exclusive license to Springer Nature Singapore Pte Ltd. 2021
R. M. Singari et al. (eds.), *Advances in Manufacturing and Industrial Engineering*, Lecture Notes in Mechanical Engineering, https://doi.org/10.1007/978-981-15-8542-5_48

due to its efficiency to provide wide taxonomy of results of invigorative inserts from CT-imaged tissue imitation [2].

2 Materials

2.1 Polymers and Composites

Polymers and composites are accounted as most basic material utilized in 3D printing. Polymers are prepared in various 3D printing methods. Polymers for 3D printing are originated as thermoplastic fibers, responsive monomers, tar, or powders.

Photopolymer-based framework gives accuracy, slight layers, and fine exactness. Further advancements utilizing new pitches have developed quality and heat obstruction. Thermoplastic polymers, for example, acrylonitrile–butadiene–styrene copolymers (ABS) [3], poly-carbonate [4], and polylactic corrosive [5] could be prepared by dealing different 3D printing strategies.

Different 3D printing procedures are accessible for manufacturing polymers in particular: SLA, SLS, FDM, inkjet printing, bioprinting. FDM is greatly utilized in creation of polymers, composites, and thermoplastic with less liquefying focuses [4]. Nonetheless, eco-accommodating polymer materials with great substantial characteristics are of significant worry for FDM because ordinary business polymers for AM, for example, ABS and PLA, do not consider the necessary yields. ABS has great mechanical characteristics; however, it discharges a smell during handling which is unsavory though PLA is eco-friendly yet has not good mechanical characteristics [5].

2.2 Metals

Metals give extraordinary opportunity to assembling complex geometries with the 3D printing techniques contrasted with customary assembling strategies. Specifically, multi-practical parts can be created to give answers for basic, defensive building, and protection issues simultaneously. Two fundamental parameters of 3D printing of metals are kind of crude material and vitality source used to deliver parts. The information crude material gave as metallic powder or wire which is dissolved utilizing a vitality source, for example, a laser or an electron bar [1, 6]. The material which is dissolved is change into layer by layer to shape a strong part. The most generally utilized strategies for added substance assembling of metals are PBF and DED, yet there are some different methods that have been as of late created for printing metals, for example, fastener streaming [7], cold splashing [8], grinding mix welding [9], direct metal composing [10], and diode-based procedures [11]. These procedures can accomplish greater exactness or rapidity. Fundamental delegate of PBF process is selective laser sintering (SLS), selective laser melting (SLM), direct metal laser sintering (DMLS), and electron beam melting (EBM). Warm vitality specifically combines the powder bed in the PBF-based procedure. DMLS process is an added substance assembling or rapid prototyping (RP) process that utilizes powder metal and advanced force laser for sintering. This technique is utilized for creating high dense parts however to accomplish gas or weight snugness; post-treatment is regularly requisite. SLS and DMLS processes are reasonably same; however, the thing that matters is in the SLS procedure polymers or covered metal powder is utilized then again DMLS utilizes not covered pre-alloyed powder of metal as the sintering material [12]. The EBM innovation utilizes vacuum in which warmed powder bed of metal is then softened and shaped like layer utilizing an electron shaft vitality source same as electron bar welding [13].

2.3 Concrete

The enthusiasm of 3D printing is developing quickly, so its application is extended to development industry. 3D printing empowers 3D strong task to be printed from a computerized pattern by settlement of progressive layers of material. A comparable technique to inkjet printing referred to as shape making as the principle innovation for added substance assembling of building structure has been created. This technique utilizes high weight expulsion of solid glue utilizing bigger spout. A trowel-like contraption is joined to print-head to give a smooth completion [14].

In conventional development to accomplish the ideal state of structure by flow of cement stream in a predefined formwork, then again, it is finished by expelling concrete in alternate layer form. In 3D concrete printing (3DCP), no frame work is required to form the solid layers, yet the crisp properties of printable material become significant [15].

Conventional cement cannot legitimately utilize. For limiting the twisting in the middle of the layers, a very nearly zero droop yet full capable cement is required. Another significant property of crisp cement is thixotropy. It is characterized as an abatement in thickness when shear is applied; in the wake of expelling the shear, continuous recuperation happens in new material. Since in 3D printing, solid material should be siphoned, and simultaneously it increases enough solidarity to convey the heap from the following layer. It tends to be said that thixotropy is significant property (Fig. 1).

2.4 Ceramics

3D printing of earthenware production is currently become the most recent pattern on the grounds that the capacity to create fired segments of complex shape with utilizing 3D printing is a lot simpler than customary techniques including infusion forming,

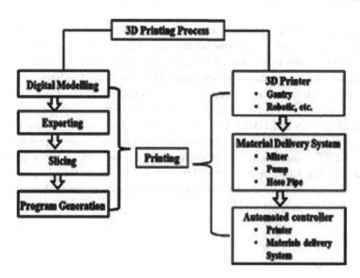

Fig. 1 Process of 3D printing used in construction industry [13]

kick the bucket squeezing, tape casting, gel throwing, and so forth. Machining of fired segments is amazingly troublesome as a result of their extraordinary hardness and fragility. The prolog to 3D printing into the assembling of earthenware particles segments diminishes the issues and difficulties which are looking during conventional techniques. AM has become a fundamental strategy for assembling of cutting-edge earthenware production for biomaterials and tissue designing, for example, platforms for bones and teeth [16]. An incredible assortment of 3D printing techniques has been explicitly produced for clay fabricating as indicated by the information type of earthenware. These techniques can for the most part be ordered into slurry-based, powder-based, and mass strong-based strategies. Slurry-based fired 3D printing innovations for the most part include fluid or semi-fluid frameworks scattered with fine artistic either as inks or glues, contingent upon the strong stacking and thickness of the framework. The slurry-based earthenware is manufactured by using additive techniques, i.e., photo-polymerization, inkjet printing, or expulsion. Photograph polymerization incorporates methods, for example, 'stereolithography'(SL) and its subordinates such as 'Digital light Preparing'(DLP) and 'Two-Photon Polymerization'(TPP), alongside 'inkjet-based printing'(IJP), and expulsion-based 'Direct Ink Composing' (DIW) [17].

3 Application

3.1 Biomaterial

Biomaterials are generally utilized for tissue designing, biomedicine, biochip, and medication discharge gadgets. 3D printing of biomaterials called bioprints is utilized for delivering complex 3D practical living tissues. Some biomaterials have been generally utilized in bioprints are bio-ceramics and metallic biomaterials. For instance, calcium phosphate biomaterials containing tricalcium phosphate (TCP), hydroxyapatite (HA), bioactive glass, and bioglass have great similarity, and creation is like the bone and teeth; in this way, these materials are utilized in the creation of permeable frameworks for bone and tissue recovery.

Three classifications of biomaterial include biopolymers for advance goals bioprints of hydrogels for cell printing tissue designing frameworks, and multimaterial bioprints [16]. Biopolymer is generally utilized in 3D printing in light of their great manufacturability. Different biopolymers including poly-L-lactide (PLLA), polycaprolactone (PCL), polylactic acid (PLA), and their combination are utilized to fabricate manufactured tissue designing platforms with electro–hydro-dynamic (EHD) printing which gives printed filaments of an appropriate scales about 800 nm–40 μm that is near the size of living cells [18]. Platform creation utilizing 3D EHD bioprinting and printed fiber have great power over fiber plans and proper interior pores, which improve cell in growth and had the option to produce complex huge tissues with bended geometries and microscale sinewy structure [16].

3.2 Metamaterial

3D printing is proficient to create complex parts without the necessity of molds, complex assembling methodology, unique installations, or apparatuses. Essentially, 3D printing was utilized for prototyping application, yet now, scientists have started to concentrate in the metalloid stage under huge non-equilibrium condition in appropriation of inward pressure or execution control of the part. For instance, the shape and execution of manufactured segment can be constrained by maintaining laser bar force and rate of motion during printing. In this way, AM is being changed from prototype innovation to shape and execution controllable technology [19].

Nature has made numerous materials and structures that are light in weight yet with high quality, for example, bamboo and human bones, by gaining from regular things to make elite material and items. This elite material is light weight with high quality and internal plan with void and honeycomb sandwich structure to accomplish high explicit quality and solidness. By utilizing 3D printing, this kind of material is created with negligible materials. 3D printing gives hardware practically slope material and structures of superior, and the ideal plan of such structure is called metamaterial [19].

Wang et al. [20] clarified the manufacture of auxetic metamaterial of double material utilizing 3D printing.

3.3 Aerospace

AM strategies are utilized for aviation parts since it gives following qualities:

The airplane business utilizes progressed and expensive materials, for example, titanium composites, nickel-based superalloys, and high-quality steel amalgams at high rate of heat which is very difficult to produce and make lot of wastage (~95%) [21]. 3D printing limits the waste materials (around 10–20%) [22]. The avionic business is portrayed by the creation of little clusters of altered parts. 3D printing is more advantageous financially than customary methods for little clusters as it does not require costly hardware. Planes have a longer operating existence of as long as 30 years. Stored components cause a remarkable expense in stock [23], yet 3D printing is fit for assembling parts on request, in this way lessening the support time.

Aviation parts should be light weighted with high quality and rigidity-to-weight proportions to diminish expenses and discharges [24]. Both metals and non-metals (e.g., metamaterials) and components of aviation can be fabricated or fixed utilizing AM, for example, air motor segments, turbine sharp edges and warmth exchangers. Non-metal AM techniques, for example, stereolithography, multi-stream displaying and melded statement demonstrating (FDM), are utilized for the quick prototyping of components and assembling apparatuses which made up of plastics, composite materials, earthenware production. AM advancements produce complex aviation leaves behind high quality and low weight.

3.4 Construction Industries

Mechanized structure development utilizing 3D printing has increased developing lately. 3D printing is used to build structures that are necessary, such as structures with irregular geometries and void structures. Subsequently, it is re-risk because of their competence to manufacture with more precision and used for separate plan potential results. Khoshnevis [25] created innovation for the robotized advancement of structures for aeronautical application. In 2014, WinSun, a structural sector in China, printed majority of houses in Shanghai (Fig. 2) [26]. Because of the size, 3D printers confined utilization of innovation in construction firm. In any case, a 3D printer of dimension of length (150 m), width (10 m), and height (6.6 m) was utilized, which is consuming concrete and glass fiber. Chunk of difficulties looked by WinSun all through the span of venture incorporates affair including frailty, reconciliation of structure, and aberrant printing [26]. Lim et al. [27, 28] depicted three enormous strategies which are appropriate for development business, depending on material used and the procedure used. Nadal suggested two principle techniques of

Fig. 2 House constructed using AM by WinSun

3D printing according to working condition: (1) connect crane methods and (2) strategies tantamount to CC. In any case, some problems faced during these techniques are conducted off-stage, e.g., less yield of material, not precise result; in general, more work is concentrated. Hager et al. [28] used a method such as contour crafting (CC) innovation which utilizes thermoplastics, cementous material, and fired items.

4 Conclusions

The principle advantages of 3D printing are mass customization, opportunity of plan, and the capacity to print complex structure with least waste. A survey of materials can be utilized in 3D printing for print items and the utilizations of 3D imprinting in different fields. Material utilized in 3D printing is fibers, wire, powder, glue, sheets, and inks. The most normally utilized material is polymers.

AM appropriation in aeronautic trade experienced few issues, e.g., constrained materials with the addition of significant expense and conflicting nature of 3D printed parts. AM modernization is abnormal in development industry in light of its significant expenditure, and less mechanical implementation varies with customized processes. 3D printing requires greater advancement so as to accomplish elite and mechanical quality as the conventional strategies in large-scale manufacturing at greater expense and lower speed. The innovative work worldwide would bring about a quick changes customary strategy to assembling to 3D printing.

References

1. Haleem A, Javaid M (2019) 3D printed medical parts with different materials using additive manufacturing. Clin Epidemiol Global Health
2. Ngo TD et al (2018) Additive manufacturing (3D printing): a review of materials, methods, applications and challenges. Compos B Eng 143:172–196
3. Postiglione G et al (2015) Conductive 3D microstructures by direct 3D printing of polymer/carbon nanotube nanocomposites via liquid deposition modeling. Compos A Appl Sci Manuf 76:110–114
4. Yang J-U, Cho JH, Yoo MJ (2017) Selective metallization on copper aluminate composite via laser direct structuring technology. Compos B Eng 110:361–367
5. Zhuang Y et al (2017) 3D–printing of materials with anisotropic heat distribution using conductive polylactic acid composites. Mater Des 126:135–140
6. Gratton A (2012) Comparison of mechanical, metallurgical properties of 17–4PH stainless steel between direct metal laser sintering (DMLS) and traditional manufacturing methods. 2012 NCUR
7. Bai Y, Williams CB (2015) An exploration of binder jetting of copper. Rapid Prototyping J
8. Sova A et al (2013) Potential of cold gas dynamic spray as additive manufacturing technology. Int J Adv Manuf Technol 69(9–12):2269–2278
9. Chen W et al (2017) Direct metal writing: controlling the rheology through microstructure. Appl Phys Lett 110(9):094104
10. Matthews MJ et al (2017) Diode-based additive manufacturing of metals using an optically-addressable light valve. Opt Express 25(10):11788–11800
11. Kuo C-C et al (2016) Preparation of starch/acrylonitrile-butadiene-styrene copolymers (ABS) biomass alloys and their feasible evaluation for 3D printing applications. Compos B Eng 86:36–39
12. Herderick E (2011) Additive manufacturing of metals: a review. Mater Sci Technol 1413
13. Bhavar V, Kattire P, Patil V, Khot S, Gujar K, Singh R (2017) A review on powder bed fusion technology of metal AM 2017. https://doi.org/10.1201/9781315119106-15
14. Khayat KH, Assaad JJ (2006) Effect of w/cm and high-range water-reducing admixture on formwork pressure and thixotropy of self-consolidating concrete. ACI Mater J 103(3):186
15. Wen Y et al (2017) 3D printed porous ceramic scaffolds for bone tissue engineering: a review. Biomater Sci 5(9):1690–1698
16. Chen Z et al (2019) 3D printing of ceramics: a review. J Eur Ceram Soc 39(4):661–687
17. Chang J et al (2018) Advanced material strategies for next-generation additive manufacturing. Materials 11(1):166
18. Zhang YS, Khademhosseini A (2017) Advances in engineering hydrogels. Science 356(6337):eaaf3627
19. Farina I et al (2016) On the reinforcement of cement mortars through 3D printed polymeric and metallic fibers. Compos B Eng 90:76–85
20. Singh R, Singh S, Fraternali F (2016) Development of in-house composite wire based feed stock filaments of fused deposition modelling for wear-resistant materials and structures. Compos B Eng 98:244–249
21. Shim J-H et al (2012) Bioprinting of a mechanically enhanced three-dimensional dual cell-laden construct for osteochondral tissue engineering using a multi-head tissue/organ building system. J Micromech Microeng 22(8):085014
22. Visser J et al (2015) Reinforcement of hydrogels using three-dimensionally printed microfibres. Nat Commun 6(1):1–10
23. Lu B, Li D, Tian X (2015) Development trends in additive manufacturing and 3D printing. Engineering 1(1):85–89
24. Wang K et al (2015) Designable dual-material auxetic metamaterials using three-dimensional printing. Mater Des 67:159–164
25. Kumar LJ, Nair CK (2017) Current trends of additive manufacturing in the aerospace industry. In: Advances in 3d printing & additive manufacturing technologies. Springer, pp 39–54

26. Wu P, Wang J, Wang X (2016) A critical review of the use of 3-D printing in the construction industry. Autom Constr 68:21–31
27. Khajavi SH, Holmström J, Partanen J (2018) Additive manufacturing in the spare parts supply chain: hub configuration and technology maturity. Rapid Prototyping J
28. Yin L et al (2018) Design and characterization of radar absorbing structure based on gradient-refractive-index metamaterials. Compos B Eng 132:178–187

Effect of Infill Percentage on Vibration Characteristic of 3D-Printed Structure

Pradeep Kumar Yadav, Abhishek, Kamal Singh, and Jitendra Bhaskar

Abstract 3D printing techniques are used to fabricate the structures for prototyping and for customized products. Fused deposition modeling (FDM) technique is commonly used due to low cost and simplicity. Evaluation of mechanical characteristics of the printed structure is very essential for best utilization. Mechanical characteristics printed structure depends upon various printing parameters. Among various geometrical parameters, the infill percentage is very important. This paper presents the study of the vibration and flexural behavior of beam fabricated using fused deposition modeling (FDM) using polylactic (PLA) thermoplastic material. In this study, we observed the effect of infill percentage on mechanical characteristics under static and dynamic conditions.

Keywords Fused deposition modeling (FDM) · Natural frequency · Damping ratio

1 Introduction

3-D printing or additive manufacturing (AM) is that the procedure for creating a three-dimensional solid object or model around any shape from a 3D model or other electronic information sources basically through additive processes during which progressive layers of material are laid down under some logical controller or unit [1, 2]. American Society for Testing and Materials (ASTM) grouped AM into various classes [3, 4]. Among various 3D printing methods, fused deposition modeling (FDM) is a reasonable and most generally used method for 3D printing as compared to other methods. It is also called fused filament fabrication (FFF). It was developed by 'S. Scott Crump' in 1980, and it had been popularized in 1990 [2, 4, 5]. 3D filaments of thermoplastic materials are used to fabricate products using FDM technique [1, 4]. Thermoplastic materials in the form of filament are fed from a large coil into a moving heated extruder head where it is melted and is forced out of the

---

P. K. Yadav (✉) · Abhishek · K. Singh · J. Bhaskar
Department of Mechanical Engineering, Harcourt Butler Technical University Kanpur, Kanpur, India
e-mail: Pradeepyadav596@gmail.com

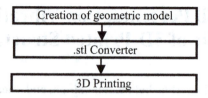

Fig. 1 Flow diagram of printing process

nozzle through the extruder for depositing on the heated bed layer by layer. For 3D printing, geometrical modeling of the structure is prepared, and then it is converted into .stl file for preparing g-codes. This has been shown in the block diagram (Fig. 1).

Support structures are required for printing hanging structures. Support structures are printed using the same material or different materials [2, 3]. 3D printer deposits a removable material (dual nozzle system in case of different support material) that acts as a supporting structure. Support structures are removed after completing the printing of the structure. There are many parameters that affect the features of the 3D-printed model. Following are the parameters which play important role in modifying the mechanical properties of the 3D-printed structure [1, 6]:

Infill: This is the quantification of material for the printing of a solid model. It is a value usually represented in % [2]. The value of 0% is hollow whereas 100% is solid. Infill % changes the distribution of material inside the 3D print and changes the density of material on macroscopically.

Number of Shells: This is referred as amount of printed shells per unit area of print on each layer. Effective density and strength of 3D-printed object are decided by the higher value of the amount of shells [6]. A higher number of shells make the printed outside walls denser.

Layer Height: Layer height is height of each layer that is decided for printing the model. It is directly related to the surface finish and printing time [2, 3, 7].

Raster angle: Angle at which infill takes place as shown in Fig. 2 [5].

In this research paper, an effort has been made to observe the effect of infill % on mechanical performances in terms of static characteristic and dynamic characteristic.

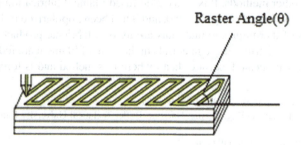

Fig. 2 Raster angle of 3D printing

2 Methodology

Two steps procedure has been adopted in this paper for evaluating the role of infill % for mechanical properties of 3D-printed structure. The first step was to print the structure using various infill % [8], and then the second step was to evaluate the performance using a vibration test and flexural deflections [7]. The detailed information on printing strategy and experimentation for performance evaluations are given below:

2.1 3D Printing of Structure

3D printing consists of geometric modeling of structure and 3D printing. Details are given below:

Geometrical Modeling Geometric model of a rectangular (100 mm × 10 mm × 3 mm) shaped cantilever was created by software CREO 2.0. Geometrical model file was converted into a standard triangular language (STL) format. This design was imported into a 3D printing machine by using the Repetier-Host software package which generates G-Code for the design model [7, 9] as shown in Fig. 3:

Printing: Specification of the 3D printer taken for conducting experiments is as following [2, 6]:

- Extruder Type: Direct with MK8 gearing
- Type: Cartesian

A safe level of printing parameters was taken for pilot experiments which were as below [2, 10]:

The above parameters were adopted from a research paper for experimenting with a secure range [5]. Polylactic acid (PLA) the material was used to print the structure. Samples were printed using different infill percentages of 20%, 40% and 60%. The printing process of the rectangular beam is shown in Fig. 4.

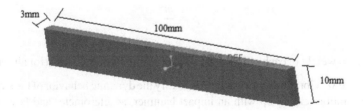

Fig. 3 Isometric view of rectangular beam

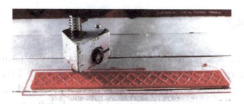

Fig. 4 3D printing with rectangular infill

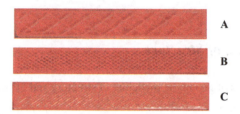

Fig. 5 Samples of different infill %

Verification of infill percentage It was desired to verify the infill % printed by 3D printer. Three samples of 20%, 40% and 60% infill were printed to ensure the infill % (Fig. 5). For verification, weight of each sample was taken from electronic balance. Appropriate volume of printed part was calculated according to infill %. Density was calculated and found that density was same for all sample according to their mass and respective volumes, i.e., (Fig. 5).

2.2 Experimentations

Experiments were performed to evaluate the performances. Details are given below:

Macroscopic Density The density of each sample is calculated by using this formula [2, 5]:

$$\rho = \frac{m}{v}$$

where m = weight of each sample, and v = 100 mm * 10 mm * 3 mm (for all sample).

Dynamic test: Vibration test was used to identify the dynamic behavior of the structure [11]. Vibration test set-up with an impact hammer, accelerometer and laser sensor was used for the experiment. In this experiment, an impact hammer was used to excite the beam, and the response was shown using 'impulse response software' [5].

Natural frequency and damping ratio were measured using a laser sensor, as shown in Fig. 6. Damping (ξ) defines the energy dissipation properties of a material

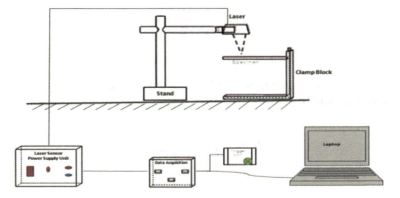

Fig. 6 Schematic of experimental setup for vibration test

or framework under cyclic stress [3–5] (Fig. 7).

Damped frequency (ω_d)

$$\omega_d = \frac{2\pi}{t}$$

Natural frequency (ω_n)

$$\omega_n = \omega_d \bigg/ \sqrt{(1-\xi^2)}$$

Logarithmic decrement (δ)

$$\delta = \frac{1}{N} \ln \frac{x_i}{x_{i+N}}$$

Damping ratio (ξ)

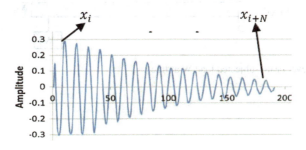

Fig. 7 Logarithmic decay under free vibration

$$\xi = \frac{\delta}{\sqrt{(2\pi)^2 + \delta^2}}$$

where $N = 1, 2, 3 \ldots\ldots$

Effective elastic modulus was calculated from natural frequency using the following relation [2]:

$$E = \frac{\rho A L^4}{I} \left[\frac{2\pi f}{(1.875)^2} \right]^2$$

where

$$I = \frac{bh^3}{12}$$

where E — elastic modulus of specimen, ρ = density of specimen, A = cross sectional area of specimen, L = length of specimen, I = moment of inertia, and h = beam thickness.

Flexural test Cantilever beam with various infill % has applied a load for deflecting in the flexural situation as shown in Fig. 8 [7, 12].

$$\text{Flexural stiffness,} \quad \frac{P}{\delta} = \frac{Ebh^3}{4L^3}$$

$$\text{Elastic modulus,} \quad E = \frac{\text{Flexural stiffness} \times 4L^3}{bh^3}$$

where P is load applied and 'δ' deflection. Flexural stiffness and effective modulus were also calculated [8]. Flexural stiffness and effective modulus of 3D printed structures were also calculated using the above formulas.

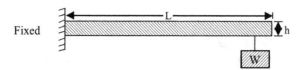

Fig. 8 Cantilever beam subjected load W at free end

3 Results and Discussions

Responses under free vibration were recorded in the time domain for all three infill % and given as in Figs. 9, 10 and 11.

Natural frequencies and damping ratio were obtained as in Table 1.

From the results, a clear-cut increase of natural frequencies was observed. This is due to an increase in infill % which increases the effective density of printed structure and increasing effective elastic modulus. The damping ratio was maximum in the lowest infill % sample. It is due to more unfilled pockets for allowing the deformation inside the sample for energy dissipation [13]. Table 2 shows that as we increase the density of the sample, the elastic modulus of the sample increases proportionally.

Table 3 shows that a 60% infilled sample has minimum deflection on applying the same load because it has maximum flexural stiffness.

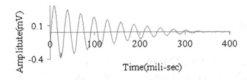

Fig. 9 Response under free vibration for sample A (20% infill)

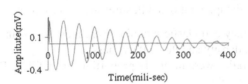

Fig. 10 Response under free vibration for sample B (40% infill)

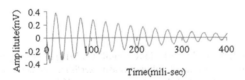

Fig. 11 Response under free vibration for sample C (60% infill)

Table 1 Natural frequency and damping ratio

Sample	Infill %	Density (g/cm^3)	Natural freq. (Hz)	Damping ratio
A	20	0.633	31.90	0.044
B	40	0.766	35.34	0.035
C	60	0.866	39.58	0.020

Table 2 Effective elastic modulus calculated using natural frequency

Samples	Infill %	Effective elastic modulus (GPa)
A	20	0.274
B	40	0.407
C	60	0.577

Table 3 Flexural stiffness and elastic modulus

Sample	Deflection (mm)	Flexural stiffness (N/mm)
A	2.47	0.4145
B	2.32	0.4413
C	2.29	0.4471

4 Conclusions

FDM technique was used for 3D printing the test samples with varying the infill % from 20 to 60. Mechanical characteristics under static and dynamic were evaluated. Vibration test was performed for dynamic test, and flexural test was performed for static test. Following conclusions have been made in view of role of infill % on the natural frequency, damping ratio and flexural stiffness as given below:

1. Natural frequency and elastic modulus were increased with infill % due to an increase in the density and flexural stiffness.
2. The damping ratios were inversely proportional to infill %. Damping ratio was higher for low infill %.

Reference

1. Avella M, Cocca M, Avolio R, Castaldo R, Gentile G (2018) PLA (poly lactic acid) based plasticized nanocomposites: Effect of polymer/plasticizer/filler interactions on the time evolution of properties. Compos Part B (2018). https://doi.org/10.1016/j.compositesb.2018.07.011
2. Srivastava A, Bhaskar J (2018) Development and performance analysis of fused deposition modeling based 3D printer. In: 7th international and 28th all India manufacturing technology design and research Conference (AIMTDR 2018)
3. Galantuccia LM, Bodib I, Kacanib J, Lavecchiaa F (2015) Analysis of dimensional performance for a 3D open-source printer. CIRP 28
4. Li Y, Linke BS (2017) Cost sustainability & surface roughness quality; A comprehensive analysis of products made with personal 3-D printers. CIRP J Manuf Sci Technol
5. Vaezi M, Chua CK (2011) Effects of every layer thickness and binder saturation level parameters on the 3-D printing process. Int J Adv Manuf Technol 53:275–284. https://doi.org/10.1007/s00170-010-2821-1

6. Galantucci LM, Bodi I, Kacani J, Lavecchia F (2015) Analysis of dimensional performance for a 3-D open-source printer based on fused deposition modeling (FDM) technique. In: 3rd CIRP global web conference
7. Kun K (2016) Reconstruction & development of a 3-D printer using fused deposition modeling technology. In: ICMEM, Slovakia
8. Singh K, Bhaskar J (2018) Numerical & experimental study of vibrational behaviour of glass & carbon fiber reinforced hybrid composite
9. Krishna Chaitanya S, Madhava Reddy K, Sai Naga Sri Harsh (2015) Vibration properties of 3-D printed/rapid prototype parts
10. Chaidas D, Kitsakis K et al (2016) The impact of temp. changing on surface roughness of FDM process. IOP Conf Ser Mater Sci Eng 161:01203
11. Mohammed Raffic N, Ganesh Babu K, Manoj Kannan M, Arul Mani G, Nandhu Krishnan R, George M (2017) Effect of FDM process parameters on vibration properties of PET-G &ABS plastic. Int J Mech Prod Eng 3(1)
12. Hong GS, Tlegenov Y, Lu WF (2018) Nozzle condition monitoring in 3-D printing. Robot Comput Integr Manuf 54:45–55
13. Tumbleston JR, Shirvanyants D (2015) Continuous liquid interface production of 3D objects. CIRP J Manuf Sci Tech

Study of Slender Carbon Fiber-Reinforced Columns Filled with Concrete

Utkarsh Roy, Shubham Khurana, Pratikshit Arora, and Vipin

Abstract Slender reinforced concrete columns are used for extension of the life of column used in construction. This is done to enable the preservation of the original structure for a longer time, with a greater level of reliability and at a relatively low maintenance cost. The behavior of these columns depends on various parameters such as types of fibers and matrix, fiber orientation angle, thickness of fiber, slenderness ratio and concrete strength. In the present study, the influence of fiber orientation angle on buckling load capacity, equivalent stress and equivalent strain is investigated to find the optimum angle of orientation. Analysis is performed for the critical buckling load on the carbon fiber-reinforced polymer column filled with M-20 concrete.

Keywords Composites · Carbon fiber · Reinforced concrete column · Buckling

1 Introduction

The requirement of increased strength and durability of slender load-bearing members in the construction industry creates a need for research about the same. In place of conventional steel tubular reinforcement, fiber-reinforced polymer composite columns are more preferred in the structural strengthening or retrofitting of columns. The application is found in the construction of buildings and bridges, especially in coastal areas.

New and developing composites like fiber-reinforced polymers (FRPs) have numerous advantages which include higher strength-to-weight ratios, ease of manufacture, ease of installation and durability (corrosion and water resistance). Their only major disadvantage is their high cost. The most widely used composites for static structural strengthening purposes are carbon fiber composites with epoxy resin-hardener.

U. Roy (✉) · S. Khurana · P. Arora · Vipin
Delhi Technological University, New Delhi 110042, India
e-mail: roy.utkarsh98@gmail.com

Two types of carbon fiber-reinforced polymers (CFRPs) are employed for the toughening of columns or structural members. Firstly, there are polymers formed which are in strip form and are manufactured by the near-surface mounted reinforcement (NSMR) method. The other type of polymers includes those in confined sheet form which can be reshaped as stirrups or continuously like a spiral. The more these polymers contain or confine the lateral deformation, greater is their strength.

In this study, fiber-reinforced polymer columns filled with concrete are first analyzed using finite element method on ANSYS 15.0 software. Equivalent stress, equivalent strain and critical buckling load are evaluated for various fiber orientation angles. A specimen for the optimum angle of orientation is then fabricated and tested for critical buckling load on a universal compressive testing machine. The concrete used is of grade M20 as per the specifications given by IS 10262. The fiber orientation angle of $+\theta°/-\theta°$ is represented simply as $\theta°$.

2 Literature Review

Amir Mirmiran was the first to research in the area of concrete-filled fiber-reinforced polymer tubes (CFFT) [1]; the conclusion of his study is that the strength of the column reduces when the slenderness ratio is increased.

Pan, Xu and Hu examined slender elliptical concrete pillars encased with FRP [2]. While these types of columns differ from CFFTs, the bolstering impact drops with an escalation in the slenderness ratio as well as the initial end eccentricity.

Tao and Yu investigated the confinement of RC columns with unidirectional carbon fiber as well as bidirectional fiber polymer jacket [3]. For columns with slenderness ratio of about 70, the ultimate strength is measured and found to be very close to that of unstrengthened ones. When there is prominent bending, the longitudinal fibers become further efficient.

A miniature system analysis of Bisby and Fitzwilliam [4] included columns encased or wrapped with carbon fiber in a longitudinal direction as well as in a hoop direction. There was only a marginal increase in load-carrying ability of slender columns wrapped in a hoop direction. Longitudinal reinforced polymer wraps enhance the performance of slender concrete-filled columns and thereby help in attainment of more eminent strength and capacity.

A. R. Rahai also conducted an experiment to study about concrete columns enclosed with carbon fiber polymer (CFRP) composites while also taking into consideration various parameters like fiber orientation angle and thickness of wrapped carbon fiber [5]. The results obtained further helped in the development of these columns.

A study on the performance of short concrete-filled steel tubular (CFST) columns was carried out by Muhammad Naseem [6]. The columns were placed under axial compressive loading. To further estimate the ultimate axial compressive load, an equation was also developed by M. Naseem for square CFST columns.

A prominent finite element software analysis was done for the first time by Raghu [7]. The research elaborated the performance of circular concrete-filled columns subjected to axial load.

Slender RC columns with varying slenderness ratio ranging from 30 to 125 were studied by Min Yu in his research [8]. The axial load-bearing capacity of these steel tube columns was predicted for different cross-sectional profiles. The research was performed on the same.

Won-Kee Hong conducted both the experiment and the analysis of circular as well as square cross-sectional slender concrete columns confined by carbon fiber composites while the load applied being axial [9].

To facilitate further extensive applications of fiber-reinforced concrete-filled columns or structural members, it is important to undertake more experiments and research in this field so that we are able to devise new methods for the strengthening of these composite columns.

3 Methodology

When a structural member is subjected to compressive stresses, buckling is seen to occur. This buckling is seen as a sideways bend or deflection of a structural member (a column or a strut). The buckling load may even be lesser than the load at which the same material fails or the compressive strength load. As the applied load on a column is increased, the column will ultimately become unstable and will buckle. Further loading at this point will cause substantial and unpredictable damage or deformations which may lead to a complete loss of the column's load-carrying capacity.

To determine the optimum angle of orientation, simulation is done for different angles of orientation using ANSYS 16.0 software. Experimental research is then carried out for the optimum angle of orientation.

3.1 Composite Column

A composite column is a combination of multiple structural components that are used together to exhibit the desirable structural properties. An advantage of fabricating parts with fiber-reinforced composites is the ability to tailor the stiffness and strength of the structure to specific design loads. A common example involves long composite columns, in which hoop stresses are double the axial stress. First introduced over 40 years ago, netting analysis is still used to determine the optimum fiber orientation for many columns. As the name implies, netting analysis assumes that load is carried only by the fibers. The optimum fiber angle is found from a stress transformation in which the axial and hoop stresses produce only fiber direction stress in the material coordinate system as:

$$\text{Tan}^{-1}(1) = 45° \tag{1}$$

Verifying the accuracy of an optimum fiber angle requires the application of a biaxial stress state. The biaxial stress state is achieved by applying internal or external pressure to induce hoop stress while applying an axial load to achieve axial stress.

3.2 Column Specifications

Carbon fiber polymer is used as reinforcement while the matrix is a mixture of epoxy and hardener. The transversely isotropic carbon fiber T-700 which has five independent elastic constants is used. The matrix used is a mix of Araldite LY 1564 (resin) and Aradur 22962 (hardener). The specificatons of the column are given in Table 1 while the properties of carbon fiber and epoxy-hardender formulation are shown in Tables 2 and 3 respectively.

The resulting stiffness matrix is developed from the properties of carbon fiber and epoxy-hardener and is used as an input parameter for the finite element analysis.

Table 1 Specifications of composite column

S. No.	Dimension	Value
1	Internal diameter	40 mm
2	External diameter	45 mm
3	Thickness	2.5 mm
4	Length	500 mm
5	Concrete grade	M20

Table 2 Elastic properties of carbon fiber T-700

S. No.	Property	Value
1	Young's modulus:	
	In fiber direction	230 GPa
	In transverse direction	6.43 GPa
2	Major Poisson's ratio	0.28
3	Minor Poisson's ratio	0.34
4	Shear modulus	6.0 GPa

Table 3 Properties of epoxy-hardener formulation

S. No.	Property	Value
1	Tensile strength	80 MPa
2	Tensile modulus	3000 MPa
3	Poisson's ratio	0.34
4	Shear modulus	6.0 GPa

3.3 Experimental Setup

The experiment is done for a composite column with fiber orientation angle of 45°.

Manufacturing Process. Firstly, a huge number of fiber rovings (bundles of fiber bound together like a pipe) are pulled from a series of wound reels into a solution of liquid resin, hardener, required catalyst and other pigments. The tension in these fibers is controlled by the lateral and transverse guides which are located between the wound reels and the resin-hardener solution or bath. The fibers, once impregnated with the resin solution, are wound around a pipe or mandrel via lateral movement. The more the number of lateral movements, more is the thickness achieved by the column. After the winding process is complete, the yarns are cut. Curing of filament wound around the mandrel is done by a heated mandrel and an infrared lamp. The mandrel is then removed by cutting or sawing the edges, and the finished filament wound composite tube is de-molded.

Filling of Concrete. Concrete is composed of sand and aggregates bonded together in a fluid cement paste that hardens over time. For this experiment, M20 grade cement (conforming to IS 10262) is used. The concrete used is assumed to be homogenous. The properties of the concrete are shown in Table 4.

Crushed angular aggregates are used with maximum nominal aggregate size 20 mm. The water–cement ratio is 0.55 with minimum and maximum cement content being 300 kg/m^3 and 450 kg/m^3, respectively. The exposure condition is taken as moderate.

End Supports. To achieve the fixed end conditions, two collars of mild steel are inserted at the two ends of the column and sealed using epoxy resin. The fixed end condition results in the reduction of the effective length from original column length to half the length. The effective length thereby reduces to 250 mm. The filling of concrete and the attachment of end supports can be seen in Fig 1.

Table 4 Properties of M20 concrete

S. No.	Property	Value
1	Compressive strength	20 MPa
2	Modulus of elasticity	16.946 GPa
3	Poisson's ratio	0.18

Fig. 1 Filling of concrete and attachment of mild steel end supports

Buckling Test. The test for critical buckling load is done on a universal compression testing machine which is specifically configured to assess the static compressive strength attributes of various materials, products and their components (Fig. 2).

The critical buckling load for this compressive test is found to be 80.1 kN.

Fig. 2 Buckled column

3.4 Finite Element Analysis

The finite element analysis for the composite column is done using ANSYS 15.0 software. ANSYS mechanical is preferred due to its robustness, accuracy and computational turnaround time. Static structural system is employed for running this analysis. The solver settings were modified to provide the most accurate results for pressure simulation. For 3D analysis, the geometrical model with two collars at both the ends of the column (fixed at both ends) was imported in ANSYS. The input parameter used is the stiffness matrix obtained from the properties of carbon fiber and epoxy.

The results from the ANSYS simulation are tabulated in Table 5. They are found to be in accordance with the experimental result, thereby validating the input parameters.

The maximum critical buckling load found by simulation is 84.2 kN at 45° angle of orientation. This also has the least equivalent stress and least equivalent strain.

The results of this simulation are plotted on the graph and shown in Figs. 3, 4 and 5.

Table 5 Simulation results for different angles of orientation

Angle (°)	Equivalent stress (MPa)	Equivalent strain (mm/mm)	Axial deformation (mm)	Lateral deformation (mm)	Critical buckling load (kN)
15	176.2028	0.0135	2.2083	0.8141	81.4057
30	141.0946	0.0120	2.1881	0.8103	82.2285
45	135.7523	0.0110	2.1745	0.8028	84.1933
60	146.1563	0.0124	2.1892	0.8097	82.3192
75	176.5237	0.0135	2.2087	0.8145	81.4133

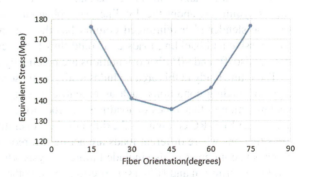

Fig. 3 Line graph of equivalent stress (MPa) versus fiber angle (degrees)

Fig. 4 Line graph of equivalent strain versus fiber angle (degrees)

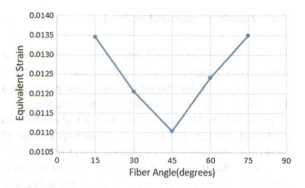

Fig. 5 Line graph of critical buckling load (kN) versus fiber angle (degrees)

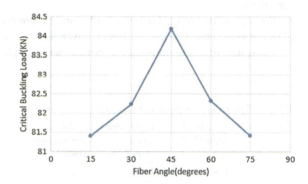

4 Conclusion

The result from this research is that the equivalent stress is minimum and the critical buckling load is maximum for 45° as shown in Table 1. The graph generated was almost symmetric about the 45° line. This phenomenon was observed because at 45°, the slender fiber-reinforced column has the maximum strength to bear both compressive and buckling loads and could thus be generated in a lower equivalent stress. Also, 0° and 90° fiber orientations are practically unfeasible.

The future study in this area includes testing of these columns under several other types of loading conditions such as eccentric loading, shear loading and seismic loading among other types of loadings in order to find a winding angle at which the slender FRP-RC columns will exhibit the best operational characteristics in real-life situations where different loads are acting. Moreover, the future study of these composite columns can also include failure analysis while being subjected to Tsai–Wu failure criterion and Hill–Tsai criterion among others.

References

1. Mirmiran A, Shahawy M, Beitleman T (2001) Slenderness limit for hybrid FRP-concrete columns. J Compos Constr 5(1):26–34
2. Pan JL, Xu T, Hu ZJ (2007) Experimental investigation of load carrying capacity of the slender reinforced concrete columns wrapped with FRP. Constr Build Mater 21(11):1991–1996
3. Tao Z, Han LH (2007) Behaviour of fire-exposed concrete-filled steel tubular beam columns repaired with CFRP wraps. Thin-Walled Struct 45(1):63–76
4. Fitzwilliam J, Bisby LA (2010) Slenderness effects on circular CFRP confined reinforced concrete columns. J Compos Constr 14(3):280–288
5. Rahai AR, Sadeghian P, Ehsani MR (2008) Experimental behavior of concrete cylinders confined with CFRP composites. In: Proceedings of the 14th World Conference on Earthquake Engineering, pp A-X. China
6. Baig MN, Fan J, Nie J (2006) Strength of concrete filled steel tubular columns. Tsinghua Sci Technol 11(6):657–666
7. Raghu KS, Ramesh Babu E (2013) Buckling behavior of concrete filled steel tube under finite element method. Int J Emerg Trends Eng Dev 4:145–161
8. Yu M, Zha X, Ye J, Li Y (2013) A unified formulation for circle and polygon concrete-filled steel tube columns under axial compression. Eng Struct 49:1–10
9. Hong WK, Kim HC (2004) Behavior of concrete columns confined by carbon composite tubes. Can J Civ Eng 31(2):178–188

Factors Affecting Import Demand in India: A Principal Component Analysis Framework

Khyati Kathuria and Nand Kumar

Abstract The paper investigates the principal factors affecting the import demand in India using annual time series data for the period 1995–2017. The estimation results were obtained using R software. The empirical results from the principal component analysis reveal strong evidence of high correlations among the variables considered in the study. GDP and relative prices are found to be the principal factors that explain the maximum variation in import demand.

Keywords PCA · Import demand · Exports · Final consumption expenditure · Real effective exchange rate · Relative prices · India

JEL Classification F41 · B17 · B41 · C51

1 Introduction

Examination of the long-run import demand function is one of the foremost elementary tasks in the field of International Economics nowadays. In this backdrop, Tang [14] explains that it is essential to comprehend the determinants that are responsible for the presence of a stable long-run cointegrating relation among the variables involved in an import demand function. For India, Dutta and Ahmed [3] reinvestigated the behavior of imports and argued that there was empirical evidence of the presence of a cointegrating relationship among the variables employed in the aggregate import demand function during the period 1971–1995.

India had been delineated as an 'import-substituting country par excellence' and remained a shielded economy for a significantly long time. Dutta and Ahmed [3] mentions that preceding the 1990s, her import regime was commanded by a highly

---

K. Kathuria · N. Kumar (✉)
Delhi Technological University, New Delhi, India
e-mail: nandkumar@dce.ac.in

K. Kathuria
e-mail: Kathuria.khyati19@gmail.com

© The Author(s), under exclusive license to Springer Nature Singapore Pte Ltd. 2021
R. M. Singari et al. (eds.), *Advances in Manufacturing and Industrial Engineering*,
Lecture Notes in Mechanical Engineering,
https://doi.org/10.1007/978-981-15-8542-5_51

protectionist import tariff structure and quantitative restrictions on imports. The World Bank counted India in the inventory of 'strongly inward-oriented' nations, implying that the general motivation structure earnestly bolstered creation for the local market. However, as per Dean, Desai and Riedel [2], the Indian economy has been encountering impressive changes since 1991. For all the intents and purposes every domain of the economy has been opened to both local and foreign private ventures, import licensing restrictions on intermediaries and capital goods have been by and large disposed off, tariffs have been fundamentally reduced and full convertibility of foreign exchange earnings for current account transactions has been established [3].

In 1991–92, due to import restrictions, imports fell down drastically, but later in the second half of 1991–92, as the restrictions were eased, they started increasing after which they surged owing to an increase in crude oil prices. As shown in the graph Fig. 1 till 1997–98, exports and imports were almost matching, but after that, the growth of imports was higher than the growth of exports in India. Given such a rise in import growth, it becomes vital for the policymakers to realize whether short-run disequilibrium in the import sector is wiped out in the long run through reforms in the sector. Unless policymakers comprehend the components influencing import demand in India and their reaction to economic reforms, in the long run, they would be unable to make steady policy prescription on imports that would guarantee necessary investment and output expansion. Additionally, a cognizance of the behavior of import demand of a large emerging economy like India is central for designing trade policies. Imports are a key piece of worldwide trade, and import of capital goods specifically is crucial to financial development. Imported capital goods directly influence investment which in turn builds up the vehicle of economic expansion. Consequently, knowledge of import demand behavior is basic for significant import forecasts and exchange rate policy formulation.

Subsequently, this empirical inquiry investigates the principal factors that affect the import demand in India for the period 1995–2017. The study may be distinguished

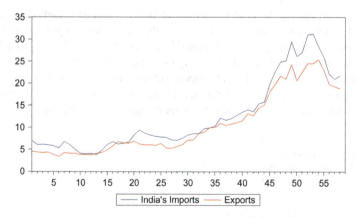

Fig. 1 India's imports and exports

from the previous studies in the sense that the previous studies have focused on the estimation of the import demand function without measuring and checking the principal factors out of a pool of potential factors that may impact the demand for imports in the country. Therefore, the present study fills the gap in the literature by conducting a principal component analysis (PCA) to identify the principal variables that impact the demand for imports.

The remaining study is structured as follows: Sect. 2 describes the overview of import demand literature, Sect. 3 specifies sources of data and methodology, Sect. 4 discusses the estimation results, and Sect. 5 concludes the study.

2 Literature Survey

There exists a surplus of the literature on the estimation of import demand function; however, in this section, we tend to review solely the studies that have used disaggregated import demand function variable-wise.

Gafar [6] estimated the disaggregated traditional import demand function using the log-linear specification to test whether appropriate exchange rate policies were necessary and sufficient for BOP adjustment, whether they must be accompanied by wage restraints, reduction in government spending and monetary contractions. The test was conducted to verify the validity of the traditional import demand specification. Stability of trade relationship was tested using a dummy variable pre- and post-oil price hike in 1974. Finally, it was tested whether disaggregation in the import demand function produced superior results. However, the results evidenced that the exchange rate could be used to correct for BOP disequilibrium. Disaggregation did not produce any superior results and the traditional import demand function performed satisfactorily.

Tang [13] investigated aggregate import demand function of China for 1970–99 by molding the traditional specification for import demand function and considered four descriptions of domestic activity variable specifically gross domestic product (GDP), GDP-Exports, National cash flow [16] and final expenditure components (decomposed GDP). Cointegration bounds testing approach by Pesaran and Johansen Juselius were used. Results indicated a long-run equilibrium relationship among the volume of imports demanded, domestic real activity and relative prices.

Tang [14] extended his 2003 study that supported the absence of long-term cointegrating relationship among aggregate imports, real income and relative price of imports of Japan. This result was attributed to omitted-variable bias. To fill this gap, financial variables, particularly govt. bond yield, lending rate, bank credit, share prices data from 1973:1 to 2000:2, were incorporated in conjunction with national income, the relative price of imports and time trend variable in import demand function of Japan. Bounds test and Johansen's multivariate test indicated long-run cointegration among the variables for Japan. Financial variables were found insignificant but the establishment of cointegration due to their inclusion proved implication of monetary policies on accelerating Japanese import demand at least at aggregate levels.

Narayan and Narayan [10] estimated the import demand model for Fiji by disaggregating the real income variable into total consumption, investment expenditure and export expenditure variables along with traditional relative prices over 1970–2000 and applied cointegration bounds testing procedure and estimated the ARDL model to obtain short-run and long-run elasticities. Cointegration was evident among variables in import demand model.

Alias and Tang [1] examined the long-run relationship between Malaysian aggregate imports and final demand expenditure components and relative prices using Johansen multivariate cointegration analysis. An error correction model was also prepared using annual data for 1970–1988. Final demand expenditure components and relative prices were found to be cointegrated with aggregate import demand both in the short run and in the long run.

Tang [13] applied Xu's [16] import demand function and reinvestigated the presence of cointegrating relation in Japan's aggregate import demand function from 1973–2000. National cash flow (GDP-I-G-EX), relative prices (ratio of import price to domestic price) and time trend were thought of as important determinants of Japanese import demand. To check for cointegration, single equation approaches of Engle [4] and Pesaran et al. [11] and system approach of Johansen (1988) were used. Results suggested an absence of cointegration among variables entering the import demand function, thus providing conclusive evidence of unstable aggregate import demand function of Japan. Gozgar [7] re-estimated aggregated and disaggregated import demand functions for China over Jan 1993–Sept 2012 using quarterly data of 6 groups of primary and manufactured goods including the period effects of 2008–09. A measure of perception of risk- SKEW index (Skewness index of Chicago Board options), the exchange rate was also included as control variables. Dynamic OLS technique is used to obtain long term parameters. To check the robustness of the parameters cointegration is checked through the Hansen test, park test, ARDL model regressions. GDP was found to be the main determinant of total and disaggregated import demand. Results indicated an absence of aggregation bias for import demand in China.

Hu et al [8] estimated China's Import demand function for steel products using monthly data for the period 1996–2004. Effects of trade liberalization, China's economic activities & the real effective exchange rate of the Chinese economy on China's demand for steel imports were examined. China's imports were found to be cointegrated with its economic activities and real exchange rate tested using the Johansen and Juselius cointegration test.

Fukumoto [5] estimated the disaggregate import demand functions of 3 basic classes of good namely Capital goods, Intermediate goods, final consumption goods using data from the period 1988–2005. Relative import prices and different macroeconomic variables namely GDP, disposable income, aggregate consumption, aggregate investment, aggregate exports are adopted for these 3 different classes of imports. According to bounds testing approach cointegration was found evident and according to ARDL short-run price elasticities are inelastic whereas domestic macroeconomic variables were elastic.

Kim et al [9] estimated the future soybean import demand from South Korea, China and Japan from the USA and examined the major determinants of import demand from 2011–17 using Ordinary Least Square method. Soybean import demand, world soybean price, Exchange rate, GDP per capita and WTO participation were considered as dependent and independent variables respectively. It was found that US soybean exporters must expand their share in the Chinese soybean market while due to the price sensitivity in the South Korean market the exporters must try to control the production and transportation cost to remain competitive.

Sinha [12] focused on the impact of developmental variables namely infrastructure, human resources, resources, openness, production and market on India's imports using principal component analysis, composite index and panel regression model under Heckscher Ohlin International Trade Theory for 1990–2013. Indian imports were mainly determined by resource and openness and therefore should be given due importance while formulating India's import policy.

Tang [13] contributed to the empirical literature by testing for the stability of Japan's aggregate import demand function by applying the rolling windows technique to the bounds testing approach to cointegration and estimating the long-run income elasticity and the relative price elasticity via least squares estimator. Japan's aggregate import demand function was not stable over the sample period 1973:1 to 2007:2 i.e. cointegration prevailed for certain periods (windows) and disappeared for other (windows) periods.

3 Data and Methodology

3.1 Data

The study uses annual data from 1995–2017 on final consumption expenditure, gross capital formation, real effective exchange rate, exports, CPI and import price index of goods and services sourced from World Bank (World Development Indicator). In spite of the fact that the usage of quarterly data by increasing the sample size and by highlighting quarterly fluctuations would improve the results of cointegration, we are not able to use quarterly data as it is not accessible for an adequately extensive stretch for all the factors that appear in the import demand function.

3.2 Methodology

Principal component analysis (PCA) is a multivariate statistical technique in which there is no distinction between dependent and independent variables. In PCA, all variables under investigation are analyzed together to extract the underlined factors. It is a data reduction method. It reduces a large number of variables resulting in data complexity to a few manageable factors. These factors explain most part of the variation of the original set of data. The basic principle behind the application of factor analysis is that the initial set of variables should be highly correlated. This test is carried out by using the Barttlet test of sphericity which tests for the significance of the correlation matrix. Another condition requires fulfillment of the Kaiser–Meyer–Olkin (KMO) statistics which takes a value between 0 and 1.

The first and foremost step is to decide how many factors are to be extracted from the given set of data. This can be accomplished by Kaiser–Guttman method which decides the number of factors to be extracted on the basis of the eigenvalue more than 1.

The second step in the factor analysis is the rotation of the initial factor solutions. Generally used method is varimax rotation which maximizes the variance of the loadings within each factor. Once this is done, a cutoff point on the factor loading is selected. Generally, it is greater than 0.5. All the variables attached to the factor are used for naming the factors. A variable which appears in one factor should not be present in the other factor. The total variance explained by all the factors taken together remains the same after rotation. The communalities also remain unchanged.

The R software was employed to obtain the results.

4 Results

The results indicate that a principal component analysis can be applied to the set of given data as the values of the correlations between the variables shown in the correlation matrix Table 1 are high. There are two factors having eigenvalue greater than 1 resulting from the analysis, i.e., the component matrix shown in Table 2 explaining 91% of the variations in the entire dataset. The first two factors explain the maximum variation of the entire dataset which can also be seen from the screen plot and the line plot as shown in Figs. 2 and 3. On the basis of the communalities Table 4 derived from the extracted factor loadings Table 3, both the factors explain the respective percentage of the variation or the information content of the respective variable. The percentage of the variation explained by first and second factors is 47% and 40.7%, respectively, after varimax rotation is performed Table 5. Using cutoff point of 0.65 for factor loading, we find that factor 1 comprising variables FCE, GCF and exports can be named as GDP variables. Factor 2 comprising import price and CPI can be named as price variable which can be entered in the model as relative price variable, i.e., by taking the ratio of import price to CPI. One interesting point to note is that REER variable which as expected by economic theory is not a part of any of the factors and its factor loading value is also very low Table 5. However, this can also be verified by checking for the significance of the REER variable in the regression analysis. But the rationale behind such a behavior of the REER variable may be attributed to the fact that India's imports include items of necessity, and

Table 1 Correlation matrix

	Log(QuantityM)	Log(REER)	Log(FCE)	Log(GCF)	Log(Exports)	Log(Import P)	Log(CPI)
Log(QuantityM)	1.000	0.496	−0.946	0.872	0.990	0.936	0.775
Log(REER)	0.496	1.000	−0.519	0.409	0.499	0.593	0.572
Log(FCE)	−0.946	−0.519	1.000	−0.931	−0.940	−0.864	−0.666
Log(GCF)	0.872	0.409	−0.931	1.000	0.835	0.702	0.444
Log(Exports)	0.990	0.499	−0.940	0.835	1.000	0.939	0.801
Log(Import P)	0.936	0.593	−0.866	0.702	0.939	1.000	0.935
Log(CPI)	0.775	0.572	−0.666	0.444	0.801	0.935	1.000

Table 2 Eigenvalues (component matrix)

	Comp.1	Comp.2	Comp.3	Comp.4	Comp.5	Comp.6	Comp.7
Standard Deviation	2.359	0.898	0.728	0.218	0.184	0.113	0.035
Proportion of Variance	0.795	0.115	0.075	0.006	0.004	0.001	0.0001
Cumulative Proportion	0.795	0.910	0.986	0.993	0.997	0.999	1.000

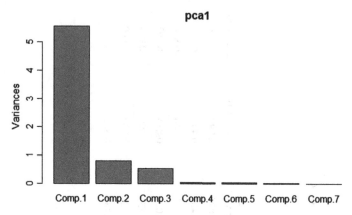

Fig. 2 Scree plot of the component matrix

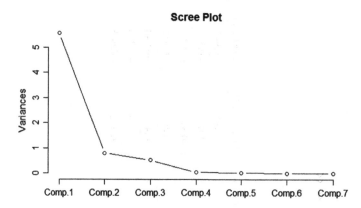

Fig. 3 Line plot of the component matrix

therefore, REER cannot explain the variation in the demand for imports in India (Fig. 4).

Table 3 Extracted factor loadings component matrix

Loadings	Comp.1	Comp.2
Log(QuantityM)	0.415	0.166
Log(REER)	0.266	−0.658
Log(FCE)	−0.404	−0.256
Log(GCF)	0.359	0.487
Log(Exports)	0.414	0.124
Log(Import P)	0.409	-0.163
Log(CPI)	0.354	-0.441

Table 4 Communalities

	Communalities
Log(QuantityM)	0.199
Log(REER)	0.503
Log(FCE)	0.228
Log(GCF)	0.366
Log(Exports)	0.186
Log(Import P)	0.193
Log(CPI)	0.319

Table 5 Rotated component matrix

Loadings	Comp.1	Comp.2
Log(QuantityM)	0.809	0.584
Log(REER)	0.245	0.530
Log(FCE)	−0.845	−0.460
Log(GCF)	0.932	0.206
Log(Exports)	0.775	0.619
Log(Import P)	0.569	0.816
Log(CPI)	0.265	0.962
Sum of Squared Loadings	3.29	2.85
Proportion Var	0.47	0.407
Cumulative Var	0.47	0.877

5 Conclusion

The paper investigates the principal factors affecting the import demand in India using annual time series data for the period 1995–2017. The study explains the factors affecting India's import demand function and provides a comprehensive analysis of

Fig. 4 Biplot of the component matrix

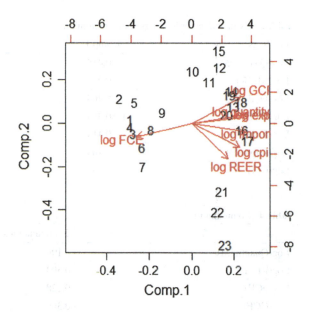

India's import behavior. The empirical results from the principal component analysis reveal strong evidence of high correlations among the variables considered in the study. The estimated results further reveal that there are two principal factors that explain the maximum variation in import demand. Using cutoff point of 0.65 for factor loading, the study finds that factor 1 comprising variables FCE, GCF and exports can be named as GDP variables. Factor 2 comprising import price and CPI can be named as price variable which can be entered in the model as relative price variable, i.e., by taking the ratio of import price to CPI.

References

1. Arize AC, Walker J (1992) A reexamination of Japan's aggregate import demand function: an application of the Engle and Granger two-step procedure. Int Econ J 6(2):41–55
2. Dean JM, Desai S, Riedel J (1994) Trade policy reform in developing countries since 1985: a review of the evidence. The World Bank
3. Dutta D, Ahmed N (1999) An aggregate import demand function for Bangladesh: a cointegration approach. Appl Econ 31(4):465–472
4. Engle RF, Granger CW (1987) Co-integration and error correction: representation, estimation, and testing. Econometrica: J Econ Soc 251–276
5. Fukumoto M (2012) Estimation of China's disaggregate import demand functions. China Econ Rev 23(2):434–444
6. Gafar JS (1988) The determinants of import demand in Trinidad and Tobago: 1967–84. Appl Econ 20(3):303–313
7. Gozgar G (2014) Aggregated and disaggregated import demand in China: an empirical study. Econ Model 43:1–8

8. Hu X, Ping H, Xie C, Hu X (2008) Globalization and China's iron and steel industry: Modelling China's demand for steel importation. J Chin Econ Foreign Trade Stud 1(1):62–74
9. Kim CM, Lee E (2014) Soybean Import Demand Analysis in East Asia: Korea, China, and Japan. Manage Rev Int J 9(1):4
10. Narayan S, Narayan PK (2005) An empirical analysis of Fiji's import demand function. J Econ Stud 32(2):158–168
11. Pesaran MH, Shin Y, Smith RJ (2001) Bounds testing approaches to the analysis of level relationships. J Appl Econ 16(3):289–326
12. Sinha MK (2016) An analysis of determinants of India's import: panel regression approach. FIIB Bus Rev 5(3):52–61
13. Tang TC (2003c) Japanese aggregate import demand function: reassessment from the 'bounds' testing approach. Japan World Econ 15(4):419–436
14. Tang TC (2004) Does financial variable (s) explain the Japanese aggregate import demand? A cointegration analysis. Appl Econ Lett 11(12):775–780
15. Tang TC (2008) The demand for imports in Japan: a review article. Int J Bus Soc 9(1):53
16. Xu QF, Jiang CX (2002) An analysis of correlation relationship between external trade and economic growth [J]. Rorecast (2):3

Theoretical and Statistical Analysis of Inventory and Warehouse Management in Supply Chain Management—A Case Study on Small-Scale Industries

Mahesh R. Latte and Channappa M. Javalagi

Abstract The main function of the supply chain is to satisfy the customer by making use of all the parties such as the manufacturers and suppliers, transporters, warehouses, retailers, and even customer themselves. Within small-scale organization, the supply chain includes all those functions which are involved in receiving and filling a customer request. These functions include, but are not limited to, desired new product development, innovative and cost-effective marketing, systematic operations, well-planned distribution, wisely managed finance, and dedicated customer service. The present case study is the analysis of supply chain management (SCM) in small-scale functional industries, located at two industrial estates, namely Gokul Shirgaon MIDC and Shiroli MIDC in Kolhapur, Maharashtra. These operational units were divided majorly into two parts, manufacturing and logistics, and the study is mainly focused on theoretical & statistical analysis of Inventory management & Warehouse management of small-scale industries. For an effective supply chain model, every company has adopted its own supply chain strategy on the basis of their needs, capacity, and available resources. So, to analyse the SCM in small-scale industry, the relevant information from these industries was collected both from primary and secondary sources. And the relevant information is statistically analysed and verified statistically by using Minitab software, and conclusion was drawn.

Keywords Inventory management · Warehouse management

M. R. Latte
K.L.E. College of Engineering and Technology, Chikodi, Belgavi, India
e-mail: latthemahesh@gmail.com

C. M. Javalagi (✉)
Basaveshwar Engineering College, Bagalkot, India
e-mail: cmjavalagi@gmail.com

1 Supply Chain Scenario in India

India is the fourth principal country in terms of purchasing power parity (PPP) and constitutes one of the best growing markets in the world. Globalization of businesses, infrastructural bottlenecks, rising ambiguity of supply chain networks, reduction of product life cycles and rise of product variety have strained Indian firms to look beyond their four walls. Indian small-scale firms face issues related to choosing and working with the right supply chain partners, fostering trust between them and designing the right system of gauging performance [1]. There are millions of small retail outlets in India. The small size of Indian retail outlet restricts the amount of inventory they can hold. Thus, the only way for manufactures to keep transportation cost near to the ground is to bring jam-packed truck loads of product close to the market and then share out it locally using smaller vehicles. In India, most conventional supply chain operates is multi-echelon distribution model in which the supplies move from the supplier and manufactures through a series of small and large stockholders and mediators to reach the customer which are not always connected. The dynamics of such supply chain differs quite significantly from the IT—it enables supply chain serving well-developed and well-connected urban markets through the systematize retailing [2]. The present case study includes the analysis of supply chain management in small-scale industries with a view to improve the performance of these industries. While doing analysis, it is been observed that there are many numbers of factors which directly or indirectly affects the supply chain; so to study systematically further, the literature survey of five factors was considered significant in the performance of SCM, and they are inventory management, trust, commitment and cooperation between the vendors and suppliers, transport management, communication and information sharing between the vendors and other related companies and warehouse management in small industries. In this paper, we will be discussing majorly of two factors inventory management and warehouse management. The industries, namely Akhilesh Industries, Kalakruti steel furniture's Pvt. Ltd. Mahalaxmi Tubes, Kamala Plastics, Mogane Machine Works, Ratnaprabha Industries, Saraswati Machine Works in Kolhapur area, and some more warehouse-related small-scale industries, were chosen for the study and analysis purpose.

2 Inventory Management

In today's progressively more competitive market place, end users are demanding for more quality products with reduced costs and with less lead time. Due to difference is supply and demand, there is disparity in supply chain and the disparity is intentional at a retail store where inventory is held in anticipation of future demand. In these instances, inventory is held to lessen the price or enhance the level of product availability. In supply chain, inventory is mostly affected by the resources held, responsiveness and the cost. High levels of inventory in an apparel supply chain improve responsiveness but also leave the supply chain at risk, lowering profit margins. A higher level of inventory also facilitates a reduction in production and transportation costs but increases holding costs. Low levels of inventory are most desirable for effective supply chain but may result in loosing of customers. (Unavailability of products may be a possibility & when the customers want to buy). In general, managers should aim to reduce inventory in such a way that it does not increase cost or reduce responsiveness [3]. To understand the inventory management in small-scale industries, questionnaires were prepared and survey is done with the owner of the business and production and planning managers and consolidated data is as below.

2.1 Overall Analysis of Inventory Management

It is been found that various parameters affect the company's inventory management while doing survey in the Shiroli MIDC and Gokul Shirgaon MIDC area. Most of the industries are vendors of the large-scale industries like Ghatge-Patil industries, Manugraph Pvt. Ltd., Kirloskar Oil engines and Menon and Menon industries. These industries place an order in range of 50–200 qty or as per requirement. Mainly, the small-scale firms place the order in 15–25 days. Industries follow JIT manufacturing technique, and they make inventory as per the need of the batch, and tolerances are given for defects or poor quality of the raw material. It is observed that industries only carry the inventory of standard parts like nut-bolts, tools, inspection gauges, dies, maintenance equipments and spanners, and more. If order is taken from the customer, then for most of the small-scale suppliers as they do not carry any inventory, they need to wait for the inventory and procurement of raw material. They are reluctant to invest their capital in inventory it is costly and risky. Because of this, company needs more time to place an order, and due to this, they fail to take further orders as their inventory management is very poor.

2.2 Industry-Wise Data

Factors	Description						
	Kalakruti steel furniture's Pvt. Ltd	Akhilesh Industries	Mahalaxmi Tubes	Kamala Plastics	Mogane Machine Works	Ratnaprabha Industries	Saraswati Machine Works
Raw Material Procurement Method	Yes	Yes	Yes	Yes	Poor	Fairly good	No
Batch quantity	Order-wise	100	Continues production	Mass production	300	50–100	50
Manufacturing Time	18 days	8 days	2 h for 1 ton	24 h production	8 days	8–10 days	8 days
Time to place order	21 days	10 days	As per order	Daily dispatch	10 days	15 days	10 days
Inventory carried or not	Yes	Yes	Yes	Yes	No	Only standard parts	No
Inventory carrying techniques	ABC & JIT techniques	ABC, VED, JIT techniques	VED technique	VED techniques	JIT technique	JIT technique	JIT technique

Some of the companies are making exceptions and implementing new technology and software for inventory management like vector flow and B2-B. Customers place an order through software with the support of Internet. Company can plan their work well before 2 months so they can make proper arrangements of raw material, and as customers place the order with detailed specifications, they can achieve the commitment in very short interval. Software provides efficiency in inventory management. From above observations, discussions with the process owners and analysis with them, the following suggestions were found to improve inventory management and for better supply chain model especially in small-scale industries.

- Statistical tools like ABC analysis, VED analysis, ERP method, MRP methods, and EOQ techniques should be used for inventory management.
- Software like Vector flow, B2-B, and SAP should be used for inventory management.
- Joint inventory forecasting is made at least 3 months in advance, and this would optimize production capacity, cost reduction and would result in enhanced profitability.
- Manufacturing firms must timely examine the firm's financial performance related to inventory, sales, gross margins, inventory turn-over, and inventory management index.
- Company needs to invest some capital in the inventory and should focus to make safety stock and buffer stock of required materials.
- To ensure ready availability of the stock, warehousing facilities within and outside the state could be enhanced by utilizing the services of private warehousing agencies.
- For improving the business, success managers of small-scale industries should be educated and trained for reducing the complexities involved in planning, executing, and controlling a supply chain network.
- Sales forecasting, MRP, and sometimes physical inventory planning help the small-scale industries from the problem like high inventory and high customer service. And it helps to produce staged benefits in customer service with low inventory, no matter how complex a firm's network is.
- Industries should propose booking or incentive programme to customers that could include reduced wholesale prices for high volume purchases to reduce inventory holding and maintenance cost.
- Firms should coordinate inventory among multiple branches both for purchases and sales among different stores and warehouse locations to cater customer's needs and attention.

Now, the same theoretical analysis is been verified through the statistical tool Minitab, and it is as below. For analysis based on the survey and brainstorming sessions with the process owners, we have selected six points which will be prominent for the inventory management, and they are potential saving by the industries, service level required for the customers, reduce the lead time, having control on price fluctuation and statistical tools or software's significance in inventory management.

2.3 Statistical Analysis of Inventory

(1) *Descriptive Statistics*

	Mean	Std. deviation	Analysis N
Potential_saving	3.6667	1.21106	6
Service_level	4.0000	0.63246	6
Reduce_time	4.0000	0.89443	6
Price_fluctuation	3.6667	0.51640	6
Statistical_tools	4.0000	0.89443	6
Firm_size	4.0000	0.63246	6

Here, as you observe that the factors such as service level, reduction of time, using of statistical tool for inventory and firm quantity of inventory play significant role as they are having highest mean. So, the theoretical analysis and statistical analysis give favourable result.

(2) *Communalities*

	Initial	Extraction
Potential_saving	1.000	0.784
Service_level	1.000	0.791
Reduce_time	1.000	0.868
Price_fluctuation	1.000	0.759
Statistical_tools	1.000	0.868
Firm_size	1.000	0.791

Communality value should be more than 0.5, and our all selected factors are showing values more than 0.5. So, suggestions drawn with respect to these factors are satisfactory.

(3) *Total Variance Explained*

The percentage variance obtained after the analysis is the same points which are important and have significance in overall factor analysis. So, we have obtained correct variance values for all the points we have selected for analysis (Fig. 1).

Component	Initial eigenvalues			Extraction sums of squared loadings		
	Total	% of variance	Cumulative %	Total	% of variance	Cumulative %
Potential_saving	4.861	81.025	81.025	4.861	81.025	81.025
Service_level	0.720	12.005	93.030			
Reduce_time	0.278	4.638	97.668			
Price_fluctuation	0.140	2.332	100.000			

(continued)

(continued)

Component	Initial eigenvalues			Extraction sums of squared loadings		
	Total	% of variance	Cumulative %	Total	% of variance	Cumulative %
Statistical_tools	−2.584E−017	−4.307E−016	100.000			
Firm_size	−3.323E−016	−5.538E−015	100.000			

The screen plots show the principal components analysis and factor analysis to visually calculate which components or factors explain most of the unpredictability in the data [4]. It is the point at which graph flattens. The eigen values less than 1 is insignificant so the factors corresponding to those values are also insignificant in the study; so in our case, potential saving of money by employing and implementing technologies is significant.

3 Warehouse Management

Warehousing management is defined as "the direct control of handling equipment producing movement and storage of loads without the need for operators or drivers". It includes equipments such as automated storage and a retrieval system, automated guided vehicles (AGVs) but excludes technology where warehouse operators are still necessary. Warehouses are the concluding point in the supply chain for order assembly, dispatch to the customer and value-added services, represent approximately 20–24% of entire logistics costs and are crucial to the achievement of customer service levels. In supply chain, warehouses act as the nodes where customer orders are assembled and dispatched. Failing to customer service levels can have significant impacts on company's sales and profits, market share, brand switching competitive capabilities, and picking efficiency. Warehousing management in supply chain can be improved by potential improvements in productivity, order accuracy, reduced space requirements, increased volume capacity, control of inventory, and increased customer service [5]. To understand the warehouse management in small-scale industries, questionnaires were prepared, and survey is done with the owner of the business and production and planning managers and consolidated data is as below.

3.1 Industry-Wise Performance

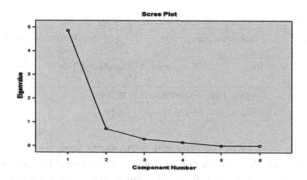

Fig. 1 Screen plot of warehouse management

Factors	Description				
	TCI Supply Chain Solution	VRL Logistics	V-Trans Logistics	V-Express Logistics	Jay Mata Di Cold Storage
Warehouse management system	Good	Good	Good	Good	Good
Methods used for storage of goods	Racks, pallets	Racks, pallets	Racks, pallets	Wooden box, racks, pallets	Wooden box, racks, pallets
Warehouse management policy	FIFO	FIFO	FIFO	FIFO	FIFO, LIFO
Warehouse activity used by organization	Assembly facility, trans-shipment facility, return good depot	Assembly facility, trans-shipment facility, return good depot	Assembly facility, trans-shipment facility, return good depot	Assembly facility, trans-shipment facility, return good depot	Cold storage service, dairy product cold, storage, fruit cold storage, vegetable cold storage, industrial cold storage
Automation or advanced technology used by organization	No	Mobile tab (for paper less	ERP	ERP	AGV in warehouse

(continued)

(continued)

Factors	Description				
	TCI Supply Chain Solution	VRL Logistics	V-Trans Logistics	V-Express Logistics	Jay Mata Di Cold Storage
Factors influencing warehouse operation	Area of warehouse, location, type of goods	Area of warehouse, location, type of goods	Area of warehouse, location, type of goods	Area of warehouse, location, type of goods	Types of product, area of cold storage, location

3.2 Overall Analysis of Warehouse Management

Warehouses are the last point in the supply chain for order assembly, value-added services and dispatch to the customer and are critical to the achievement of customer service levels. TCI Supply Chain Solution, VRL Logistics, V-Trans Logistics-Express Logistics, and Jay Mata Di Cold Storage are the companies located in Shiroli & Gokul Shirgaon MIDC at Kolhapur, Maharashtra, India, that we have selected for our analysis purpose. As all have different products for warehousing, we have taken survey on common points such as warehouse planning strategies, storages policies and strategies, automation in warehousing and technologies any if used for warehousing. On the basis of analysis, the following points were found to enhance warehousing management in supply chain management of small manufacturing firms:

- Small firms should be encouraged to adopt advanced warehousing management system as it can handle multi-stockroom inventories, leads to efficient space utilization and flexibility of arrangement, provides ready availability of stocks, outperforms competitors on customer service, and leads to minimization of material deterioration and pilferage.
- Managers must be educated in warehousing planning & control as it provides complete storage to various items, helps in distribution of goods economically, meets the demands of consuming departments and also assists in building goodwill & inviting business.
- A central industrial warehouse could be constructed and maintained by the government where small-scale industries could stock their finished products during slack period and obtain funds to run industries against it.
- A shared warehousing facility by firms producing similar products is constructed for meeting market demand and absorbing uncertainties. The task of warehousing could be outsourced to warehousing agencies specialized in multi-stock storing. Modern warehousing techniques especially for cement, insecticides, pesticides, and emulsifiers firms are made.

- Seminars and workshops conducted by DIC will help in the warehouse management planning, and this enhances the storage capacity, cost reduction, inventory accuracy, and all this in turn increases customer service.
- Use Enterprise Resource Planning (ERP) system for warehouse system.

3.3 Statistical Analysis of Warehouse Management

(1) *Descriptive Statistics*

Here, as you observe that the factors such as warehouse activity play significant role as it is having the highest mean. So, the theoretical analysis and statistical analysis give favourable result.

(2) *Communalities*

	Initial	Extraction
Warehouse_planning	1.000	0.978
Proper_storage	1.000	0.978
Warehouse_policy	1.000	0.950
Warehouse_activity	1.000	1.000
Automization	1.000	0.953
Location_structure	1.000	0.909

Communality value should be more than 0.5, and our all selected factors are showing values more than 0.5. So, suggestions drawn with respect to these factors are satisfactory.

(3) *Total Variance Explained*

Component	Initial eigenvalues			Extraction sums of squared loadings		
	Total	% of variance	Cumulative %	Total	% of variance	Cumulative %
Warehouse planning	4.56	76.113	76.113	76.113	3.920	4.56
Proper storage	1.20	96.125	20.013	96.125	3.332	1.20
Warehouse policy	0.181	3.023	99.149			
Warehouse activity	0.051	0.851	100.000			
Automization	1.37	−2.295E−017	100.000			
Warehouse planning	1.99	−3.316E−016	100.000			

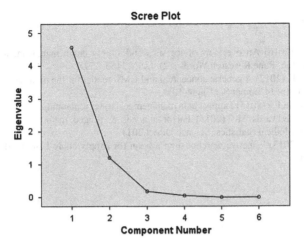

Fig. 2 Screen plot of warehouse management

The percentage variance obtained after the analysis is the same points which are important and have significance in overall factor analysis. So, we have obtained correct variance values for all the points we have selected for analysis (Fig. 2). A screen plot displays the eigenvalues associated with a component or factor in descending order versus the number of the component or factor. The screen plots show the principal components analysis and factor analysis to visually assess which components or factors explain most of the variability in the data [4]. It is the point at which graph flattens. The point below 1 in eigenvalue should not be considered which means that they are not so much significant; so in our case, planning of warehouse is significant compared to automation and policies. The significant factor is proper planning of warehouse location, transportation ways, vehicles, etc.

4 Conclusion

The present case study is done with a view to improve the supply chain performance in small-scale industries, and in this paper, we have discussed extensively of inventory management and warehouse management, and we have found from our analysis that in inventory management, use of new technologies and proper implementation of it is to improve the performance in small-scale industries, and in warehouse management, we have found that proper planning of warehouse is most significant factor for improving the performance of supply chain management.

References

1. Malhotra N (2016) An overview of logistics and supply chain management in India. Int J Interdiscip Stud, Pune Research World 1(2). ISSN 2455-359X
2. Ech-Cheikh H (2012) A generic conceptual and UML model for the multi-echelon distribution supply chain. Int J Comput Sci Eng 4(10)
3. Chopra S, Meindl P (2015) Supply chain management strategy, planning, and operation. Pearsons
4. Woods CM, Edwards MC (2011) Factor analysis & related methods. Essential Statistical Methods for Medical Statistics, Science Direct 2011
5. Chalotra V (2013) Effective warehousing: a boon for supply chain Laurels. IFSMRC 01(01), Jan-Jun 2013

Evaluation of Separation Efficiency of a Cyclone-Type Oil Separator

Ujjwal Suri, Shraman Das, Utkarsh Garg, and B. B. Arora

Abstract Oil separators play an important role in screw chillers for preventing oil circulation in the system and providing continuous oil return to the compressor crankcase. The present study intends to evaluate the performance of a cyclone-type oil separator for a water-cooled screw chiller having a cooling capacity of 245 TR. Operation parameters are calculated on the basis of AHRI standard conditions. Taking these parameters as inputs, the performance is first evaluated using an analytical mathematical model. Subsequently, computational fluid dynamics simulations are conducted in ANSYS Fluent. Results obtained using both methodologies are compared and analyzed.

Keywords Cyclone-type oil separator · Screw chiller · Refrigerant-oil mixture

1 Introduction

In refrigeration devices, oil is used as a lubricant and helps prevent the compressor from seizing. It is essential that these oils are present at the required places for greasing. However, it is inevitable that they are sucked into the compressor and circulated throughout the refrigeration system. This causes significant changes to the properties of the refrigerant [1]. It also could cause significant damage to the refrigeration system. There have been several studies about the mixing and separation characteristics of organic oils from refrigerants [2].

Chillers are machines that cause cooling by removing heat from the liquid by vapor compression or absorption cycles. Screw chillers are commonly used for large-scale refrigeration and air conditioning applications. They consist of semi-hermetic screw compressors which are more suitable for lower refrigeration loads and partial loads in comparison with centrifugal compressors [3]. At the discharge of a screw compressor, some percentage of lubricating oil leaves the compressor crankcase along with the refrigerant and may get circulated through the refrigerant system. Circulation of

U. Suri (✉) · S. Das · U. Garg · B. B. Arora
Delhi Technological University, New Delhi, Delhi 110042, India
e-mail: ujjwalsuri99@gmail.com

lubricating oil in the system may have several negative effects including mechanical breakdown, decrease in heat exchanger efficiency and modification of the physical and chemical properties of the refrigerant. There are significant effects to the compressor in particular [4]. In modern screw compressor equipped chillers, it has been found that an effective method of preventing lubricating oil from flowing out is the cyclone method [5]. Cyclone separators have been found to be widely applicable. It is based on gravity and vortex generation to separate particles generally from gaseous streams [6].

Study of separator efficiency of oil-gas cyclone separators can be seen in the works of Gao et al. [7]. However, the proposed present paper is intended to determine the same using a sophisticated analytical algorithm based on Monte Carlo simulations for a different refrigerant gas in a more industry-applicable separator using the RNG k-ε turbulence model instead of Reynolds stress model and compare the obtained data with CFD results acquired using the discrete phase method.

2 Methodology

In the present work, the operation of cyclone-type oil separator is evaluated at standard AHRI conditions for the selected chiller.

2.1 Refrigerant Mass Flow Rate Calculations

The heat transfer can be calculated from the following equation:

$$Q = \eta * (m * C * \Delta T + m * L) \tag{1}$$

where Q is heat transfer in kW, η is condenser heat transfer efficiency, m is mass flow rate, C is sensible heat capacity, L is latent heat capacity and ΔT is degree of superheat.

For R-134a, at 35.6 °C condenser temperature, C is 1.35 kJ/kg K and L is 168 kJ/kg. The cooling capacity of the selected chiller is 245 TR which is equivalent to 861.6 kW. Furthermore, from test data, it was found that ΔT is 11 K and $\eta = 90.7\%$.

Calculating from these obtained values, it was found that the

$$\text{Refrigerant mass flow rate}(m) = 5.19 \, \text{kg/s} \tag{2}$$

From test data, it is found that oil mass flow rate is

$$\text{Oil mass flow rate}(m) = 0.95 \, \text{kg/s} \tag{3}$$

Table 1 Input parameters for analytical model

Parameter	Case I (kg/s)	Case II (kg/s)	Case III (kg/s)
Refrigerant mass flow rate	5.19	5.19	6.82
Oil mass flow rate	0.95	2.31	2.31

Table 2 Results from analytical model

Parameter	Case I	Case II	Case III
Percentage of escaped particles	1.68	0.52	0.111
Oil separator efficiency	98.32	99.48	99.24

Now, these values are utilized in the analytical and CFD models to obtain the separation efficiency.

3 Analytical Model

The analytical model was developed on the lines of earlier work by Murakami et al. [8]. It considers two stages of separation, namely centrifugal and gravity separation. Particle distances from inlet centerline and particle diameters were initiated in a range of random values. The value of separation efficiency was then determined by employing Monte Carlo methods [9]. A distribution of 10,000 particles was initialized at the inlet of the cyclone-oil separator and the number of particles separated by both the centrifugal and gravity separation methods was determined. This simulation was implemented in the present work using a Python program. The mass flow rates of refrigerant and oil were taken as input variables, and the dimensions and distribution of the particles were then determined by the program as functions of these variables. Subsequently, the program determines the separation efficiency. Tables 1 and 2 tabulate the values obtained by this analytical method.

4 Computational Fluid Dynamics Simulation

A computational fluid dynamics (CFD) simulation has been performed on a 3D CAD model of the cyclone separator for flow visualization and determination of oil separation efficiency. The CFD simulation was conducted using ANSYS Fluent.

Firstly, a 3D CAD model was developed using Dassault SolidWorks software. Then, the geometry cleanup was performed on the 3D CAD model of the cyclone separator. All components such as valves and brackets support structures were removed from the model. The purpose of this step is to eliminate unnecessary surfaces that increase meshing and computation time and do not contribute to the simulation results.

Subsequently, the CAD model was imported into ANSYS Workbench. Now, the RNG k-ε model was used for the simulation as it is known to be highly accurate for swirl-type flows [10].

4.1 Meshing

The meshing strategy used is proximity and curvature. This is chosen because the body surface contains primarily of curved surfaces and the proximity of the body to the floor of the domain is also very small.

The minimum size of elements is taken as 0.5 mm and maximum size as 3 mm. It has been found by Seon et al. [11] that the Y-plus at the walls should be within the range of 1 and 10. So, a Y-plus of 10 is chosen to keep the mesh within the turbulence model range and right levels of refinement.

Inflation layers are added around the body surface. The size of the first layer is kept as 0.69 mm in thickness. The size is chosen to keep the Y-plus around 10, taking a reference length as the diameter of the separator. The number of layers added is 10 with a growth rate of 1.2.

4.2 Fluid Modeling

As detailed in an earlier section, the RNG k-ε model was chosen as it produces relatively higher accuracy predictions for swirl flows. For modeling of the oil phase, discrete phase model was selected [12]. Second-order discretization was used for pressure. Generally, the discrete phase model (DPM) is employed for the simulation of either a fluid or solid particle which is dispersed in a fluid phase. A key assumption that is made in this model is that a relatively low fraction of the volume is occupied by the discrete phase. In the oil separator case, the oil volume ratio of the oil-refrigerant mixture in the discharge pipe is estimated to be less than 20%. Therefore, the condition of oil mist at the compressor discharge pipe satisfies the assumption of the discrete phase model. To simulate the movement of oil droplets, we utilize the Euler–Lagrangian approach. The vapor phase is treated as a continuum by the solution of the Reynolds-averaged Navier–Stokes equations. Simultaneously, the discrete phase is calculated by tracking all the generated droplets through the calculated flow field. Momentum exchange can take place between the discrete phase and the fluid phase. The discrete phase is introduced into the simulation by the definition of an injection at the inlet surface of the test section. The internal volume of the test section is extracted as the flow region for the refrigerant-oil mixture.

4.3 Boundary Conditions

The boundary conditions, in accordance with the specific locations, are presented in Fig. 4 The boundary condition at the entrance of the separator was set to mass flow, and the mass flux used was 5.19 kg/s. The working fluid, R-134a, has a viscosity of 1.3×10^{-5} Pa s and a density value of 44.0 kg/m, based on 845 kPa pressure test data, 80 °C which was the exit condition of the compressor. The flux of the oil particle was set to 0.95 kg/s and the density value of 937 kg/m^3. The average size of the lubricant particle was set to 10^{-5} m. The droplet size distribution is given between 5 and 50 μm and the size distribution was assumed as the Rosin-Rammler distribution [11]. The wall surfaces of the geometry were assigned the trap boundary condition for the purpose of the present CFD simulation. In this boundary condition, the calculations for the trajectory of the particle are terminated and the particle is recorded to be trapped. This boundary condition helps effectively model the deposition of the particles on these surfaces.

4.4 Solver Parameters

The simulation employs SIMPLE as the pressure-velocity coupling scheme and uses second order scheme for pressure discretization and second-order upwind scheme for momentum discretization to obtain highly accurate results [10]. The gradient scheme used was least squares cell based.

4.5 Solution

The solution was performed with a convergence criterion of 10^{-6} for residuals. It was found that this criterion is reached anywhere between 3500 and 3800 iterations. The DPM model shows results in the form of parcels of particles rather than the number of particles. The total number of parcels in the model based on distribution and size is 3590.

4.6 Results

The separation efficiency has been calculated on the basis of number of parcels that were separated or 'trapped' and the number that escaped. The simulation was run for 120,000 DPM iterations in order to produce a complete result. Results are tabulated in Tables 3 and 4.

Table 3 Input parameters for CFD simulation

Parameter	Case I (kg/s)	Case II (kg/s)	Case III (kg/s)
Refrigerant mass flow rate	5.19	5.19	6.82
Oil mass flow rate	0.95	2.31	2.31

Table 4 Results from CFD simulation

Parameter	Case I	Case II	Case III
Total number of oil parcels injected	3590	3590	3590
Number of oil parcels trapped by separator	3583	3587	3586
Number of oil parcels escaped from separator	7	3	4
Percentage of escaped particles	0.195	0.083	0.111
Oil separator efficiency	99.81	99.91	99.88

The representative streamlines of refrigerant and discrete oil particle flow are illustrated in Fig. 1. As can be observed, the majority of oil particles settle at the bottom of the separator below the baffle plate, while the refrigerant is released from the outlet. Furthermore, refrigerant flow is severely retarded beneath the baffle plate (Fig. 2).

Figure 3 illustrates particle tracks of all the oil particles. It can be seen that a very small fraction of the particles escape.

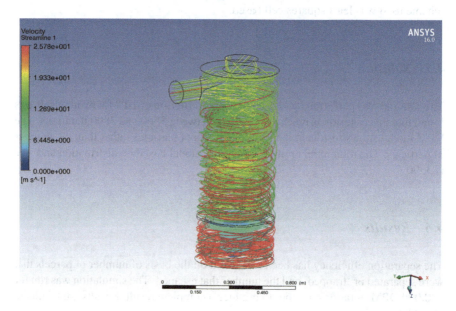

Fig. 1 Streamlines of refrigerant and oil particles

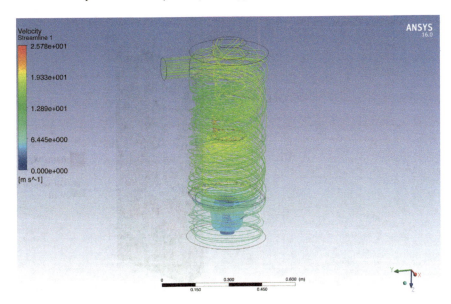

Fig. 2 Streamlines of refrigerant only

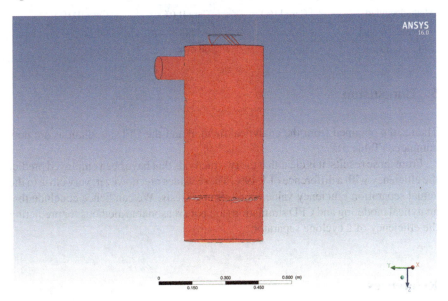

Fig. 3 Oil particle tracks

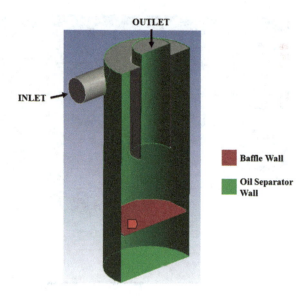

Fig. 4 Boundary Conditions

Table 5 Comparison of separation efficiency obtained

Method applied	Case I (%)	Case II (%)	Case III (%)
Analytical model	98.32	99.48	99.24
CFD simulation	99.81	99.91%	99.88

5 Conclusion

The results obtained from the analytical method and the CFD simulations are now compared (Table 5).

From these results, it is clear that the two methods that have been employed predict the efficiency with a difference of 1.49%. These values are, however, very close to the actual separation efficiency value found through tests. We can hence conclude that analytical modeling and CFD simulations are both reasonable methods for predicting the efficiency of a cyclone separator.

References

1. Youbi-idrissi M, Bonjour J, Marvillet C, Meunier F (2003) Impact of refrigerant—oil solubility on an evaporator performance working with R-407C. Int J Refrig 26:284–292
2. Cooper KW, Mount AG (1972) Oil circulation—its effect on compressor capacity, theory and experiment. In: International compressor engineering conference, Purdue University
3. Kim HS, Kang GH, Yoon PH, Sa YC, Chung BY, Kim MS (2016) Flow characteristics of refrigerant and oil mixture in an oil separator. Int J Refrig 70:206–218

4. Fukuta M, Yanagisawa T, Omura M, Ogi Y (2005) Mixing and separation characteristics of isobutane with refrigeration oil. Int J Refrig 28:997–1005
5. Liu H, Xu J, Zhang J, Sun H, Zhang J, Wu Y (2012) Oil/water separation in a liquid-liquid cylindrical cyclone. J Hydrodyn Ser B 24(1):116–123
6. Erdal FM, Shirazi SA, Shoham O, Kouba GE (1997) CFD simulation of single-phase and two-phase flow in gas-liquid cylindrical cyclone separators. SPE J 2(04):436–446
7. Gao X, Zhao Y, Feng J, Chang Y, Peng X (2012) The research on the performance of oil-gas cyclone separators in oil injected compressor systems with considering the breakup of oil droplets. In: International compressor engineering conference, Purdue University
8. Murakami H, Wakamoto OM (2006) Performance prediction of a cyclone oil separator. In: International refrigeration and air conditioning conference, Purdue University
9. Granovskii BL, Ermakov SM (1977) The monte carlo method. J Sov Math 7(2):161–192
10. Feng J, Chang Y, Peng XY, Qu Z (2008) Investigation of the oil-gas separation in a horizontal separator for oil-injected compressor units. Proc Inst Mech Eng Part A: J Power Energy 222(4):403–412
11. Seon G, Ahn J (2016) Design of the inlet-port of the cyclone-type oil separator using CFD. Int J Air-Cond Refrig 24(04):1–8
12. Xu J, Hrnjak P (2018) Flow visualization and CFD simulation of impinging oil separator for compressors. In: International compressor engineering conference, Purdue University

Energy Analysis of Double Evaporator Ammonia Water Vapour Absorption Refrigeration System

Deepak Panwar and Akhilesh Arora

Abstract An exhaustive analysis of double evaporator ammonia water vapour absorption refrigeration system (DEVARS) based on energy analysis has been carried out in this communication. To carry out analysis, a programme is made in the engineering equation solver (EES) software. The coefficient of performance (COP) is calculated at various operating conditions to study the effect of variation of evaporators, generators, absorbers and condenser temperatures and generators pressure on the COP. It is observed that by using DEVARS instead of using simple VARS, there is an increase of 11.11% COP when the load on both the evaporators is equal, i.e. 10 TR, and there is an increase of 22.22% COP when load on high temperature evaporator is 20 TR, and low temperature evaporator operates at no load. It is also observed that the variation in the temperature of absorber-2 has more effect on the COP of the system than the variation in absorber-1 and condenser temperatures. The study shows that NH_3–H_2O DEVARS is more promising than simple VAR systems with single evaporator in a wide operating range of evaporator temperatures.

Keywords Energy · Double evaporator VARS · Ammonia water · Parametric investigation

Nomenclature

COP	Coefficient of performance (Non-dimensional)
DEVARS	Double evaporator vapour absorption refrigeration system
EES	Engineering equation solver
f_1	Solution circulation ratio in circuit 1 (Non-dimensional)
f_2	Solution circulation ratio in circuit 2 (Non-dimensional)

D. Panwar · A. Arora (✉)
Delhi Technological University, New Delhi, Delhi 110042, India
e-mail: akhilesharora@dce.ac.in

D. Panwar
e-mail: deepakpanwar1297@gmail.com

h	Enthalpy per unit mass (kJ/kg)
$\dot{m}$	Rate of mass flow (kg/s)
P	Pressure (bar)
$\dot{Q}$	Rate of heat transfer (kW)
s	Specific entropy (kJ/kg K)
SEV	Solution expansion valve
SHX	Solution heat exchanger
T	Temperature (K)
TV	Throttle valve
$\dot{W}$	Rate of work transfer (kW)
X	Ammonia mass fraction in the solution (dimensionless)

Greek Letters

ϵ	Effectiveness of heat exchanger
v	Specific volume (m^3/kg)
$\sum$	Represents sum of

Subscripts

1, 2, 3, ..., 25	Represent state points in the equations or represents components
$a_{1,2}$	Absorber-1 and 2
c	Condenser
e	Evaporator, exit
$e_{1,2}$	Evaporator-1 and 2
$g_{1,2}$	Generator-1 and 2
h	High
i	Inlet
l	Low
$p_{1,2}$	Solution pump-1 and 2
r	Refrigerant
rec	Rectifier
ss	Strong solution (strong in ammonia)
she	Solution heat exchanger
ws	Weak solution (weak in ammonia

1 Introduction

The problem of environment pollution and world energy scarcity has been increasing seriously. VAR systems with the use of environmental-friendly refrigerants like ammonia, Li-Br, etc. have been substantially noticed in recent years. Moreover, heating energy for these systems can be supplied from renewable energy sources such as solar, geothermal and waste heat. Some industries like pharmaceutical, paper industries and pelagic seas vessels require sub-zero temperature for storing products along with comfort for peoples working in these industries required above zero temperature for air conditioning purpose. To fulfil these purposes, a suitable combination of refrigerants and working cycle is required.

A number of researches had been made to increase the performance of the VARS. Wang et al. [1] used molecular simulations dynamic modelling of the double-effect absorption cycles to reduce the system complexity by avoiding the use of rectification column. Liang et al. [2] added Li-Br for performance increment of $NH_3–H_2O$ absorption cooling system. It can likewise build the concentration of ammonia in the vapour stream, in this manner expanding energy utilization. Through joining the cycle of electrodialysis in this system, the working temperature in the generator is diminished altogether, and the proposed system COP is expanded contrasted with the typical $NH_3–H_2O$. Vasudev et al. [3] used a fluid-like ethylene glycol which is used as a latent energy storage material in a model $NH_3–H_2O$ installed in the evaporator to test its transient temperature effects. Experimentally, it has been shown that ethylene glycol significantly reduces this frequency, thereby protecting the performance of the products stored inside. Aprile et al. [4] broke down a gas-driven $NH_3–H_2O$ heat pump at full burden activity by differing high-temperature water and partial load operation examined by diminishing gas contribution to half of the full load value. Results indicated that partial load output could be improved if the solution mass flow rate was actively regulated. Swarnkar et al. [5] run the system with refrigerant ammonia and various ionic liquids as well as water as solvent and cosolvent. The results showed that when water is used as a cosolvent results in a significant reduction in the circulation ratio and therefore in the size of the device, whereas the COP decreases marginally.

Jia et al. [6] proposed a single-stage balanced type $NH_3–H_2O$ absorption-reabsorption heat pump (ARHP) cycle eliminates the need of the rectifier and relying on a single solution circulation pump when it is based on the unbalanced ARHP cycle. The cycle will accomplish inside mass and species preservation by increasing the concentration difference between ammonia and water solution at different levels of pressure. A model obtained feasible value of high pressure/low pressure (P_h/P_l) to enable the cycle. The maximum value of COP obtained is 1.51 corresponds to heat supply temperature of 316.4 K, (P_h/P_l) pair value of 1.50/0.48 MPa and heat source temperature of 368 K.

Jawahar et al. [7] analyzed $NH_3–H_2O$ absorption cooling system to recover total internal heat from streams by using pinch point analysis thus increasing the efficiency of the system. The proposed cycle's coefficient of performance (COP) was found to be 17–56% higher than that of a traditional cycle in terms of operating conditions. Horuz [8] concluded that VAR systems using Li-Br water used as a refrigerant provided better performance than ammonia water but water-lithium bromide had danger of crystallization and impossibility of using water as a refrigerant at very low temperatures. Toppi et al. [9] numerically evaluated semi-GAX cycle at working conditions reasonable for a low-temperature driven cooling application. The COP came to fruition to be immovably influenced by the split ratio, which chooses the middle pressure and the probability to achieve the GAX impact. The best suitable air temperature which allows a circulation ratio underneath 15 is 40 °C, with chilled water at 7/12 °C and driving temperature of 90 °C. Wu et al. [10] analysed that a temperature of 130 °C at inlet of generator, the evaporator bay temperature decreases from −5 to −25 °C and the COP of heat pump drops from 1.513 to 1.372, while the heating capacity declined from 77.26 kW to 47.11 kW; moreover, the connections between compressor assisted absorption heat pump (CAHP) and ordinary absorption heat pump (AHP) exhibited that CAHP can widen as far as possible uttermost spans of evaporator gulf temperature from −10 to −25 °C, and CAHP can improve the heating capacity by around 55.5–85.0% in any occasion, when AHP can work commonly.

On the basis of the literature survey, it is understood that a lot of research work has been done on the simple NH_3–H_2O VARS cycle to increase performance of the system. The researchers have worked on single evaporator system and no literature is found on multi-evaporator in VARS. Multi-evaporator system is required where two types of applications at different temperatures are required. For examples, air conditioning usually requires temperatures in the range of 5°–10 °C whereas refrigeration requires temperatures lower than 0 °C. Hence, in the present work, efforts have been made to work with two evaporators in which one have at low temperature and others at high temperature to increase the performance of the system. Since some research work has been done on the multi-evaporator vapour compression refrigeration system, but no efforts have been made to analyse the research work on the multi-evaporator (double evaporator) ammonia water vapour absorption refrigeration system. Further, the effect of cooling load of either evaporator on the performance of system is scantly available. Also, the effect of both generators, absorber and condenser temperature on the performance of DEVARS is not done.

In this communication, a detailed energy analysis of multi-evaporator ammonia water vapour absorption refrigeration system is performed and analysed. The coefficient of performance is determined at different conditions to think about the impact of cooling load on both evaporators, effect of temperature of generators, absorbers and condenser on them; also the effect of absorber and generator pressure on the performance is also studied.

2 System Description

The main components of Double Evaporator Vapour Absorber Refrigeration System (DEVARS) are divided into two circuits, namely high-pressure circuit and low-pressure circuit illustrated in Fig. 1 are as follows. The subscript 1 and 2 denote the components of high-pressure and low-pressure circuit, respectively.

- Condenser
- Throttle valve (TV-1 and TV-2)
- High and low pressure evaporator, i.e. (Evaporator-1 and Evaporator-2)
- High and low pressure absorber, i.e. (Absorber-1 and Absorber-2)
- Solution pump (Pump-1 and Pump-2)
- Solution expansion valve (SEV-1 and SEV-2)
- Solution heat exchanger (SHX-1 and SHX-2)
- High- and low-pressure generator, i.e. (Generator-1 and generator-2).

In the high-pressure circuit, the saturated strong solution (strong in ammonia) is forced to high pressure (state point 2) and then passed through a solution heat exchanger (SHX-1) to recuperate the heat internally. In the generator, outside heat is provided which bubbles off the refrigerant and the remaining saturated weak solution (state point 4) after giving its heat to the strong solution by using SHX, expanded to low pressure in solution expansion valve flows back to the absorber. The vapours leaving the generator (state point 7) consists of ammonia and water vapours. At the end of throttling process, water presents in the ammonia chokes the system, which

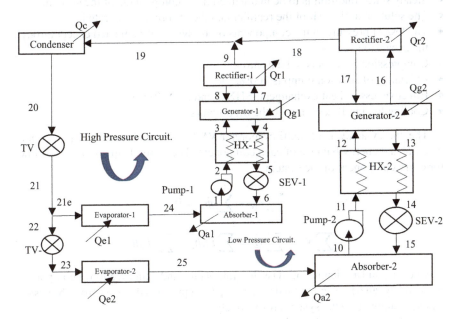

Fig. 1 Schematic diagram of double evaporator NH_3–H_2O vapour absorption refrigeration system

hampers the performance of the system so it is to be rectified in the rectifier after leaving the generator. The vapours at the exit of rectifier (state point 9) consists of almost 99.9% pure ammonia vapours and remaining condensed water in the rectifier flows back to the generator.

In low-pressure circuit, all the processes occur similarly as that of high-pressure circuit. The ammonia vapours leaving from low-pressure circuit, i.e. rectifier-2 (state point 18) mixes with ammonia vapours (state point 9) which then flows through the condenser tubes giving out heat to the surrounding (state point 20). The temperature at inlet of evaporator-1 is controlled by the amount of pressure reduced in the throttling valve-1 (TV-1). The liquid refrigerant takes heat from high temperature cooling space evaporates into saturated vapour, and this saturated vapour mixes with the weak solution in the absorber-1. This completes the high-pressure cycle.

The remaining saturated liquid (state point 22) reduced to low pressure (state point 23) in throttle valve-2 (TV-2) depending upon the design temperature of refrigerated space. The liquid refrigerant takes heat in evaporator-2 and is converted into vapour and then mixed with weak solution in absorber-2. This completes the cycle in low-pressure circuit.

3 Thermodynamic Analysis

To carry out the analysis, some assumptions are made which are as follows:

- Steady state condition is to be maintained in all components of the system.
- The solution at the exit of the rectifier and the absorber is saturated.
- The loss of pressure in the components of the system is neglected except through the expansion valves.
- Compression process is to be isentropic.
- Efficiency of solution pump is 100%.
- Effectiveness of heat exchangers (SHX-1 and SHX-2) is 0.8.
- Cooling load is 10 kW in both the evaporators.

The energy analysis of multi-evaporator VAR cycles includes the application of material and energy balance of each component. The general equation is as follows for every component of the system.

$$\sum \dot{m}_i - \sum \dot{m}_e = 0 \quad (1)$$

$$\sum \dot{W} - \sum \dot{Q} = \sum \dot{m}_i h_i - \sum \dot{m}_e h_e \quad (2)$$

where $\dot{m}_i$ and $\dot{m}_e$ denote the mass flow rate at inlet and outlet section of component, respectively; also first and second term in Eq. (2) represents the work from the system and heat supplied to the system, respectively.

The thermodynamic analysis of each section of the system is as follows.

In steady state, the mass and energy balance across the components of the system give the following equations.

3.1 Evaporator

$$\dot{m}_{21e} = \dot{m}_{24} = \dot{m}_{r1} \text{ and } \dot{m}_{23} = \dot{m}_{25} = \dot{m}_{r2} \tag{3}$$

$$\dot{Q}_{e1} = \dot{m}_{r1}(h_{24} - h_{21}) \tag{4}$$

$$\dot{Q}_{e2} = \dot{m}_{r2}(h_{25} - h_{23}) \tag{5}$$

3.2 Absorber

$$\dot{Q}_{a1} + \dot{m}_{ss1}h_1 = \dot{m}_{r1}h_{24} + \dot{m}_{ws1}h_6 \tag{6}$$

$$\dot{Q}_{a2} + \dot{m}_{ss2}h_{10} = \dot{m}_{r2}h_{25} + \dot{m}_{ws2}h_{15} \tag{7}$$

3.3 Solution Pump

$$\dot{W}_{p1} = \dot{m}_{ss1}(h_2 - h_1) = v_1(P_{h1} - P_{l1}) \tag{8}$$

$$\dot{W}_{p2} = \dot{m}_{ss2}(h_{11} - h_{10}) = v_{10}(P_{h2} - P_{l2}) \tag{9}$$

3.4 Generator

$$\dot{Q}_{g1} + \dot{m}_{ss1}h_3 + \dot{m}_8 h_8 = \dot{m}_7 h_7 + \dot{m}_{ws1}h_4 \tag{10}$$

$$\dot{Q}_{g2} + \dot{m}_{ss2}h_{12} + \dot{m}_{17}h_{17} = \dot{m}_{16}h_{16} + \dot{m}_{ws2}h_{13} \tag{11}$$

3.5 Rectifier

$$\dot{m}_8 + \dot{m}_9 = \dot{m}_7 \qquad (12)$$

$$\dot{m}_8 X_8 + \dot{m}_9 X_9 = \dot{m}_7 X_7 \qquad (13)$$

$$\dot{Q}_{r1} + \dot{m}_8 h_8 + \dot{m}_9 h_9 = \dot{m}_7 h_7 \qquad (14)$$

$$\dot{m}_{17} + \dot{m}_{18} = \dot{m}_{16} \qquad (15)$$

$$\dot{m}_{17} X_{17} + \dot{m}_{18} X_{18} = \dot{m}_{16} X_{16} \qquad (16)$$

$$\dot{Q}_{r2} + \dot{m}_{17} h_{17} + \dot{m}_{18} h_{18} = \dot{m}_{16} h_{16} \qquad (17)$$

In steady state, the mass coming out from the either of the rectifier is equal to the mass which is entered the absorber of that particular circuit, i.e. high-pressure circuit and low pressure-circuit which is represented by Eqs. 18 and 19, respectively.

$$\dot{m}_{r1} = \dot{m}_9 \qquad (18)$$

$$\dot{m}_{r2} = \dot{m}_{18} \qquad (19)$$

These refrigerant flows from either rectifier mixed before going to the condenser so combined mass flow rate, material and energy balance (state point 19) is given by

$$\dot{m}_r = \dot{m}_{r1} + \dot{m}_{r2} \qquad (20)$$

$$\dot{m}_9 X_9 + \dot{m}_{18} X_{18} = \dot{m}_{19} X_{19} \qquad (21)$$

$$\dot{m}_9 h_9 + \dot{m}_{18} h_{18} = \dot{m}_{19} h_{19} \qquad (22)$$

where

$$X_r = X_{19} \qquad (23)$$

$$\dot{m}_r = \dot{m}_{19} \qquad (24)$$

3.6 Solution Heat Exchanger

Suppose heat exchanger effectiveness is ϵ and the enthalpy (state point 5) is calculated by the relation

$$\epsilon = (h_4 - h_5)/(h_4 - h_{m5}) \tag{25}$$

where h_{m5} is the minimum enthalpy reached (state point 5) when temperature at state point (5) is equal to the temperature at (state point 2). Similarly,

$$\epsilon = (h_{13} - h_{14})/(h_{13} - h_{m14}) \tag{26}$$

where h_{m14} is the minimum enthalpy (state point 14) when temperature (state point 14) is equal to temperature (state point 11).

The enthalpy (state point 5 and 12) is calculated by the following relation

$$(f_2 - 1)(h_{13} - h_{14}) = f_2(h_{12} - h_{11}) \tag{27}$$

$$(f_1 - 1)(h_4 - h_5) = f_1(h_3 - h_2) \tag{28}$$

where f_1 and f_2 are the solution circulation ratio in high- and low-pressure circuit, respectively.

Which is calculated by the following relation

$$f_1 = \dot{m}_{ss1}/\dot{m}_{r1} = (X_r - X_{ws1})/(X_{ss1} - X_{ws1}) \tag{29}$$

$$f_2 = \dot{m}_{ss2}/\dot{m}_{r2} = (X_r - X_{ws1})/(X_{ss1} - X_{ws2}) \tag{30}$$

And the mass of weak solution is calculated by the following relation

$$f_1 - 1 = \dot{m}_{ws1}/\dot{m}_{r1} \tag{31}$$

$$f_2 - 1 = \dot{m}_{ws2}/\dot{m}_{r2} \tag{32}$$

3.7 Condenser

$$\dot{Q}_c = \dot{m}_r(h_{19} - h_{20}) \tag{33}$$

3.8 Throttle Valve

Since, during throttling process enthalpy remained constant. Therefore,

$$h_{20} = h_{21} \tag{34}$$

$$h_{22} = h_{23} \tag{35}$$

Coefficient of performance (COP) is the ratio of desired effect to the energy input. The expression for COP of simple and double evaporator ammonia water VARS is given below:

$$\text{COP}_{\text{VARS}} = \frac{\dot{Q}_e}{(\dot{Q}_g + \dot{W}_p)} \tag{36}$$

$$\text{COP}_{\text{DEVARS}} = \frac{\dot{Q}_{e1} + \dot{Q}_{e2}}{(\dot{Q}_{g1} + \dot{Q}_{g2} + \dot{W}_{p1} + \dot{W}_{p2})} \tag{37}$$

4 Results and Discussion

A programme has been made in Engineering Equation Solver (EES) programming (Klein and Alvarado 2016) to achieve vitality examination of double evaporator NH_3–H_2O VARS. The DEVARS cycle operating on high evaporator temperature called high-pressure cycle and cycle operating on low evaporator temperature called low-pressure cycle.

4.1 Simulation Validation

So as to confirm the present model, the after effects of vitality investigation of present work have been contrasted and the accessible numerical information determined by Herold et al. [11] for the simple VARS cycle in the present work when the cooling load in the evaporator-2 is put to zero. Comparison of heat transfer in various components is shown in Table 1, and the difference in heat duty in all the components are under 1% when compared to Herold et al. [11] as shown in Table 1. The values of COP and SCR in the present case are 0.442 and 7.034, respectively, and these values also match with the results of Herold et al. [11].

Table 1 Examination of the consequences of the energy analysis of present work with the numerical information given in Herold et al. [11]

Components	Keith. E. Herold $\dot{Q}$(kW)	Present work $\dot{Q}$(kW)	Difference %
Generator	329.2	330.2	0.303
Absorber	275.8	275.7	−0.036
Condenser	157.2	158.6	0.88
Evaporator	147.2	147.3	0.03
Rectifier	45.0	45.33	0.727
Solution heat Exchanger	256.0	257.0	0.389
Pump	1.5	1.489	−0.728
SCR (dimensionless)	7.025	7.034	0.127
COP (dimensionless)	0.445	0.442	−0.678

Parameters: $T_{eo} = -10\,°C$, $T_c = T_a = 40\,°C$, $X_r = 0.999634$, $\dot{M}_{ss} = 1$ kg/s, difference in mass fraction of two solution stream $= 0.10$, fluid leaving evaporator at vapour quality $= 0.975$, heat exchanger effectiveness $= 0.8$, $Q_{e2} = 0$

4.2 Effect of Generator Temperature on COP of System

The variation in COP and solution circulation ratio with the temperature of generator-1 and generator-2 is shown in Fig. 2. At constant temperature of generator-2 (120 °C), as the temperature of generator-1 increases, the COP of the system first increases

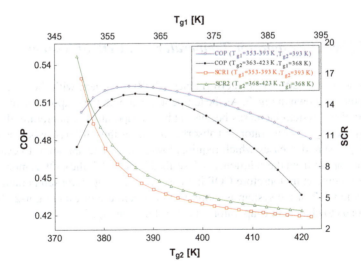

Fig. 2 Variation of COP and SCR with generator-1 and 2 temperature [i.e. at $T_{g1} = 368$ K (T_{g2} vary) and $T_{g2} = 393$ K (T_{g1} vary)]

reaching maximum value of 0.525 at the temperature of 85 °C and then COP decreases as the temperature further increases; similarly, at constant temperature of generator-1 (95 °C), as the generator-2 temperature increases up to certain temperature (115 °C) COP increases up to a value of 0.515, after that point COP diminishes as the temperature increments further. The purpose behind expanding COP up to some temperature is due to generation of more ammonia vapours. Due to which the concentration of ammonia in weak solution exiting from the generator decreases. The decrease in the ammonia concentration of the solution will decrease the solution circulation ratio in the particular circuit which will decrease the heat duty in the components of the circuit (e.g. generator-1 and generator-2) which increases the COP due to reduction in the heat supplied to the generator. But after certain temperature, increase in temperature of generator increases the average temperature of rectifier, condenser and absorber which increase the irreversibility of these components thus increase in COP due to increase in temperature is balanced by the decrease in COP due to increase in irreversibility of components; hence, after some particular, the net effect is the reduction in COP of the system; also the variation of COP of system is more pronounced of changing in generator-2 temperature, i.e. T_{g2}. This is due to the higher temperature of the generator-2 in comparison to generator-1 which brings higher irreversibility in components in comparison to generator-1.

The solution circulation ratio (f_1 and f_2) defined in Eqs. 29 and 30 for circuit 1 and 2, respectively, decreases as the temperature of generator increases. This happens because, as the temperature of generator increases, the ammonia mass fraction of weak solution increases; thus, at constant absorber temperature, the difference in mass fraction of strong and weak solution decrease therefore SCR decreases as shown in Fig. 2.

4.3 Effect of Evaporator Temperature on COP of System

The variation of coefficient of performance (COP) of system with the evaporators temperature is shown in Fig. 3. As the temperature of both the evaporators increases, the COP of the system increases because at high evaporators temperature, the pressure (P_{l1} and P_{l2}) in evaporator and absorber increases thus difference between high and low pressure decreases which requires low-generator temperature to evaporate ammonia vapour from the solution. As described in Sect. 4.2, the COP is more at low generator temperature therefore COP increases at high evaporator temperature. The variation in COP of the system is more when temperature of evaporator-2 changes corresponds to change in evaporator-1 is also shown in Fig. 3.

Energy Analysis of Double Evaporator Ammonia Water Vapour ...

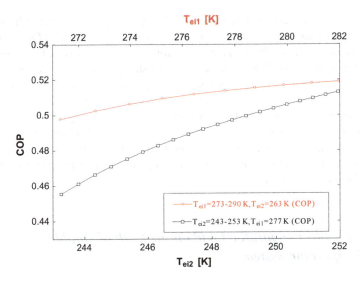

Fig. 3 Variation of COP with evaporator-1&2 temperature [i.e. at $T_{ei1} = 277\,\text{K}(T_{ei2}$ vary) and $T_{ei2} = 353\,\text{K}(T_{ei1}$ vary)]

4.4 Effect of Absorber Temperature on the COP of System

The effect of absorber temperature on the COP of the system is shown in Fig. 4. As absorber temperature increases, the COP of the system decreases. This is due to fact

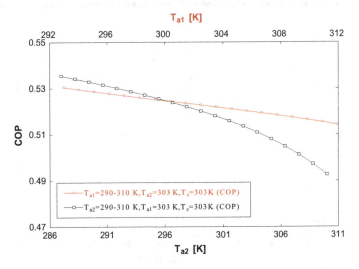

Fig. 4 Variation of COP with absorber-1 and 2 temperature [i.e. at $T_c = 303$ K and $T_{a1} = 303\,\text{K}(T_{a2}$ vary) and $T_{a2} = 303\,\text{K}(T_{a1}$ vary)]

that as the absorber temperature increases, the SCR in the solution circuit increases because ammonia mass fraction of strong solution decreases therefore difference between ammonia mass fraction between strong and weak solution decreases which will increase SCR. As SCR increases, mass of strong and weak solution increases which increase heat duty of generator; hence, at constant evaporator load, COP of the system decreases. But the effect of absorber-2 has been more pronounced on the performance of the system as that of absorber-1 as shown in Fig. 4 because pressure in absorber-2 is lower in comparison to absorber-1 thus difference in pressure in generator-2 and absorber-2 is more, and as discussed in Sect. 4.2, the effect of high temperature of generator is more therefore variation in COP is more in absorber-2 in comparison to absorber-1.

4.5 Effect of Evaporator Load on the COP of Double Evaporator System

Figure 5 shows the variation of evaporator load on the COP of the system. As the load on evaporator-1 increases keeping load on evaporator-2 constant, the value of COP increases and the COP decreases when the load on the evaporator-2 increases keeping load on evaporator-1 constant; thus, it can be concluded that the performance or COP of the system increases as the load is distributed on both the evaporator instead of working on single evaporator-2. The reason for this is that by distributing load on both the evaporators, the mass flow rate of refrigerant is distributed to low- and high-pressure circuit, and for total constant load on evaporator, the total heat transfer to generator-1 and generator-2 decreases because lesser amount of heat is required to

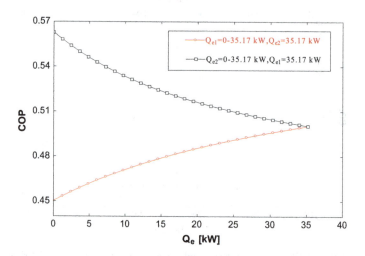

Fig. 5 Variation of COP with the load on evaporator-1 and 2 [i.e. at $Q_{e1} = 10\,\text{TR}(Q_{e2}$ vary) and $Q_{e2} = 10\,\text{TR}(Q_{e1}$ vary)]

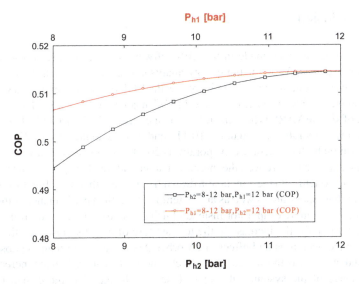

Fig. 6 Variation of COP with generator-1 and 2 pressure [i.e. at $P_{h1} = 10\,\text{bar}(P_{h2}$ vary) and $P_{h2} = 10\,\text{bar}(P_{h1}$ vary)]

evaporate ammonia vapour in generator-1, i.e. in high-pressure circuit corresponds to generator-2, i.e. low-pressure circuit as discussed in detail in Sect. 4.2.

4.6 Effect of Generator Pressure on the COP of the System

Figure 6 shows the variation of both the generator pressure, i.e. P_{h1} and P_{h2} on the COP of system. At constant generator-2 pressure of 12 bar, as generator-1 pressure vary, the COP of system first increases up to some pressure, and then, it becomes constant, and the maximum value is obtained at the pressure of approximately 12 bar; similarly, at constant generator-1 pressure of 12 bar, as the generator-2 pressure vary, the COP of the system first increases, and then, it becomes constant because as generator pressure increases, the temperature of both generators also increases, and the variation of COP with generator temperature is explained in detail in Sect. 4.2; therefore, similar pattern is observed but the effect of varying generator-2 pressure is more on the COP of the system than varying the generator-1 pressure because generator-2 temperature has more effect on COP than generator-1 as discussed in Sect. 4.2.

5 Conclusion

A programme has been made in the EES software to envisage the performance of ammonia water DEVARS using energy analysis. It is observed that the increase in temperature of both the generators increases the COP of the system up to a generator temperature of 90 °C in generator-1 and 115 °C in generator-2. The COP of DEVARS instead of simple VARS is higher, and it is an increase of 11.11% COP when the load on both the evaporators is equal, i.e. 10 TR, and there is an increase of 22.22% COP when load on high temperature evaporator is 20 TR and low temperature evaporator operates at no load; moreover, the increase in temperature of evaporator increases the COP of the system, and the conclusion drawn from here is that if cooling load is available at high temperature as well as at low temperature, then performance is increased by using two or more evaporator instead of using single evaporator at lower temperature; additionally, the increase in the temperature of both the absorber, the system performance decreases and effect of absorber-2 is more severe than the absorber-1 and condenser; also the increase in effectiveness of heat exchanger increases the performance of the system but effect of heat exchanger-2 is more than the heat exchanger-1.

References

1. Wang M, Becker TM, Schouten BA, Vlugt TJ, Ferreira CA (2018) Ammonia/ionic liquid-based double-effect vapor absorption refrigeration cycles driven by waste heat for cooling in fishing vessels. Energy Convers Manage 174:824–843
2. Liang Y, Li S, Yue X, Zhang X (2017) Analysis of NH_3–H_2O–LiBr absorption refrigeration integrated with an electrodialysis device. Appl Therm Eng 115:134–140
3. Vasudev V, Dondapati RS (2017) Experimental and exergy analysis of ammonia/water absorption system using ethylene glycol [$C_2H_4(OH)_2$] in the evaporator. Energy Procedia 109:401–408
4. Aprile M, Scoccia R, Toppi T, Guerra M, Motta M (2016) Modelling and experimental analysis of a GAX NH_3–H_2O gas-driven absorption heat pump. Int J Refrig 66:145–155
5. Swarnkar SK, Srinivasa Murthy S, Gardas RL, Venkatarathnam G (2014) Performance of a vapour absorption refrigeration system operating with ionic liquid-ammonia combination with water as cosolvent. Appl Thermal Eng 72:250–257
6. Jia T, Dai E, Dai Y (2019) Thermodynamic analysis and optimization of a balanced-type single-stage NH_3–H_2O absorption-resorption heat pump cycle for residential heating application. Energy. https://doi.org/10.1016/j.energy2019.01.002
7. Jawahar CP, Raj B, Saravanan R (2010) Thermodynamic studies on NH_3–H_2O absorption cooling system using pinch point approach. Int J Refrig 33:1377–1385
8. Horuz I (1998) A comparison between NH_3–H_2O and Li-Br & H_2O solution in vapor absorption refrigeration system. Int Comm Heat Mass Transfer 25(5):711–721
9. Toppi T, Aprile M, Guerra M, Motta M (2016) Numerical investigation on semi-GAX NH_3–H_2O absorption cycles. Int J Refrig 66:169–180
10. Wu W, Wang B, Shang S, Shi W, Li X (2016) Experimental investigation on NH_3–H_2O compression-assisted absorption heat pump (CAHP) for low temperature heating in colder conditions. Int J Refrig 67:109–124
11. Herold KE, Radermacher R, Klein SA (2016) Absorption chillers and heat pumps. CRC Press